INSTRUCTOR'S SOLUTIONS MANUAL

to accompany

Physics Fifth Edition

John D. Cutnell
Kenneth W. Johnson

Southern Illinois University at Carbondale

Volume 2

Chapters 18 – 32

John Wiley & Sons, Inc.

New York • Chichester • Weinheim • Brisbane • Singapore • Toronto

To order books or for customer service call 1-800-CALL-WILEY (225-5945).

Copyright © 2001 by John Wiley & Sons, Inc.

Excerpts from this work may be reproduced by instructors for distribution on a not-for-profit basis for testing or instructional purposes only to students enrolled in courses for which the textbook has been adopted. *Any other reproduction or translation of this work beyond that permitted by Sections 107 or 108 of the 1976 United States Copyright Act without the permission of the copyright owner is unlawful. Requests for permission or further information should be addressed to the Permissions Department, John Wiley & Sons, Inc., 605 Third Avenue, New York, NY 10158-0012.*

ISBN 0-471-35584-4

Printed in the United States of America

10 9 8 7 6 5 4 3 2 1

Printed and bound by Bradford & Bigelow, Inc.

PREFACE for VOLUME 2

This volume contains the complete solutions to the problems and conceptual questions for chapters 18 through 32. These chapters include electricity and magnetism, light and optics, and modern physics. The solutions for chapters 1 through 17 [mechanics (including fluids), thermal physics, and wave motion] are contained in Volume 1.

The manual is organized so that the solutions to the problems are in the front part of the book (part A), while those for the conceptual questions are in the back (part B).

An electronic version of this manual is also available on the Instructor Resource CD. The files are available in two formats: Microsoft Word, and PDF files of each individual solution.

There are two types of icons at the beginning of some problems. The [SSM] and [WWW] icons indicate, respectively, that the solution is also available in the Student Solutions Manual and on the World Wide Web at http://www.wiley.com/college/cutnell.

Note to adopters regarding The Instructor's Solutions Manual to accompany *Physics, 5e* by Cutnell & Johnson

Thank you very much for adopting PHYSICS, 5e. We are pleased to be able to provide you with a variety of support material to help you in the teaching of your course. Please note that all this material—including the Instructor's Solutions Manual—is copyrighted by John Wiley & Sons, Inc. and is explicitly intended for use only at your institution.

Please note that our providing these solutions does not carry with it permission to distribute them beyond your institution. Before putting any of the solutions on a web site, we ask that you request formal permission from us to do so. Please write to: Permissions Department, John Wiley & Sons, Inc., 605 Third Avenue, New York, NY 10158-0012. In most cases, we will grant such permission PROVIDED THAT THE WEB SITE IS PASSWORD PROTECTED.

Our goal is to prevent students from other campuses from being able to access your solutions. We trust that you can understand how that might undermine the efforts of your colleagues at other institutions.

We appreciate your support and understanding in this matter.

CONTENTS for VOLUME 2

PART A: PROBLEMS

Chapter 18	663	Chapter 26	938
Chapter 19	708	Chapter 27	994
Chapter 20	741	Chapter 28	1024
Chapter 21	799	Chapter 29	1049
Chapter 22	834	Chapter 30	1070
Chapter 23	865	Chapter 31	1097
Chapter 24	891	Chapter 32	1123
Chapter 25	918		

PART B: CONCEPTUAL QUESTIONS

Chapter 18	99	Chapter 26	155
Chapter 19	110	Chapter 27	168
Chapter 20	119	Chapter 28	176
Chapter 21	126	Chapter 29	181
Chapter 22	135	Chapter 30	186
Chapter 23	141	Chapter 31	191
Chapter 24	146	Chapter 32	197
Chapter 25	150		

CHAPTER 18 ELECTRIC FORCES AND ELECTRIC FIELDS

PROBLEMS

1. **SSM REASONING AND SOLUTION** The charge on a single electron is -1.60×10^{-19} C. In order to give a neutral silver dollar a charge of +2.4 μC, we must remove an amount of negative charge equal to -2.4 μC. This corresponds to

$$(-2.4 \times 10^{-6} \text{ C})\left(\frac{1 \text{ electron}}{-1.60 \times 10^{-19} \text{ C}}\right) = \boxed{1.5 \times 10^{13} \text{ electrons}}$$

2. **REASONING AND SOLUTION** The total charge to be removed is -5.0 µC. The number of electrons corresponding to this charge is

$$N = (-5.0 \times 10^{-6} \text{ C})/(-1.60 \times 10^{-19} \text{ C}) = \boxed{3.1 \times 10^{13}}$$

3. **REASONING AND SOLUTION** The total charge of the electrons is

$$q = N(-e) = (6.0 \times 10^{13})(-1.60 \times 10^{-19} \text{ C})$$

$$q = -9.6 \times 10^{-6} \text{ C} = -9.6 \text{ µC}$$

The net charge on the sphere is, therefore,

$$q_{net} = +8.0 \text{ µC} - 9.6 \text{ µC} = \boxed{-1.6 \text{ µC}}$$

4. **REASONING AND SOLUTION** Object A is negative because of an *excess* of electrons, while object B is positive because of a *deficiency* of electrons. The mass difference between the charged objects arises because the mass of A is greater due to the additional electrons, while the mass of object B is less due to the loss of electrons. Since $q = Ne$, where e is the magnitude of the charge on one electron, the number of excess electrons on object A is

$$N = \frac{q}{e} = \frac{3.0 \times 10^{-6} \text{ C}}{1.60 \times 10^{-19} \text{ C}} = 1.9 \times 10^{13}$$

This corresponds to an increase in mass given by

664 ELECTRIC FORCES AND ELECTRIC FIELDS

$$\Delta m = (1.9 \times 10^{13} \text{ electrons}) \left(\frac{9.11 \times 10^{-31} \text{ kg}}{1 \text{ electron}} \right) = 1.7 \times 10^{-17} \text{ kg}$$

Since both objects carry charge of the same magnitude, N is also equal to the number of electrons lost by object B. Hence, the mass of B is reduced by the amount Δm. If M is the mass of either object when they are electrically neutral, then the mass of each charged object is:

$$M_A = M + \Delta m \qquad \text{while} \qquad M_B = M - \Delta m$$

The mass difference between the charged objects is, therefore,

$$\Delta M = M_A - M_B = (M + \Delta m) - (M - \Delta m)$$

$$= 2\Delta m = 2(1.7 \times 10^{-17} \text{ kg}) = \boxed{3.4 \times 10^{-17} \text{ kg}}$$

From the discussion above, $\boxed{\text{object A has the larger mass}}$.

---

5. **SSM REASONING** Identical conducting spheres equalize their charge upon touching. When spheres A and B touch, an amount of charge $+q$, flows from A and instantaneously neutralizes the $-q$ charge on B leaving B momentarily neutral. Then, the remaining amount of charge, equal to $+4q$, is equally split between A and B, leaving A and B each with equal amounts of charge $+2q$. Sphere C is initially neutral, so when A and C touch, the $+2q$ on A splits equally to give $+q$ on A and $+q$ on C. When B and C touch, the $+2q$ on B and the $+q$ on C combine to give a total charge of $+3q$, which is then equally divided between the spheres B and C; thus, B and C are each left with an amount of charge $+1.5q$.

SOLUTION Taking note of the initial values given in the problem statement, and summarizing the final results determined in the *Reasoning* above, we conclude the following:

a. Sphere C ends up with an amount of charge equal to $\boxed{+1.5q}$.

b. The charges on the three spheres before they were touched, are, according to the problem statement, $+5q$ on sphere A, $-q$ on sphere B, and zero charge on sphere C. Thus, the total charge on the spheres is $+5q - q + 0 = \boxed{+4q}$.

c. The charges on the spheres after they are touched are $+q$ on sphere A, $+1.5q$ on sphere B, and $+1.5q$ on sphere C. Thus, the total charge on the spheres is $+q + 1.5q + 1.5q = \boxed{+4q}$.

Chapter 18 Problems 665

6. **REASONING**
a. The number N of electrons is 10 times the number of water molecules in 1 liter of water. The number of water molecules is equal to the number n of moles of water molecules times Avogadro's number N_A: $N = 10 n N_A$.

b. The net charge of all the electrons is equal to the number of electrons times the change on one electron.

SOLUTION
a. The number N of water molecules is equal to $10 n N_A$, where n is the number of moles of water molecules and N_A is Avogadro's number. The number of moles is equal to the mass m of 1 liter of water divided by the mass per mole of water. The mass of water is equal to its density ρ times the volume, as expressed by Equation 11.1. Thus, the number of electrons is

$$N = 10 \, n \, N_A = 10 \left(\frac{m}{18.0 \text{ g/mol}} \right) N_A = 10 \left(\frac{\rho V}{18.0 \text{ g/mol}} \right) N_A$$

$$= 10 \left[\frac{(1000 \text{ kg/m}^3)(1.00 \times 10^{-3} \text{ m}^3)\left(\frac{1000 \text{ g}}{1 \text{ kg}}\right)}{18.0 \text{ g/mol}} \right] (6.022 \times 10^{23} \text{ mol}^{-1})$$

$$= \boxed{3.35 \times 10^{26} \text{ electrons}}$$

b. The net charge Q of all the electrons is equal to the number of electrons times the change on one electron: $Q = (3.35 \times 10^{26})(-1.60 \times 10^{-19} \text{ C}) = \boxed{-5.36 \times 10^7 \text{ C}}$.

---

7. **SSM WWW REASONING** Initially, the two spheres are neutral. Since negative charge is removed from the sphere which loses electrons, it then carries a net positive charge. Furthermore, the neutral sphere to which the electrons are added is then negatively charged. Once the charge is transferred, there exists an electrostatic force on each of the two spheres, the magnitude of which is given by Coulomb's law (Equation 18.1), $F = k q_1 q_2 / r^2$.

SOLUTION
a. Since each electron carries a charge of -1.60×10^{-19} C, the amount of negative charge removed from the first sphere is

$$(3.0 \times 10^{13} \text{ electrons}) \left(\frac{1.60 \times 10^{-19} \text{ C}}{1 \text{ electron}} \right) = 4.8 \times 10^{-6} \text{ C}$$

666 ELECTRIC FORCES AND ELECTRIC FIELDS

Thus, the first sphere carries a charge $+4.8 \times 10^{-6}$ C, while the second sphere carries a charge -4.8×10^{-6} C. The magnitude of the electrostatic force that acts on each sphere is, therefore,

$$F = \frac{kq_1 q_2}{r^2} = \frac{(8.99 \times 10^9 \text{ N} \cdot \text{m}^2/\text{C}^2)(4.8 \times 10^{-6} \text{ C})^2}{(0.50 \text{ m})^2} = \boxed{0.83 \text{ N}}$$

b. Since the spheres carry charges of opposite sign, the force is $\boxed{\text{attractive}}$.

---

8. **REASONING AND SOLUTION** The electric force exerted on each proton can be obtained from Coulomb's law

$$F = \frac{kq_1 q_2}{r^2} = \frac{(8.99 \times 10^9 \text{ N} \cdot \text{m}^2/\text{C}^2)(1.60 \times 10^{-19} \text{ C})(1.60 \times 10^{-19} \text{ C})}{(3.0 \times 10^{-15} \text{ m})^2} = \boxed{26 \text{ N}}$$

---

9. **REASONING AND SOLUTION** The electrostatic forces decreases with the square of the distance separating the charges. If this distance is increased by a factor of 5 then the force will decrease by a factor of 25. The new force is, then,

$$F = \frac{3.5 \text{ N}}{25} = \boxed{0.14 \text{ N}}$$

---

10. **REASONING** The magnitude of the electrostatic force that acts on particle 1 is given by Coulomb's law as $F = k q_1 q_2 / r^2$. This equation can be used to find the magnitude q_2 of the charge.

SOLUTION Solving Coulomb's law for the magnitude q_2 of the charge gives

$$q_2 = \frac{F r^2}{k q_1} = \frac{(3.4 \text{ N})(0.26 \text{ m})^2}{(9.0 \times 10^9 \text{ N} \cdot \text{m}^2/\text{C}^2)(3.5 \times 10^{-6} \text{ C})} = \boxed{7.3 \times 10^{-6} \text{ C}} \qquad (18.1)$$

Since q_1 is positive and experiences an attractive force, the charge q_2 must be $\boxed{\text{negative}}$.

11. **SSM** *REASONING AND SOLUTION* The net electrostatic force on charge 3 at $x = +3.0$ m is the vector sum of the forces on charge 3 due to the other two charges, 1 and 2. According to Coulomb's law (Equation 18.1), the magnitude of the force on charge 3 due to charge 1 is

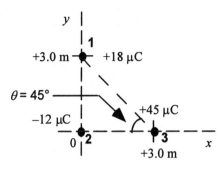

Figure 1

$$F_{13} = \frac{kq_1 q_3}{r_{13}^2}$$

where the distance between charges 1 and 3 is r_{13}.

According to the Pythagorean theorem, $r_{13}^2 = x^2 + y^2$. Therefore,

$$F_{13} = \frac{(8.99 \times 10^9 \text{ N} \cdot \text{m}^2/\text{C}^2)(18 \times 10^{-6} \text{ C})(45 \times 10^{-6} \text{ C})}{(3.0 \text{ m})^2 + (3.0 \text{ m})^2} = 0.405 \text{ N}$$

Charges 1 and 3 are equidistant from the origin, so that $\theta = 45°$ (see Figure 1). Since charges 1 and 3 are both positive, the force on charge 3 due to charge 1 is repulsive and along the line that connects them, as shown in Figure 2. The components of F_{13} are:

$$F_{13x} = F_{13} \cos 45° = 0.286 \text{ N} \quad \text{and} \quad F_{13y} = -F_{13} \sin 45° = -0.286 \text{ N}$$

The second force on charge 3 is the attractive force (opposite signs) due to its interaction with charge 2 located at the origin. The magnitude of the force on charge 3 due to charge 2 is, according to Coulomb's law,

$$F_{23} = \frac{kq_2 q_3}{r_{23}^2} = \frac{kq_2 q_3}{x^2}$$

$$= \frac{(8.99 \times 10^9 \text{ N} \cdot \text{m}^2/\text{C}^2)(12 \times 10^{-6} \text{ C})(45 \times 10^{-6} \text{ C})}{(3.0 \text{ m})^2}$$

$$= 0.539 \text{ N}$$

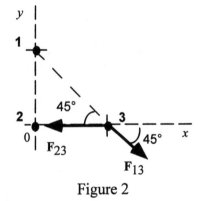

Figure 2

Since charges 2 and 3 have opposite signs, they attract each other, and charge 3 experiences a force to the left as shown in Figure 2. Taking up and to the right as the positive directions, we have

$$F_{3x} = F_{13x} + F_{23x} = +0.286 \text{ N} - 0.539 \text{ N} = -0.253 \text{ N}$$

$$F_{3y} = F_{13y} = -0.286 \text{ N}$$

Using the Pythagorean theorem, we find the magnitude of F_3 to be

$$F_3 = \sqrt{F_{3x}^2 + F_{3y}^2} = \sqrt{(-0.253 \text{ N})^2 + (-0.286 \text{ N})^2} = \boxed{0.38 \text{ N}}$$

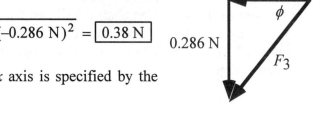

Figure 3

The direction of F_3 relative to the $-x$ axis is specified by the angle ϕ, where

$$\phi = \tan^{-1}\left(\frac{0.286 \text{ N}}{0.253 \text{ N}}\right) = \boxed{49° \text{ below the } -x \text{ axis}}$$

---

12. **REASONING AND SOLUTION** Calculate the magnitude of each force acting on the center charge. Using Coulomb's law, we can write

$$F_{43} = \frac{kq_4q_3}{r_{43}^2} = \frac{(8.99 \times 10^9 \text{ N} \cdot \text{m}^2/\text{C}^2)(4.00 \times 10^{-6} \text{ C})(3.00 \times 10^{-6} \text{ C})}{(0.100 \text{ m})^2}$$

$$= 10.8 \text{ N} \quad \text{(toward the south)}$$

$$F_{53} = \frac{kq_5q_3}{r_{53}^2} = \frac{(8.99 \times 10^9 \text{ N} \cdot \text{m}^2/\text{C}^2)(5.00 \times 10^{-6} \text{ C})(3.00 \times 10^{-6} \text{ C})}{(0.100 \text{ m})^2}$$

$$= 13.5 \text{ N} \quad \text{(toward the east)}$$

Adding F_{43} and F_{53} vectorially, we have

$$F = \sqrt{F_{43}^2 + F_{53}^2} = \sqrt{(10.8 \text{ N})^2 + (13.5 \text{ N})^2} = \boxed{17.3 \text{ N}}$$

$$\theta = \tan^{-1}\left(\frac{F_{43}}{F_{53}}\right) = \tan^{-1}\left(\frac{10.8 \text{ N}}{13.5 \text{ N}}\right) = \boxed{38.7° \text{ S of E}}$$

---

13. **REASONING AND SOLUTION** Place the $-9.0 \text{ }\mu\text{C}$ charge at the apex of the triangle (north), the $+8.0 \text{ }\mu\text{C}$ charge at the lower left corner (west), and the $+2.0 \text{ }\mu\text{C}$ charge at the lower right (east). Find the forces on the $+2.0 \text{ }\mu\text{C}$ charge due to the $-9.0 \text{ }\mu\text{C}$ and $+8.0 \text{ }\mu\text{C}$ charges.

$$F_{92} = \frac{kq_9q_2}{r_{92}^2} = \frac{(8.99 \times 10^9 \text{ N} \cdot \text{m}^2/\text{C}^2)(9.0 \times 10^{-6} \text{ C})(2.0 \times 10^{-6} \text{ C})}{(0.15 \text{ m})^2}$$

$$F_{92} = 7.2 \text{ N, at } 60.0° \text{ N of W}$$

The components are

$$F_{92x} = F_{92} \cos 60.0° = 3.6 \text{ N (due west)}$$

$$F_{92y} = F_{92} \sin 60.0° = 6.2 \text{ N (due north)}$$

Similarly,

$$F_{82} = \frac{kq_8 q_2}{r_{82}^2} = \frac{(8.99 \times 10^9 \text{ N} \cdot \text{m}^2/\text{C}^2)(8.0 \times 10^{-6} \text{ C})(2.0 \times 10^{-6} \text{ C})}{(0.15 \text{ m})^2}$$

$F_{92} = 6.4$ N, directed due east

The resultant force on the +2.0 µC charge is obtained from

$$F_2 = \sqrt{F_x^2 + F_y^2} = \sqrt{(6.4 \text{ N} - 3.6 \text{ N})^2 + (6.2 \text{ N})^2} = \boxed{6.8 \text{ N}}$$

14. **REASONING**
   a. The magnitude of the electrostatic force that acts on each sphere is given by Coulomb's law as $F = kq_1 q_2 / r^2$, where q_1 and q_2 are the magnitudes of the charges, and r is the distance between the centers of the spheres.

   b. When the spheres are brought into contact, the net charge after contact and separation must be equal to the net charge before contact. Since the spheres are identical, the charge on each after being separated is one-half the net charge. Coulomb's law can be applied again to determine the magnitude of the electrostatic force that each sphere experiences.

SOLUTION
a. The magnitude of the force that each sphere experiences is given by Coulomb's law as:

$$F = \frac{kq_1 q_2}{r^2} = \frac{(8.99 \times 10^9 \text{ N} \cdot \text{m}^2/\text{C}^2)(20.0 \times 10^{-6} \text{ C})(50.0 \times 10^{-6} \text{ C})}{(2.50 \times 10^{-2} \text{ m})^2} = \boxed{1.44 \times 10^4 \text{ N}}$$

Because the charges have opposite signs, the force is $\boxed{\text{attractive}}$.

b. The net charge on the spheres is $-20.0 \text{ } \mu\text{C} + 50.0 \text{ } \mu\text{C} = +30.0 \text{ } \mu\text{C}$. When the spheres are brought into contact, the net charge after contact and separation must be equal to the net charge before contact, or $+30.0 \text{ } \mu\text{C}$. Since the spheres are identical, the charge on each after being separated is one-half the net charge, so $q_1 = q_2 = +15.0 \text{ } \mu\text{C}$. The electrostatic force that acts on each sphere is now

$$F = \frac{kq_1q_2}{r^2} = \frac{(8.99\times 10^9 \text{ N}\cdot\text{m}^2/\text{C}^2)(15.0\times 10^{-6} \text{ C})(15.0\times 10^{-6} \text{ C})}{(2.50\times 10^{-2} \text{ m})^2} = \boxed{3.24\times 10^3 \text{ N}}$$

Since the charges now have the same signs, the force is $\boxed{\text{repulsive}}$.

---

15. **SSM** **WWW** *REASONING* Each particle will experience an electric force due to the presence of the other charge. According to Coulomb's law (Equation 18.1), the magnitude of the force felt by each particle can be calculated from $F = kq_1q_2/r^2$, where q_1 and q_2 are the respective charges on particles 1 and 2 and r is the distance between them. According to Newton's second law, the magnitude of the force experienced by each particle is given by $F = ma$, where a is the acceleration of the particle.

SOLUTION
a. Since the two particles have identical positive charges, $q_1 = q_2 = q$, and we have, using the data for particle 1,

$$\frac{kq^2}{r^2} = m_1 a_1$$

Solving for q, we find that

$$q = \sqrt{\frac{m_1 a_1 r^2}{k}} = \sqrt{\frac{(6.00\times 10^{-6} \text{ kg})(4.60\times 10^3 \text{ m/s}^2)(2.60\times 10^{-2} \text{ m})^2}{8.99\times 10^9 \text{ N}\cdot\text{m}^2/\text{C}^2}} = \boxed{4.56\times 10^{-8} \text{ C}}$$

b. Since each particle experiences a force of the same magnitude (From Newton's third law), we can write $F_1 = F_2$, or $m_1 a_1 = m_2 a_2$. Solving this expression for the mass m_2 of particle 2, we have

$$m_2 = \frac{m_1 a_1}{a_2} = \frac{(6.00\times 10^{-6} \text{ kg})(4.60\times 10^3 \text{ m/s}^2)}{8.50\times 10^3 \text{ m/s}^2} = \boxed{3.25\times 10^{-6} \text{ kg}}$$

16. **REASONING** The unknown charge q must be positive. To see why, consider the unknown charge at the upper right corner in the drawing at the right (the unknown charge at the lower left corner could also be used). Three forces act on this charge: (1) F_O is the repulsive force due to the other unknown charge on the opposite corner, (2) F_B is the attractive force due to the negative charge at the lower right corner, and (3) F_L is the attractive force due to the negative charge at the upper left corner. These three forces add to give a net force of zero. The unknown charge can not be negative, because then F_B and F_L would have directions opposite to those shown in the shown in the drawing, and the sum of F_O, F_B, and F_L

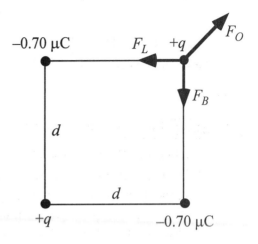

could not be zero. We note that the magnitudes of F_B and F_L are equal according to Coulomb's law (Equation 18.1), since the sides of the square have equal lengths and the charge magnitudes are q and 0.70 µC in each case. We also note that the directions of F_B and F_L are perpendicular. Thus, the resultant of F_B and F_L is given by the Pythagorean theorem and points along the diagonal of the square, directly opposite to the direction of F_O. Since the vector sum of F_O, F_B, and F_L is zero, the magnitude of the resultant of F_B and F_L must equal the magnitude of F_O, and it is with this fact in mind that we begin our solution.

SOLUTION Using the Pythagorean to express the magnitude of the resultant of F_B and F_L, which is equal to F_O, we have

$$\sqrt{F_B^2 + F_L^2} = F_O \tag{1}$$

Coulomb's law indicates that

$$F_B = F_L = \frac{k(0.70 \times 10^{-6}\ \text{C})q}{d^2} \quad \text{and} \quad F_O = \frac{kq^2}{\left(\sqrt{2}\,d\right)^2} = \frac{kq^2}{2d^2}$$

where we have used d for the length of a side of the square and the fact that the diagonal of the square has a length of $\sqrt{d^2 + d^2} = \sqrt{2}\,d$. Substituting these expressions for F_O, F_B, and F_L into Equation (1), we find

$$\sqrt{\left[\frac{k(0.70 \times 10^{-6}\ \text{C})q}{d^2}\right]^2 + \left[\frac{k(0.70 \times 10^{-6}\ \text{C})q}{d^2}\right]^2} = \sqrt{2\left[\frac{k(0.70 \times 10^{-6}\ \text{C})q}{d^2}\right]^2} = \frac{kq^2}{2d^2}$$

Simplifying this result shows that

$$\frac{\sqrt{2}\,k(0.70\times 10^{-6}\text{ C})q}{d^2} = \frac{kq^2}{2d^2} \quad \text{or} \quad q = 2\sqrt{2}(0.70\times 10^{-6}\text{ C}) = \boxed{+2.0\,\mu\text{C}}$$

---

17. **REASONING AND SOLUTION** The new force acting on each of the charges A and B is

$$F = kq_Aq_B/r^2$$

where

$$q_A + q_B = 0$$

since the net charge on the system is zero. We want the net force to equal 45.0 N, and can write

$$F = kq_A^2/r^2$$

so

$$q_A = r\sqrt{\frac{F}{k}} = (0.0200\text{ m})\sqrt{\frac{45.0\text{ N}}{8.99\times 10^9\text{ N}\cdot\text{m}^2/\text{C}^2}}$$

$$= 1.41\times 10^{-6}\text{ C}$$

Therefore, we need to remove $-0.41\,\mu\text{C}$ of charge from A and give it to B. The number of electrons will then be

$$N = q/(-e) = (-0.41\times 10^{-6}\text{ C})/(-1.60\times 10^{-19}\text{ C}) = \boxed{2.6\times 10^{12}}.$$

---

18. **REASONING**
a. There are two electrostatic forces that act on q_1; that due to q_2 and that due to q_3. The magnitudes of these forces can be found by using Coulomb's law. The magnitude and direction of the net force that acts on q_1 can be determined by using the method of vector components.

b. According to Newton's second law, Equation 4.2b, the acceleration of q_1 is equal to the net force divided by its mass. However, there is only one force acting on it, so this force is the net force.

SOLUTION
a. The magnitude F_{12} of the force exerted on q_1 by q_2 is given by Coulomb's law, Equation 18.1, where the distance is specified in the drawing:

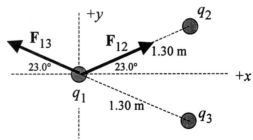

$$F_{12} = \frac{k q_1 q_2}{r_{12}^2} = \frac{(8.99 \times 10^9 \text{ N} \cdot \text{m}^2/\text{C}^2)(8.00 \times 10^{-6} \text{ C})(5.00 \times 10^{-6} \text{ C})}{(1.30 \text{ m})^2} = 0.213 \text{ N}$$

Since the magnitudes of the charges and the distances are the same, the magnitude of $\mathbf{F}_{13}$ is the same as the magnitude of $\mathbf{F}_{12}$, or $F_{13} = 0.213$ N. From the drawing it can be seen that the x-components of the two forces cancel, so we need only to calculate the y components of the forces.

Force	y component
$\mathbf{F}_{12}$	$+F_{12} \sin 23.0° = +(0.213 \text{ N}) \sin 23.0° = +0.0832 \text{ N}$
$\mathbf{F}_{13}$	$+F_{13} \sin 23.0° = +(0.213 \text{ N}) \sin 23.0° = +0.0832 \text{ N}$
$\mathbf{F}$	$F_y = +0.166 \text{ N}$

Thus, the net force is $\boxed{\mathbf{F} = +0.166 \text{ N (directed along the +y axis)}}$.

b. According to Newton's second law, Equation 4.2b, the acceleration of q_1 is equal to the net force divided by its mass. However, there is only one force acting on it, so this force is the net force:

$$\mathbf{a} = \frac{\mathbf{F}}{m} = \frac{+0.166 \text{ N}}{1.50 \times 10^{-3} \text{ kg}} = \boxed{+111 \text{ m/s}^2}$$

where the plus sign indicates that $\boxed{\text{the acceleration is along the +y axis}}$.

---

19. **SSM** *REASONING AND SOLUTION* Before the spheres have been charged, they exert no forces on each other. After the spheres are charged, each sphere experiences a repulsive force F due to the charge on the other sphere, according to Coulomb's law (Equation 18.1). Therefore, since each sphere has the same charge, the magnitude F of this force is

$$F = \frac{k q_1 q_2}{r^2} = \frac{(8.99 \times 10^9 \text{ N} \cdot \text{m}^2/\text{C}^2)(1.60 \times 10^{-6} \text{ C})^2}{(0.100 \text{ m})^2} = 2.30 \text{ N}$$

674   ELECTRIC FORCES AND ELECTRIC FIELDS

The repulsive force on each sphere compresses the spring to which it is attached. The magnitude of this repulsive force is related to the amount of compression by Equation 10.1: $F = kx$. Therefore, solving for k, we find

$$k = \frac{F}{x} = \frac{2.30 \text{ N}}{0.0250 \text{ m}} = \boxed{92.0 \text{ N/m}}$$

---

20. **REASONING AND SOLUTION** The magnitude of the force exerted on each charge is given by Coulomb's law, $F = kq_1q_2/r^2$. In this case we also have the constraint that

$$q_1 + q_2 = q = 9.00 \text{ μC} \quad \text{or} \quad q_1 = q - q_2$$

Using this in Coulomb's law, we can eliminate q_1 and solve for q_2, i.e.,

$$F = k(q - q_2)q_2/r^2$$

gives

$$q_2^2 - qq_2 + Fr^2/k = 0$$

where $F = 8.00 \times 10^{-3}$ N, and $r = 3.00$ m. Solving the above quadratic equation for q_2 we obtain;

$$q_2 = \boxed{1.00 \times 10^{-6} \text{ C}} \quad \text{and} \quad q_1 = \boxed{8.00 \times 10^{-6} \text{ C}}$$

---

21. **REASONING** This is a problem that deals with motion in a circle of radius r. As Chapter 5 discusses, a centripetal force acts on the plane to keep it on its circular path. The centripetal force F_c is the name given to the net force that acts on the plane in the radial direction and points toward the center of the circle. When there are no electric charges present, only the tension in the guideline supplies this force, and it has a value T_{max} at the moment the line breaks. However, when there is a charge of $+q$ on the plane and a charge of $-q$ on the guideline at the center of the circle, there are two contributions to the centripetal force. One is the electrostatic force of attraction between the charges and is given by Coulomb's law (Equation 18.1) as $F = kq^2/r^2$. The other is the tension T_{max}, which is characteristic of the rope and has the same value as when no charges are present. Whether or not charges are present, the centripetal force is equal to the mass m times the centripetal acceleration, according to Newton's second law and stated in Equation 5.3, $F_c = mv^2/r$. In this expression v is the speed of the plane. Since we are given information about the plane's kinetic energy, we will use the definition of kinetic energy, which is $KE = mv^2/2$, according to Equation 6.2.

SOLUTION From the definition of kinetic energy, we see that $mv^2 = 2(KE)$, so that Equation 5.3 for the centripetal force becomes

$$F_c = \frac{mv^2}{r} = \frac{2(KE)}{r}$$

Applying this result to the situations with and without the charges, we get

$$\underbrace{T_{max} + \frac{kq^2}{r^2}}_{\text{Centripetal force}} = \frac{2(KE)_{charged}}{r} \quad (1) \qquad \underbrace{T_{max}}_{\text{Centripetal force}} = \frac{2(KE)_{uncharged}}{r} \quad (2)$$

Subtracting Equation (2) from Equation (1) eliminates T_{max} and gives

$$\frac{kq^2}{r^2} = \frac{2\left[(KE)_{charged} - (KE)_{uncharged}\right]}{r}$$

Solving for q gives

$$q = \sqrt{\frac{2r\left[(KE)_{charged} - (KE)_{uncharged}\right]}{k}} = \sqrt{\frac{2(3.0 \text{ m})(51.8 \text{ J} - 50.0 \text{ J})}{8.99 \times 10^9 \text{ N} \cdot \text{m}^2/\text{C}^2}} = \boxed{3.5 \times 10^{-5} \text{ C}}$$

---

22. **REASONING AND SOLUTION** Assume that before the objects are touched that the left object has a negative charge of magnitude q_1 and the right object has a charge q_2. The force between them then has a magnitude of

$$F = kq_1q_2/r^2$$

After touching the charge on each object is the same and of magnitude $(q_2 - q_1)/2$. The magnitude of the force between the objects is now

$$F = k[(q_2 - q_1)/2]^2/r^2$$

Equating the equations and rearranging gives

$$q_2^2 - 6q_1q_2 + q_1^2 = 0$$

The solutions to this quadratic equation are

676   ELECTRIC FORCES AND ELECTRIC FIELDS

$$q_1 = 5.58 \ \mu C, \ q_2 = 0.957 \ \mu C \quad \text{and} \quad q_1 = 0.957 \ \mu C, \ q_2 = 5.58 \ \mu C$$

The charge, q_1, was assumed negative, so the possible solutions are

$\boxed{-5.58 \ \mu C \text{ on left}, \ +0.957 \ \mu C \text{ on right}}$ and $\boxed{-0.957 \ \mu C \text{ on left}, \ +5.58 \ \mu C \text{ on right}}$

---

23. **SSM** **REASONING** The charged insulator experiences an electric force due to the presence of the charged sphere shown in the drawing in the text. The forces acting on the insulator are the downward force of gravity (i.e., its weight, $W = mg$), the electrostatic force $F = kq_1q_2/r^2$ (see Coulomb's law, Equation 18.1) pulling to the right, and the tension T in the wire pulling up and to the left at an angle θ with respect to the vertical as shown in the drawing in the problem statement. We can analyze the forces to determine the desired quantities θ and T.

SOLUTION.
a. We can see from the diagram that

$$T_x = F \quad \text{which gives} \quad T \sin\theta = kq_1q_2/r^2$$

and

$$T_y = W \quad \text{which gives} \quad T \cos\theta = mg$$

Dividing the first equation by the second yields $(T\sin\theta)/(T\cos\theta) = \tan\theta = kq_1q_2/(mgr^2)$. Solving for θ, we find that

$$\theta = \tan^{-1}\left(\frac{kq_1q_2}{mgr^2}\right)$$

$$= \tan^{-1}\left[\frac{(8.99 \times 10^9 \ \text{N} \cdot \text{m}^2/\text{C}^2)(0.600 \times 10^{-6} \ \text{C})(0.900 \times 10^{-6} \ \text{C})}{(8.00 \times 10^{-2} \ \text{kg})(9.80 \ \text{m/s}^2)(0.150 \ \text{m})^2}\right] = \boxed{15.4°}$$

b. Since $T\cos\theta = mg$, the tension can be obtained as follows:

$$T = \frac{mg}{\cos\theta} = \frac{(8.00 \times 10^{-2} \ \text{kg})(9.80 \ \text{m/s}^2)}{\cos 15.4°} = \boxed{0.813 \ \text{N}}$$

---

24. **REASONING AND SOLUTION**
a. To find the charge on each ball we first need to determine the electric force acting on each ball. This can be done by noting that each thread makes an angle of 18° with respect to the vertical.

$$F_e = mg \tan 18° = (8.0 \times 10^{-4} \text{ kg})(9.80 \text{ m/s}^2) \tan 18° = 2.547 \times 10^{-3} \text{ N}$$

We also know that

$$F_e = kq_1q_2/r^2 = kq^2/r^2$$

where $r = 2(0.25 \text{ m}) \sin 18° = 0.1545$ m. Now

$$q = r\sqrt{\frac{F_e}{k}} = (0.1545 \text{ m})\sqrt{\frac{2.547 \times 10^{-3} \text{ N}}{8.99 \times 10^9 \text{ N} \cdot \text{m}^2/\text{C}^2}} = \boxed{8.2 \times 10^{-8} \text{ C}}$$

b. The tension is due to the combination of the weight of the ball and the electric force, the two being perpendicular to one another. The tension is therefore,

$$T = \sqrt{(mg)^2 + F_e^2}$$

$$= \sqrt{\left[(8.0 \times 10^{-4} \text{ kg})(9.80 \text{ m/s}^2)\right]^2 + (2.547 \times 10^{-3} \text{ N})^2} = \boxed{8.2 \times 10^{-3} \text{ N}}$$

---

25. **SOLUTION** Knowing the electric field at a spot allows us to calculate the force that acts on a charge placed at that spot, without knowing the nature of the object producing the field. This is possible because the electric field is defined as $\mathbf{E} = \mathbf{F}/q_0$, according to Equation 18.2. This equation can be solved directly for the force $\mathbf{F}$, if the field $\mathbf{E}$ and charge q_0 are known.

SOLUTION Using Equation 18.2, we find that the force has a magnitude of

$$F = Eq = (260\,000 \text{ N/C})(7.0 \times 10^{-6} \text{ C}) = 1.8 \text{ N}$$

If the charge were positive, the direction of the force would be due west, the same as the direction of the field. But the charge is negative, so the force points in the opposite direction or due east. Thus, the force on the charge is $\boxed{1.8 \text{ N due east}}$.

678   ELECTRIC FORCES AND ELECTRIC FIELDS

26. **REASONING AND SOLUTION** The electric field lines must originate on the positive charges and terminate on the negative charges. They cannot cross one another. Furthermore, the number of field lines beginning or ending on any charge must be proportional to the magnitude of the charge. If 10 electric field lines leave the +5q charge, then six lines must originate from the +3q charge, and eight lines must end on each –4q charge. The drawing shows the electric field lines that meet these criteria.

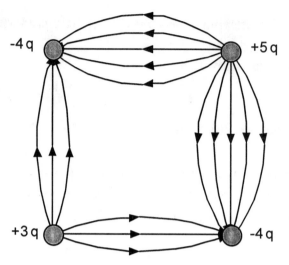

27. **SSM REASONING AND SOLUTION** The force **F** exerted on a charge q_0 placed in an electric field **E** can be determined from Equation 18.2, the definition of electric field ($\mathbf{E} = \mathbf{F}/q_0$). Writing this in terms of magnitudes, and taking due east as the positive direction, we have, solving for F,

$$F = q_0 E = (+3.0 \times 10^{-5} \text{ C})(+15\,000 \text{ N/C}) = \boxed{+0.45 \text{ N}}$$

where the plus sign indicates that the force on the charge is directed $\boxed{\text{due east}}$.

28. **REASONING**

a. The drawing shows the two point charges q_1 and q_2. Point A is located at $x = 0$ cm, and point B is at $x = +6.0$ cm.

Since q_1 is positive, the electric field points away from it. At point A, the electric field E_1 points to the left, in the $-x$ direction. Since q_2 is negative, the electric field points toward it. At point A, the electric field E_2 points to the right, in the $+x$ direction. The net electric field is $E = -E_1 + E_2$. We can use Equation 18.3, $E = kq/r^2$, to find the magnitude of the electric field due to each point charge.

b. The drawing shows the electric field produced by the charges q_1 and q_2 at point B, which is located at $x = +6.0$ cm.

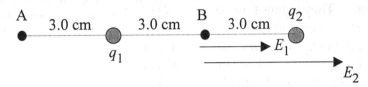

Since q_1 is positive, the electric field points away from it. At point B, the electric field points to the right, in the $+x$ direction. Since q_2 is negative, the electric field points toward it. At point B, the electric field points to the right, in the $+x$ direction. The net electric field is $E = +E_1 + E_2$.

SOLUTION
a. The net electric field at the origin (point A) is $E = -E_1 + E_2$:

$$E = -E_1 + E_2 = \frac{-kq_1}{r_1^2} + \frac{kq_2}{r_2^2}$$

$$= \frac{-(8.99 \times 10^9 \text{ N} \cdot \text{m}^2/\text{C}^2)(8.5 \times 10^{-6} \text{ C})}{(3.0 \times 10^{-2} \text{ m})^2} + \frac{(8.99 \times 10^9 \text{ N} \cdot \text{m}^2/\text{C}^2)(21 \times 10^{-6} \text{ C})}{(9.0 \times 10^{-2} \text{ m})^2}$$

$$= \boxed{-6.2 \times 10^7 \text{ N/C}}$$

The minus sign tells us that the net electric field points along the $-x$ axis.

b. The net electric field at $x = +6.0$ cm (point B) is $E = E_1 + E_2$:

$$E = E_1 + E_2 = \frac{kq_1}{r_1^2} + \frac{kq_2}{r_2^2}$$

$$= \frac{(8.99 \times 10^9 \text{ N} \cdot \text{m}^2/\text{C}^2)(8.5 \times 10^{-6} \text{ C})}{(3.0 \times 10^{-2} \text{ m})^2} + \frac{(8.99 \times 10^9 \text{ N} \cdot \text{m}^2/\text{C}^2)(21 \times 10^{-6} \text{ C})}{(3.0 \times 10^{-2} \text{ m})^2}$$

$$= \boxed{+2.9 \times 10^8 \text{ N/C}}$$

The plus sign tells us that the net electric field points along the $+x$ axis.

680 ELECTRIC FORCES AND ELECTRIC FIELDS

29. ***REASONING AND SOLUTION***
a. In order for the field to be zero, the point cannot be between the two charges. Instead, it must be located on the line between the two charges on the side of the positive charge and away from the negative charge. If x is the distance from the positive charge to the point in question, then the negative charge is at a distance (3.0 m + x) meters from this point. For the field to be zero here we have

$$kq_-/(3.0 \text{ m} + x)^2 = kq_+/x^2$$

Substituting and solving for x yields (the units have been suppressed for convenience),

$$(12 \times 10^{-6})x^2 - 6.0(4.0 \times 10^{-6})x - 9.0(4.0 \times 10^{-6}) = 0$$

which we can solve for x using the quadratic equation. We obtain $\boxed{x = 3.0 \text{ m}}$.

b. Since the field is zero at this point, the force acting on a charge at that point would be $\boxed{0 \text{ N}}$.

---

30. ***REASONING AND SOLUTION*** The electric field due to a point charge is

$$E = kq/r^2 \qquad \text{so that} \qquad kq = Er^2$$

Since the charge is the same at either location, we can write

$$E_1 r_1^2 = E_2 r_2^2$$

Solving for E_2 we obtain

$$E_2 = E_1(r_1^2/r_2^2) = (160 \text{ N/C})[(0.15 \text{ m})^2/(0.45 \text{ m})^2] = \boxed{18 \text{ N/C}}$$

---

31. $\boxed{\text{SSM}}$ ***REASONING*** Before the 3.0-μC point charge q is introduced into the region, the region contains a uniform electric field **E** of magnitude 1.6×10^4 N/C. After the 3.0-μC charge is introduced into the region, the net electric field changes. In addition to the uniform electric field **E**, the region will also contain the electric field $\mathbf{E}_q$ due to the point charge q. The field at any point in the region is the vector sum of **E** and $\mathbf{E}_q$. The field $\mathbf{E}_q$ is radial as discussed in the text, and its magnitude at any distance r from the charge q is given by Equation 18.3, $E_q = kq/r^2$. There will be one point P in the region where the net electric field $\mathbf{E}_{net}$ is zero. This point is located where the field **E** has the same magnitude and points in the direction opposite to the field $\mathbf{E}_q$. We will use this reasoning to find the distance r_0 from the charge q to the point P.

SOLUTION Let us assume that the field **E** points to the right and that the charge q is *negative* (the problem is done the same way if q is positive, although then the relative positions of P and q will be reversed). Since q is negative, its electric field is radially *inward* (i.e., toward q); therefore, in order for the field $\mathbf{E}_q$ to point in the opposite direction to **E**, the charge q will have to be to the left of the point P where $\mathbf{E}_{net}$ is zero, as shown in the following drawing.

From Equation 18.3, $E_q = kq/r_0^2$, and solving for the distance r_0, we have $r_0 = \sqrt{kq/E_q}$. Since the magnitude E_q must be equal to the magnitude of E at the point P, we have

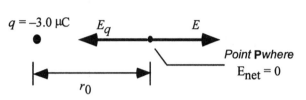

$$r_0 = \sqrt{\frac{kq}{E}} = \sqrt{\frac{(8.99 \times 10^9 \text{ Nm}^2/\text{C}^2)(3.0 \times 10^{-6} \text{ C})}{1.6 \times 10^4 \text{ N/C}}} = \boxed{1.3 \text{ m}}$$

---

32. **REASONING AND SOLUTION** The electric field is defined by Equation 18.2: $\mathbf{E} = \mathbf{F}/q_0$. Thus, the magnitude of the force exerted on a charge q in an electric field of magnitude E is given by

$$F = qE \qquad (1)$$

The magnitude of the electric field can be determined from its x and y components by using the Pythagorean theorem:

$$E = \sqrt{E_x^2 + E_y^2} = \sqrt{(6.00 \times 10^3 \text{ N/C})^2 + (8.00 \times 10^3 \text{ N/C})^2} = 1.00 \times 10^4 \text{ N/C}$$

a. From Equation (1) above, the magnitude of the force on the charge is

$$F = (7.50 \times 10^{-6} \text{ C})(1.00 \times 10^4 \text{ N/C}) = \boxed{7.5 \times 10^{-2} \text{ N}}$$

b. From the defining equation for the electric field, it follows that the direction of the force on a charge is the same as the direction of the field, provided that the charge is positive. Thus, the angle that the force makes with the x axis is given by

$$\theta = \tan^{-1}\left(\frac{E_y}{E_x}\right) = \tan^{-1}\left(\frac{8.00 \times 10^3 \text{ N/C}}{6.00 \times 10^3 \text{ N/C}}\right) = \boxed{53.1°}$$

682   ELECTRIC FORCES AND ELECTRIC FIELDS

33. **REASONING AND SOLUTION** The electric field between the plates of a parallel plate capacitor is given by

$$E = \frac{q}{\varepsilon_0 A} \tag{18.4}$$

Solving for the area of the plate, A, yields

$$A = \frac{q}{\varepsilon_0 E} = \frac{0.15 \times 10^{-6} \text{ C}}{\left[8.85 \times 10^{-12} \text{ C}^2/(\text{N} \cdot \text{m}^2)\right]\left(2.4 \times 10^5 \text{ N/C}\right)} = \boxed{7.1 \times 10^{-2} \text{ m}^2}$$

---

34. **REASONING** The two charges lying on the x axis produce no net electric field at the coordinate origin. This is because they have identical charges, are located the same distance from the origin, and produce electric fields that point in opposite directions. The electric field produced by q_3 at the origin points away from the charge, or along the $-y$ direction. The electric field produced by q_4 at the origin points toward the charge, or along the $+y$ direction. The net electric field is, then, $E = -E_3 + E_4$, where E_3 and E_4 can be determined by using Equation 18.3.

SOLUTION The net electric field at the origin is

$$E = -E_3 + E_4 = \frac{-kq_3}{r_3^2} + \frac{kq_4}{r_4^2}$$

$$= \frac{-\left(8.99 \times 10^9 \text{ N} \cdot \text{m}^2/\text{C}^2\right)\left(3.0 \times 10^{-6} \text{ C}\right)}{\left(5.0 \times 10^{-2} \text{ m}\right)^2} + \frac{\left(8.99 \times 10^9 \text{ N} \cdot \text{m}^2/\text{C}^2\right)\left(8.0 \times 10^{-6} \text{ C}\right)}{\left(7.0 \times 10^{-2} \text{ m}\right)^2}$$

$$= \boxed{+3.9 \times 10^6 \text{ N/C}}$$

The plus sign indicates that $\boxed{\text{the net electric field points along the +y direction}}$.

---

35. **SSM REASONING** Since the charged droplet (charge = q) is suspended motionless in the electric field **E**, the net force on the droplet must be zero. There are two forces that act on the droplet, the force of gravity **W** = $m\mathbf{g}$, and the electric force **F** = $q\mathbf{E}$ due to the electric field. Since the net force on the droplet is zero, we conclude that $mg = qE$. We can use this reasoning to determine the sign and the magnitude of the charge on the droplet.

SOLUTION

a. Since the net force on the droplet is zero, and the weight W points downward, the electric force F = qE must point upward. Since the electric field points upward, the excess charge on the droplet must be positive in order for the force F to point upward.

b. The excess charge on the droplet is, using the expression $mg = qE$,

$$q = \frac{mg}{E} = \frac{(3.50 \times 10^{-9} \text{ kg})(9.80 \text{ m/s}^2)}{8480 \text{ N/C}} = 4.04 \times 10^{-12} \text{ C}$$

The charge on a proton is 1.60×10^{-19} C, so the excess number of protons is

$$(4.04 \times 10^{-12} \text{ C})\left(\frac{1 \text{ proton}}{1.60 \times 10^{-19} \text{ C}}\right) = \boxed{2.53 \times 10^7 \text{ protons}}$$

---

36. **REASONING AND SOLUTION** The figure at the right shows the configuration given in text Figure 18.22b. The electric field at the center of the rectangle is the resultant of the electric fields at the center due to each of the four charges. As discussed in Conceptual Example 11, the magnitudes of the electric field at the center due to each of the four charges are equal. However, the fields produced by the charges in corners 1 and 3 are in opposite directions. Since they have the same magnitudes, they combine to give zero resultant.

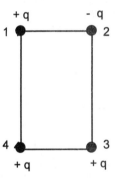

Figure 1

The fields produced by the charges in corners 2 and 4 point in the same direction (toward corner 2).

Thus, $E_C = E_{C2} + E_{C4}$, where E_C is the magnitude of the electric field at the center of the rectangle, and E_{C2} and E_{C4} are the magnitudes of the electric field at the center due to the charges in corners 2 and 4 respectively. Since both E_{C2} and E_{C4} have the same magnitude, we have $E_C = 2 E_{C2}$.

The distance r, from any of the charges to the center of the rectangle, can be found using the Pythagorean theorem:

$$d = \sqrt{(3.00 \text{ cm})^2 + (5.00 \text{ cm})^2} = 5.83 \text{ cm}$$

Therefore, $r = \dfrac{d}{2} = 2.92 \text{ cm} = 2.92 \times 10^{-2} \text{ m}$

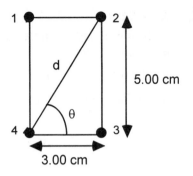

Figure 2

The electric field at the center has a magnitude of

$$E_C = 2E_{C2} = \frac{2kq_2}{r^2} = \frac{2(8.99 \times 10^9 \text{ N} \cdot \text{m}^2/\text{C}^2)(8.60 \times 10^{-12} \text{ C})}{(2.92 \times 10^{-2} \text{ m})^2} = \boxed{1.81 \times 10^2 \text{ N/C}}$$

The figure at the right shows the configuration given in text Figure 18.22c. All four charges contribute a non-zero component to the electric field at the center of the rectangle. As discussed in Conceptual Example 11, the contribution from the charges in corners 2 and 4 point toward corner 2 and the contribution from the charges in corners 1 and 3 point toward corner 1.

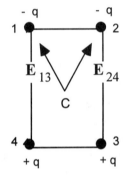

Notice also, the magnitudes of E_{24} and E_{13} are equal, and, from the first part of this problem, we can know that

$$E_{24} = E_{13} = 1.81 \times 10^2 \text{ N/C}$$

Figure 3

The electric field at the center of the rectangle is the vector sum of $\mathbf{E}_{24}$ and $\mathbf{E}_{13}$. The x components of $\mathbf{E}_{24}$ and $\mathbf{E}_{13}$ are equal in magnitude and opposite in direction; hence

$$(E_{13})_x - (E_{24})_x = 0$$

Therefore,
$$E_C = (E_{13})_y + (E_{24})_y = 2(E_{13})_y = 2(E_{13})\sin\theta$$

From Figure 2 above,
$$\sin\theta = \frac{5.00 \text{ cm}}{d} = \frac{5.00 \text{ cm}}{5.83 \text{ cm}} = 0.858$$

and
$$E_C = 2(E_{13})\sin\theta = 2(1.81 \times 10^2 \text{ N/C})(0.858) = \boxed{3.11 \times 10^2 \text{ N/C}}$$

37. **REASONING AND SOLUTION** The average force $\overline{F}$ on the proton can be determined from the impulse-momentum theorem (Equation 7.4):

$$\overline{F}\Delta t = \Delta(mv)$$

$$\overline{F} = \frac{mv - mv_0}{\Delta t} = \frac{(5.0 \times 10^{-23} \text{ kg} \cdot \text{m/s}) - (1.5 \times 10^{-23} \text{ kg} \cdot \text{m/s})}{6.3 \times 10^{-6} \text{ s}} = 5.6 \times 10^{-18} \text{ N}$$

From the definition of electric field: $E = \dfrac{\overline{F}}{q_0}$, we have

$$E = \frac{\overline{F}}{q} = \frac{5.6 \times 10^{-18} \text{ N}}{1.60 \times 10^{-19} \text{ C}} = \boxed{35 \text{ N/C}}$$

---

38. **REASONING AND SOLUTION** From kinematics, $v_y^2 = v_{0y}^2 + 2a_y y$. Since the electron starts from rest, $v_{0y} = 0$ m/s. The acceleration of the proton is given by

$$a_y = \frac{F}{m} = \frac{eE}{m}$$

where e and m are the electron's charge magnitude and mass, respectively, and E is the magnitude of the electric field. The magnitude of the electric field between the plates of a parallel plate capacitor is $E = \sigma/\varepsilon_0$, where σ is the magnitude of the charge per unit area on each plate. Thus, $a_y = e\sigma/(m\varepsilon_0)$. Combining this expression for a with the kinematics equation we have

$$v_y^2 = 2\left(\frac{e\sigma}{m\varepsilon_0}\right) y$$

Solving for v_y gives

$$v_y = \sqrt{\frac{2e\sigma y}{m\varepsilon_0}} = \sqrt{\frac{2(1.60 \times 10^{-19} \text{ C})(1.8 \times 10^{-7} \text{ C/m}^2)(1.5 \times 10^{-2} \text{ m})}{(9.11 \times 10^{-31} \text{ kg})[8.85 \times 10^{-12} \text{ C}^2/(\text{N} \cdot \text{m}^2)]}} = \boxed{1.0 \times 10^7 \text{ m/s}}$$

686 ELECTRIC FORCES AND ELECTRIC FIELDS

39. **SSM** **WWW** *REASONING* The figure shows the arrangement of the three charges. Let $\mathbf{E}_q$ represent the electric field at the empty corner due to the $-q$ charge. Furthermore, let $\mathbf{E}_1$ and $\mathbf{E}_2$ be the electric fields at the empty corner due to charges $+q_1$ and $+q_2$, respectively.

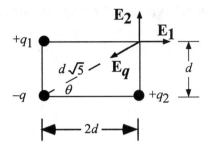

According to the Pythagorean theorem, the distance from the charge $-q$ to the empty corner along the diagonal is given by $\sqrt{(2d)^2 + d^2} = \sqrt{5d^2} = d\sqrt{5}$. The magnitude of each electric field is given by Equation 18.3, $E = kq/r^2$. Thus, the magnitudes of each of the electric fields at the empty corner are given as follows: $E_q = kq/r^2 = kq/(d\sqrt{5})^2 = kq/(5d^2)$, since the length of the diagonal is $d\sqrt{5}$; $E_1 = kq_1/(4d^2)$; and $E_2 = kq_2/d^2$. The angle θ that the diagonal makes with the horizontal is $\theta = \tan^{-1}(d/2d) = 26.57°$. Since the net electric field E_{net} at the empty corner is zero, the horizontal component of the net field must be zero, and we have

$$E_1 - E_q \cos 26.57° = 0 \quad \text{or} \quad kq_1/(4d^2) - kq(\cos 26.57°)/(5d^2) = 0$$

Similarly, the vertical component of the net field must be zero, and we have

$$E_2 - E_q \sin 26.57° = 0 \quad \text{or} \quad kq_2/d^2 - kq(\sin 26.57°)/(5d^2) = 0$$

These last two expressions can be solved for the charges q_1 and q_2.

SOLUTION Solving the last two expressions for the charges q_1 and q_2, we have

$$q_1 = \tfrac{4}{5} q \, \cos 26.57° = \boxed{0.716 \, q}$$

$$q_2 = \tfrac{1}{5} q \, \sin 26.57° = \boxed{0.0895 \, q}$$

---

40. *REASONING* The magnitude E of the electric field is defined as the magnitude F of the electric force exerted on a small test charge divided by the charge: $E = F/q$. According to Newton's second law, Equation 4.2, the net force acting on an object is equal to its mass m times its acceleration a. Since there is only one force acting on the object, it is the net force. Thus, the magnitude of the electric field can be written as

$$E = \frac{F}{q} = \frac{ma}{q}$$

The acceleration is related to the initial and final velocities, v_0 and v, and the time t through Equation 2.4, as $a = \dfrac{v - v_0}{t}$. Substituting this expression for a into the one above for E gives

$$E = \frac{ma}{q} = \frac{m\left(\dfrac{v - v_0}{t}\right)}{q} = \frac{m(v - v_0)}{qt}$$

SOLUTION The magnitude E of the electric field is

$$E = \frac{m(v - v_0)}{qt} = \frac{(9.0 \times 10^{-5} \text{ kg})(2.0 \times 10^3 \text{ m/s} - 0 \text{ m/s})}{(7.5 \times 10^{-6} \text{ C})(0.96 \text{ s})} = \boxed{2.5 \times 10^4 \text{ N/C}}$$

---

41. **REASONING AND SOLUTION** The net field at the center of the circle is E_n, and since E_1 and E_2 are perpendicular to one another, we see that

$$E_1 = E_n \cos 20.0° \quad \text{and} \quad E_2 = E_n \sin 20.0°$$

Solving for E_n in each equation and equating the two expressions gives

$$E_1/(\cos 20.0°) = E_2/(\sin 20.0°)$$

Using $E = kq/r^2$ we have

$$q_1/(\cos 20.0°) = q_2/(\sin 20.0°)$$

so

$$(q_2/q_1) = (\sin 20.0°/\cos 20.0°) = \boxed{0.364}$$

---

42. **REASONING AND SOLUTION** From two-dimensional kinematics, taking the entry point as the origin, we have

$$x = v_{0x} t + \tfrac{1}{2} a_x t^2 \qquad (1)$$

$$y = v_{0y} t + \tfrac{1}{2} a_y t^2 \qquad (2)$$

There is no acceleration in the x-direction, so $a_x = 0 \text{ m/s}^2$. Initially, the electron travels in the +x direction, so $v_{0y} = 0 \text{ m/s}$. Solving Equation (1) for t and substituting into Equation (2) gives:

$$y = \frac{1}{2}a_y\left(\frac{x}{v_{0x}}\right)^2 \qquad (3)$$

From Newton's second law and the definition of electric field we have

$$a_y = \frac{F}{m} = \frac{eE}{m}$$

Thus, Equation (3) becomes

$$y = \frac{1}{2}\left(\frac{eE}{m}\right)\left(\frac{x}{v_{0x}}\right)^2$$

where e is the magnitude of the electron's charge and m is its mass. Solving for E gives

$$E = \frac{2my}{e}\left(\frac{v_{0x}}{x}\right)^2 = \frac{2(9.11 \times 10^{-31}\text{ kg})(1.50 \times 10^{-3}\text{ m})}{(1.60 \times 10^{-19}\text{ C})}\left(\frac{7.00 \times 10^6\text{ m/s}}{2.00 \times 10^{-2}\text{ m}}\right)^2 = \boxed{2.09 \times 10^3 \text{ N/C}}$$

---

43. **SSM** **REASONING** The net electric field at point P in Figure 1 is the vector sum of the fields $\mathbf{E}_+$ and $\mathbf{E}_-$, which are due, respectively, to the charges $+q$ and $-q$. These fields are shown in Figure 2.

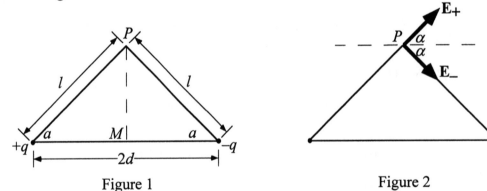

Figure 1   Figure 2

According to Equation 18.3, the magnitudes of the fields E_+ and E_- are the same, since the triangle is an isosceles triangle with equal sides of length ℓ. Therefore, $E_+ = E_- = kq/\ell^2$. The vertical components of these two fields cancel, while the horizontal components reinforce, leading to a total field at point P that is horizontal and has a magnitude of

$$E_P = E_+ \cos\alpha + E_- \cos\alpha = 2\left(\frac{kq}{\ell^2}\right)\cos\alpha$$

At point M in Figure 1, both $\mathbf{E}_+$ and $\mathbf{E}_-$ are horizontal and point to the right. Again using Equation 18.3, we find

$$E_M = E_+ + E_- = \frac{kq}{\ell^2} + \frac{kq}{\ell^2} = \frac{2kq}{\ell^2}$$

Since $E_M/E_P = 9.0$, we have

$$\frac{E_M}{E_P} = \frac{2kq/d^2}{2kq(\cos\alpha)/\ell^2} = \frac{1}{(\cos\alpha)d^2/\ell^2} = 9.0$$

But from Figure 1, we can see that $d/\ell = \cos\alpha$. Thus, it follows that

$$\frac{1}{\cos^3\alpha} = 9.0 \quad \text{or} \quad \cos\alpha = \sqrt[3]{1/9.0} = 0.48$$

The value for α is, then, $\alpha = \cos^{-1}(0.48) = \boxed{61°}$.

---

44. **REASONING AND SOLUTION** The electric field due to a parallel plate capacitor is given by $E = \sigma/\varepsilon_0$ (Equation 18.4).

a. The induced charge density σ is, therefore,

$$\sigma = \varepsilon_0 E = [8.85 \times 10^{-12} \text{ C}^2/(\text{N·m}^2)](480 \text{ N/C}) = \boxed{4.2 \times 10^{-9} \text{ C/m}^2}$$

b. The area of one face of the circular coin is

$$A = \pi r^2 = \pi (0.019 \text{ m})^2$$

The total charge is

$$q = \sigma A = \boxed{4.8 \times 10^{-12} \text{ C}}$$

---

45. **REASONING AND SOLUTION** Since the thread makes an angle of 30.0° with the vertical, it can be seen that the electric force on the ball, F_e, and the gravitational force, mg, are related by

$$F_e = mg \tan 30.0°$$

The force F_e is due to the charged ball being in the electric field of the parallel plate capacitor. That is,

$$F_e = Eq$$

where q is the ball's charge and E is the field due to the plates, given by

690 ELECTRIC FORCES AND ELECTRIC FIELDS

$$E = Q/(\varepsilon_0 A) \quad (18.4)$$

Combining equations gives

$$mg \tan 30.0° = qQ/(\varepsilon_0 A)$$

Solving for Q yields

$$Q = \frac{\varepsilon_0 A mg \tan 30.0°}{q}$$

$$= \frac{\left[8.85 \times 10^{-12}\ C^2/(N \cdot m^2)\right](0.0150\ m^2)(6.50 \times 10^{-3}\ kg)(9.80\ m/s^2) \tan 30.0°}{0.150 \times 10^{-6}\ C}$$

$$= \boxed{3.25 \times 10^{-8}\ C}$$

---

46. **REASONING AND SOLUTION** Gauss' Law is given by text Equation 18.7: $\Phi_E = \dfrac{Q}{\varepsilon_0}$, where Q is the *net charge* enclosed by the Gaussian surface.

a. $\Phi_E = \dfrac{3.5 \times 10^{-6}\ C}{8.85 \times 10^{-12}\ C^2/(N \cdot m^2)} = \boxed{4.0 \times 10^5\ N \cdot m^2/C}$

b. $\Phi_E = \dfrac{-2.3 \times 10^{-6}\ C}{8.85 \times 10^{-12}\ C^2/(N \cdot m^2)} = \boxed{-2.6 \times 10^5\ N \cdot m^2/C}$

c. $\Phi_E = \dfrac{(3.5 \times 10^{-6}\ C) + (-2.3 \times 10^{-6}\ C)}{8.85 \times 10^{-12}\ C^2/(N \cdot m^2)} = \boxed{1.4 \times 10^5\ N \cdot m^2/C}$

---

47. **SSM** *REASONING* As discussed in Section 18.9, the electric flux Φ_E through a surface is equal to the component of the electric field that is normal to the surface multiplied by the area of the surface, $\Phi_E = E_\perp A$, where $E_\perp$ is the component of **E** that is normal to the surface of area A. We can use this expression and the figure in the text to determine the flux through the two surfaces.

SOLUTION

a. The flux through surface 1 is

$$(\Phi_E)_1 = (E \cos 35°)A_1 = (250 \text{ N/C})(\cos 35°)(1.7 \text{ m}^2) = \boxed{350 \text{ N} \cdot \text{m}^2/\text{C}}$$

b. Similarly, the flux through surface 2 is

$$(\Phi_E)_2 = (E \cos 55°)A_2 = (250 \text{ N/C})(\cos 55°)(3.2 \text{ m}^2) = \boxed{460 \text{ N} \cdot \text{m}^2/\text{C}}$$

---

48. **REASONING AND SOLUTION** The maximum possible flux occurs when the electric field is parallel to the normal of the rectangular surface (that is, when the angle between the direction of the field and the direction of the normal is zero). Then

$$\Phi_E = (E \cos \phi)A = (580 \text{ N/C})(\cos 0°)(0.16 \text{ m})(0.38 \text{ m}) = \boxed{35 \text{ N} \cdot \text{m}^2/\text{C}}$$

---

49. **REASONING AND SOLUTION**

a. In all three cases, the net charge enclosed by the surface is the same, because the net charge enclosed by each surface is the same; therefore, by Gauss' Law, the electric flux through the surfaces described in parts (a) through (c) is the same:

$$\Phi_E = \frac{Q}{\varepsilon_0} = \frac{2.0 \times 10^{-6} \text{ C}}{8.85 \times 10^{-12} \text{ C}^2/(\text{N} \cdot \text{m}^2)} = \boxed{2.3 \times 10^5 \text{ N} \cdot \text{m}^2/\text{C}}$$

b. $\Phi_E = \boxed{2.3 \times 10^5 \text{ N} \cdot \text{m}^2/\text{C}}$

c. $\Phi_E = \boxed{2.3 \times 10^5 \text{ N} \cdot \text{m}^2/\text{C}}$

---

50. **REASONING AND SOLUTION** Since the electric field is uniform, its magnitude and direction are the same at each point on the wall. The angle ϕ between the electric field and the normal to the wall is 35°. Therefore, the electric flux is

$$\Phi_E = (E \cos \phi) A = (150 \text{ N/C})(\cos 35°)[(5.9 \text{ m})(2.5 \text{ m})] = \boxed{1.8 \times 10^3 \text{ N} \cdot \text{m}^2/\text{C}}$$

51. **SSM** *REASONING* The electric flux through each face of the cube is given by $\Phi_E = (E\cos\phi)A$ (see Section 18.9) where E is the magnitude of the electric field at the face, A is the area of the face, and ϕ is the angle between the electric field and the outward normal of that face. We can use this expression to calculate the electric flux Φ_E through each of the six faces of the cube.

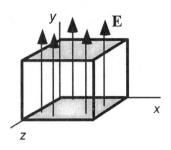

SOLUTION
a. On the bottom face of the cube, the outward normal points parallel to the $-y$ axis, in the opposite direction to the electric field, and $\phi = 180°$. Therefore,

$$(\Phi_E)_{bottom} = (1500 \text{ N/C})(\cos 180°)(0.20 \text{ m})^2 = \boxed{-6.0 \times 10^1 \text{ N} \cdot \text{m}^2/\text{C}}$$

On the top face of the cube, the outward normal points parallel to the $+y$ axis, and $\phi = 0.0°$. The electric flux is, therefore,

$$(\Phi_E)_{top} = (1500 \text{ N/C})(\cos 0.0°)(0.20 \text{ m})^2 = \boxed{+6.0 \times 10^1 \text{ N} \cdot \text{m}^2/\text{C}}$$

On each of the other four faces, the outward normals are perpendicular to the direction of the electric field, so $\phi = 90°$. So for each of the four side faces,

$$(F_E)_{sides} = (1500 \text{ N/C})(\cos 90°)(0.20 \text{ m})^2 = \boxed{0 \text{ N} \cdot \text{m}^2/\text{C}}$$

b. The total flux through the cube is

$$(\Phi_E)_{total} = (\Phi_E)_{top} + (\Phi_E)_{bottom} + (\Phi_E)_{side\,1} + (\Phi_E)_{side\,2} + (\Phi_E)_{side\,3} + (\Phi_E)_{side\,4}$$

Therefore,

$$(\Phi_E)_{total} = (+6.0 \times 10^1 \text{ N} \cdot \text{m}^2/\text{C}) + (-6.0 \times 10^1 \text{ N} \cdot \text{m}^2/\text{C}) + 0 + 0 + 0 + 0 = \boxed{0 \text{ N} \cdot \text{m}^2/\text{C}}$$

---

52. *REASONING AND SOLUTION* Since both charge distributions are uniformly spread over concentric spherical shells, the electric field possesses spherical symmetry. Gauss' law can be used to determine the magnitude of the electric field, provided we choose spherical Gaussian surfaces (concentric with the spherical shells) to evaluate the electric flux. To find the magnitude of the electric field at any distance r from the center of the spherical shells, we construct a spherical Gaussian surface of radius r. The electric flux through this Gaussian surface is

$$\Phi_E = \Sigma(E \cos \phi)\Delta A$$

Because the charge distributions have spherical symmetry, we expect the electric field to be directed radially. That is, the electric field is everywhere perpendicular to the Gaussian surface. Thus, for any surface element, ϕ will be 0 or 180°. Furthermore, since the charge distribution possesses spherical symmetry, we expect the electric field to be uniform in magnitude over any sphere concentric with the shells. Thus, E is constant over any Gaussian surface concentric with the shells. Then, (E cos ϕ) can be factored out of the summation.

$$\Phi_E = \Sigma(E \cos \phi)\Delta A = (E \cos \phi)\Sigma \Delta A$$

where $\Sigma \Delta A$ is the sum of the area elements that make up the Gaussian surface. This sum must equal the surface area of the Gaussian surface or

$$\Phi_E = (E \cos \phi)\Sigma \Delta A = (E \cos \phi)(4\pi r^2)$$

where r is the radius of the Gaussian surface. From Gauss' law this becomes

$$(E \cos \phi)(4\pi r^2) = \frac{Q}{\varepsilon_0} \tag{1}$$

where Q is the net charge enclosed by the Gaussian surface.

a. **r = 0.20 m**

The Gaussian surface encloses both shells. The net charge enclosed is

$(+5.1 \times 10^{-6}$ C$) + (-1.6 \times 10^{-6}$ C$) = +3.5 \times 10^{-6}$ C

Since the net charge is positive, E will be radially outward for all points on the Gaussian surface, and $\phi = 0.0°$ for all elements on the Gaussian surface.

Solving Equation (1) for E gives

$$E = \frac{Q}{\varepsilon_0 (4\pi r^2)} = \frac{3.5 \times 10^{-6} \text{ C}}{[8.85 \times 10^{-12} \text{ C}^2/(\text{N} \cdot \text{m}^2)][4\pi(0.20 \text{ m})^2]} = \boxed{7.9 \times 10^5 \text{ N/C}}$$

The direction of **E** is radially outward , because the net charge within the Gaussian surface is positive.

b. r = 0.10 m

The Gaussian surface encloses only the inner shell. The net charge enclosed is

$$Q = -1.6 \times 10^{-6} \text{ C}$$

Since the net charge is negative, E will be radially inward for all points on the Gaussian surface, and $\phi = 180°$ for all elements on the Gaussian surface.

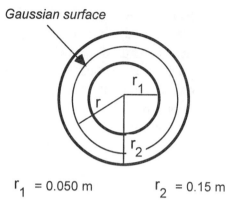

$r_1 = 0.050$ m $r_2 = 0.15$ m

Solving Equation (1) for E gives

$$E = -\frac{Q}{\varepsilon_0(4\pi r^2)} = -\frac{(-1.6 \times 10^{-6} \text{ C})}{[8.85 \times 10^{-12} \text{ C}^2/(\text{N} \cdot \text{m}^2)][4\pi(0.10 \text{ m})^2]} = \boxed{1.4 \times 10^6 \text{ N/C}}$$

The direction of **E** is $\boxed{\text{radially inward}}$, because the net charge within the Gaussian surface is negative.

c. r = 0.025 m

The net charge enclosed by the Gaussian surface is zero. This implies that the net electric flux is zero, so the electric field is either a constant or zero everywhere within the Gaussian surface. However, an electric field does not exist within the Gaussian surface, because there are only negative charges on the shell of radius r_1, so electric field lines cannot originate from any place on this shell. Thus, $\boxed{E = 0 \text{ N/C}}$ in this region.

---

53. **REASONING** Because the charge is distributed uniformly along the straight wire, the electric field is directed radially outward, as the end view of the wire given below illustrates.

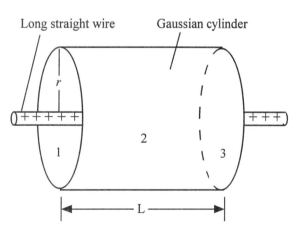
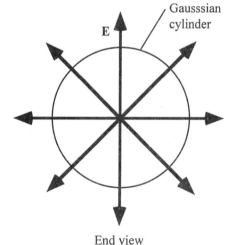

End view

And because of symmetry, the magnitude of the electric field is the same at all points equidistant from the wire. In this situation we will use a Gaussian surface that is a cylinder concentric with the wire. The drawing shows that this cylinder is composed of three parts, the two flat ends (1 and 3) and the curved wall (2). We will evaluate the electric flux for this three-part surface and then set it equal to Q/ε_0 (Gauss' law) to find the magnitude of the electric field.

SOLUTION Surfaces 1 and 3 – the flat ends of the cylinder – are parallel to the electric field, so $\cos\phi = \cos 90° = 0$. Thus, there is no flux through these two surfaces: $\Phi_1 = \Phi_3 = 0 \text{ N·m}^2/\text{C}$.

Surface 2 – the curved wall – is everywhere perpendicular to the electric field **E**, so $\cos\phi = \cos 0° = 1$. Furthermore, the magnitude E of the electric field is the same for all points on this surface, so it can be factored outside the summation in Equation 18.6:

$$\Phi_2 = \Sigma(E\cos 0°)\Delta A = E\Sigma A$$

The area ΣA of this surface is just the circumference $2\pi r$ of the cylinder times its length L: $\Sigma A = (2\pi r)L$. The electric flux through the entire cylinder is, then,

$$\Phi_E = \Phi_1 + \Phi_2 + \Phi_3 = 0 + E(2\pi rL) + 0 = E(2\pi rL)$$

Following Gauss' law, we set Φ_E equal to Q/ε_0, where Q is the net charge inside the Gaussian cylinder: $E(2\pi rL) = Q/\varepsilon_0$. The ratio Q/L is the charge per unit length of the wire and is known as the linear charge density λ: $\lambda = Q/L$. Solving for E, we find that

$$\boxed{E = \frac{Q/L}{2\pi\varepsilon_0 r} = \frac{\lambda}{2\pi\varepsilon_0 r}}$$

---

54. **REASONING AND SOLUTION** The electric field lines must originate on the positive charges and terminate on the negative charge. They cannot cross one another. Furthermore, the number of field lines beginning or terminating on any charge must be proportional to the magnitude of the charge. Thus, for every field line that leaves the charge +q, two field lines must leave the charge +2q. These three lines must terminate on the -3q charge. If the sketch is to have six field lines, two of them must originate on +q, and four of them must originate on the charge +2q.

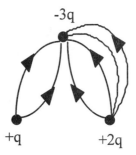

696 ELECTRIC FORCES AND ELECTRIC FIELDS

55. **SSM REASONING AND SOLUTION** The magnitude of the force of attraction between the charges is given by Coulomb's law (Equation 18.1): $F = kq_1q_2/r^2$, where q_1 and q_2 are the magnitudes of the charges and r is the separation of the charges. Let F_A and F_B represent the magnitudes of the forces between the charges when the separations are r_A and $r_B = r_A/9$, respectively. Then

$$\frac{F_B}{F_A} = \frac{kq_1q_2/r_B^2}{kq_1q_2/r_A^2} = \left(\frac{r_A}{r_B}\right)^2 = \left(\frac{r_A}{r_A/9}\right)^2 = (9)^2 = 81$$

Therefore, we can conclude that $F_B = 81 F_A = (81)(1.5 \text{ N}) = \boxed{120 \text{ N}}$.

---

56. **REASONING AND SOLUTION** The $+2q$ of charge initially on the sphere lies entirely on the outer surface. When the $+q$ charge is placed inside of the sphere, then a $\boxed{-q}$ charge will still be induced on the interior of the sphere. An additional $+q$ will appear on the outer surface, giving a net charge of $\boxed{+3q}$.

---

57. **REASONING** The drawing at the right shows the set-up. The force on the $+q$ charge at the origin due to the other $+q$ charge is given by Coulomb's law (Equation 18.1), as is the force due to the $+2q$ charge. These two forces point to the left, since each is repulsive. The sum of the two is twice the force on the $+q$ charge at the origin due to the other $+q$ charge alone.

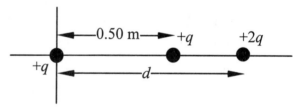

SOLUTION Applying Coulomb's law, we have

$$\underbrace{\frac{k(q)(q)}{(0.50 \text{ m})^2}}_{\substack{\text{Force due to } +q \\ \text{charge at } x = +0.50 \text{ m}}} + \underbrace{\frac{k(2q)(q)}{(d)^2}}_{\substack{\text{Force due to } +2q \\ \text{charge at } x = +d}} = \underbrace{2\frac{k(q)(q)}{(0.50 \text{ m})^2}}_{\substack{\text{Twice the force due to} \\ +q \text{ charge at } x = +0.50 \text{ m}}}$$

Rearranging this result and solving for d give

$$\frac{k(2q)(q)}{(d)^2} = \frac{k(q)(q)}{(0.50 \text{ m})^2} \quad \text{or} \quad d^2 = 2(0.50 \text{ m})^2 \quad \text{or} \quad d = \pm 0.71 \text{ m}$$

We reject the negative root, because a negative value for d would locate the $+2q$ charge to the left of the origin. Then, the two forces acting on the charge at the origin would have different directions, contrary to the statement of the problem. Therefore, the $+2q$ charge is located at a position of $\boxed{x = +0.71 \text{ m}}$.

---

58. **REASONING AND SOLUTION**
a. The electrostatic force acting on q_2 is the vector sum of the forces due to charges q_1 and q_3, i.e., $\mathbf{F}_2 = \mathbf{F}_{12} + \mathbf{F}_{32}$. We can write for the magnitude of each force,

$$F_{12} = \frac{kq_1q_2}{r^2} = \frac{(8.99 \times 10^9 \text{ N} \cdot \text{m}^2/\text{C}^2)(25 \times 10^{-6} \text{ C})(11 \times 10^{-6} \text{ C})}{(2.0 \text{ m})^2} = 0.62 \text{ N}$$

The direction of the force is to the right (+x direction).

$$F_{32} = \frac{kq_3q_2}{r^2} = \frac{(8.99 \times 10^9 \text{ N} \cdot \text{m}^2/\text{C}^2)(45 \times 10^{-6} \text{ C})(11 \times 10^{-6} \text{ C})}{(1.5 \text{ m})^2} = 2.0 \text{ N}$$

The direction of the force is to the left (–x direction).

Since F_{32} is larger than F_{12}, the net force is $\boxed{\text{to the left (–x direction)}}$ and has a magnitude of

$$F_2 = 2.0 \text{ N} - 0.62 \text{ N} = \boxed{1.4 \text{ N}}$$

b. If q_2 were negative, the net force would still be 1.4 N, but would be in the +x direction instead. This is because F_{12} would now be to the left (since q_1 attracts q_2), with the same magnitude, and F_{32} would act to the right. The resulting force $F_2 = \boxed{1.4 \text{ N}}$ would now be $\boxed{\text{to the right (+x direction)}}$.

---

59. **SSM** **WWW** **REASONING** Two forces act on the charged ball (charge q); they are the downward force of gravity $m\mathbf{g}$ and the electric force $\mathbf{F}$ due to the presence of the charge q in the electric field $\mathbf{E}$. In order for the ball to float, these two forces must be equal in magnitude and opposite in direction, so that the net force on the ball is zero (Newton's second law). Therefore, $\mathbf{F}$ must point upward, which we will take as the positive direction. According to Equation 18.2, $\mathbf{F} = q\mathbf{E}$. Since the charge q is negative, the electric field $\mathbf{E}$ must point downward, as the product $q\mathbf{E}$ in the expression $\mathbf{F} = q\mathbf{E}$ must be positive, since the force $\mathbf{F}$ points upward. The magnitudes of the two forces must be equal, so that $mg = qE$. This expression can be solved for E.

SOLUTION The magnitude of the electric field E is

$$E = \frac{mg}{q} = \frac{(0.012 \text{ kg})(9.80 \text{ m/s}^2)}{18 \times 10^{-6} \text{ C}} = \boxed{6.5 \times 10^3 \text{ N/C}}$$

As discussed in the reasoning, this electric field points $\boxed{\text{downward}}$.

---

60. **REASONING** The drawing at the right shows the set-up. Here, the electric field **E** points along the +y axis and applies a force of +**F** to the +q charge and a force of −**F** to the −q charge, where q = 8.0 μC denotes the magnitude of each charge. Each force has the same magnitude of F = Eq, according to Equation 18.2. The torque is measured as discussed in Section 9.1. According to Equation 9.1, the torque produced by each force has a magnitude given by the magnitude of the force times the lever arm, which is the perpendicular distance between the point of application of the force and the axis of rotation. In the drawing the z axis is the axis of rotation and is midway between the ends of the rod. Thus, the lever arm for each force is half the length L of the rod or L/2, and the magnitude of the torque produced by each force is $\tau = (Eq)(L/2)$.

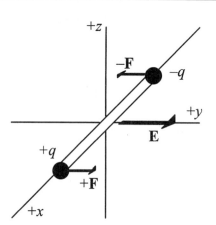

SOLUTION The +**F** and the −**F** force each cause the rod to rotate in the same sense about the z axis. Therefore, the torques from these forces reinforce one another. Using $\tau = (Eq)(L/2)$ for each torque, we find that the magnitude of the net torque is

$$\tau = Eq\left(\frac{L}{2}\right) + Eq\left(\frac{L}{2}\right) = EqL = (5.0 \times 10^3 \text{ N/C})(8.0 \times 10^{-6} \text{ C})(4.0 \text{ m}) = \boxed{0.16 \text{ N} \cdot \text{m}}$$

---

61. **SSM** **REASONING AND SOLUTION**

a. Since the gravitational force between the spheres is one of attraction and the electrostatic force must balance it, the electric force must be one of repulsion. Therefore, the charges must have $\boxed{\text{the same algebraic signs, both positive or both negative}}$.

b. There are two forces that act on each sphere; they are the gravitational attraction F_G of one sphere for the other, and the repulsive electric force F_E of one sphere on the other. From the problem statement, we know that these two forces balance each other, so that $F_G = F_E$. The magnitude of F_G is given by Newton's law of gravitation (Equation 4.3: $F_G = Gm_1m_2/r^2$), while the magnitude of F_E is given by Coulomb's law (Equation 18.1: $F_E = kq_1q_2/r^2$). Therefore, we have $Gm_1m_2/r^2 = kq_1q_2/r^2$, and since the spheres have the same mass and carry charges of the same magnitude, $Gm^2/r^2 = kq^2/r^2$. Solving for q, we find

$$q = m\sqrt{\frac{G}{k}} = (2.0 \times 10^{-6} \text{ kg})\sqrt{\frac{6.67 \times 10^{-11} \text{ N} \cdot \text{m}^2/\text{kg}^2}{8.99 \times 10^9 \text{ N} \cdot \text{m}^2/\text{C}^2}} = \boxed{1.7 \times 10^{-16} \text{ C}}$$

62. **REASONING AND SOLUTION** The force on q_1 due to q_2 is given by Coulomb's law:

$$F_{12} = \frac{kq_1q_2}{r_{12}^2} \quad (1)$$

The force on q_1 due to the electric field of the capacitor is given by

$$F_{1C} = q_1 E_C = q_1\left(\frac{\sigma}{\varepsilon_0}\right) \quad (2)$$

Equating the right hand sides of Equations (1) and (2) above gives

$$\frac{kq_1q_2}{r_{12}^2} = q_1\left(\frac{\sigma}{\varepsilon_0}\right)$$

Solving for r_{12} gives

$$r_{12} = \sqrt{\frac{\varepsilon_0 k q_2}{\sigma}}$$

$$= \sqrt{\frac{[8.85 \times 10^{-12} \text{ C}^2/(\text{N} \cdot \text{m}^2)](8.99 \times 10^9 \text{ N} \cdot \text{m}^2/\text{C}^2)(5.00 \times 10^{-6} \text{ C})}{(1.30 \times 10^{-4} \text{ C/m}^2)}} = \boxed{5.53 \times 10^{-2} \text{ m}}$$

63. **REASONING AND SOLUTION** We need to find the electric force, and hence the electric field, at the point in question. The net force is

$$\mathbf{F} = m\mathbf{a} = (2.0 \times 10^{-3} \text{ kg})(3.5 \times 10^3 \text{ m/s}^2) = 7.0 \text{ N} \quad (+x \text{ direction})$$

The electric field is

$$E = F/q = (7.0 \text{ N})/(25 \times 10^{-6} \text{ C}) = \boxed{2.8 \times 10^5 \text{ N/C}}$$

Since the charge is negative, the field is directed opposite to the direction of the force. Therefore, the electric field is along the $\boxed{-x \text{ direction}}$.

700 ELECTRIC FORCES AND ELECTRIC FIELDS

64. **REASONING AND SOLUTION**
a. Since the spring is stretched, the electric force must be a repulsion. Therefore, the charges must be the same polarity (both positive or both negative) .

b. The force needed to stretch the spring is $F = k_{spring} x$, which is provided by the electric force given by Coulomb's law.

$$\frac{kq^2}{r^2} = k_{spring} x$$

$$q = \sqrt{\frac{k_{spring} x r^2}{k}} = \sqrt{\frac{(220 \text{ N/m})(0.020 \text{ m})(0.34 \text{ m})^2}{8.99 \times 10^9 \text{ N} \cdot \text{m}^2 / \text{C}^2}} = \boxed{7.5 \times 10^{-6} \text{ C}}$$

---

65. **SSM REASONING** The charge at A experiences a force $\mathbf{F_A}$ given by $\mathbf{F_A} = q_A \mathbf{E_A}$, where $\mathbf{E_A}$ is the net electric field at A. The net electric field $\mathbf{E_A}$ is the vector sum of the fields due to the presence of the other two charges. In order for the net force on the charge at corner A to point along the vertical, the fourth charge that is placed at the empty corner, must produce an electric field $\mathbf{E_q}$ at A with an x-component that is equal in magnitude and opposite in direction to the x-component of $\mathbf{E_A}$. The x-component of $\mathbf{E_A}$ is due solely to the +3.0 μC charge in the lower right corner. According to Equation 18.3, then, $E_{Ax} = kq_{right}/(16d^2)$, where q_{right} is the charge in the lower right corner. Since q_{right} is positive, E_{Ax} points to the left. The x component of the field $\mathbf{E_q}$ must, therefore, be equal to $E_{qx} = kq_{right}/(16d^2)$ and point to the right. We will now use this expression to solve for q.

SOLUTION According to Equation 18.3, the electric field $\mathbf{E_q}$ has a magnitude that is given by

$$E_q = kq/r^2 = kq/(d^2 + 16d^2) = kq/(17d^2)$$

Furthermore, this field must point along the diagonal from the charge q_A toward the empty corner, as shown in the following drawing. Therefore, the charge placed at the empty corner must be negative.

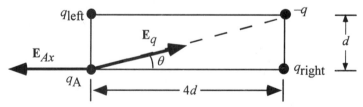

The angle θ is given by $\theta = \tan^{-1}[d/(4d)] = \tan^{-1}(0.250) = 14.0°$. Using the arguments developed in the reasoning, we have $E_{qx} = kq_{right}/(16d^2)$, or

$$E_{qx} = E_q \cos\theta = \frac{kq}{17d^2}\cos\theta = \frac{kq_{\text{right}}}{16d^2}$$

Solving for the charge magnitude q, we obtain

$$q = \frac{q_{\text{right}}}{\cos\theta}\left(\frac{17}{16}\right) = \left(\frac{3.0\times 10^{-6}\text{ C}}{\cos 14.0°}\right)\left(\frac{17}{16}\right) = 3.3\times 10^{-6}\text{ C}$$

Thus, the charge at the empty corner is $\boxed{-3.3\times 10^{-6}\text{ C}}$.

---

66. **REASONING** The proton (charge = $+e = 1.60\times 10^{-19}$ C and mass = $m = 1.67\times 10^{-27}$ kg) moves in the direction of the electric field because of the force that the field applies to the proton. This force does work and thereby changes the proton's kinetic energy. According to the work-energy theorem (Equation 6.3), the work done causes the proton's kinetic energy to change. Kinetic energy is $mv^2/2$, where v is the speed. The work-energy theorem will involve the final speed of the proton and we will use it to obtain that speed.

SOLUTION According to Equation 6.3, the work-energy theorem is $W = \text{KE}_f - \text{KE}_0$, where W is the work done by the net external force that acts on the proton and KE_f and KE_0 are, respectively, the final and initial kinetic energies. The force applied by the electric field **E** is the only force acting, so it is the net force and, according to Equation 18.2, has a magnitude of $F = Ee$. The direction of the force is in the same direction as the electric field, since the proton has a positive charge. Since the motion is in the direction of the field and the force, the work done by the force is given by Equation 6.1 as $W = Fs = Ees$, where s is the distance traveled. Thus, the work-energy theorem can be stated as follows:

$$W = \text{KE}_f - \text{KE}_0 \qquad \text{or} \qquad Ees = \frac{1}{2}mv_f^2 - \frac{1}{2}mv_0^2$$

Solving this result for the final speed v_f, we find

$$v_f = \sqrt{\frac{2Ees}{m} + v_0^2}$$

$$= \sqrt{\frac{2(2.3\times 10^3 \text{ N/C})(1.60\times 10^{-19}\text{ C})(2.0\times 10^{-3}\text{ m})}{1.67\times 10^{-27}\text{ kg}} + (2.5\times 10^4 \text{ m/s})^2}$$

$$= \boxed{3.9\times 10^4 \text{ m/s}}$$

702 ELECTRIC FORCES AND ELECTRIC FIELDS

67. **REASONING AND SOLUTION** In order for the net force on any charge to be directed inward toward the center of the square, the charges must be placed with alternate + and − signs on each successive corner. The force on any charge due to an adjacent charge is

$$F = \frac{kq^2}{r^2} = \frac{(8.99 \times 10^9 \text{ N} \cdot \text{m}^2/\text{C}^2)(2.0 \times 10^{-6} \text{ C})^2}{(0.30 \text{ m})^2} = 0.40 \text{ N}$$

The forces due to two adjacent charges are perpendicular to one another and produce a resultant force

$$F' = \sqrt{2F^2} = \sqrt{2(0.40 \text{ N})^2}$$

$$F' = 0.57 \text{ N}.$$

The force due to the diagonal charge is

$$F'' = kq^2/(2r^2) = 0.20 \text{ N}$$

since the diagonal distance is $r\sqrt{2}$. The force F'' is directed opposite to F' (since the diagonal charges are of the same sign). Therefore, the net force acting on any of the charges is directed inward and has a magnitude

$$F_{net} = F' - F'' = 0.57 \text{ N} - 0.20 \text{ N} = \boxed{0.37 \text{ N}}$$

---

68. **CONCEPT QUESTIONS**
a. The conservation of electric charge states that, during any process, the net electric charge of an isolated system remains constant (is conserved). Therefore, the net charge $(q_1 + q_2)$ on the two spheres before they touch is the same as the net charge after they touch.

b. When the two identical spheres touch, the net charge will spread out equally over both of them. When the spheres are separated, the charge on each is the same.

SOLUTION
a. Since the final charge on each sphere is +5.0 µC, the final net charge on both spheres is 2(+5.0 µC) = +10.0 µC. The initial net charge must also be +10.0 µC. The only spheres whose net charge is +10.0 µC are

$$\boxed{\text{B } (q_B = -2.0 \ \mu\text{C}) \text{ and D } (q_D = +12.0 \ \mu\text{C})}.$$

b. Since the final charge on each sphere is +3.0 µC, the final net charge on the three spheres is 3(+3.0 µC) = +9.0 µC. The initial net charge must also be +9.0 µC. The only spheres whose net charge is +9.0 µC are

A ($q_A = -8.0\ \mu C$), C ($q_C = +5.0\ \mu C$) and D ($q_D = +12.0\ \mu C$).

c. Since the final charge on a given sphere in part (b) is +3.0 µC, we would have to add −3.0 µC to make it electrically neutral. Since the charge on an electron is -1.6×10^{-19} C, the number of electrons that would have to be added is

$$\text{Number of electrons} = \frac{-3.0 \times 10^{-6}\ \text{C}}{-1.6 \times 10^{-19}\ \text{C}} = \boxed{1.9 \times 10^{13}}$$

69. **CONCEPT QUESTIONS**
a. The electrical force that each charge exerts on the middle charge is shown in the drawing below. $\mathbf{F}_{21}$ is the force exerted on 2 by 1, and $\mathbf{F}_{23}$ is the force exerted on 2 by 3. Each force has the same magnitude, because the charges have the same magnitude and the distances are equal.

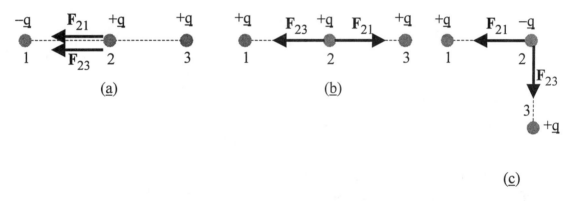

b. The net electric force **F** that acts on 2 is shown in the diagrams below.

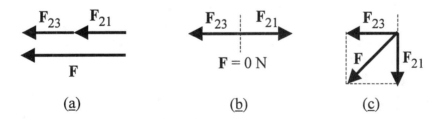

It can be seen from the diagrams that the largest electric force occurs in (a), followed by (c), and then by (b).

SOLUTION The magnitude F_{21} of the force exerted on 2 by 1 is the same as the magnitude F_{23} of the force exerted on 2 by 3, since the magnitudes of the charges are the same and the distances are the same. Coulomb's law gives the magnitudes as

704 ELECTRIC FORCES AND ELECTRIC FIELDS

$$F_{21} = F_{23} = \frac{kqq}{r^2}$$

$$= \frac{(8.99 \times 10^9 \ \text{N} \cdot \text{m}^2/\text{C}^2)(8.6 \times 10^{-6} \ \text{C})(8.6 \times 10^{-6} \ \text{C})}{(3.8 \times 10^{-3} \ \text{m})^2} = 4.6 \times 10^4 \ \text{N}$$

In part (a) of the drawing, both $\mathbf{F}_{21}$ and $\mathbf{F}_{23}$ point to the left, so the net force has a magnitude of

$$F = 2F_{12} = 2(4.6 \times 10^4 \ \text{N}) = \boxed{9.2 \times 10^4 \ \text{N}}$$

In part (b) of the drawing, $\mathbf{F}_{21}$ and $\mathbf{F}_{23}$ point in opposite directions, so the net force has a magnitude of $\boxed{0 \ \text{N}}$.

In part (c) the magnitude can be obtained from the Pythagorean theorem:

$$F = \sqrt{F_{21}^2 + F_{23}^2} = \sqrt{(4.6 \times 10^4 \ \text{N})^2 + (4.6 \times 10^4 \ \text{N})^2} = \boxed{6.5 \times 10^4 \ \text{N}}$$

---

70. **CONCEPT QUESTIONS**
a. The gravitational force is an attractive force. To neutralize this force, the electrical force must be a repulsive force. Therefore, the charges must both be positive or both negative.

b. Newton's law of gravitation, Equation 4.3, states that the gravitational force depends inversely on the square of the distance between the earth and the moon. Coulomb's law, Equation 18.1 states that the electrical force also depends inversely on the square of the distance. When these two forces are added together to give a zero net force, the distance can be algebraically eliminated. Thus, we do not need to know the distance between the two bodies.

SOLUTION Since the repulsive electrical force neutralizes the attractive gravitational force, the magnitudes of the two forces are equal:

$$\underbrace{\frac{kqq}{r^2}}_{\substack{\text{Electrical} \\ \text{force,} \\ \text{Equation 18.1}}} = \underbrace{\frac{GM_e M_m}{r^2}}_{\substack{\text{Gravitational} \\ \text{force,} \\ \text{Equation 4.3}}}$$

Solving this equation for the magnitude q of the charge on either body, we find

$$q = \sqrt{\frac{GM_e M_m}{k}} = \sqrt{\frac{\left(6.67 \times 10^{-11} \dfrac{\text{N} \cdot \text{m}^2}{\text{kg}^2}\right)\left(5.98 \times 10^{24} \text{ kg}\right)\left(7.35 \times 10^{22} \text{ kg}\right)}{8.99 \times 10^9 \dfrac{\text{N} \cdot \text{m}^2}{\text{C}^2}}} = \boxed{5.71 \times 10^{13} \text{ C}}$$

71. **CONCEPT QUESTIONS**
a. The magnitude of the electric field is obtained by dividing the magnitude of the force (obtained from the meter) by the magnitude of the charge. Since the charge is positive, the direction of the electric field is the same as the direction of the force.

b. As in part (a), the magnitude of the electric field is obtained by dividing the magnitude of the force by the magnitude of the charge. Since the charge is negative, however, the direction of the force (as indicated by the meter) is opposite to the direction of the electric field. Thus, the direction of the electric field is opposite to that of the force.

SOLUTION
a. According to Equation 18.2, the magnitude of the electric field is

$$E = \frac{F}{q} = \frac{40.0 \ \mu\text{N}}{20.0 \ \mu\text{C}} = \boxed{2.0 \text{ N/C}}$$

As mentioned in the answer to Concept Question (a), the direction of the electric field is the same as the direction of the force, or $\boxed{\text{due east}}$.

b. The magnitude of the electric field is

$$E = \frac{F}{q} = \frac{20.0 \ \mu\text{N}}{10.0 \ \mu\text{C}} = \boxed{2.0 \text{ N/C}}$$

Since the charge is negative, the direction of the electric field is opposite to the direction of the force, or $\boxed{\text{due east}}$. Thus, the electric fields in parts (a) and (b) are the same.

72. **CONCEPT QUESTIONS**
a. The electric field produced by a charge points away from a positive charge and toward a negative charge. Therefore, the electric field $\mathbf{E}_{+2}$ produced by the +2.0 μC charge points away from it, and the electric fields $\mathbf{E}_{-3}$ and $\mathbf{E}_{-5}$ produced by the −3.0 μC and −5.0 μC charges point toward them (see the drawing). The magnitude of the electric field produced by a point charge is given by Equation 18.3 as $E = kq/r^2$. Since the distance from each charge to the origin is the same, the magnitude of the electric field is proportional only to the magnitude q of the charge. Thus, the x component $\mathbf{E}_x$ of the net electric field is proportional

to 5.0 μC (2.0 μC + 3.0 μC). Since only one of the charges produces an electric field in the y direction, the y component E_y of the net electric field is proportional to the magnitude of this charge, or 5.0 μC. Thus, the x and y components are equal, as indicated in the right drawing, where the net electric field **E** is also shown.

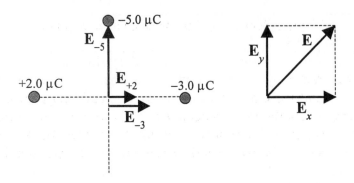

b. Using the same arguments as in part (a), the electric field produced by the four charges are shown in the left drawing. These fields also produce the same net electric field **E** as in part (a), as indicated in the right drawing.

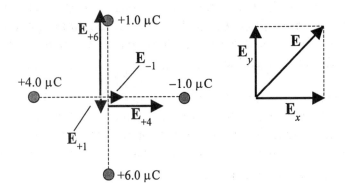

SOLUTION
a. The net electric field in the x direction is

$$E_x = \frac{(8.99 \times 10^9 \text{ N} \cdot \text{m}^2/\text{C}^2)(2.0 \times 10^{-6} \text{ C})}{(0.061 \text{ m})^2} + \frac{(8.99 \times 10^9 \text{ N} \cdot \text{m}^2/\text{C}^2)(3.0 \times 10^{-6} \text{ C})}{(0.061 \text{ m})^2}$$

$$= 1.2 \times 10^7 \text{ N/C}$$

The net electric field in the y direction is

$$E_y = \frac{(8.99 \times 10^9 \text{ N} \cdot \text{m}^2/\text{C}^2)(5.0 \times 10^{-6} \text{ C})}{(0.061 \text{ m})^2} = 1.2 \times 10^7 \text{ N/C}$$

The magnitude of the net electric field is

$$E = \sqrt{E_x^2 + E_y^2} = \sqrt{(1.2 \times 10^7 \text{ N/C})^2 + (1.2 \times 10^7 \text{ N/C})^2} = \boxed{1.7 \times 10^7 \text{ N/C}}$$

b. The magnitude of the net electric field is the same as in part (a).

---

73. **CONCEPT QUESTIONS**
a. Since the proton and the electron have the same charge magnitude e, the electric force that each experiences has the same magnitude. The directions are different, however. The proton, being positive, experiences a force in the same direction as the electric field (due east). The electron, being negative, experiences a force in the opposite direction (due west).

b. Newton's second law indicates that the direction of the acceleration is the same as the direction of the net force, which, in this case, is the electric force. The proton's acceleration is in the same direction (due east) as the electric field. The electron's acceleration is in the opposite direction (due west) as the electric field.

c. Newton's second law indicates that the magnitude of the acceleration is equal to the magnitude of the electric force divided by the mass. Although the proton and electron experience the same force magnitude, they have different masses. Thus, they have accelerations of different magnitudes.

SOLUTION According to Newton's second law, Equation 4.2, the acceleration a of an object is equal to the net force divided by the object's mass m. In this situation there is only one force, the electric force F, so it is the net force. According to Equation 18.2, the electric force is equal to the product of the charge and the electric field, or $F = q_0 E$. Thus, the magnitude of the acceleration can be written as

$$a = \frac{F}{m} = \frac{q_0 E}{m}$$

The magnitude of the acceleration of the electron is

$$a = \frac{q_0 E}{m} = \frac{(1.60 \times 10^{-19} \text{ C})(8.0 \times 10^4 \text{ N/C})}{9.11 \times 10^{-31} \text{ kg}} = \boxed{1.4 \times 10^{16} \text{ m/s}^2}$$

The magnitude of the acceleration of the proton is

$$a = \frac{q_0 E}{m} = \frac{(1.60 \times 10^{-19} \text{ C})(8.0 \times 10^4 \text{ N/C})}{1.67 \times 10^{-27} \text{ kg}} = \boxed{7.7 \times 10^{12} \text{ m/s}^2}$$

CHAPTER 19 | ELECTRIC POTENTIAL ENERGY AND THE ELECTRIC POTENTIAL

PROBLEMS

1. **SSM** *REASONING AND SOLUTION* Combining Equations 19.1 and 19.3, we have

$$W_{AB} = \text{EPE}_A - \text{EPE}_B = q_0(V_A - V_B) = (+1.6 \times 10^{-19} \text{ C})(0.070 \text{ V}) = \boxed{1.1 \times 10^{-20} \text{ J}}$$

2. *REASONING* Equation 19.1 indicates that the work done by the electric force as the particle moves from point A to point B is $W_{AB} = \text{EPE}_A - \text{EPE}_B$. For motion through a distance s along the line of action of a constant force of magnitude F, the work is given by Equation 6.1 as either $+Fs$ (if the force and the displacement have the same direction) or $-Fs$ (if the force and the displacement have opposite directions). Here, $\text{EPE}_A - \text{EPE}_B$ is given to be positive, so we can conclude that the work is $W_{AB} = +Fs$ and that the force points in the direction of the motion from point A to point B. The electric field is given by Equation 18.2 as $\mathbf{E} = \mathbf{F}/q_0$, where q_0 is the charge.

SOLUTION a. Using Equation 19.1 and the fact that $W_{AB} = +Fs$, we find

$$W_{AB} = +Fs = \text{EPE}_A - \text{EPE}_B$$

$$F = \frac{\text{EPE}_A - \text{EPE}_B}{s} = \frac{9.0 \times 10^{-4} \text{ J}}{0.20 \text{ m}} = \boxed{4.5 \times 10^{-3} \text{ N}}$$

As discussed in the reasoning, the direction of the force is $\boxed{\text{from } A \text{ toward } B}$.

b. From Equation 18.2, we find that the electric field has a magnitude of

$$E = \frac{F}{q_0} = \frac{4.5 \times 10^{-3} \text{ N}}{1.5 \times 10^{-6} \text{ C}} = \boxed{3.0 \times 10^3 \text{ N/C}}$$

The direction is the same as that of the force on the positive charge, namely $\boxed{\text{from } A \text{ toward } B}$.

3. *REASONING* The number N of electrons that jump from your hand (point A) to the door knob (point B) is equal to the total charge q that jumps divided by the charge $-e$ of one electron: $N = q/(-e)$, where $e = 1.6 \times 10^{-19}$ C. We can determine q by using Equation 19.4,

which relates the work W_{AB} done by the electric force to the difference in electric potentials, $V_B - V_A$, and the charge. The difference in potentials is given as $V_B - V_A = 2.0 \times 10^4$ V.

SOLUTION The number of electrons that jumps from your hand to the door knob is

$$N = \frac{q}{-e} = \frac{\frac{-W_{AB}}{V_B - V_A}}{-e} = \frac{\frac{-1.5 \times 10^{-7} \text{ J}}{2.0 \times 10^4 \text{ V}}}{-1.6 \times 10^{-19} \text{ C}} = \boxed{4.7 \times 10^7}$$

---

4. **REASONING AND SOLUTION**

a. According to Equation 19.4, the work done by the electric force as the electron goes from point A (the cathode) to point B (the anode) is

$$W_{AB} = -q(V_B - V_A) = -(-1.6 \times 10^{-19} \text{ C})(+125\,000 \text{ V}) = \boxed{+2.00 \times 10^{-14} \text{ J}}$$

b. The only force that acts on the electron is the conservative electric force. Therefore, the total energy of the electron is conserved as it moves from point A to point B:

$$\underbrace{\tfrac{1}{2}mv_A^2 + \text{EPE}_A}_{\text{Total energy at point }A} = \underbrace{\tfrac{1}{2}mv_B^2 + \text{EPE}_B}_{\text{Total energy at point }B}$$

Since the electron starts from rest, $v_A = 0$. The electric potential V is related to the electric potential energy EPE by $V = \text{EPE}/q$ (see Equation 19.3). With these changes, the equation above gives the kinetic energy of the electron at point B (the anode) to be

$$\tfrac{1}{2}mv_B^2 = -\text{EPE}_B + \text{EPE}_A$$

$$= -q(V_B - V_A) = -(-1.60 \times 10^{-19} \text{ C})(125\,000 \text{ V}) = \boxed{2.00 \times 10^{-14} \text{ J}}$$

---

5. [SSM] **REASONING** The only force acting on the moving electron is the conservative electric force. Therefore, the total energy of the electron (the sum of the kinetic energy KE and the electric potential energy EPE) remains constant throughout the trajectory of the electron. Let the subscripts A and B refer to the initial and final positions, respectively, of the electron. Then,

$$\tfrac{1}{2}mv_A^2 + \text{EPE}_A = \tfrac{1}{2}mv_B^2 + \text{EPE}_B$$

Solving for v_B gives

$$v_B = \sqrt{v_A^2 - \frac{2}{m}(\text{EPE}_B - \text{EPE}_A)}$$

Since the electron starts from rest, $v_A = 0$. The difference in potential energies is related to the difference in potentials by Equation 19.4, $\text{EPE}_B - \text{EPE}_A = q(V_B - V_A)$.

SOLUTION The speed v_B of the electron just before it reaches the screen is

$$v_B = \sqrt{-\frac{2q}{m}(V_B - V_A)} = \sqrt{-\frac{2(-1.6 \times 10^{-19}\text{ C})}{9.11 \times 10^{-31}\text{ kg}}(25\,000\text{ V})} = \boxed{9.4 \times 10^7 \text{ m/s}}$$

---

6. ***REASONING AND SOLUTION***
a. The only force that acts on the particle is the conservative electric force. Therefore, the total energy of the particle is conserved as it moves from point A to point B:

$$\underbrace{\tfrac{1}{2}mv_A^2 + \text{EPE}_A}_{\text{Total energy at point } A} = \underbrace{\tfrac{1}{2}mv_B^2 + \text{EPE}_B}_{\text{Total energy at point } B}$$

Since the particle starts from rest, $v_A = 0$. The electric potential V is related to the electric potential energy EPE by $V = \text{EPE}/q$ (see Equation 19.3). With these changes, we can solve the equation above for the potential difference $V_B - V_A$:

$$V_B - V_A = \frac{1}{q}\left(0 - \tfrac{1}{2}mv_B^2\right) = \frac{1}{-1.5 \times 10^{-6}\text{ C}}\left[0 - \tfrac{1}{2}(2.5 \times 10^{-6}\text{ kg})(42\text{ m/s})^2\right] = \boxed{1500\text{ V}}$$

b. $\boxed{\text{Point } B}$ is at the higher potential, because a negative charge accelerates from a lower potential to a higher potential.

---

7. ***REASONING AND SOLUTION*** The power rating P is defined as the work W_{AB} done by the battery divided by the time t,

$$P = \frac{W_{AB}}{t}$$

The work done by the electric force as the charge moves from point A (the positive terminal), through the electric motor, and to point B (the negative terminal) is

$$W_{AB} = q(V_A - V_B) = (1300\text{ C})(320\text{ V}) = 4.2 \times 10^5\text{ J} \tag{19.4}$$

The power rating is

$$P = \frac{W_{AB}}{t} = \frac{4.2 \times 10^5\text{ J}}{8.0\text{ s}} = 5.2 \times 10^4\text{ W}$$

Since 746 W = 1 hp, the minimum horsepower rating of the car is

$$(5.20 \times 10^4 \text{ W}) \frac{1 \text{ hp}}{746 \text{ W}} = \boxed{7.0 \times 10^1 \text{ hp}}$$

---

8. **REASONING** According to Equation 19.4, the energy available is $\Delta(\text{EPE}) = q_0 \Delta V$, where q_0 is the charge and ΔV is the potential difference between the battery terminals, that is, 12 V. The energy (heat) need to boil m kilograms of water is given by Equation 12.5 as $Q = mL_v$, where $L_v = 22.6 \times 10^5$ J/kg is the latent heat of vaporization of water (see Table 12.3).

SOLUTION Using Equations 19.4 and 12.5, we obtain

$$\Delta(\text{EPE}) = mL_v = q_0 \Delta V$$

$$m = \frac{q_0 \Delta V}{L_v} = \frac{(7.5 \times 10^5 \text{ C})(12 \text{ V})}{22.6 \times 10^5 \text{ J/kg}} = \boxed{4.0 \text{ kg}}$$

---

9. [SSM] [WWW] **REASONING** The only force acting on the moving charge is the conservative electric force. Therefore, the total energy of the charge remains constant. Applying the principle of conservation of energy between locations A and B, we obtain

$$\tfrac{1}{2}mv_A^2 + \text{EPE}_A = \tfrac{1}{2}mv_B^2 + \text{EPE}_B$$

Since the charged particle starts from rest, $v_A = 0$. The difference in potential energies is related to the difference in potentials by Equation 19.4, $\text{EPE}_B - \text{EPE}_A = q(V_B - V_A)$. Thus, we have

$$q(V_A - V_B) = \tfrac{1}{2}mv_B^2 \tag{1}$$

Similarly, applying the conservation of energy between locations C and B gives

$$q(V_C - V_B) = \tfrac{1}{2}m(2v_B)^2 \tag{2}$$

Dividing Equation (1) by Equation (2) yields

$$\frac{V_A - V_B}{V_C - V_B} = \frac{1}{4}$$

This expression can be solved for V_B.

712 ELECTRIC POTENTIAL ENERGY AND THE ELECTRIC POTENTIAL

SOLUTION Solving for V_B, we find that

$$V_B = \frac{4V_A - V_C}{3} = \frac{4(452\text{ V}) - 791\text{ V}}{3} = \boxed{339\text{ V}}$$

---

10. **REASONING** The gravitational and electric forces are conservative forces, so the total energy of the particle remains constant as it moves from point A to point B. Recall from Equation 6.5 that the gravitational potential energy (GPE) of a particle of mass m is GPE = mgh, where h is the height of the particle above the earth's surface. The conservation of energy is written as

$$\tfrac{1}{2}mv_A^2 + mgh_A + \text{EPE}_A = \tfrac{1}{2}mv_B^2 + mgh_B + \text{EPE}_B \tag{1}$$

We will use this equation several times to determine the initial speed v_A of the negatively charged particle.

SOLUTION
When the negatively charged particle is thrown upward, it attains a maximum height of h. For this particle we have:

$v_A = ?$ $\qquad\qquad\qquad\qquad$ $v_B = 0$ (at maximum height)
$\text{EPE}_A = (-q)V_A$ $\qquad\qquad$ $\text{EPE}_B = (-q)V_B$
$h_A = 0$ (ground level) $\qquad\;\;$ $h_B = h$ (the maximum height)

Solving the conservation of energy equation, Equation (1), for v_A and substituting in the data above gives

$$v_A = \sqrt{\frac{2}{m}\left[mgh + (-q)(V_B - V_A)\right]} \tag{2}$$

Equation (2) cannot be solved as it stands because the height h and the potential difference $(V_B - V_A)$ are not known. We now make use of the fact that a positively charged particle, when thrown straight upward with an initial speed of 30.0 m/s, also reaches the maximum height h. For this particle we have:

$v_A = 30.0$ m/s $\qquad\qquad\qquad$ $v_B = 0$ (at maximum height)
$\text{EPE}_A = (+q)V_A$ $\qquad\qquad$ $\text{EPE}_B = (+q)V_B$
$h_A = 0$ (ground level) $\qquad\;\;$ $h_B = h$

Solving the conservation of energy equation, Equation (1), for the potential difference $(V_B - V_A)$ and substituting in the data above gives

$$V_B - V_A = \frac{1}{+q}\left[\tfrac{1}{2}m(30.0 \text{ m/s})^2 - mgh\right] \qquad (3)$$

Substituting Equation (3) into Equation (2) gives, after some algebraic simplifications,

$$v_A = \sqrt{4gh - (30.0 \text{ m/s})^2} \qquad (4)$$

Equation (4) cannot be solved because the height h is still unknown. We now make use of the fact that the uncharged particle, when thrown straight upward with an initial speed of 25.0 m/s, also reaches the maximum height h. For this particle we have:

$v_A = 25.0$ m/s $\qquad v_B = 0$ (at maximum height)
$EPE_A = qV_A = 0$ (since $q = 0$) $\qquad EPE_B = qV_B = 0$ (since $q = 0$)
$h_A = 0$ (ground level) $\qquad h_A = h$

Solving Equation (1) with this data for the maximum height h yields

$$h = \frac{(25.0 \text{ m/s})^2}{2g} = \frac{(25.0 \text{ m/s})^2}{2(9.80 \text{ m/s}^2)} = 31.9 \text{ m}$$

Substituting $h = 31.9$ m into Equation (4) gives $v_A = \boxed{18.7 \text{ m/s}}$.

---

11. **SSM** **REASONING AND SOLUTION** The electric potential V at a distance r from a point charge q is given by Equation 19.6, $V = kq/r$. Solving this expression for q, we find that

$$q = \frac{rV}{k} = \frac{(0.25 \text{ m})(+130 \text{ V})}{9.0 \times 10^9 \text{ N} \cdot \text{m}^2/\text{C}^2} = \boxed{+3.6 \times 10^{-9} \text{ C}}$$

---

12. **REASONING** The potential of each charge q at a distance r away is given by Equation 19.6 as $V = kq/r$. By applying this expression to each charge, we will be able to find the desired ratio, because the distances are given for each charge.

SOLUTION According to Equation 19.6, the potentials of each charge are

$$V_A = \frac{kq_A}{r_A} \quad \text{and} \quad V_B = \frac{kq_B}{r_B}$$

Since we know that $V_A = V_B$, it follows that

$$\frac{kq_A}{r_A} = \frac{kq_B}{r_B} \quad \text{or} \quad \frac{q_B}{q_A} = \frac{r_B}{r_A} = \frac{0.43 \text{ m}}{0.18 \text{ m}} = \boxed{2.4}$$

714 ELECTRIC POTENTIAL ENERGY AND THE ELECTRIC POTENTIAL

13. **REASONING** The potential V at a distance r from a proton is $V = k(+e)/r$ (see Equation 19.6), where $+e$ is the charge of the proton. When an electron ($q = -e$) is placed at a distance r from the proton, the electric potential energy is EPE $= -eV$, as per Equation 19.3.

SOLUTION The difference in the electric potential energies when the electron and proton are separated by $r_{final} = 5.29 \times 10^{-11}$ m and when they are very far apart ($r_{initial} = \infty$) is

$$\text{EPE}_{final} - \text{EPE}_{initial} = \frac{(-e)ke}{r_{final}} - \frac{(-e)ke}{r_{initial}}$$

$$= -(8.99 \times 10^9 \text{ N} \cdot \text{m}^2/\text{C}^2)(1.60 \times 10^{-19} \text{ C})^2$$

$$\times \left(\frac{1}{5.29 \times 10^{-11} \text{ m}} - \frac{1}{\infty} \right) = \boxed{-4.35 \times 10^{-18} \text{ J}}$$

14. **REASONING AND SOLUTION** The potential at point A is

$$V_A = \frac{kq}{r_A} = \frac{(8.99 \times 10^9 \text{ N} \cdot \text{m}^2/\text{C}^2)(-3.00 \times 10^{-8} \text{ C})}{2.00 \text{ m}} = -135 \text{ V}$$

Similarly, the potential at point B is $V_B = kq/r_B = -89.9$ V

The difference in the potentials is

$$V_B - V_A = \boxed{+45 \text{ V}}$$

$$\boxed{\text{Point } B \text{ is at the higher potential.}}$$

15. **SSM WWW** *REASONING* Initially, suppose that one charge is at C and the other charge is held fixed at B. The charge at C is then moved to position A. According to Equation 19.4, the work W_{CA} done by the electric force as the charge moves from C to A is $W_{CA} = q(V_C - V_A)$, where, from Equation 19.6, $V_C = kq/d$ and $V_A = kq/r$. From the figure at the right we see that $d = \sqrt{r^2 + r^2} = \sqrt{2}r$. Therefore, we find that

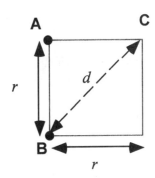

$$W_{CA} = q\left(\frac{kq}{\sqrt{2}r} - \frac{kq}{r}\right) = \frac{kq^2}{r}\left(\frac{1}{\sqrt{2}} - 1\right)$$

SOLUTION Substituting values, we obtain

$$W_{CA} = \frac{(8.99 \times 10^9 \text{ N} \cdot \text{m}^2/\text{C}^2)(3.0 \times 10^{-6} \text{ C})^2}{0.500 \text{ m}}\left(\frac{1}{\sqrt{2}} - 1\right) = \boxed{-4.7 \times 10^{-2} \text{ J}}$$

---

16. *REASONING AND SOLUTION* The initial and final electric potential energies of the configuration are, respectively,

$$\text{EPE}_A = \frac{kq_1q_2}{r_A} \quad \text{and} \quad \text{EPE}_B = \frac{kq_1q_2}{r_B}$$

Since $\text{EPE}_B = 2\,\text{EPE}_A$, we have

$$\frac{kq_1q_2}{r_B} = 2\frac{kq_1q_2}{r_A}$$

or

$$r_B = \frac{r_A}{2} = \frac{0.74 \text{ m}}{2} = \boxed{0.37 \text{ m}}.$$

---

17. *REASONING AND SOLUTION* Let s be the length of the side of the square and Q be the value of the unknown charge. The potential at either of the vacant corners is

$$V = 0 = \frac{k(9q)}{s} + \frac{k(-8q)}{s} + \frac{kQ}{s/\sqrt{2}}$$

so

$$\boxed{Q = \frac{-q}{\sqrt{2}}}$$

716 ELECTRIC POTENTIAL ENERGY AND THE ELECTRIC POTENTIAL

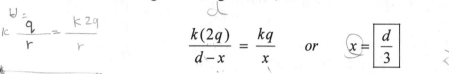

18. **REASONING AND SOLUTION** Let the first spot where the potential is zero be a distance x to the left of the negative charge. Then,

$$\frac{k(2q)}{d-x} = \frac{kq}{x} \quad \text{or} \quad \boxed{x = \frac{d}{3}}$$

Let the second spot where the potential is zero be a distance x to the right of the negative charge. Then,

$$\frac{k(2q)}{d+x} = \frac{kq}{x} \quad \text{or} \quad \boxed{x = d}$$

---

19. **SSM REASONING** The only force acting on the moving charge is the conservative electric force. Therefore, the sum of the kinetic energy KE and the electric potential energy EPE is the same at points A and B:

$$\tfrac{1}{2}mv_A^2 + \text{EPE}_A = \tfrac{1}{2}mv_B^2 + \text{EPE}_B$$

Since the particle comes to rest at B, $v_B = 0$. Combining Equations 19.3 and 19.6, we have

$$\text{EPE}_A = qV_A = q\left(\frac{kq_1}{d}\right)$$

and

$$\text{EPE}_B = qV_B = q\left(\frac{kq_1}{r}\right)$$

where d is the initial distance between the fixed charge and the moving charged particle, and r is the distance between the charged particles after the moving charge has stopped. Therefore, the expression for the conservation of energy becomes

$$\tfrac{1}{2}mv_A^2 + \frac{kqq_1}{d} = \frac{kqq_1}{r}$$

This expression can be solved for r. Once r is known, the distance that the charged particle moves can be determined.

SOLUTION Solving the expression above for r gives

$$r = \frac{kqq_1}{\tfrac{1}{2}mv_A^2 + \dfrac{kqq_1}{d}}$$

$$= \frac{(8.99 \times 10^9 \text{ N}\cdot\text{m}^2/\text{C}^2)(-8.00 \times 10^{-6} \text{ C})(-3.00 \times 10^{-6} \text{ C})}{\tfrac{1}{2}(7.20 \times 10^{-3} \text{ kg})(65.0 \text{ m/s})^2 + \dfrac{(8.99 \times 10^9 \text{ N}\cdot\text{m}^2/\text{C}^2)(-8.00 \times 10^{-6} \text{ C})(-3.00 \times 10^{-6} \text{ C})}{0.0450 \text{ m}}}$$

$$= 0.0108 \text{ m}$$

Therefore, the charge moves a distance of $0.0450 \text{ m} - 0.0108 \text{ m} = \boxed{0.0342 \text{ m}}$.

20. **REASONING** It will not matter in what order the group is assembled. For convenience, we will assemble the group from one end of the line to the other. The potential energy of each charge added to the group will be determined and the four values added together to get the total. At each step, the electric potential energy of an added charge q_0 is given by Equation 19.3 as $EPE = q_0 V_{Total}$, where V_{Total} is the potential at the point where the added charge is placed. The potential V_{Total} will be determined by adding together the contributions from the charges previously put in position, each according to Equation 19.6 ($V = kq/r$).

SOLUTION The first charge added to the group has no electric potential energy, since the spot where it goes has a total potential of $V_{Total} = 0$ J, there being no charges in the vicinity to create it:

$$EPE_1 = q_0 V_{Total} = (2.0 \times 10^{-6} \text{ C})(0 \text{ V}) = 0 \text{ J}$$

The second charge experiences a total potential that is created by the first charge:

$$V_{Total} = \frac{kq}{r} = \frac{(8.99 \times 10^9 \text{ N}\cdot\text{m}^2/\text{C}^2)(+2.0 \times 10^{-6} \text{ C})}{0.40 \text{ m}} = 4.5 \times 10^4 \text{ V}$$

$$EPE_2 = q_0 V_{Total} = (2.0 \times 10^{-6} \text{ C})(4.5 \times 10^4 \text{ V}) = 0.090 \text{ J}$$

The third charge experiences a total potential that is created by the first and the second charges:

$$V_{Total} = \frac{(8.99 \times 10^9 \text{ N}\cdot\text{m}^2/\text{C}^2)(+2.0 \times 10^{-6} \text{ C})}{0.80 \text{ m}}$$

$$+ \frac{(8.99 \times 10^9 \text{ N}\cdot\text{m}^2/\text{C}^2)(+2.0 \times 10^{-6} \text{ C})}{0.40 \text{ m}} = 6.7 \times 10^4 \text{ V}$$

$$EPE_3 = q_0 V_{Total} = (2.0 \times 10^{-6} \text{ C})(6.7 \times 10^4 \text{ V}) = 0.13 \text{ J}$$

The fourth charge experiences a total potential that is created by the first, second, and third charges:

$$V_{Total} = \frac{(8.99 \times 10^9 \text{ N} \cdot \text{m}^2/\text{C}^2)(+2.0 \times 10^{-6} \text{ C})}{1.2 \text{ m}}$$

$$+ \frac{(8.99 \times 10^9 \text{ N} \cdot \text{m}^2/\text{C}^2)(+2.0 \times 10^{-6} \text{ C})}{0.80 \text{ m}}$$

$$+ \frac{(8.99 \times 10^9 \text{ N} \cdot \text{m}^2/\text{C}^2)(+2.0 \times 10^{-6} \text{ C})}{0.40 \text{ m}} = 8.2 \times 10^4 \text{ V}$$

$$EPE_4 = q_0 V_{Total} = (2.0 \times 10^{-6} \text{ C})(8.2 \times 10^4 \text{ V}) = 0.16 \text{ J}$$

The total electric potential energy of the group is

$$EPE_{Total} = EPE_1 + EPE_2 + EPE_3 + EPE_4$$

$$= 0 \text{ J} + 0.090 \text{ J} + 0.13 \text{ J} + 0.16 \text{ J} = \boxed{0.38 \text{ J}}$$

---

21. **REASONING** The only force acting on each proton is the conservative electric force. Therefore, the total energy (kinetic energy plus electric potential energy) is conserved at all points along the motion. For two points, A and B, the conservation of energy is

$$\underbrace{\tfrac{1}{2}mv_A^2 + \tfrac{1}{2}mv_A^2}_{\text{Initial kinetic energy of the two protons}} + \underbrace{EPE_A}_{\substack{\text{Initial electric} \\ \text{potential} \\ \text{energy}}} = \underbrace{\tfrac{1}{2}mv_B^2 + \tfrac{1}{2}mv_B^2}_{\text{Final kinetic energy of the two protons}} + \underbrace{EPE_B}_{\substack{\text{Final electric} \\ \text{potential} \\ \text{energy}}}$$

The electric potential energy of two protons (charge = $+e$) that are separated by a distance r is $EPE = ke^2/r$. By substituting this expression for EPE into the conservation of energy, we will be able to determine the distance of closest approach.

SOLUTION When the protons are very far apart ($r_A = \infty$), so that $EPE_A = 0$. At the distance r_B of closest approach, the speed of each proton is momentarily zero ($v_B = 0$). With these substitutions, the conservation of energy equation reduces to

$$\tfrac{1}{2}mv_A^2 + \tfrac{1}{2}mv_A^2 = \frac{ke^2}{r_B}$$

Solving for r_B, the distance of closest approach is

$$r_B = \frac{ke^2}{mv_A^2} = \frac{(9.0 \times 10^9 \text{ N} \cdot \text{m}^2/\text{C}^2)(1.6 \times 10^{-19} \text{ C})^2}{(1.67 \times 10^{-27} \text{ kg})(1.5 \times 10^6 \text{ m/s})^2} = \boxed{6.1 \times 10^{-14} \text{ m}}$$

22. **REASONING AND SOLUTION** The figure at the right shows two identical charges, q, fixed to diagonally opposite corners of a square. The potential at corner A is caused by the presence of the two charges. It is given by

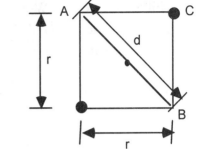

$$(V_A)_0 = \frac{kq}{r} + \frac{kq}{r} = \frac{2kq}{r}$$

Since both charges are the same distance from corner B, this is equal to the potential at corner B as well. If a third charge, q_3, is placed at the center of the square, the potential at corner A (as well as corner B) becomes

$$(V_A)_f = \frac{kq}{r} + \frac{kq}{r} + \frac{kq_3}{(d/2)}$$

We can find d by applying the Pythagorean theorem to the geometry in the figure above:

$$d = \sqrt{r^2 + r^2} = \sqrt{2}\,r \qquad \text{so that} \qquad \frac{d}{2} = \frac{r}{\sqrt{2}}$$

The expression for $(V_A)_f$ becomes

$$(V_A)_f = \frac{kq}{r} + \frac{kq}{r} + \frac{kq_3\sqrt{2}}{r}.$$

From the problem statement, we know that the addition of q_3 causes the potential at A and B to change sign without changing magnitude.

In other words, $(V_A)_f = -(V_A)_0$, or

$$\frac{kq}{r} + \frac{kq}{r} + \frac{kq_3\sqrt{2}}{r} = -\frac{2kq}{r}$$

Solving for q_3 gives

$$q_3 = -(2\sqrt{2})q = -(2\sqrt{2})(1.7 \times 10^{-6}\ \text{C}) = \boxed{-4.8 \times 10^{-6}\ \text{C}}$$

23. **[SSM] REASONING** Initially, the three charges are infinitely far apart. We will proceed as in Example 8 by adding charges to the triangle, one at a time, and determining the electric potential energy at each step. According to Equation 19.3, the electric potential energy EPE is the product of the charge q and the electric potential V at the spot where the charge is placed, EPE = qV. The total electric potential energy of the group is the sum of the energies of each step in assembling the group.

720 ELECTRIC POTENTIAL ENERGY AND THE ELECTRIC POTENTIAL

SOLUTION Let the corners of the triangle be numbered clockwise as 1, 2 and 3, starting with the top corner. When the first charge ($q_1 = 8.00$ μC) is placed at a corner 1, the charge has no electric potential energy, $EPE_1 = 0$. This is because the electric potential V_1 produced by the other two charges at corner 1 is zero, since they are infinitely far away.

Once the 8.00-μC charge is in place, the electric potential V_2 that it creates at corner 2 is

$$V_2 = \frac{kq_1}{r_{21}}$$

where $r_{21} = 5.00$ m is the distance between corners 1 and 2, and $q_1 = 8.00$ μC. When the 20.0-μC charge is placed at corner 2, its electric potential energy EPE_2 is

$$EPE_2 = q_2 V_2 = q_2 \left(\frac{kq_1}{r_{21}}\right)$$

$$= (20.0 \times 10^{-6} \text{ C}) \left[\frac{(8.99 \times 10^9 \text{ N} \cdot \text{m}^2/\text{C}^2)(8.00 \times 10^{-6} \text{ C})}{5.00 \text{ m}}\right] = 0.288 \text{ J}$$

The electric potential V_3 at the remaining empty corner is the sum of the potentials due to the two charges that are already in place on corners 1 and 2:

$$V_3 = \frac{kq_1}{r_{31}} + \frac{kq_2}{r_{32}}$$

where $q_1 = 8.00$ μC, $r_{31} = 3.00$ m, $q_2 = 20.0$ μC, and $r_{32} = 4.00$ m. When the third charge ($q_3 = -15.0$ μC) is placed at corner 3, its electric potential energy EPE_3 is

$$EPE_3 = q_3 V_3 = q_3 \left(\frac{kq_1}{r_{31}} + \frac{kq_2}{r_{32}}\right) = q_3 k \left(\frac{q_1}{r_{31}} + \frac{q_2}{r_{32}}\right)$$

$$= (-15.0 \times 10^{-6} \text{ C})(8.99 \times 10^9 \text{ N} \cdot \text{m}^2/\text{C}^2) \left(\frac{8.00 \times 10^{-6} \text{ C}}{3.00 \text{ m}} + \frac{20.0 \times 10^{-6} \text{ C}}{4.00 \text{ m}}\right) = -1.034 \text{ J}$$

The electric potential energy of the entire array is given by

$$EPE = EPE_1 + EPE_2 + EPE_3 = 0 + 0.288 \text{ J} + (-1.034 \text{ J}) = \boxed{-0.746 \text{ J}}$$

24. **REASONING AND SOLUTION**
   a. Let d be the distance between the charges. The potential at the point $x_1 = 4.00$ cm to the left of the negative charge is

$$V = 0 = \frac{kq_1}{d-x_1} - \frac{kq_2}{x_1}$$

which gives

$$\frac{q_1}{q_2} = \frac{d}{x_1} - 1 \qquad (1)$$

Similarly, at the point $x_2 = 7.00$ cm to the right of the negative charge we have

$$V = 0 = \frac{kq_1}{x_2 + d} - \frac{kq_2}{x_2}$$

which gives

$$\frac{q_1}{q_2} = \frac{d}{x_2} + 1 \qquad (2)$$

Equating Equations (1) and (2) and solving for d gives $d = \boxed{0.187 \text{ m}}$.

   b. Using the above value for d in Equation (1) yields $\frac{q_1}{q_2} = \boxed{3.67}$.

---

25. **REASONING AND SOLUTION** The electrical potential energy of the group of charges is

$$\text{EPE} = kq_1q_2/d + kq_1q_3/(2d) + kq_2q_3/d = 0$$

so

$$q_1q_2 + (1/2)q_1q_3 + q_2q_3 = 0$$

   a. If $q_1 = q_2 = q$, then

$$q + (1/2)q_3 + q_3 = 0 \quad \text{or} \quad \boxed{q_3 = -\tfrac{2}{3}q}$$

   b. If $q_1 = q$ and $q_2 = -q$ then

$$-q + (1/2)q_3 - q_3 = 0 \quad \text{or} \quad \boxed{q_3 = -2q}$$

---

26. **REASONING** The only force acting on each particle is the conservative electric force. Therefore, the total energy (kinetic energy plus electric potential energy) is conserved as the particles move apart. In addition, the net external force acting on the system of two particles is zero (the electric force that each particle exerts on the other is an internal force). Thus, the total linear momentum of the system is also conserved. We will use the conservation of energy and the conservation of linear momentum to find the final speed of each particle.

722 ELECTRIC POTENTIAL ENERGY AND THE ELECTRIC POTENTIAL

SOLUTION For two points, A and B, along the motion, the conservation of energy is

$$\underbrace{\tfrac{1}{2}mv_{1,A}^2 + \tfrac{1}{2}mv_{2,A}^2}_{\text{Initial kinetic energy of the two particles}} + \underbrace{\frac{kq_1q_2}{r_A}}_{\text{Initial electric potential energy}} = \underbrace{\tfrac{1}{2}mv_{1,B}^2 + \tfrac{1}{2}mv_{2,B}^2}_{\text{Final kinetic energy of the two particles}} + \underbrace{\frac{kq_1q_2}{r_B}}_{\text{Final electric potential energy}}$$

Setting $v_{1,A} = v_{2,A} = 0$ since the particles are initially at rest, and letting $r_B = \tfrac{1}{2}r_A$, the conservation of energy equation becomes

$$\tfrac{1}{2}mv_{1,B}^2 + \tfrac{1}{2}mv_{2,B}^2 = -\frac{kq_1q_2}{r_A} \qquad (1)$$

This equation cannot be solved for $v_{1,B}$ because the final speed $v_{2,B}$ of the second particle is not known. To find this speed, we will use the conservation of linear momentum:

$$\underbrace{mv_{1,A} + mv_{2,A}}_{\text{Initial linear momentum}} = \underbrace{mv_{1,B} + mv_{2,B}}_{\text{Final linear momentum}}$$

Setting $v_{1,A} = v_{2,A} = 0$ and solving for $v_{2,B}$ gives $v_{2,B} = -v_{1,B}$. Substituting this result into Equation (1) and solving for $v_{1,B}$ yields

$$v_{1,B} = \sqrt{\frac{-kq_1q_2}{mr_A}}$$

$$= \sqrt{\frac{-(9.0 \times 10^9 \text{ N} \cdot \text{m}^2/\text{C}^2)(+5.0 \times 10^{-6} \text{ C})(-5.0 \times 10^{-6} \text{ C})}{(6.0 \times 10^{-3} \text{ kg})(0.80 \text{ m})}} = \boxed{6.8 \text{ m/s}}$$

This is also the speed of $v_{2,B}$.

---

27. **SSM REASONING AND SOLUTION** Since all points on the equipotential surface are the same distance r from the point charge, the potential is given by Equation 19.6,

$$V = \frac{kq}{r} = \frac{(8.99 \times 10^9 \text{ N} \cdot \text{m}^2/\text{C}^2)(3.0 \times 10^{-7} \text{ C})}{0.15 \text{ m}} = \boxed{18\,000 \text{ V}}$$

28. **REASONING** The potential of a point charge is given by Equation 19.6 ($V = kq/r$). This expression can be solved directly for the charge q.

    SOLUTION Using Equation 19.6, we find

    $$V = \frac{kq}{r} \quad \text{or} \quad q = \frac{rV}{k} = \frac{(0.18 \text{ m})(+36 \text{ V})}{8.99 \times 10^9 \text{ N} \cdot \text{m}^2/\text{C}^2} = \boxed{+7.2 \times 10^{-10} \text{ C}}$$

29. **REASONING** The magnitude E of the electric field is given by Equation 19.7 (without the minus sign) as $E = \frac{\Delta V}{\Delta s}$, where ΔV is the potential difference between the two metal conductors of the spark plug, and Δs is the distance between the two conductors. We can use this relation to find ΔV.

    SOLUTION The potential difference between the conductors is

    $$\Delta V = E \Delta s = (4.7 \times 10^7 \text{ V/m})(0.75 \times 10^{-3} \text{ m}) = \boxed{3.5 \times 10^4 \text{ V}}$$

30. **REASONING AND SOLUTION** From Equation 19.7 we know that $E = -\frac{\Delta V}{\Delta s}$, where ΔV is the potential difference between the two surfaces of the membrane, and Δs is the distance between them. If A is a point on the positive surface and B is a point on the negative surface, then $\Delta V = V_A - V_B = 0.070$ V. The electric field between the surfaces is

    $$E = -\frac{\Delta V}{\Delta s} = -\frac{V_B - V_A}{\Delta s} = \frac{V_A - V_B}{\Delta s} = \frac{0.070 \text{ V}}{8.0 \times 10^{-9} \text{ m}} = \boxed{8.8 \times 10^6 \text{ V/m}}$$

31. **SSM REASONING AND SOLUTION** Equation 19.7 gives the result directly:

    $$E = -\frac{\Delta V}{\Delta s} = -\frac{V_B - V_A}{\Delta s} = -\frac{28 \text{ V} - 95 \text{ V}}{0.016 \text{ m}} = \boxed{4.2 \times 10^3 \text{ V/m}}$$

    The electric field points from high potential to low potential. Thus, it points from $\boxed{A \text{ to } B}$.

32. **REASONING AND SOLUTION** Let E_1 represent the magnitude of the electric field on the first equipotential surface, and E_2 represent the magnitude of the electric field when it has shrunk to one-half of its initial value. Then

$$E_1 = \frac{kq}{r_1^2} \quad \text{and} \quad E_2 = \frac{kq}{r_2^2}$$

Since $E_2 = \tfrac{1}{2} E_1$, we have

$$\frac{kq}{r_2^2} = \frac{1}{2} \frac{kq}{r_1^2} \quad \text{so that} \quad \frac{r_2}{r_1} = \sqrt{2}$$

Thus, the electric field shrinks to one-half its original value at a distance of $r_2 = \sqrt{2}\, r_1$.

The potential difference between the first equipotential surface and the locus of points where the electric field has dropped to one-half of its original magnitude is

$$\Delta V = \frac{kq}{r_2} - \frac{kq}{r_1} = \frac{kq}{r_1}\left(\frac{1}{\sqrt{2}} - 1\right)$$

$$= \frac{(8.99 \times 10^9 \text{ N} \cdot \text{m}^2/\text{C}^2)(2.00 \times 10^{-6} \text{ C})}{1.60 \text{ m}}\left(\frac{1}{\sqrt{2}} - 1\right) = -3290 \text{ V}$$

Since the potential difference between two successive surfaces is 1.00×10^3 V, we have

$$\frac{\text{Magnitude of } \Delta V \text{ from } r_1 \text{ to } r_2}{\Delta V \text{ between successive equipotential surfaces}} = \frac{3290 \text{ V}}{1.00 \times 10^3 \text{ V}} = 3.29$$

Therefore, the number of equipotential surfaces crossed in going radially outward from r_1 to r_2 is $\boxed{3}$.

---

33. **REASONING AND SOLUTION** The drawing shows the electric field and the three points, A, B, and C, around the point P, which we take as the origin. We choose the upward direction as being positive. Thus, $E = -3.0 \times 10^3$ V/m, since the electric field points straight down.

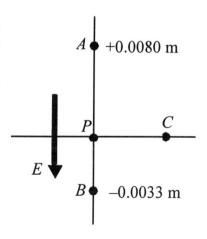

a. The electric potential at point A can be determined from Equation 19.7 as

$$E = -\frac{\Delta V}{\Delta s} = -\frac{V_P - V_A}{\Delta s} = -\frac{135 \text{ V} - V_A}{0 - 8.0 \times 10^{-3} \text{ m}}$$

$$\underbrace{}_{-3.0 \times 10^3 \text{ V/m}}$$

so that $\boxed{V_A = 159 \text{ V}}$.

b. The electric potential at point B is found in a similar manner:

$$E = -\frac{\Delta V}{\Delta s} = -\frac{V_B - V_P}{\Delta s} = -\frac{V_B - 135 \text{ V}}{-3.3 \times 10^{-3} \text{ m} - 0}$$

$$\underbrace{}_{-3.0 \times 10^3 \text{ V/m}}$$

so that $\boxed{V_B = 125 \text{ V}}$

c. $\Delta V = 0$, since the path is perpendicular to the electric field, and no work is done in moving a charge along such a path: $\boxed{V_C = 135 \text{ V}}$

---

34. **REASONING AND SOLUTION** Let point A be on the x-axis where the potential is 515 V. Let point B be on the x-axis where the potential is 495 V. From Equation 19.7, the electric field is

$$E = -\frac{\Delta V}{\Delta s} = -\frac{V_B - V_A}{\Delta s} = -\frac{495 \text{ V} - 515 \text{ V}}{-2(6.0 \times 10^{-3} \text{ m})} = -1.7 \times 10^3 \text{ V/m}$$

The magnitude of the electric field is $\boxed{1.7 \times 10^3 \text{ V/m}}$. Since the electric field is negative, it points to the $\boxed{\text{left}}$, from the high toward the low potential.

---

35. **SSM** **REASONING** The total energy of the particle on the equipotential surface A is $E_A = \frac{1}{2}mv_A^2 + qV_A$. Similarly, the total energy of the particle when it reaches equipotential surface B is $E_B = \frac{1}{2}mv_B^2 + qV_B$. According to Equation 6.8, the work W done by the outside force is equal to the final total energy E_B minus the initial total energy E_A, $W = E_B - E_A$.

SOLUTION The work done by the outside force in moving the particle from A to B is,

$$W = E_B - E_A = \left(\tfrac{1}{2}mv_B^2 + qV_B\right) - \left(\tfrac{1}{2}mv_A^2 + qV_A\right) = \tfrac{1}{2}m(v_B^2 - v_A^2) + q(V_B - V_A)$$

$$= \tfrac{1}{2}(5.00 \times 10^{-2} \text{ kg})\left[(3.00 \text{ m/s})^2 - (2.00 \text{ m/s})^2\right]$$

$$+ (4.00 \times 10^{-5} \text{ C})[7850 \text{ V} - 5650 \text{ V}] = \boxed{0.213 \text{ J}}$$

726 ELECTRIC POTENTIAL ENERGY AND THE ELECTRIC POTENTIAL

36. **REASONING AND SOLUTION** The capacitance, voltage, and charge are related by

$$V = \frac{q}{C} = \frac{7.2 \times 10^{-5} \text{ C}}{6.0 \times 10^{-6} \text{ F}} = \boxed{12 \text{ V}} \qquad (19.8)$$

37. [SSM] **REASONING** According to Equation 19.11, the energy stored in a capacitor with capacitance C and potential V across its plates is Energy $= \frac{1}{2}CV^2$.

    SOLUTION Therefore, solving Equation 19.11 for V, we have

$$V = \sqrt{\frac{2(\text{Energy})}{C}} = \sqrt{\frac{2(73 \text{ J})}{120 \times 10^{-6} \text{ F}}} = \boxed{1.1 \times 10^3 \text{ V}}$$

38. **REASONING AND SOLUTION** The capacitance is $C = q_0/V_0 = q/V$. The new charge q is, therefore,

$$q = \frac{q_0 V}{V_0} = \frac{(5.3 \times 10^{-5} \text{ C})(9.0 \text{ V})}{6.0 \text{ V}} = \boxed{8.0 \times 10^{-5} \text{ C}}$$

39. [SSM] **REASONING** The charge that resides on the outer surface of the cell membrane is $q = CV$, according to Equation 19.8. Before we can use this expression, however, we must first determine the capacitance of the membrane. If we assume that the cell membrane behaves like a parallel plate capacitor filled with a dielectric, Equation 19.10 ($C = \kappa \varepsilon_0 A/d$) applies as well.

    SOLUTION The capacitance of the cell membrane is

$$C = \frac{\kappa \varepsilon_0 A}{d} = \frac{(5.0)(8.85 \times 10^{-12} \text{ F/m})(5.0 \times 10^{-9} \text{ m}^2)}{1.0 \times 10^{-8} \text{ m}} = 2.2 \times 10^{-11} \text{ F}$$

a. The charge on the outer surface of the membrane is, therefore,

$$q = CV = (2.2 \times 10^{-11} \text{ F})(60.0 \times 10^{-3} \text{ V}) = \boxed{1.3 \times 10^{-12} \text{ C}}$$

b. If the charge in part (a) is due to K$^+$ ions with charge $+e$ ($e = 1.6 \times 10^{-19}$ C), the number of ions present on the outer surface of the membrane is

$$\frac{\text{Number of}}{\text{K}^+ \text{ ions}} = \frac{1.3 \times 10^{-12} \text{ C}}{1.6 \times 10^{-19} \text{ C}} = \boxed{8.1 \times 10^6}$$

40. **REASONING AND SOLUTION** The total charge transferred is given by

$$q = CV = (2.5 \times 10^{-8} \text{ F})(450 \text{ V})$$

The number of electrons transferred is, then,

$$\text{Number of electrons} = \frac{q}{e} = \frac{(2.5 \times 10^{-8} \text{ F})(450 \text{ V})}{1.60 \times 10^{-19} \text{ C}} = \boxed{7.0 \times 10^{13}}$$

41. **REASONING AND SOLUTION**
a. We know that

$$\text{Energy} = \tfrac{1}{2}CV^2 = \tfrac{1}{2}(750 \times 10^{-6} \text{ F})(330 \text{ V})^2 = \boxed{41 \text{ J}}$$

b. The power developed by the flash is

$$P = \frac{\text{Energy}}{\text{Time}} = \frac{41 \text{ J}}{5.0 \times 10^{-3} \text{ s}} = \boxed{8200 \text{ W}}$$

42. **REASONING** Equation 19.11 (Energy = $CV^2/2$) gives the energy stored by a capacitor. Applying this expression to each capacitor will allow us to determine the voltage ratio, since the two energies are given.

SOLUTION Applying Equation 19.11 to each capacitor gives

$$(\text{Energy})_A = \tfrac{1}{2}CV_A^2 \quad \text{and} \quad (\text{Energy})_B = \tfrac{1}{2}CV_B^2$$

Dividing the A-equation by the B-equation, we find

$$\frac{(\text{Energy})_A}{(\text{Energy})_B} = \frac{\tfrac{1}{2}CV_A^2}{\tfrac{1}{2}CV_B^2} \quad \text{or} \quad \frac{(\text{Energy})_A}{(\text{Energy})_B} = \frac{V_A^2}{V_B^2}$$

$$V_A = V_B \sqrt{\frac{(\text{Energy})_A}{(\text{Energy})_B}} = (12 \text{ V})\sqrt{\frac{310 \text{ J}}{34 \text{ J}}} = \boxed{36 \text{ V}}$$

43. [SSM] **REASONING** According to Equation 19.11, the energy stored in a capacitor with a capacitance C and potential V across its plates is Energy = $\tfrac{1}{2}CV^2$. Once we determine how much energy is required to operate a 75-W light bulb for one minute, we can then use the expression for the energy to solve for V.

728 ELECTRIC POTENTIAL ENERGY AND THE ELECTRIC POTENTIAL

SOLUTION The energy stored in the capacitor, which is equal to the energy required to operate a 75-W bulb for one minute (= 60 s), is

$$\text{Energy} = Pt = (75 \text{ W})(60 \text{ s}) = 4500 \text{ J}$$

Therefore, solving Equation 19.11 for V, we have

$$V = \sqrt{\frac{2(\text{Energy})}{C}} = \sqrt{\frac{2(4500 \text{ J})}{3.3 \text{ F}}} = \boxed{52 \text{ V}}$$

---

44. *REASONING AND SOLUTION* The electric energy stored in the region between the metal spheres is

$$\text{Energy} = \tfrac{1}{2} \kappa \varepsilon_0 E^2 (\text{Volume}) \tag{19.12}$$

However, the magnitude E of the electric field is related to the potential difference ΔV between the spheres and the distance Δs between them via Equation 19.7, $E = \Delta V / \Delta s$. Thus,

$$\text{Energy} = \tfrac{1}{2} \kappa \varepsilon_0 \left(\frac{\Delta V}{\Delta s}\right)^2 (\text{Volume})$$

$$= \tfrac{1}{2}(2.1)\left(8.85 \times 10^{-12} \frac{\text{C}^2}{\text{N}\cdot\text{m}^2}\right)\left(\frac{3.0 \text{ V}}{0.0020 \text{ m}}\right)^2 \underbrace{\left[4\pi(0.1500 \text{ m})^2 (0.0020 \text{ m})\right]}_{\text{Volume}}$$

$$= \boxed{1.2 \times 10^{-8} \text{ J}}$$

---

45. *REASONING AND SOLUTION* From Equations 19.9 and 18.4, we have

$$A = \frac{q}{\kappa \varepsilon_0 E} = \frac{1.7 \times 10^{-7} \text{ C}}{(3.5)\left(8.85 \times 10^{-12} \frac{\text{C}^2}{\text{N}\cdot\text{m}^2}\right)(1.4 \times 10^7 \text{ N/C})} = 3.9 \times 10^{-4} \text{ m}^2$$

Since $A = \pi r^2$, the radius of the plates is

$$r = \sqrt{\frac{A}{\pi}} = \sqrt{\frac{3.9 \times 10^{-4} \text{ m}^2}{\pi}} = \boxed{1.1 \times 10^{-2} \text{ m}}$$

46. **REASONING AND SOLUTION** The charge on the empty capacitor is $q_0 = C_0 V_0$. With the dielectric in place, the charge remains the same. However, the new capacitance is $C = \kappa C_0$ and the new voltage is V. Thus,

$$q_0 = CV = \kappa C_0 V = C_0 V_0$$

Solving for the new voltage yields

$$V = V_0/\kappa = (12.0 \text{ V})/2.8 = 4.3 \text{ V}$$

The potential difference is $12.0 - 4.3 = \boxed{7.7 \text{ V}}$. The change in potential is a $\boxed{\text{decrease}}$.

---

47. **SSM WWW REASONING** If we assume that the motion of the proton and the electron is horizontal in the $+x$ direction, the motion of the proton is determined by Equation 2.8, $x = v_0 t + \frac{1}{2} a_p t^2$, where x is the distance traveled by the proton, v_0 is its initial speed, and a_p is its acceleration. If the distance between the capacitor places is d, then this relation becomes $\frac{1}{2} d = v_0 t + \frac{1}{2} a_p t^2$, or

$$d = 2 v_0 t + a_p t^2 \qquad (1)$$

We can solve Equation (1) for the initial speed v_0 of the proton, but, first, we must determine the time t and the acceleration a_p of the proton. Since the proton strikes the negative plate at the same instant the electron strikes the positive plate, we can use the motion of the electron to determine the time t.

For the electron, $\frac{1}{2} d = \frac{1}{2} a_e t^2$, where we have taken into account the fact that the electron is released from rest. Solving this expression for t we have $t = \sqrt{d/a_e}$. Substituting this expression into Equation (1), we have

$$d = 2 v_0 \sqrt{\frac{d}{a_e}} + \left(\frac{a_p}{a_e}\right) d \qquad (2)$$

The accelerations can be found by noting that the magnitudes of the forces on the electron and proton are equal, since these particles have the same magnitude of charge. The force on the electron is $F = eE = eV/d$, and the acceleration of the electron is, therefore,

$$a_e = \frac{F}{m_e} = \frac{eV}{m_e d} \qquad (3)$$

Newton's second law requires that $m_e a_e = m_p a_p$, so that

730 ELECTRIC POTENTIAL ENERGY AND THE ELECTRIC POTENTIAL

$$\frac{a_p}{a_e} = \frac{m_e}{m_p} \quad (4)$$

Combining Equations (2), (3) and (4) leads to the following expression for v_0, the initial speed of the proton:

$$v_0 = \frac{1}{2}\left(1 - \frac{m_e}{m_p}\right)\sqrt{\frac{eV}{m_e}}$$

SOLUTION Substituting values into the expression above, we find

$$v_0 = \frac{1}{2}\left(1 - \frac{9.11\times 10^{-31}\text{ kg}}{1.67\times 10^{-27}\text{ kg}}\right)\sqrt{\frac{(1.6\times 10^{-19}\text{ C})(175\text{ V})}{9.11\times 10^{-31}\text{ kg}}} = \boxed{2.77\times 10^6\text{ m/s}}$$

---

48. **REASONING AND SOLUTION** The electron and proton experience forces of the same magnitude in the same electric field, since they have the same magnitude of charge. Newton's second law gives

$$m_p a_p = m_e a_e$$

where a_p and a_e are the accelerations of the proton and the electron. Kinematics gives

$$a_p = 2x_p/t^2 \quad \text{and} \quad a_e = 2x_e/t^2$$

so

$$x_p = \frac{m_e}{m_p}x_e = \frac{9.11\times 10^{-31}\text{ kg}}{1.67\times 10^{-27}\text{ kg}}(0.0400\text{ m}) = \boxed{2.18\times 10^{-5}\text{ m}}$$

---

49. **REASONING AND SOLUTION** First, we need to find the magnitude of the charge. Since $V = kq/r$, we have

$$q = \frac{Vr}{k} = \frac{(164\text{ V})(0.20\text{ m})}{9.0\times 10^9\text{ N}\cdot\text{m}^2/\text{C}^2} = 3.6\times 10^{-9}\text{ C}$$

Thus, at a distance of 0.80 m the potential would be

$$V = \frac{kq}{r} = \frac{(9.0\times 10^9\text{ N}\cdot\text{m}^2/\text{C}^2)(3.6\times 10^{-9}\text{ C})}{0.80\text{ m}} = \boxed{41\text{ V}}$$

50. ***REASONING AND SOLUTION***

a. The change in the electric potential energy is

$$\text{EPE}_A - \text{EPE}_B = W_{AB} = \boxed{5.80 \times 10^{-3} \text{ J}}$$

b. The potential difference between the points is

$$V_A - V_B = \frac{\text{EPE}_A - \text{EPE}_B}{q} = \frac{5.80 \times 10^{-3} \text{ J}}{1.80 \times 10^{-4} \text{ C}} = \boxed{32.2 \text{ V}}$$

c. $\boxed{\text{Point } A}$ has the higher potential.

---

51. $\boxed{\text{SSM}}$ ***REASONING AND SOLUTION*** Equation 19.10 gives the capacitance for a parallel plate capacitor filled with a dielectric of constant κ: $C = \kappa \varepsilon_0 A / d$. Solving for κ, we have

$$\kappa = \frac{Cd}{\varepsilon_0 A} = \frac{(7.0 \times 10^{-6} \text{ F})(1.0 \times 10^{-5} \text{ m})}{(8.85 \times 10^{-12} \text{ F/m})(1.5 \text{ m}^2)} = \boxed{5.3}$$

---

52. ***REASONING AND SOLUTION*** The only force that acts on the α-particle is the conservative electric force. Therefore, the total energy of the α-particle is conserved as it moves from point A to point B:

$$\underbrace{\tfrac{1}{2} m v_A^2 + \text{EPE}_A}_{\text{Total energy at point } A} = \underbrace{\tfrac{1}{2} m v_B^2 + \text{EPE}_B}_{\text{Total energy at point } B}$$

Since the α-particle starts from rest, $v_A = 0$. The electric potential V is related to the electric potential energy EPE by $V = \text{EPE}/q$ (see Equation 19.3). With these changes, the equation above gives the kinetic energy of the α-particle at point B to be

$$\tfrac{1}{2} m v_B^2 = \text{EPE}_A - \text{EPE}_B = q(V_A - V_B)$$

Since an α-particle contains two protons, its charge is $q = 2e = 3.2 \times 10^{-19}$ C. Thus, the kinetic energy (in electron-volts) is

$$\tfrac{1}{2} m v_B^2 = q(V_A - V_B) = (3.2 \times 10^{-19} \text{ C})[+250 \text{ V} - (-150 \text{ V})]$$

$$= 1.28 \times 10^{-16} \text{ J} \left(\frac{1.0 \text{ eV}}{1.6 \times 10^{-19} \text{ J}} \right) = \boxed{8.0 \times 10^2 \text{ eV}}$$

732 ELECTRIC POTENTIAL ENERGY AND THE ELECTRIC POTENTIAL

53. **REASONING AND SOLUTION** The capacitance is given by

$$C = \frac{k\varepsilon_0 A}{d} = \frac{5(8.85 \times 10^{-12} \text{ F/m})(5 \times 10^{-6} \text{ m}^2)}{1 \times 10^{-8} \text{ m}} = \boxed{2 \times 10^{-8} \text{ F}}$$

54. **REASONING AND SOLUTION** The initial and final separations of the charges are the same. Since the electric potential energy depends only on the separation of the charges, there is no change in the electric potential energy, and, hence, $\boxed{\text{no work is done}}$.

55. **SSM WWW REASONING AND SOLUTION** As described in the problem statement, the charges jump between your hand and a doorknob. If we assume that the electric field is uniform, Equation 19.7 applies, and we have

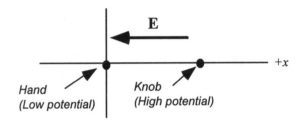

$$E = -\frac{\Delta V}{\Delta s} = -\frac{V_{\text{knob}} - V_{\text{hand}}}{\Delta s}$$

Therefore, solving for the potential difference between your hand and the doorknob, we have

$$V_{\text{knob}} - V_{\text{hand}} = -E\Delta s = -(-3.0 \times 10^6 \text{ N/C})(3.0 \times 10^{-3} \text{ m}) = \boxed{+9.0 \times 10^3 \text{ V}}$$

56. **REASONING** The electric field E that exists between two points in space is, according to Equation 19.7, proportional to the electric potential difference ΔV between the points divided by the distance Δx between them: $E = -\Delta V/\Delta x$.

SOLUTION
a. The electric field in the region from A to B is

$$E = -\frac{\Delta V}{\Delta x} = -\frac{5.0 \text{ V} - 5.0 \text{ V}}{0.20 \text{ m} - 0 \text{ m}} = \boxed{0 \text{ V/m}}$$

b. The electric field in the region from B to C is

$$E = -\frac{\Delta V}{\Delta x} = -\frac{3.0 \text{ V} - 5.0 \text{ V}}{0.40 \text{ m} - 0.20 \text{ m}} = \boxed{1.0 \times 10^1 \text{ V/m}}$$

c. The electric field in the region from C to D is

$$E = -\frac{\Delta V}{\Delta x} = -\frac{1.0 \text{ V} - 3.0 \text{ V}}{0.80 \text{ m} - 0.40 \text{ m}} = \boxed{5.0 \text{ V/m}}$$

---

57. **REASONING AND SOLUTION** The potential difference is constant so $V_0 = q_0/C_0 = q/C = q/(\kappa C_0)$. Thus,

$$q = kq_0 = kC_0V_0 = (4.0)(2.7 \times 10^{-6} \text{ F})(12 \text{ V}) = 1.3 \times 10^{-4} \text{ C}$$

The original charge is $q_0 = C_0V_0 = 3.2 \times 10^{-5}$ C. The surface charge Q is, therefore,

$$Q = q - q_0 = \boxed{1.0 \times 10^{-4} \text{ C}}$$

---

58. **REASONING** The only force acting on each particle is the conservative electric force. Therefore, the total energy (kinetic energy plus electric potential energy) is conserved as the particles move apart. In addition, the net external force acting on the system of two particles is zero (the electric force that each particle exerts on the other is an internal force). Thus, the total linear momentum of the system is also conserved. We will use the conservation of energy and the conservation of linear momentum to find the initial separation of the particles.

SOLUTION For two points, A and B, along the motion, the conservation of energy is

$$\underbrace{\tfrac{1}{2}m_1v_{1,A}^2 + \tfrac{1}{2}m_2v_{2,A}^2}_{\text{Initial kinetic energy of the two particles}} + \underbrace{\frac{kq_1q_2}{r_A}}_{\text{Initial electric potential energy}} = \underbrace{\tfrac{1}{2}m_1v_{1,B}^2 + \tfrac{1}{2}m_2v_{2,B}^2}_{\text{Final kinetic energy of the two particles}} + \underbrace{\frac{kq_1q_2}{r_B}}_{\text{Final electric potential energy}}$$

Solving this equation for $1/r_A$ and setting $v_{1,A} = v_{2,A} = 0$ since the particles are initially at rest, we obtain

$$\frac{1}{r_A} = \frac{1}{r_B} + \frac{1}{kq_1q_2}\left(\tfrac{1}{2}m_1v_{1,B}^2 + \tfrac{1}{2}m_2v_{2,B}^2\right) \quad (1)$$

This equation cannot be solved for the initial separation r_A, because the final speed $v_{2,B}$ of the second particle is not known. To find this speed, we will use the conservation of linear momentum:

$$\underbrace{m_1v_{1,A} + m_2v_{2,A}}_{\text{Initial linear momentum}} = \underbrace{m_1v_{1,B} + m_2v_{2,B}}_{\text{Final linear momentum}}$$

734 ELECTRIC POTENTIAL ENERGY AND THE ELECTRIC POTENTIAL

Setting $v_{1,A} = v_{2,A} = 0$ and solving for $v_{2,B}$ gives

$$v_{2,B} = -\frac{m_1}{m_2}v_{1,B} = -\frac{3.00 \times 10^{-3} \text{ kg}}{6.00 \times 10^{-3} \text{ kg}}(125 \text{ m/s}) = -62.5 \text{ m/s}$$

Substituting this value for $v_{2,B}$ into Equation (1) yields

$$\frac{1}{r_A} = \frac{1}{0.100 \text{ m}} + \frac{1}{(8.99 \times 10^9 \text{ N} \cdot \text{m}^2/\text{C}^2)(8.00 \times 10^{-6} \text{ C})^2}$$

$$\times \left[\tfrac{1}{2}(3.00 \times 10^{-3} \text{ kg})(125 \text{ m/s})^2 + \tfrac{1}{2}(6.00 \times 10^{-3} \text{ kg})(-62.5 \text{ m/s})^2 \right]$$

$$\boxed{r_A = 1.41 \times 10^{-2} \text{ m}}$$

---

59. **SSM REASONING** According to Equation 19.10, the capacitance of a parallel plate capacitor filled with a dielectric is $C = \kappa \varepsilon_0 A/d$, where κ is the dielectric constant, A is the area of one plate, and d is the distance between the plates.

From the definition of capacitance (Equation 19.8), $q = CV$. Thus, using Equation 19.10, we see that the charge q on a parallel plate capacitor that contains a dielectric is given by $q = (\kappa \varepsilon_0 A/d)V$. Since each dielectric occupies one-half of the volume between the plates, the area of each plate in contact with each material is $A/2$. Thus,

$$q_1 = \frac{\kappa_1 \varepsilon_0 (A/2)}{d}V = \frac{\kappa_1 \varepsilon_0 A}{2d}V \quad \text{and} \quad q_2 = \frac{\kappa_2 \varepsilon_0 (A/2)}{d}V = \frac{\kappa_2 \varepsilon_0 A}{2d}V$$

According to the problem statement, the total charge stored by the capacitor is

$$q_1 + q_2 = CV \tag{1}$$

where q_1 and q_2 are the charges on the plates in contact with dielectrics 1 and 2, respectively. Using the expressions for q_1 and q_2 above, Equation (1) becomes

$$CV = \frac{\kappa_1 \varepsilon_0 A}{2d}V + \frac{\kappa_2 \varepsilon_0 A}{2d}V = \frac{\kappa_1 \varepsilon_0 A + \kappa_2 \varepsilon_0 A}{2d}V = \frac{(\kappa_1 + \kappa_2)\varepsilon_0 A}{2d}V$$

This expression can be solved for C.

SOLUTION Solving for C, we obtain $\boxed{C = \dfrac{\varepsilon_0 A(\kappa_1 + \kappa_2)}{2d}}$

60. **REASONING AND SOLUTION** The information about the electric field requires that

$$kq_2/(1.00 \text{ m})^2 = kq_1/(4.00 \text{ m})^2 \quad \text{so} \quad q_2 = (1/16.0)\, q_1$$

Let x be the distance of the zero-potential point from the negative charge. Then

$$kq_1/(d+x) = kq_2/x$$

if the point is to the right of q_2 and

$$kq_1/(d-x) = kq_2/x$$

if the point is to the left of q_2. Solving for x gives

$$\boxed{x = 0.200 \text{ m to the right of the negative charge}}$$
$$\boxed{x = 0.176 \text{ m to the left of the negative charge}}$$

61. **CONCEPT QUESTIONS** a. The total potential at either of the empty corners is the sum of the individual potentials created by each of the charges. According to Equation 19.6 ($V = kq/r$), greater positive charges q and smaller distances r lead to greater potentials. Thus, the potential at corner A is greater than at corner B, because corner A is closer to the larger charge.

b. According to Equation 19.3, the electric potential energy of the third charge is $\text{EPE} = q_3 V_{\text{Total}}$, where V_{Total} is the total potential at the point where the third charge is placed. Since the charges q_1 and q_2 are both positive, V_{Total} is also positive. Therefore, since q_3 is negative, Equation 19.3 indicates that the electric potential energy is negative.

c. According to Equation 19.3, the electric potential energy of the third charge is $\text{EPE} = q_3 V_{\text{Total}}$, where V_{Total} is greater at corner A than at corner B. Therefore, the electric potential energy has a greater magnitude at corner A.

SOLUTION We begin by noting that each side of the square has a length L, so that the diagonal has a length of $\sqrt{2}L$, according to the Pythagorean theorem. Using Equation 19.6, we can express the total potential at corners A and B as follows:

$$V_{\text{Total, A}} = \frac{kq_2}{L} + \frac{kq_1}{\sqrt{2}L} \quad \text{and} \quad V_{\text{Total, B}} = \frac{kq_1}{L} + \frac{kq_2}{\sqrt{2}L}$$

736 ELECTRIC POTENTIAL ENERGY AND THE ELECTRIC POTENTIAL

Using Equation 19.3, we find that the electric potential energy of the third charge at each corner is

$$\text{EPE}_A = q_3 V_{\text{Total, A}} = q_3 \left(\frac{kq_2}{L} + \frac{kq_1}{\sqrt{2}L} \right) = \frac{q_3 k}{L} \left(q_2 + \frac{q_1}{\sqrt{2}} \right)$$

$$= \frac{(-6.0 \times 10^{-9} \text{ C})(8.99 \times 10^9 \text{ N} \cdot \text{m}^2/\text{C}^2)}{0.25 \text{ m}} \left(+4.0 \times 10^{-9} \text{ C} + \frac{+1.5 \times 10^{-9} \text{ C}}{\sqrt{2}} \right)$$

$$= \boxed{-1.1 \times 10^{-6} \text{ J}}$$

$$\text{EPE}_B = q_3 V_{\text{Total, B}} = q_3 \left(\frac{kq_1}{L} + \frac{kq_2}{\sqrt{2}L} \right) = \frac{q_3 k}{L} \left(q_1 + \frac{q_2}{\sqrt{2}} \right)$$

$$= \frac{(-6.0 \times 10^{-9} \text{ C})(8.99 \times 10^9 \text{ N} \cdot \text{m}^2/\text{C}^2)}{0.25 \text{ m}} \left(+1.5 \times 10^{-9} \text{ C} + \frac{+4.0 \times 10^{-9} \text{ C}}{\sqrt{2}} \right)$$

$$= \boxed{-0.93 \times 10^{-6} \text{ J}}$$

As expected the electric potential energy at corner A has a greater magnitude.

---

62. **CONCEPT QUESTIONS** a. The potential from the positive charge is positive, while the potential from the negative charge is negative at the spot in question. The magnitude of the positive potential must be equal to the magnitude of the negative potential. Only then will the algebraic sum of the two potentials give zero for the total potential.

b. The drawing shows the arrangement of the charges. The two charges and the spot in question form a right triangle. Therefore, the distance between the positive charge and the zero-potential spot is the hypotenuse of the right triangle and, according to the Pythagorean theorem, is greater than L.

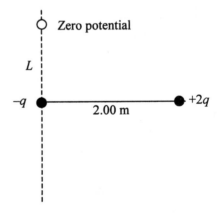

c. There are two spots on the dashed line where the total potential is zero. One spot is shown in the drawing. The other spot is on the dashed line a distance L below the negative charge.

SOLUTION We begin by noting that the distance between the positive charge and the zero-potential spot is given by the Pythagorean theorem as $\sqrt{L^2 + (2.00 \text{ m})^2}$. With this in mind and using Equation 19.6 ($V = kq/r$), we write the total potential at the spot in question as follows:

$$V_{Total} = \frac{k(-q)}{L} + \frac{k(+2q)}{\sqrt{L^2 + (2.00 \text{ m})^2}} = 0 \quad \text{or} \quad \frac{1}{L} = \frac{+2}{\sqrt{L^2 + (2.00 \text{ m})^2}}$$

Squaring both sides of this result gives

$$\frac{1}{L^2} = \frac{4}{L^2 + (2.00 \text{ m})^2} \quad \text{or} \quad L^2 + (2.00 \text{ m})^2 = 4L^2$$

Solving for L, we obtain

$$3L^2 = (2.00 \text{ m})^2 \quad \text{or} \quad L = \frac{2.00 \text{ m}}{\sqrt{3}} = \boxed{1.15 \text{ m}}$$

63. **CONCEPT QUESTIONS** a. Conservation of energy requires that as each particle gains kinetic energy, it must lose electric potential energy. The loss in one type of energy appears as a gain in the other type.

b. The electric potential difference ΔV experienced by the electron has the same magnitude as the electric potential difference experienced by the proton. Moreover, the charge q_0 on either particle has the same magnitude. According to Equation 19.4 ($\Delta \text{EPE} = q_0 \Delta V$), therefore, the losses in EPE for the electron and the proton are the same. Conservation of energy, then, dictates that the electron gains the same amount of kinetic energy as does the proton.

c. Both particles start from rest, so the gain in kinetic energy is equal to the final kinetic energies in each case, which are the same for each particle. However, kinetic energy is $mv^2/2$, and the mass of the electron is much less than the mass of the proton (see the inside of the front cover). Since the kinetic energies are the same, the speed v_e of the electron must, then, be greater than the speed v_p of the proton.

SOLUTION Let us assume that the proton accelerates from point A to point B. According to the energy-conservation principle (including only kinetic and electric potential energies), we have

$$\underbrace{\text{KE}_{p,B} + \text{EPE}_{p,B}}_{\text{Final total energy}} = \underbrace{\text{KE}_{p,A} + \text{EPE}_{p,A}}_{\text{Initial total energy}} \quad \text{or} \quad \text{KE}_{p,B} = \text{EPE}_{p,A} - \text{EPE}_{p,B}$$

Here we have used the fact that the initial kinetic energy is zero since the proton starts from rest. According to Equation 19.3, the electric potential energy of a charge q_0 is $\text{EPE} = q_0 V$, where V is the potential experienced by the charge. We can, therefore, write the final kinetic energy of the proton as follows:

$$\text{KE}_{p,B} = \text{EPE}_{p,A} - \text{EPE}_{p,B} = q_p(V_A - V_B) \tag{1}$$

The electron has a negative charge, so it accelerates in the direction opposite to that of the proton, or from point B to point A. Energy conservation applied to the electron gives

$$\underbrace{\text{KE}_{e,A} + \text{EPE}_{e,A}}_{\text{Final total energy}} = \underbrace{\text{KE}_{e,B} + \text{EPE}_{e,B}}_{\text{Initial total energy}} \quad \text{or} \quad \text{KE}_{e,A} = \text{EPE}_{e,B} - \text{EPE}_{e,A}$$

Here we have used the fact that the initial kinetic energy is zero since the electron starts from rest. Again using Equation 19.3, we can write the final kinetic energy of the electron as follows:

$$\text{KE}_{e,A} = \text{EPE}_{e,B} - \text{EPE}_{e,A} = q_e(V_B - V_A)$$

But an electron has a negative charge that is equal in magnitude to the charge on a proton, so $q_e = -q_p$. With this substitution, we can write the kinetic energy of the electron as

$$\text{KE}_{e,A} = \text{EPE}_{e,B} - \text{EPE}_{e,A} = -q_p(V_B - V_A) = q_p(V_A - V_B) \tag{2}$$

Comparing Equations (1) and (2), we can see that

$$\text{KE}_{p,B} = \text{KE}_{e,A} \quad \text{or} \quad \tfrac{1}{2}m_p v_p^2 = \tfrac{1}{2}m_e v_e^2$$

Solving for the ratio v_e/v_p and referring to the inside of the front cover for the masses of the electron and the proton, we obtain

$$\frac{v_e}{v_p} = \sqrt{\frac{m_p}{m_e}} = \sqrt{\frac{1.67 \times 10^{-27} \text{ kg}}{9.11 \times 10^{-31} \text{ kg}}} = \boxed{42.8}$$

As expected, the speed of the electron is greater than the speed of the proton.

---

64. **CONCEPT QUESTIONS** a. Since the electric force does negative work, the electric force must point opposite to the displacement of the test charge.

b. The point charge is positive, so it exerts an outward-directed force on the positive test charge. Since the force and the displacement of the test charge have opposite directions, the displacement must be directed inward toward the point charge. Therefore, the radius r_B is less than the radius r_A.

SOLUTION According to Equation 19.4, we have

$$V_B - V_A = \frac{-W_{AB}}{q_0}$$

The potential of a point charge is given by Equation 19.6 as $V = kq/r$. With this substitution for V_B and V_A, we have

$$\frac{kq}{r_B} - \frac{kq}{r_A} = \frac{-W_{AB}}{q_0} \quad \text{or} \quad \frac{kq}{r_B} = \frac{kq}{r_A} - \frac{W_{AB}}{q_0} \quad \text{or} \quad \frac{1}{r_B} = \frac{1}{r_A} - \frac{W_{AB}}{kqq_0}$$

Solving for r_B gives

$$\frac{1}{r_B} = \frac{1}{r_A} - \frac{W_{AB}}{kqq_0}$$

$$= \frac{1}{1.8 \text{ m}} - \frac{(-8.1 \times 10^{-9} \text{ J})}{(8.99 \times 10^9 \text{ N} \cdot \text{m}^2/\text{C}^2)(7.2 \times 10^{-8} \text{ C})(4.5 \times 10^{-11} \text{ C})} = 0.83 \text{ m}^{-1}$$

$$r_B = \frac{1}{0.83 \text{ m}^{-1}} = \boxed{1.2 \text{ m}}$$

As expected, r_B is less than r_A.

---

65. **CONCEPT QUESTIONS** a. Equation 19.10 gives the capacitance as $C = \kappa \varepsilon_0 A/d$, where A and d are, respectively, the plate area and separation. Other things being equal, the capacitor with the bigger plate area has the greater capacitance. The diameter of the circle equals the length of a side of the square, so the circle fits within the square. The square, therefore, has the bigger area, and the capacitor with the square plates would have the greater capacitance.

b. To make the capacitors have equal capacitances, the dielectric constant must compensate for the larger area of the square plates. Therefore, since capacitance is proportional to the dielectric constant, the capacitor with square plates must contain a dielectric material with a smaller dielectric constant. Thus, the capacitor with circular plates contains the material with the greater dielectric constant.

SOLUTION The area of the square plates is $A_{\text{square}} = L^2$, while the area of the circular plates is $A_{\text{circle}} = \pi(L/2)^2$. Using these areas and applying Equation 19.10 to each capacitor, we have

740 ELECTRIC POTENTIAL ENERGY AND THE ELECTRIC POTENTIAL

$$C = \frac{\kappa_{\text{square}} \varepsilon_0 L^2}{d} \quad \text{and} \quad C = \frac{\kappa_{\text{circle}} \varepsilon_0 \pi \left(\tfrac{1}{2} L\right)^2}{d}$$

Since the values for C are the same, we have

$$\frac{\kappa_{\text{square}} \varepsilon_0 L^2}{d} = \frac{\kappa_{\text{circle}} \varepsilon_0 \pi \left(\tfrac{1}{2} L\right)^2}{d} \quad \text{or} \quad \kappa_{\text{square}} = \frac{\kappa_{\text{circle}} \pi}{4}$$

$$\kappa_{\text{circle}} = \frac{4\kappa_{\text{square}}}{\pi} = \frac{4(3.00)}{\pi} = \boxed{3.82}$$

As expected, κ_{circle} is greater than κ_{square}.

---

66. **CONCEPT QUESTIONS** a. The energy needed to melt ice is given by Equation 12.5 as $Q = mL_f$, where m is the mass and L_f is the latent heat of fusion for water. The energy needed to boil away water is given by $Q = mL_v$, where L_v is the latent heat of vaporization for water. The latent heat of vaporization is greater than the latent heat of fusion (see Table 12.3). Therefore, it requires more energy to boil away one milligram of water than to melt one milligram of ice.

b. According to Equation 19.11 the energy stored in a capacitor is Energy = $CV^2/2$, where C is the capacitance and V is the voltage across the plates. Since the voltage is the same here for both capacitors, the capacitor storing the greater energy has the greater capacitance. Capacitor B contains more energy, since it can boil the water. Therefore, capacitor B must have the greater capacitance.

SOLUTION Using Equations 19.11 and 12.5, we find

$$(\text{Energy})_A = mL_f = \tfrac{1}{2} C_A V^2 \quad \text{and} \quad (\text{Energy})_B = mL_v = \tfrac{1}{2} C_B V^2$$

Dividing these two results gives

$$\frac{mL_v}{mL_f} = \frac{\tfrac{1}{2} C_B V^2}{\tfrac{1}{2} C_A V^2} \quad \text{or} \quad \frac{L_v}{L_f} = \frac{C_B}{C_A} \quad \text{or} \quad C_B = C_A \frac{L_v}{L_f}$$

Taking the values for the latent heats from Table 12.3, we find

$$C_B = C_A \frac{L_v}{L_f} = (9.3 \times 10^{-6} \text{ F}) \left(\frac{22.6 \times 10^5 \text{ J/kg}}{33.5 \times 10^4 \text{ J/kg}} \right) = \boxed{6.3 \times 10^{-5} \text{ F}}$$

As expected C_B is greater than C_A.

CHAPTER 20 | ELECTRIC CIRCUITS

PROBLEMS

1. **SSM REASONING** Since current is defined as charge per unit time, the current used by the portable compact disc player is equal to the charge provided by the battery pack (180 C) divided by the time in which the charge is delivered (2.0 h).

   SOLUTION The amount of current that the player uses in operation is determined from Equation 20.1:

   $$I = \frac{\Delta q}{\Delta t} = \frac{180 \text{ C}}{2.0 \text{ h}} \underbrace{\left(\frac{1.0 \text{ h}}{3600 \text{ s}}\right)}_{\text{Converts hours to seconds}} = \boxed{0.025 \text{ A}}$$

2. **REASONING** The current I is defined in Equation 20.1 as the amount of charge Δq per unit of time Δt that flows in a wire. Therefore, the amount of charge is the product of the current and the time interval. The number of electrons is equal to the charge that flows divided by the magnitude of the charge on an electron.

   SOLUTION
   a. The amount of charge that flows is

   $$\Delta q = I\,\Delta t = (18 \text{ A})(2.0 \times 10^{-3} \text{ s}) = \boxed{3.6 \times 10^{-2} \text{ C}}$$

   b. The number of electrons N is equal to the amount of charge divided by e, the magnitude of the charge on an electron.

   $$N = \frac{\Delta q}{e} = \frac{3.6 \times 10^{-2} \text{ C}}{1.60 \times 10^{-19} \text{ C}} = \boxed{2.3 \times 10^{17}}$$

3. **REASONING AND SOLUTION** We know that $V = IR$. Therefore,

   $$I = \frac{V}{R} = \frac{120 \text{ V}}{580 \text{ }\Omega} = \boxed{0.21 \text{ A}}$$

742   ELECTRIC CIRCUITS

4. **REASONING AND SOLUTION** Ohm's law gives

$$I = \frac{V}{R} = \frac{120 \text{ V}}{14 \text{ }\Omega} = \boxed{8.6 \text{ A}}$$

5. **SSM** **REASONING AND SOLUTION** Ohm's law (Equation 20.2, $V = IR$) gives the result directly

$$I = \frac{V}{R} = \frac{240 \text{ V}}{11 \text{ }\Omega} = \boxed{22 \text{ A}}$$

6. **REASONING AND SOLUTION** First determine the total charge delivered to the battery using Equation 20.1:

$$\Delta q = I \Delta t = (6.0 \text{ A})(5.0 \text{ h})[(3600 \text{ s})/(1 \text{ h})] = 1.1 \times 10^5 \text{ C}$$

To find the energy delivered to the battery, multiply this charge by the energy per unit charge (i.e., the voltage) to get

$$\text{Energy} = (\Delta q)V = (1.1 \times 10^5 \text{ C})(12 \text{ V}) = \boxed{1.3 \times 10^6 \text{ J}}$$

7. **REASONING** As discussed in Section 20.1, the voltage gives the energy per unit charge. Thus, we can determine the energy delivered to the toaster by multiplying the voltage V by the charge Δq that flows during a time Δt of one minute. The charge can be obtained by solving Equation 20.1, $I = (\Delta q)/(\Delta t)$, since the current I can be obtained from Ohm's law.

    SOLUTION Remembering that voltage is energy per unit charge, we have

    $$\text{Energy} = V \Delta q$$

    Solving Equation 20.1 for Δq gives $\Delta q = I \Delta t$, which can be substituted in the previous result to give

    $$\text{Energy} = V \Delta q = VI \Delta t$$

    According to Ohm's law (Equation 20.2), the current is $I = V/R$, which can be substituted in the energy expression to show that

    $$\text{Energy} = VI \Delta t = V\left(\frac{V}{R}\right)\Delta t = \frac{V^2 \Delta t}{R} = \frac{(120 \text{ V})^2 (60 \text{ s})}{14 \text{ }\Omega} = \boxed{6.2 \times 10^4 \text{ J}}$$

8. **REASONING AND SOLUTION**
   a. The total charge that can be delivered is

   $$\Delta q = (220 \text{ A·h})[3600 \text{ s}/(1 \text{ h})] = \boxed{7.9 \times 10^5 \text{ C}}$$

   b. The maximum current is

   $$I = (220 \text{ A·h})/[(38 \text{ min})(1 \text{ hr})/(60 \text{ min})] = \boxed{350 \text{ A}}$$

---

9. **SSM REASONING** The number N of protons that strike the target is equal to the amount of electric charge Δq striking the target divided by the charge e of a proton, $N = (\Delta q)/e$. From Equation 20.1, the amount of charge is equal to the product of the current I and the time Δt. We can combine these two relations to find the number of protons that strike the target in 15 seconds.

   The heat Q that must be supplied to change the temperature of the aluminum sample of mass m by an amount ΔT is given by Equation 12.4 as $Q = cm\Delta T$, where c is the specific heat capacity of aluminum. The heat is provided by the kinetic energy of the protons and is equal to the number of protons that strike the target times the kinetic energy per proton. Using this reasoning, we can find the change in temperature of the block for the 15 second-time interval.

   SOLUTION
   a. The number N of protons that strike the target is

   $$N = \frac{\Delta q}{e} = \frac{I \Delta t}{e} = \frac{(0.50 \times 10^{-6} \text{ A})(15 \text{ s})}{1.6 \times 10^{-19} \text{ C}} = \boxed{4.7 \times 10^{13}}$$

   b. The amount of heat Q provided by the kinetic energy of the protons is

   $$Q = (4.7 \times 10^{13} \text{ protons})(4.9 \times 10^{-12} \text{ J/proton}) = 230 \text{ J}$$

   Since $Q = cm\Delta T$ and since Table 12.2 gives the specific heat of aluminum as $c = 9.00 \times 10^2$ J/(kg·C°), the change in temperature of the block is

   $$\Delta T = \frac{Q}{cm} = \frac{230 \text{ J}}{\left(9.00 \times 10^2 \dfrac{\text{J}}{\text{kg·C°}}\right)(15 \times 10^{-3} \text{ kg})} = \boxed{17 \text{ C°}}$$

10. **REASONING AND SOLUTION** Using Equation 20.3 and the resistivity from Table 20.1, we find

$$R = \frac{\rho L}{A} = \frac{(2.82 \times 10^{-8} \ \Omega \cdot m)(10.0 \times 10^3 \ m)}{4.9 \times 10^{-4} \ m^2} = \boxed{0.58 \ \Omega}$$

---

11. [SSM] [WWW] **REASONING** The resistance of a metal wire of length L, cross-sectional area A and resistivity ρ is given by Equation 20.3: $R = \rho L / A$. Solving for A, we have $A = \rho L / R$. We can use this expression to find the ratio of the cross-sectional area of the aluminum wire to that of the copper wire.

SOLUTION Forming the ratio of the areas and using resistivity values from Table 20.1, we have

$$\frac{A_{aluminum}}{A_{copper}} = \frac{\rho_{aluminum} L / R}{\rho_{copper} L / R} = \frac{\rho_{aluminum}}{\rho_{copper}} = \frac{2.82 \times 10^{-8} \ \Omega \cdot m}{1.72 \times 10^{-8} \ \Omega \cdot m} = \boxed{1.64}$$

---

12. **REASONING AND SOLUTION** Solving Equation 20.5 for α yields

$$\alpha = \frac{(R/R_0) - 1}{T - T_0} = \frac{[(43.7 \ \Omega)/(38.0 \ \Omega)] - 1}{55 \ °C - 25 \ °C} = \boxed{0.0050 \ (C°)^{-1}}$$

---

13. **REASONING AND SOLUTION** The resistance of the cable is

$$R = \frac{V}{I} = \frac{\rho L}{A}$$

Since $A = \pi r^2$, the radius of the cable is

$$r = \sqrt{\frac{\rho L I}{\pi V}} = \sqrt{\frac{(1.72 \times 10^{-8} \ \Omega \cdot m)(0.24 \ m)(1200 \ A)}{\pi (1.6 \times 10^{-2} \ V)}} = \boxed{9.9 \times 10^{-3} \ m}$$

---

14. **REASONING** The materials in Table 20.1 are listed according to their resistivities, so we need to find the resistivity of this material. The resistivity depends on the resistance R, the length L, and the cross-sectional A of the wire (see Equation 20.3). We know the length of the wire, and the cross-sectional area can be found from the radius r as $A = \pi r^2$. The resistance depends on the voltage V and current I through the relation $R = V / I$, as expressed by Equation 20.2.

SOLUTION
The resistivity is

$$\rho = \frac{RA}{L} = \frac{R(\pi r^2)}{L} \qquad (20.3)$$

We also know that $R = V/I$, so the resistivity becomes

$$\rho = \frac{\left(\frac{V}{I}\right)(\pi r^2)}{L} = \frac{V\pi r^2}{IL} = \frac{(0.0320 \text{ V})\pi(1.03 \times 10^{-3} \text{ m})^2}{(1.35 \text{ A})(2.80 \text{ m})} = 2.82 \times 10^{-8} \; \Omega \cdot \text{m}$$

An inspection of Table 20.1 shows that the material that has this resistivity is aluminum.

---

15. **REASONING AND SOLUTION** First, we find α from the known values of T, T_0, R, and R_0. We have from Equation 20.5

$$\alpha = \frac{\dfrac{R}{R_0} - 1}{T - T_0} = \frac{\dfrac{47.6 \; \Omega}{35.0 \; \Omega} - 1}{100.0 \,°\text{C} - 20.0 \,°\text{C}} = 0.0045 \; (\text{C}°)^{-1}$$

We know that $R = R_0[1 + \alpha(T - T_0)]$, and using $R = 37.8 \; \Omega$ we find

$$T = T_0 + \frac{\dfrac{R}{R_0} - 1}{\alpha} = 20.0 \,°\text{C} + \frac{\dfrac{37.8 \; \Omega}{35.0 \; \Omega} - 1}{0.0045 \; (\text{C}°)^{-1}} = \boxed{37.8 \,°\text{C}}$$

---

16. **REASONING** The resistance R of the wire depends on its length L, so we can use Equation 20.3 to express the length in terms of the resistance:

$$L = \frac{RA}{\rho}$$

where A is the cross-sectional area of the wire and ρ is the resistivity of tungsten. The cross-sectional area can be expressed in terms of the radius r of the wire as $A = \pi r^2$. The resistance depends on the voltage V and current I through the relation $R = V/I$, Equation 20.2. Thus, the length of the wire can be expressed as

746   ELECTRIC CIRCUITS

$$L = \frac{RA}{\rho} = \frac{\left(\frac{V}{I}\right)\left(\pi r^2\right)}{\rho}$$

According to Equation 20.4, the resistivity ρ at the temperature T depends on the resistivity ρ_0 at the temperature T_0 through the relation

$$\rho = \rho_0\left[1 + \alpha(T - T_0)\right]$$

where α is the temperature coefficient of resistivity. Substituting this expression for ρ into the expression for the length of the wire, we have

$$L = \frac{\left(\frac{V}{I}\right)\left(\pi r^2\right)}{\rho_0\left[1 + \alpha(T - T_0)\right]}$$

SOLUTION The length of the wire is

$$L = \frac{\left(\frac{V}{I}\right)\left(\pi r^2\right)}{\rho_0\left[1 + \alpha(T - T_0)\right]}$$

$$= \frac{\left(\frac{120 \text{ V}}{1.5 \text{ A}}\right)\pi\left(0.075 \times 10^{-3} \text{ m}\right)^2}{\left(5.6 \times 10^{-8} \ \Omega \cdot \text{m}\right)\left\{1 + \left[4.5 \times 10^{-3} \ (\text{C}°)^{-1}\right](1320 \ °\text{C} - 20.0 \ °\text{C})\right\}} = \boxed{3.7 \text{ m}}$$

The resistivity at 20.0 °C, $\rho_0 = 5.6 \times 10^{-8} \ \Omega \cdot \text{m}$, was obtained from Table 20.1.

---

17. **SSM** **WWW** *REASONING* The resistance of a metal wire of length L, cross-sectional area A and resistivity ρ is given by Equation 20.3: $R = \rho L / A$. The volume V_2 of the new wire will be the same as the original volume V_1 of the wire, where volume is the product of length and cross-sectional area. Thus, $V_1 = V_2$ or $A_1 L_1 = A_2 L_2$. Since the new wire is three times longer than the first wire, we can write

$$A_1 L_1 = A_2 L_2 = A_2(3L_1) \quad \text{or} \quad A_2 = A_1/3$$

We can form the ratio of the resistances, use this expression for the area A_2, and find the new resistance.

SOLUTION The resistance of the new wire is determined as follows:

$$\frac{R_2}{R_1} = \frac{\rho L_2 / A_2}{\rho L_1 / A_1} = \frac{L_2 A_1}{L_1 A_2} = \frac{(3L_1) A_1}{L_1 (A_1/3)} = 9$$

Solving for R_2, we find that

$$R_2 = 9R_1 = 9(21.0\ \Omega) = \boxed{189\ \Omega}$$

---

18. **REASONING AND SOLUTION** Suppose that when the initial temperature of the wire is T_0 the resistance is R_0, and when the temperature rises to T the resistance is R. The relation between temperature and resistance is given by Equation 20.5 as $R = R_0[1 + \alpha(T - T_0)]$, where α is the temperature coefficient of resistivity. The initial and final resistances are related to the voltage and current as $R_0 = V/I_0$ and $R = V/I$, where the voltage V across the wire is the same in both cases. Substituting these values for R_0 and R into Equation 20.5 and solving for T, we arrive at

$$T = T_0 + \frac{\left(\dfrac{I_0}{I} - 1\right)}{\alpha} = 20\ ^\circ\mathrm{C} + \frac{\left(\dfrac{1.50\ \mathrm{A}}{1.30\ \mathrm{A}} - 1\right)}{4.5 \times 10^{-4}\ (\mathrm{C}^\circ)^{-1}} = \boxed{360\ ^\circ\mathrm{C}}$$

---

19. **SSM WWW** **REASONING** We will ignore any changes in length due to thermal expansion. Although the resistance of each section changes with temperature, the total resistance of the composite does not change with temperature. Therefore,

$$\underbrace{\left(R_{\text{tungsten}}\right)_0 + \left(R_{\text{carbon}}\right)_0}_{\text{At room temperature}} = \underbrace{R_{\text{tungsten}} + R_{\text{carbon}}}_{\text{At temperature } T}$$

From Equation 20.5, we know that the temperature dependence of the resistance for a wire of resistance R_0 at temperature T_0 is given by $R = R_0[1 + \alpha(T - T_0)]$, where α is the temperature coefficient of resistivity. Thus,

$$\left(R_{\text{tungsten}}\right)_0 + \left(R_{\text{carbon}}\right)_0 = \left(R_{\text{tungsten}}\right)_0 (1 + \alpha_{\text{tungsten}} \Delta T) + \left(R_{\text{carbon}}\right)_0 (1 + \alpha_{\text{carbon}} \Delta T)$$

Since ΔT is the same for each wire, this simplifies to

$$\left(R_{\text{tungsten}}\right)_0 \alpha_{\text{tungsten}} = -\left(R_{\text{carbon}}\right)_0 \alpha_{\text{carbon}} \qquad (1)$$

748   ELECTRIC CIRCUITS

This expression can be used to find the ratio of the resistances. Once this ratio is known, we can find the ratio of the lengths of the sections with the aid of Equation 20.3 ($L = RA/\rho$).

SOLUTION From Equation (1), the ratio of the resistances of the two sections of the wire is

$$\frac{(R_{tungsten})_0}{(R_{carbon})_0} = -\frac{\alpha_{carbon}}{\alpha_{tungsten}} = -\frac{-0.0005\ [(C°)^{-1}]}{0.0045\ [(C°)^{-1}]} = \frac{1}{9}$$

Thus, using Equation 20.3, we find the ratio of the tungsten and carbon lengths to be

$$\frac{L_{tungsten}}{L_{carbon}} = \frac{(R_0 A/\rho)_{tungsten}}{(R_0 A/\rho)_{carbon}} = \frac{(R_{tungsten})_0}{(R_{carbon})_0}\left(\frac{\rho_{carbon}}{\rho_{tungsten}}\right) = \left(\frac{1}{9}\right)\left(\frac{3.5\times10^{-5}\ \Omega\cdot m}{5.6\times10^{-8}\ \Omega\cdot m}\right) = \boxed{70}$$

where we have used resistivity values from Table 20.1 and the fact that the two sections have the same cross-sectional areas.

---

20. **REASONING AND SOLUTION** The mass m of the aluminum wire is equal to the density d of aluminum times the volume of the wire. The wire is cylindrical, so its volume is equal to the cross-sectional area A times the length L; $m = dAL$.

The cross-sectional area of the wire is related to its resistance R and length L by Equation 20.3; $R = \rho L/A$, where ρ is the resistivity of aluminum. Therefore, the mass of the aluminum wire can be written as

$$m = dAL = d\left(\frac{\rho L}{R}\right)L$$

The resistance R is given by Ohm's law as $R = V/I$, so the mass of the wire becomes

$$m = d\left(\frac{\rho L}{R}\right)L = \frac{d\rho L^2 I}{V}$$

$$m = \frac{(2700\ kg/m^3)(2.82\times10^{-8}\ \Omega\cdot m)(175\ m)^2(125\ A)}{0.300\ V} = \boxed{9.7\times10^2\ kg}$$

---

21. **REASONING AND SOLUTION** According to Equation 20.6, the power delivered to the battery to charge it is

$$P = IV = (19.0\ A)(12.0\ V) = \boxed{228\ W}$$

22. **REASONING AND SOLUTION**

a. According to Equation 20.6c, the resistance is

$$R = V^2/P = (12 \text{ V})^2/(33 \text{ W}) = \boxed{4.4 \text{ }\Omega}$$

b. According to Equation 20.6a, the current is

$$I = P/V = (33 \text{ W})/(12 \text{ V}) = \boxed{2.8 \text{ A}}$$

---

23. **SSM REASONING** According to Equation 6.10b, the energy used is Energy = Pt, where P is the power and t is the time. According to Equation 20.6a, the power is $P = IV$, where I is the current and V is the voltage. Thus, Energy = IVt, and we apply this result first to the drier and then to the computer.

SOLUTION The energy used by the drier is

$$\text{Energy} = Pt = IVt = (16 \text{ A})(240 \text{ V})(45 \text{ min})\underbrace{\left(\frac{60 \text{ s}}{1.00 \text{ min}}\right)}_{\text{Converts minutes to seconds}} = 1.04 \times 10^7 \text{ J}$$

For the computer, we have

$$\text{Energy} = 1.04 \times 10^7 \text{ J} = IVt = (2.7 \text{ A})(120 \text{ V})t$$

Solving for t we find

$$t = \frac{1.04 \times 10^7 \text{ J}}{(2.7 \text{ A})(120 \text{ V})} = 3.21 \times 10^4 \text{ s} = (3.21 \times 10^4 \text{ s})\left(\frac{1.00 \text{ h}}{3600 \text{ s}}\right) = \boxed{8.9 \text{ h}}$$

---

24. **REASONING** The total cost of keeping all the TVs turned on is equal to the number of TVs times the cost to keep each one on. The cost for one TV is equal to the energy it consumes times the cost per unit of energy ($0.10 per kW·h). The energy that a single set uses is, according to Equation 6.10b, the power it consumes times the time of use.

SOLUTION The total cost is

$$\text{Total cost} = (110 \text{ million sets})(\text{Cost per set})$$

$$= (110 \text{ million sets})\left[\text{Energy (in kW·h) used per set}\right]\left(\frac{\$0.10}{1 \text{ kW·h}}\right)$$

750 ELECTRIC CIRCUITS

The energy (in kW·h) used per set is the product of the power and the time, where the power is expressed in kilowatts and the time is in hours:

$$\text{Energy used per set} = Pt = (75 \text{ W})\left(\frac{1 \text{ kW}}{1000 \text{ W}}\right)(6.0 \text{ h}) \tag{6.10b}$$

The total cost of operating the TV sets is

$$\text{Total cost} = (110 \text{ million sets})\left[(75 \text{ W})\left(\frac{1 \text{ kW}}{1000 \text{ W}}\right)(6.0 \text{ h})\right]\left(\frac{\$0.10}{1 \text{ kW}\cdot\text{h}}\right) = \boxed{\$5.0 \times 10^6}$$

---

25. **REASONING AND SOLUTION** According to Equation 20.6a, we know $P = IV$, so that

$$I = P/V = (140 \text{ W})/(120 \text{ V}) = \boxed{1.2 \text{ A}}$$

---

26. **REASONING AND SOLUTION** The power delivered is $P = VI$ so

a. $\quad P_{bd} = VI_{bd} = (120 \text{ V})(11 \text{ A}) = \boxed{1300 \text{ W}}$

b. $\quad P_{vc} = VI_{vc} = (120 \text{ V})(4.0 \text{ A}) = \boxed{480 \text{ W}}$

c. The energy is $E = Pt$ so,

$$E_{bd}/E_{vc} = (P_{bd}t_{bd})/(P_{vc}t_{vc}) = (1300 \text{ W})(15 \text{ min})/[(480 \text{ W})(30.0 \text{ min})] = \boxed{1.4}$$

---

27. **SSM REASONING AND SOLUTION** As a function of temperature, the resistance of the wire is given by Equation 20.5: $R = R_0\left[1+\alpha(T-T_0)\right]$, where α is the temperature coefficient of resistivity. From Equation 20.6c, we have $P = V^2/R$. Combining these two equations, we have

$$P = \frac{V^2}{R_0\left[1+\alpha(T-T_0)\right]} = \frac{P_0}{1+\alpha(T-T_0)}$$

where $P_0 = V^2/R_0$, since the voltage is constant. But $P = \frac{1}{2}P_0$, so we find

$$\frac{P_0}{2} = \frac{P_0}{1+\alpha(T-T_0)} \qquad \text{or} \qquad 2 = 1+\alpha(T-T_0)$$

Solving for T, we find

$$T = \frac{1}{\alpha} + T_0 = \frac{1}{0.0045 \, (\text{C}°)^{-1}} + 28° = \boxed{250 \, °\text{C}}$$

---

28. **REASONING AND SOLUTION** We know that the resistance of the wire can be obtained from

$$P = V^2/R \quad \text{or} \quad R = V^2/P$$

We also know that $R = \rho L/A$. Solving for the length, noting that $A = \pi r^2$, and using $\rho = 100 \times 10^{-8} \, \Omega \cdot \text{m}$ from Table 20.1, we find

$$L = \frac{RA}{\rho} = \frac{(V^2/P)(\pi r^2)}{\rho} = \frac{V^2 \pi r^2}{\rho P} = \frac{(120 \, \text{V})^2 \pi (6.5 \times 10^{-4} \, \text{m})^2}{(100 \times 10^{-8} \, \Omega \cdot \text{m})(4.00 \times 10^2 \, \text{W})} = \boxed{50 \, \text{m}}$$

---

29. **REASONING AND SOLUTION** The electrical energy is converted to heat, so we have

$$Q = mc\Delta T = mc(T - T_0)$$

This energy is also given by

$$Q = Pt = I^2 R_{av} t = I^2 [1/2 \, (R + R_0)] t$$

where we are considering the average resistance R_{av} over the one minute interval. Also

$$R = R_0[1 + \alpha(T - T_0)]$$

We have

$$Q = I^2[1/2 \, (R + R_0)]t = I^2\{1/2 \, R_0[2 + \alpha(T - T_0)]\}t = mc(T - T_0)$$

Solving for T in the above expression yields

$$T = T_0 + \frac{\frac{R_0 I^2 t}{mc}}{1 - \frac{\alpha R_0 I^2 t}{2mc}}$$

Using $R_0 = 12 \, \Omega$, $I = 0.10 \, \text{A}$, $t = 60.0 \, \text{s}$, $m = 1.3 \times 10^{-3} \, \text{kg}$, $c = 452 \, \text{J/(kg} \cdot \text{C}°)$ from Table 12.2, $\alpha = 0.0050 \, (\text{C}°)^{-1}$, $T_0 = 20.0 \, °\text{C}$, we obtain $\boxed{T = 33 \, °\text{C}}$.

752   ELECTRIC CIRCUITS

30. **REASONING** The average power is given by Equation 20.15c as $\overline{P} = V_{rms}^2 / R$. In this expression the rms voltage V_{rms} appears. However, we seek the peak voltage V_0. The relation between the two types of voltage is given by Equation 20.13 as $V_{rms} = V_0/\sqrt{2}$, so we can obtain the peak voltage by using Equation 20.13 to substitute into Equation 20.15c.

SOLUTION Substituting V_{rms} from Equation 20.13 into Equation 20.15c gives

$$\overline{P} = \frac{V_{rms}^2}{R} = \frac{\left(V_0/\sqrt{2}\right)^2}{R} = \frac{V_0^2}{2R}$$

Solving for the peak voltage V_0 gives

$$V_0 = \sqrt{2R\overline{P}} = \sqrt{2(4.0\ \Omega)(55\ W)} = \boxed{21\ V}$$

---

31. **REASONING AND SOLUTION** The expression relating the peak current, I_0, to the rms-current, I_{rms}, is

$$I_{rms} = I_0/\sqrt{2} = (2.50\ A)/\sqrt{2} = \boxed{1.77\ A}$$

---

32. **REASONING**
a. We can obtain the frequency of the alternating current by comparing this specific expression for the current with the more general one in Equation 20.8.

b. The resistance of the light bulb is, according to Equation 20.14, equal to the rms-voltage divided by the rms-current. The rms-voltage is given, and we can obtain the rms-current by dividing the peak current by $\sqrt{2}$, as expressed by Equation 20.12.

c. The average power is given by Equation 20.15a as the product of the rms-current and the rms-voltage.

SOLUTION
a. By comparing $I = (0.707\ A)\sin\left[(314\ Hz)\,t\right]$ with the general expression (see Equation 20.8) for the current in an ac circuit, $I = I_0 \sin 2\pi f t$, we see that

$$2\pi f t = (314\ Hz)\,t \quad \text{or} \quad f = \frac{314\ Hz}{2\pi} = \boxed{50.0\ Hz}$$

b. The resistance is equal to $V_{\text{rms}}/I_{\text{rms}}$, where the rms-current is related to the peak current I_0 by $I_{\text{rms}} = I_0/\sqrt{2}$. Thus, the resistance of the light bulb is

$$R = \frac{V_{\text{rms}}}{I_{\text{rms}}} = \frac{V_{\text{rms}}}{\frac{I_0}{\sqrt{2}}} = \frac{\sqrt{2}(120.0 \text{ V})}{0.707 \text{ A}} = \boxed{2.40 \times 10^2 \ \Omega} \qquad (20.14)$$

c. The average power is the product of the rms-current and rms-voltage:

$$\overline{P} = I_{\text{rms}} V_{\text{rms}} = \left(\frac{I_0}{\sqrt{2}}\right) V_{\text{rms}} = \left(\frac{0.707 \text{ A}}{\sqrt{2}}\right)(120.0 \text{ V}) = \boxed{60.0 \text{ W}} \qquad (20.15a)$$

---

33. **SSM** *REASONING AND SOLUTION*
    a. According to Equation 20.15c, the average power consumed by the iron is

$$\overline{P} = \frac{V_{\text{rms}}^2}{R} = \frac{(120 \text{ V})^2}{16 \ \Omega} = \boxed{9.0 \times 10^2 \text{ W}}$$

b. The peak power is $P_{\text{peak}} = 2\overline{P} = \boxed{1.8 \times 10^3 \text{ W}}$.

---

34. *REASONING AND SOLUTION* The power P dissipated in the extension cord is $P = I^2 R$ (Equation 20.6b). The resistance R is related to the length L of the wire and its cross-sectional area A by Equation 20.3, $R = \rho L/A$, where ρ is the resistivity of copper. The cross-sectional area of the wire can be expressed as

$$A = \frac{\rho L}{R} = \frac{\rho I^2}{\left(\frac{P}{L}\right)}$$

where the ratio P/L is the power per unit length of copper wire that the heater produces. The wire is cylindrical, so its cross-sectional area is $A = \pi r^2$. Thus, the smallest radius of wire that can be used is

$$r = I\sqrt{\frac{\rho}{\pi\left(\frac{P}{L}\right)}} = (18 \text{ A})\sqrt{\frac{1.72 \times 10^{-8} \ \Omega \cdot \text{m}}{\pi(1.0 \text{ W}/\text{m})}} = \boxed{1.3 \times 10^{-3} \text{ m}}$$

Note that we have used 1.0 W/m as the power per unit length, rather than 2.0 W/m. This is because an extension cord is composed of two copper wires. If the maximum power per

754 ELECTRIC CIRCUITS

unit length that the extension cord itself can produce is 2.0 W/m, then each wire can produce only a maximum of 1.0 W/m.

---

35. ***REASONING AND SOLUTION*** The total amount of energy needed is

$$E = Pt = I^2Rt = (25 \text{ A})^2(5.3 \text{ }\Omega)(31 \text{ d})[(9 \text{ h})/(1 \text{ d})][(3600 \text{ s})/(1 \text{ h})] = 3.3 \times 10^9 \text{ J}$$

To find the cost, then,

$$\text{Cost} = [\$0.10/(1 \text{ kWh})](3.3 \times 10^9 \text{ J})(1 \text{ kWh})/(3.6 \times 10^6 \text{ J}) = \boxed{\$\,92}$$

---

36. ***REASONING AND SOLUTION***
a. According to Equation 20.15c,

$$P = V_{rms}^2/R = (240 \text{ V})^2/(29 \text{ }\Omega) = \boxed{2.0 \times 10^3 \text{ W}}$$

b. Assuming (3/4) of the energy is converted to heat,

$$Q = (3/4)Pt = mc\Delta T$$

Solving for t yields

$$t = (4/3)mc(T_f - T_0)/P = (4/3)(1.9 \text{ kg})[4186 \text{ J/(kg·C°)}](85 \text{ C°})/(2.0 \times 10^3 \text{ W}) = \boxed{450 \text{ s}}$$

---

37. $\boxed{\text{SSM}}$ ***REASONING*** According to Equation 6.10b, the energy supplied to the water in the form of heat is $Q = Pt$, where P is the power and t is the time. According to Equation 20.6c, the power is $P = V^2/R$, so that $Q = (V^2/R)t$, where V is the voltage and R is the resistance. But for a mass m of water with a specific heat capacity c, the heat Q is related to the change in temperature ΔT by $Q = mc\Delta T$ according to Equation 12.4. Our solution uses these two expressions for Q.

SOLUTION Combining the two expressions for Q, we have $Q = mc\Delta T = V^2t/R$. Solving for t, we have

$$t = \frac{Rcm\Delta T}{V^2} \tag{1}$$

The mass of the water can be determined from its volume V_{water}. From Equation 11.1, we have $m = \rho_{water} V_{water}$, where $\rho_{water} = 1.000 \times 10^3$ kg/m^3 is the mass density of water (see Table 11.1). Therefore, Equation (1) becomes

$$t = \frac{Rc\rho_{water}V_{water}\Delta T}{V^2} \qquad (2)$$

The value (see Table 12.2) for the specific heat of water is 4186 J/(kg·C°). The volume of water is

$$V_{water} = (52 \text{ gal})\left(\frac{3.79 \times 10^{-3} \text{ m}^3}{1.00 \text{ gal}}\right) = 0.197 \text{ m}^3$$

Substituting values into Equation (2), we obtain

$$t = \frac{(3.0 \text{ }\Omega)[4186 \text{ J/(kg·C°)}](1.000 \times 10^3 \text{ kg/m}^3)(0.197 \text{ m}^3)(53 \text{ °C} - 11 \text{ °C})}{(120 \text{ V})^2} = 7220 \text{ s}$$

$$= (7220 \text{ s})\left(\frac{1.00 \text{ h}}{3600 \text{ s}}\right) = \boxed{2.0 \text{ h}}$$

---

38. **REASONING AND SOLUTION** The energy Q_1 that is released when the water cools from an initial temperature T to a final temperature of 0.0 °C is given by Equation 12.4 as $Q_1 = cm(T - 0.0 \text{ °C})$. The energy Q_2 released when the water turns into ice at 0.0 °C is $Q_2 = mL_f$, where L_f is the latent heat of fusion for water. Since power P is energy divided by time, the power produced is

$$P = \frac{Q_1 + Q_2}{t} = \frac{cm(T - 0.0 \text{ °C}) + mL_f}{t}$$

The power produced by an electric heater is, according to Equation 20.6a, $P = IV$. Substituting this expression for P into the equation above and solving for the current I, we get

$$I = \frac{cm(T - 0.0 \text{ °C}) + mL_f}{tV}$$

$$I = \frac{(4186 \text{ J/kg·C°})(660 \text{ kg})(10.0 \text{ C°}) + (660 \text{ kg})(33.5 \times 10^4 \text{ J/kg})}{(9.0 \text{ h})\left(\frac{3600 \text{ s}}{\text{h}}\right)(240 \text{ V})} = \boxed{32 \text{ A}}$$

---

39. **SSM** **REASONING** The equivalent series resistance R_s is the sum of the resistances of the three resistors. The potential difference V can be determined from Ohm's law as $V = IR_s$.

SOLUTION
a. The equivalent resistance is

$$R_s = 25\ \Omega + 45\ \Omega + 75\ \Omega = \boxed{145\ \Omega}$$

b. The potential difference across the three resistors is

$$V = IR_s = (0.51\ \text{A})(145\ \Omega) = \boxed{74\ \text{V}}$$

---

40. **REASONING** Since the two resistors are connected in series, they are equivalent to a single equivalent resistance that is the sum of the two resistances, according to Equation 20.16. Ohm's law (Equation 20.2) can be applied with this equivalent resistance to give the battery voltage.

    SOLUTION According to Ohm's law, we find

    $$V = IR_s = I(R_1 + R_2) = (0.12\ \text{A})(47\ \Omega + 28\ \Omega) = \boxed{9.0\ \text{V}}$$

---

41. **REASONING AND SOLUTION** The equivalent resistance of the circuit is

    $$R_s = R_1 + R_2 = 36.0\ \Omega + 18.0\ \Omega = 54.0\ \Omega$$

    Ohm's law for the circuit gives $I = V/R_s = (15.0\ \text{V})/(54.0\ \Omega) = 0.278\ \text{A}$

    a. Ohm's law for R_1 gives $V_1 = (0.278\ \text{A})(36.0\ \Omega) = \boxed{10.0\ \text{V}}$

    b. Ohm's law for R_2 gives $V_2 = (0.278\ \text{A})(18.0\ \Omega) = \boxed{5.00\ \text{V}}$

---

42. **REASONING** According to Equation 20.2, the resistance R of the resistor is equal to the voltage V_R across it divided by the current I, or $R = V_R/I$. Since the resistor, the lamp, and the voltage source are in series, the voltage across the resistor is $V_R = 120.0\ \text{V} - V_L$, where V_L is the voltage across the lamp. Thus, the resistance is

    $$R = \frac{120.0\ \text{V} - V_L}{I}$$

    Since V_L is known, we need only determine the current in the circuit. Since we know the voltage V_L across the lamp and the power P dissipated by it, we can use Equation 20.6a to find the current: $I = P/V_L$. The resistance can be written as

$$R = \frac{120.0 \text{ V} - V_L}{P/V_L}$$

SOLUTION Substituting the known values for V_L and P into the equation above, the resistance is

$$R = \frac{120.0 \text{ V} - 25 \text{ V}}{(60.0 \text{ W})/(25 \text{ V})} = \boxed{4.0 \times 10^1 \text{ }\Omega}$$

---

43. **SSM** **REASONING** Using Ohm's law (Equation 20.2) we can write an expression for the voltage across the original circuit as $V = I_0 R_0$. When the additional resistor R is inserted in series, assuming that the battery remains the same, the voltage across the new combination is given by $V = I(R + R_0)$. Since V is the same in both cases, we can write $I_0 R_0 = I(R + R_0)$. This expression can be solved for R_0.

SOLUTION Solving for R_0, we have $I_0 R_0 - I R_0 = IR$ or $R_0 (I_0 - I) = IR$; therefore,

$$R_0 = \frac{IR}{I_0 - I} = \frac{(12.0 \text{ A})(8.00 \text{ }\Omega)}{15.0 \text{ A} - 12.0 \text{ A}} = \boxed{32 \text{ }\Omega}$$

---

44. **REASONING AND SOLUTION**
a. The equivalent resistance of the circuit is

$$R_s = 9.0 \text{ }\Omega + 5.0 \text{ }\Omega + 1.0 \text{ }\Omega = 15.0 \text{ }\Omega$$

The current through each of the resistors is from Ohm's law

$$I = (24 \text{ V})/(15.0 \text{ }\Omega) = \boxed{1.6 \text{ A}}$$

b. The voltage drop across the 9.0-Ω resistor is $V_1 = (1.6 \text{ A})(9.0 \text{ }\Omega) = \boxed{14 \text{ V}}$

The drop across the 5.0 Ω resistor is $V_2 = (1.6 \text{ A})(5.0 \text{ }\Omega) = \boxed{8.0 \text{ V}}$

The drop across the 1.0 Ω resistor is $V_3 = (1.6 \text{ A})(1.0 \text{ }\Omega) = \boxed{1.6 \text{ V}}$

c. The power dissipated in the 9.0 Ω resistor is

$$P_1 = I^2 R_1 = (1.6 \text{ A})^2 (9.0 \text{ }\Omega) = \boxed{23 \text{ W}}$$

758 ELECTRIC CIRCUITS

Similarly, for the 5.0 Ω resistor, $P_2 = (1.6 \text{ A})^2 (5.0 \text{ Ω}) = \boxed{13 \text{ W}}$

and for the 1.0 Ω resistor $P_3 = (1.6 \text{ A})^2 (1.0 \text{ Ω}) = \boxed{2.6 \text{ W}}$

---

45. **REASONING AND SOLUTION** The 5.0 Ω resistor can tolerate a current of

$$I = \sqrt{\frac{P_1}{R_1}} = \sqrt{\frac{20.0 \text{ W}}{5.0 \text{ Ω}}} = 2.0 \text{ A}$$

Similarly, it is found that the 30.0 Ω and 15.0 Ω resistors can tolerate currents of 0.577 A and 0.816 A, respectively. The maximum current in the circuit can, therefore, only be 0.577 A.

a. Ohm's law for the circuit gives

$$V = IR_S = (0.577 \text{ A})(50.0 \text{ Ω}) = \boxed{28.9 \text{ V}}$$

b. The battery must deliver

$$P = IV = (0.577 \text{ A})(28.9 \text{ V}) = \boxed{16.7 \text{ W}}$$

---

46. **REASONING** The answer is not 340 W + 240 W = 580 W. The reason is that each heater contributes resistance to the circuit when they are connected in series across the battery. For a series connection, the resistances add together to give the equivalent total resistance, according to Equation 20.16. Thus, the total resistance is greater than the resistance of either heater. The greater resistance means that the current from the battery is less than when either heater is present by itself. Since the power for each heater is $P = I^2 R$, according to Equation 20.6b, the smaller current means that the power delivered to an individual heater is less when both are connected than when that heater is connected alone. We approach this problem by remembering that the total power delivered to the series combination of the heaters is the power delivered to the equivalent series resistance.

SOLUTION Let the resistances of the two heaters be R_1 and R_2. Correspondingly, the powers delivered to the heaters when each is connected alone to the battery are P_1 and P_2. For the series connection, the equivalent total resistance is $R_1 + R_2$, according to Equation 20.16. Using Equation 20.6c, we can write the total power delivered to this equivalent resistance as

$$P = \frac{V^2}{R_1 + R_2} \tag{1}$$

But according to Equation 20.6c, as applied to the situations when each heater is connected by itself to the battery, we have

$$P_1 = \frac{V^2}{R_1} \quad \text{or} \quad R_1 = \frac{V^2}{P_1} \tag{2}$$

$$P_2 = \frac{V^2}{R_2} \quad \text{or} \quad R_2 = \frac{V^2}{P_2} \tag{3}$$

Substituting Equations (2) and (3) into Equation (1) gives

$$P = \frac{V^2}{\frac{V^2}{P_1} + \frac{V^2}{P_2}} = \frac{1}{\frac{1}{P_1} + \frac{1}{P_2}} = \frac{P_1 P_2}{P_1 + P_2} = \frac{(340 \text{ W})(240 \text{ W})}{340 \text{ W} + 240 \text{ W}} = \boxed{140 \text{ W}}$$

---

47. **REASONING** Ohm's law provides the basis for our solution. We will use it to express the current from the battery when both resistors are connected and when only one resistor at a time is connected. When both resistors are connected, we will use Ohm's law with the series equivalent resistance, which is $R_1 + R_2$, according to Equation 20.16. The problem statement gives values for amounts by which the current increases when one or the other resistor is removed. Thus, we will focus attention on the difference between the currents given by Ohm's law.

SOLUTION When R_2 is removed, leaving only R_1 connected, the current increases by 0.20 A. In this case, using Ohm's law to express the currents, we have

$$\underbrace{\frac{V}{R_1}}_{\substack{\text{Current given by} \\ \text{Ohm's law when} \\ \text{only } R_1 \text{ is present}}} - \underbrace{\frac{V}{R_1 + R_2}}_{\substack{\text{Current given by} \\ \text{Ohm's law when} \\ R_1 \text{ and } R_2 \text{ are present}}} = \frac{VR_2}{R_1(R_1 + R_2)} = 0.20 \text{ A} \tag{1}$$

When R_1 is removed, leaving only R_2 connected, the current increases by 0.10 A. In this case, using Ohm's law to express the currents, we have

$$\underbrace{\frac{V}{R_2}}_{\substack{\text{Current given by} \\ \text{Ohm's law when} \\ \text{only } R_2 \text{ is present}}} - \underbrace{\frac{V}{R_1 + R_2}}_{\substack{\text{Current given by} \\ \text{Ohm's law when} \\ R_1 \text{ and } R_2 \text{ are present}}} = \frac{VR_1}{R_2(R_1 + R_2)} = 0.10 \text{ A} \tag{2}$$

Multiplying Equation (1) and Equation (2), we obtain

$$\left[\frac{VR_2}{R_1(R_1+R_2)}\right]\left[\frac{VR_1}{R_2(R_1+R_2)}\right]=(0.20\text{ A})(0.10\text{ A})$$

Simplifying this result algebraically shows that

$$\frac{V^2}{(R_1+R_2)^2}=(0.20\text{ A})(0.10\text{ A}) \quad \text{or} \quad \frac{V}{R_1+R_2}=\sqrt{(0.20\text{ A})(0.10\text{ A})}=0.14\text{ A} \quad (3)$$

a. Using the result for $V/(R_1+R_2)$ from Equation (3) to substitute into Equation (1) gives

$$\frac{V}{R_1}-0.14\text{ A}=0.20\text{ A} \quad \text{or} \quad R_1=\frac{V}{0.20\text{ A}+0.14\text{ A}}=\frac{12\text{ V}}{0.20\text{ A}+0.14\text{ A}}=\boxed{35\ \Omega}$$

b. Using the result for $V/(R_1+R_2)$ from Equation (3) to substitute into Equation (2) gives

$$\frac{V}{R_2}-0.14\text{ A}=0.10\text{ A} \quad \text{or} \quad R_2=\frac{V}{0.10\text{ A}+0.14\text{ A}}=\frac{12\text{ V}}{0.10\text{ A}+0.14\text{ A}}=\boxed{5.0\times10^1\ \Omega}$$

---

48. **REASONING AND SOLUTION** The rule for combining parallel resistors is

$$\frac{1}{R_P}=\frac{1}{R_1}+\frac{1}{R_2}$$

which gives

$$\frac{1}{R_2}=\frac{1}{R_P}-\frac{1}{R_1}=\frac{1}{115\ \Omega}-\frac{1}{155\ \Omega} \quad \text{or} \quad \boxed{R_2=446\ \Omega}$$

---

49. **SSM REASONING AND SOLUTION** Since the circuit elements are in parallel, the equivalent resistance can be obtained directly from Equation 20.17:

$$\frac{1}{R_P}=\frac{1}{R_1}+\frac{1}{R_2}=\frac{1}{16\ \Omega}+\frac{1}{8.0\ \Omega} \quad \text{or} \quad \boxed{R_P=5.3\ \Omega}$$

---

50. **REASONING AND SOLUTION** The power P dissipated in a resistance R is given by Equation 20.6c as $P=V^2/R$. The resistance R_{50} of the 50.0-W filament is

$$R_{50}=\frac{V^2}{P}=\frac{(120.0\text{ V})^2}{50.0\text{ W}}=\boxed{288\ \Omega}$$

The resistance R_{100} of the 100.0-W filament is

$$R_{100} = \frac{V^2}{P} = \frac{(120.0 \text{ V})^2}{100.0 \text{ W}} = \boxed{144 \text{ }\Omega}$$

---

51. **SSM** *REASONING* Since the resistors are connected in parallel, the voltage across each one is the same and can be calculated from Ohm's Law (Equation 20.2: $V = IR$). Once the voltage across each resistor is known, Ohm's law can again be used to find the current in the second resistor. The total power consumed by the parallel combination can be found calculating the power consumed by each resistor from Equation 20.6b: $P = I^2R$. Then, the total power consumed is the sum of the power consumed by each resistor.

SOLUTION Using data for the second resistor, the voltage across the resistors is equal to

$$V = IR = (3.00 \text{ A})(64.0 \text{ }\Omega) = 192 \text{ }\Omega$$

a. The current through the 42.0-Ω resistor is

$$I = \frac{V}{R} = \frac{192 \text{ V}}{42.0 \text{ }\Omega} = \boxed{4.57 \text{ A}}$$

b. The power consumed by the 42.0-Ω resistor is

$$P = I^2 R = (4.57 \text{ A})^2 (42.0 \text{ }\Omega) = 877 \text{ W}$$

while the power consumed by the 64.0-Ω resistor is

$$P = I^2 R = (3.00 \text{ A})^2 (64.0 \text{ }\Omega) = 576 \text{ W}$$

Therefore the total power consumed by the two resistors is 877 W + 576 W = $\boxed{1450 \text{ W}}$.

---

52. *REASONING AND SOLUTION* Each piece has a resistance of 1/3 R. Then

$$1/R_p = 1/(1/3 \text{ R}) + 1/(1/3 \text{ R}) + 1/(1/3 \text{ R}) = 9/R \quad \text{or} \quad R_p = \boxed{R/9}$$

---

53. *REASONING* The total power is given by Equation 20.15c as $\overline{P} = V_{rms}^2 / R_p$, where R_p is the equivalent parallel resistance of the heater and the lamp. Since the total power and the rms voltage are known, we can use this expression to obtain the equivalent parallel resistance. This equivalent resistance is related to the individual resistances of the heater

762 ELECTRIC CIRCUITS

and the lamp via Equation 20.17, which is $R_p^{-1} = R_{heater}^{-1} + R_{lamp}^{-1}$. Since R_{heater} is given, R_{lamp} can be found once R_p is known.

SOLUTION According to Equation 20.15c, the equivalent parallel resistance is

$$R_p = \frac{V_{rms}^2}{\overline{P}}$$

Using this result in Equation 20.17 gives

$$\frac{1}{R_p} = \frac{1}{V_{rms}^2/\overline{P}} = \frac{1}{R_{heater}} + \frac{1}{R_{lamp}}$$

Rearranging this expression shows that

$$\frac{1}{R_{lamp}} = \frac{\overline{P}}{V_{rms}^2} - \frac{1}{R_{heater}} = \frac{84 \text{ W}}{(120 \text{ V})^2} - \frac{1}{6.0 \times 10^2 \text{ }\Omega} = 4.2 \times 10^{-3} \text{ }\Omega^{-1}$$

Therefore,

$$R_{lamp} = \frac{1}{4.2 \times 10^{-3} \text{ }\Omega^{-1}} = \boxed{240 \text{ }\Omega}$$

---

54. **REASONING** The electric heater and the toaster are connected in parallel with the voltage source, so each receives the same voltage as the source. The rms-current through the toaster is given by Equation 20.14, $I_{rms} = V_{rms}/R$, where V_{rms} is the rms-voltage across the toaster and R is its resistance. The total power supplied to the heater and toaster is given by Equation 20.15c as $\overline{P} = V_{rms}^2/R_P$, where R_P is the equivalent resistance of the parallel circuit.

SOLUTION

a. The voltage across the heater is the same as that of the generator, $V_{rms} = \boxed{120.0 \text{ V}}$.

b. The current through the toaster is

$$I_{rms} = \frac{V_{rms}}{R_{toaster}} = \frac{120.0 \text{ V}}{17.0 \text{ }\Omega} = \boxed{7.06 \text{ A}} \quad (20.14)$$

c. The average power supplied to the heater and toaster is $\overline{P} = V_{rms}^2/R_P$. The equivalent resistance can be obtained from Equation 20.17:

$$\frac{1}{R_p} = \frac{1}{R_{heater}} + \frac{1}{R_{toaster}}$$

The average power becomes

$$\overline{P} = \frac{V_{rms}^2}{R_p} = V_{rms}^2 \left(\frac{1}{R_{heater}} + \frac{1}{R_{toaster}} \right) = (120.0 \text{ V})^2 \left(\frac{1}{9.60 \text{ }\Omega} + \frac{1}{17.0 \text{ }\Omega} \right) = \boxed{2350 \text{ W}}$$

---

55. **SSM** *REASONING* The equivalent resistance of the three devices in parallel is R_p, and we can find the value of R_p by using our knowledge of the total power consumption of the circuit; the value of R_p can be found from Equation 20.6c, $P = V^2 / R_p$. Ohm's law (Equation 20.2, $V = IR$) can then be used to find the current through the circuit.

SOLUTION
a. The total power used by the circuit is $P = 1650 \text{ W} + 1090 \text{ W} + 1250 \text{ W} = 3990 \text{ W}$. The equivalent resistance of the circuit is

$$R_p = \frac{V^2}{P} = \frac{(120 \text{ V})^2}{3990 \text{ W}} = \boxed{3.6 \text{ }\Omega}$$

b. The total current through the circuit is

$$I = \frac{V}{R_p} = \frac{120 \text{ V}}{3.6 \text{ }\Omega} = \boxed{33 \text{ A}}$$

This current is larger than the rating of the circuit breaker; therefore, the $\boxed{\text{breaker will open}}$.

---

56. *REASONING AND SOLUTION* The resistance of the parallel combination is given by

$$1/R_p = 1/R + 1/(2.00 \text{ }\Omega)$$

and the series combination has a resistance of $R_s = R + 2.00 \text{ }\Omega$. Ohm's law gives for each case

$$I_p = V/R_p = V/R + V/(2.00 \text{ }\Omega)$$

and

$$I_s = V/R_s = V/(R + 2.00 \text{ }\Omega)$$

We know that $I_p = 5I_s$. Using this with the above equations and suppressing units yields the quadratic equation

$$R^2 - 6.0R + 4.00 = 0$$

The two solutions of this equation are $\boxed{5.24\ \Omega}$ and $\boxed{0.76\ \Omega}$.

---

57. **REASONING** Since the defogger wires are connected in parallel, the total resistance of all thirteen wires can be obtained from Equation 20.17:

$$\frac{1}{R_p} = \frac{13}{R} \quad \text{or} \quad R_p = \frac{R}{13}$$

where R is the individual resistance of one of the wires. The heat required to melt the ice is given by $Q = mL_f$, where m is the mass of the ice and L_f is the latent heat of fusion of the ice (see Section 12.8). Therefore, using Equation 20.6c, we can see that the power or energy dissipated per unit time in the wires and used to melt the ice is

$$P = \frac{V^2}{R_p} = \underbrace{\frac{mL_f}{t}}_{\text{Energy/time}} \quad \text{or} \quad \frac{V^2}{R/13} = \frac{mL_f}{t}$$

According to Equation 20.3, $R = \rho L / A$, where the length of the wire is L, its cross-sectional area is A and its resistivity is ρ; therefore, the last expression can be written

$$\frac{13V^2}{R} = \frac{13V^2}{\rho L / A} = \frac{mL_f}{t}$$

This expression can be solved for the area A.

SOLUTION Solving the above expression for A, and substituting given data, and obtaining the latent heat of fusion of water from Table 12.3, we find that

$$A = \frac{\rho L m L_f}{13 V^2 t}$$

$$= \frac{(88.0 \times 10^{-8}\ \Omega \cdot \text{m})(1.30\ \text{m})(2.10 \times 10^{-2}\ \text{kg})(33.5 \times 10^4\ \text{J/kg})}{13(12.0\ \text{V})^2 (120\ \text{s})} = \boxed{3.58 \times 10^{-8}\ \text{m}^2}$$

58. **REASONING** To find the current, we will use Ohm's law, together with the proper equivalent resistance. The coffee maker and frying pan are in series, so their equivalent resistance is given by Equation 20.16 as $R_{coffee} + R_{pan}$. This total resistance is in parallel with the resistance of the bread maker, so the equivalent resistance of the parallel combination can be obtained from Equation 20.17 as $R_p^{-1} = (R_{coffee} + R_{pan})^{-1} + R_{bread}^{-1}$.

SOLUTION Using Ohm's law and the expression developed above for R_p^{-1}, we find

$$I = \frac{V}{R_p} = V\left(\frac{1}{R_{coffee} + R_{pan}} + \frac{1}{R_{bread}}\right) = (120 \text{ V})\left(\frac{1}{14\,\Omega + 16\,\Omega} + \frac{1}{23\,\Omega}\right) = \boxed{9.2 \text{ A}}$$

---

59. **SSM** **REASONING** We will analyze the combination in parts. The 65 and 85-Ω resistors are in parallel. This parallel combination is in series with the 63-Ω resistor.

SOLUTION The equivalent resistance for the parallel combination of the 65 and 85-Ω resistors can be determined as follows, using Equation 20.17:

$$\frac{1}{R_p} = \frac{1}{65\,\Omega} + \frac{1}{85\,\Omega} = 0.027\,\Omega^{-1} \quad \text{or} \quad R_p = \frac{1}{0.027\,\Omega^{-1}} = 37\,\Omega$$

This equivalent resistance is in series with the 63-Ω resistance, the two giving the following total resistance between points A and B, according to Equation 20.16:

$$R_s = R_{AB} = 63\,\Omega + 37\,\Omega = \boxed{1.0 \times 10^2\,\Omega}$$

---

60. **REASONING AND SOLUTION** The 6.00 Ω, 5.00 Ω and 3.00 Ω resistors are in series with an equivalent resistance of $R_s = 14.0\,\Omega$.

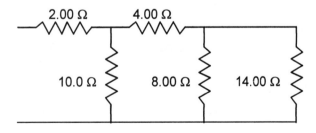

This equivalent resistor is in parallel with the 8.00-Ω resistor, so

$$1/R_p = 1/(14.0\,\Omega) + 1/(8.00\,\Omega) \quad \text{or} \quad R_p = 5.09\,\Omega$$

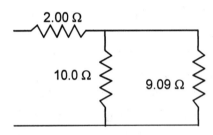

This new equivalent resistor is in series with the 4.00-Ω resistor, so $R_s' = 9.09\ \Omega$.

R_s' is in parallel with the 10.0-Ω resistor, so

$$1/R_p' = 1/(9.09\ \Omega) + 1/(10.0\ \Omega) \quad \text{or} \quad R_p' = 4.76\ \Omega$$

Finally, R_p' is in series with the 2.00-Ω, so the total equivalent resistance is $\boxed{6.76\ \Omega}$.

---

61. **SSM** *REASONING* When two or more resistors are in series, the equivalent resistance is given by Equation 20.16: $R_s = R_1 + R_2 + R_3 + \ldots$. Likewise, when resistors are in parallel, the expression to be solved to find the equivalent resistance is given by Equation 20.17: $\dfrac{1}{R_p} = \dfrac{1}{R_1} + \dfrac{1}{R_2} + \dfrac{1}{R_3} + \ldots$. We will successively apply these to the individual resistors in the figure in the text beginning with the resistors on the right side of the figure.

SOLUTION Since the 4.0-Ω and the 6.0-Ω resistors are in series, the equivalent resistance of the combination of those two resistors is 10.0 Ω. The 9.0-Ω and 8.0-Ω resistors are in parallel; their equivalent resistance is 4.24 Ω. The equivalent resistances of the parallel combination (9.0 Ω and 8.0 Ω) and the series combination (4.0 Ω and the 6.0 Ω) are in parallel; therefore, their equivalent resistance is 2.98 Ω. The 2.98-Ω combination is in series with the 3.0-Ω resistor, so that equivalent resistance is 5.98 Ω. Finally, the 5.98-Ω combination and the 20.0-Ω resistor are in parallel, so the equivalent resistance between the points A and B is $\boxed{4.6\ \Omega}$.

---

62. **REASONING** The circuit diagram is shown at the right. We can find the current in the 120.0-Ω resistor by using Ohm's law, provided that we can obtain a value for V_{AB}, the voltage between points A and B in the diagram. To find V_{AB}, we will also apply Ohm's law, this time by multiplying the current from the battery times R_{AB}, the equivalent parallel resistance between A and B. The current from the battery can be obtained by applying Ohm's law once again, now to the entire circuit and using the total equivalent resistance of the series combination of the 20.0-Ω resistor and R_{AB}. Once the current in the 120.0-Ω resistor is found, the power dissipated in it can be obtained from Equation 20.6b, $P = I^2R$.

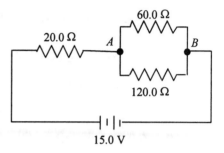

SOLUTION
a. According to Ohm's law, the current in the 120.0-Ω resistor is $I_{120} = V_{AB}/(120.0\ \Omega)$. To find V_{AB}, we note that the equivalent parallel resistance between points A and B can be obtained from Equation 20.17 as follows:

$$\frac{1}{R_{AB}} = \frac{1}{60.0\ \Omega} + \frac{1}{120.0\ \Omega} \quad \text{or} \quad R_{AB} = 40.0\ \Omega$$

This resistance of 40.0 Ω is in series with the 20.0-Ω resistance, so that, according to Equation 20.16, the total equivalent resistance connected across the battery is 40.0 Ω + 20.0 Ω = 60.0 Ω. Applying Ohm's law to the entire circuit, we can see that the current from the battery is

$$I = \frac{15.0\ \text{V}}{60.0\ \Omega} = 0.250\ \text{A}$$

Again applying Ohm's law, this time to the resistance R_{AB}, we find that

$$V_{AB} = (0.250\ \text{A})R_{AB} = (0.250\ \text{A})(40.0\ \Omega) = 10.0\ \text{V}$$

Finally, we can see that the current in the 120.0-Ω resistor is

$$I_{120} = \frac{V_{AB}}{120\ \Omega} = \frac{10.0\ \text{V}}{120\ \Omega} = \boxed{8.33 \times 10^{-2}\ \text{A}}$$

768   ELECTRIC CIRCUITS

b. The power dissipated in the 120.0-Ω resistor is given by Equation 20.6b as

$$P = I_{120}^2 R = (8.33 \times 10^{-2} \text{ A})^2 (120.0 \text{ Ω}) = \boxed{0.833 \text{ W}}$$

63. **REASONING AND SOLUTION** The resistors in the small network have an equivalent resistance

$$\frac{1}{R_p} = \frac{1}{2.0 \text{ Ω} + 1.0 \text{ Ω}} + \frac{1}{5.0 \text{ Ω} + 1.0 \text{ Ω}} \quad \text{or} \quad R_p = 2.0 \text{ Ω}$$

This resistance is in series with the 4.0-Ω resistor so the equivalent resistance of the circuit is R = 6.0 Ω. Therefore, Ohm's law gives the total current in the circuit to be

$$I = V/R = (12 \text{ V})/(6.0 \text{ Ω}) = 2.0 \text{ A}$$

This current, upon entering the parallel branch, will split in the ratios of 3:9 and 6:9, with the largest current entering the smallest resistance path. The 5.0-Ω resistor then has a current of

$$I = (3/9)(2.0 \text{ A})$$

The power dissipated in this resistor is

$$P = [(3/9)(2.0 \text{ A})]^2 (5.0 \text{ Ω}) = \boxed{2.2 \text{ W}}$$

64. **REASONING** The power P delivered to the circuit is, according to Equation 20.6c, $P = V^2 / R_{12345}$, where V is the voltage of the battery and R_{12345} is the equivalent resistance of the five-resistor circuit. The voltage and power are known, so that the equivalent resistance can be calculated. We will use our knowledge of resistors wired in series and parallel to evaluate R_{12345} in terms of the resistance R of each resistor. In this manner we will find the value for R.

SOLUTION First we note that all the resistors are equal, so $R_1 = R_2 = R_3 = R_4 = R_5 = R$. We can find the equivalent resistance R_{12345} as follows. The resistors R_3 and R_4 are in series, so the equivalent resistance R_{34} of these two is $R_{34} = R_3 + R_4 = 2R$. The resistors R_2, R_{34}, and R_5 are in parallel, and the reciprocal of the equivalent resistance R_{2345} is

$$\frac{1}{R_{2345}} = \frac{1}{R_2} + \frac{1}{R_{34}} + \frac{1}{R_5} = \frac{1}{R} + \frac{1}{2R} + \frac{1}{R} = \frac{5}{2R}$$

so $R_{2345} = 2R/5$. The resistor R_1 is in series with R_{2345}, and the equivalent resistance of this combination is the equivalent resistance of the circuit. Thus, we have

$$R_{12345} = R_1 + R_{2345} = R + \frac{2R}{5} = \frac{7R}{5}$$

The power delivered to the circuit is

$$P = \frac{V^2}{R_{12345}} = \frac{V^2}{\left(\frac{7R}{5}\right)}$$

Solving for the resistance R, we find that

$$R = \frac{5V^2}{7P} = \frac{5(45 \text{ V})^2}{7(58 \text{ W})} = \boxed{25 \text{ }\Omega}$$

---

65. **SSM** **WWW** **REASONING** Since we know that the current in the 8.00-Ω resistor is 0.500 A, we can use Ohm's law ($V = IR$) to find the voltage across the 8.00-Ω resistor. The 8.00-Ω resistor and the 16.0-Ω resistor are in parallel; therefore, the voltages across them are equal. Thus, we can also use Ohm's law to find the current through the 16.0-Ω resistor. The currents that flow through the 8.00-Ω and the 16.0-Ω resistors combine to give the total current that flows through the 20.0-Ω resistor. Similar reasoning can be used to find the current through the 9.00-Ω resistor.

SOLUTION
a. The voltage across the 8.00-Ω resistor is $V_8 = (0.500 \text{ A})(8.00 \text{ }\Omega) = 4.00 \text{ V}$. Since this is also the voltage that is across the 16.0-Ω resistor, we find that the current through the 16.0-Ω resistor is $I_{16} = (4.00 \text{ V})/(16.0 \text{ }\Omega) = 0.250 \text{ A}$. Therefore, the total current that flows through the 20.0-Ω resistor is

$$I_{20} = 0.500 \text{ A} + 0.250 \text{ A} = \boxed{0.750 \text{ A}}$$

b. The 8.00-Ω and the 16.0-Ω resistors are in parallel, so their equivalent resistance can be obtained from Equation 20.17, $\frac{1}{R_p} = \frac{1}{R_1} + \frac{1}{R_2} + \frac{1}{R_3} + ...$, and is equal to 5.33 Ω. Therefore, the equivalent resistance of the upper branch of the circuit is $R_{\text{upper}} = 5.33 \text{ }\Omega + 20.0 \text{ }\Omega = 25.3 \text{ }\Omega$, since the 5.33-$\Omega$ resistance is in series with the 20.0-Ω resistance. Using Ohm's law, we find that the voltage across the upper branch must be $V = (0.750 \text{ A})(25.3 \text{ }\Omega) = 19.0 \text{ V}$. Since the lower branch is in parallel with the upper branch, the voltage across both branches must be the same. Therefore, the current through the 9.00-Ω resistor is, from Ohm's law,

770 ELECTRIC CIRCUITS

$$I_9 = \frac{V_{lower}}{R_9} = \frac{19.0 \text{ V}}{9.00 \text{ }\Omega} = \boxed{2.11 \text{ A}}$$

---

66. **REASONING AND SOLUTION** The equivalent resistance of the initial configuration is given by

$$1/R_p = 3/R \quad \text{or} \quad R_p = R/3$$

The parallel part of the final configuration has a resistance of $R_p' = R/2$, so the total equivalent resistance is

$$R_s = R/2 + R = 3R/2$$

Now $R_s = R_p + 700 \text{ }\Omega$, so $3R/2 = R/3 + 700 \text{ }\Omega$, which gives $R = \boxed{600 \text{ }\Omega}$.

---

67. **REASONING AND SOLUTION** The terminal voltage of the battery is

$$V_T = \text{Emf} - Ir$$

where

$$I = \text{Emf}/(R + r)$$

so

$$r/(R + r) = 1 - V_T/\text{Emf} = 1 - (8.30 \text{ V})/(9.00 \text{ V}) = 0.078$$

Solving for the internal resistance gives $r = 0.078 (R + r)$, or

$$r = 0.078 \, R/(1 - 0.078) = 0.078(1.4 \text{ }\Omega)/(1 - 0.078) = \boxed{0.12 \text{ }\Omega}$$

---

68. **REASONING AND SOLUTION** Ohm's law gives

$$r = V/I = (1.5 \text{ V})/(28 \text{ A}) = \boxed{0.054 \text{ }\Omega}$$

---

69. **SSM REASONING** The terminal voltage of the battery is given by $V_{terminal} = \text{Emf} - Ir$, where r is the internal resistance of the battery. Since the terminal voltage is observed to be one-half of the emf of the battery, we have $V_{terminal} = \text{Emf}/2$ and $I = \text{Emf}/(2r)$. From Ohm's law, the equivalent resistance of the circuit is $R = \text{emf}/I = 2r$. We can also find the equivalent resistance of the circuit by considering that the identical bulbs are in parallel across the battery terminals, so that the equivalent resistance of the N bulbs is found from

$$\frac{1}{R_p} = \frac{N}{R_{bulb}} \quad \text{or} \quad R_p = \frac{R_{bulb}}{N}$$

This equivalent resistance is in series with the battery, so we find that the equivalent resistance of the circuit is

$$R = 2r = \frac{R_{bulb}}{N} + r$$

This expression can be solved for N.

SOLUTION Solving the above expression for N, we have

$$N = \frac{R_{bulb}}{2r - r} = \frac{R_{bulb}}{r} = \frac{15 \ \Omega}{0.50 \ \Omega} = \boxed{30}$$

---

70. **REASONING** According to the discussion in Section 20.9, the emf of a battery is equal to its terminal voltage V plus the voltage V_r across the internal resistance; $\text{Emf} = V + V_r$. According to Equation 20.6a, however, the voltage across the internal resistance is related to the current I and the power P dissipated by the internal resistance as $V_r = P/I$. Thus, the emf of the battery can be expressed as

$$\text{Emf} = V + \frac{P}{I}$$

SOLUTION Using the result found above, the emf of the battery is

$$\text{Emf} = V + \frac{P}{I} = 23.4 \ \text{V} + \frac{34.0 \ \text{W}}{55.0 \ \text{A}} = \boxed{24.0 \ \text{V}}$$

---

71. **REASONING AND SOLUTION** The equivalent resistance of the circuit is

$$R = 0.15 \ \Omega + 1.50 \ \Omega = 1.65 \ \Omega$$

The current in the circuit is then

$$I = (12.0 \ \text{V})/(1.65 \ \Omega) = 7.27 \ \text{A}$$

The terminal voltage is

$$V_T = 12.0 \ \text{V} - (7.27 \ \text{A})(0.15 \ \Omega) = \boxed{10.9 \ \text{V}}$$

772 ELECTRIC CIRCUITS

72. **REASONING AND SOLUTION**
a. In the first case the parallel resistance of the 75.0 Ω and the 45.0 Ω resistors have an equivalent resistance that can be calculated using Equation 20.17:

$$\frac{1}{R_p} = \frac{1}{75.0 \ \Omega} + \frac{1}{45.0 \ \Omega} \quad \text{or} \quad R_p = 28.1 \ \Omega$$

Ohm's law, Emf = IR gives Emf = (0.294 A)(28.1 Ω + r), or

$$\text{Emf} = 8.26 \ \text{V} + (0.294 \ \text{A})r \tag{1}$$

In the second case, Emf = (0.116 A)(75.0 Ω + r), or

$$\text{Emf} = 8.70 \ \text{V} + (0.116 \ \text{A})r \tag{2}$$

Multiplying Equation (1) by 0.116 A, Equation (2) by 0.294 A, and subtracting yields

Emf = $\boxed{8.99 \ \text{V}}$.

b. Substituting this result into Equation (1) and solving for r gives $r = \boxed{2.5 \ \Omega}$.

---

73. **SSM REASONING** The current I can be found by using Kirchhoff's loop rule. Once the current is known, the voltage between points A and B can be determined.

SOLUTION
a. We assume that the current is directed clockwise around the circuit. Starting at the upper-left corner and going clockwise around the circuit, we set the potential drops equal to the potential rises:

$$\underbrace{(5.0 \ \Omega)I + (27 \ \Omega)I + 10.0 \ \text{V} + (12 \ \Omega)I + (8.0 \ \Omega)I}_{\text{Potential drops}} = \underbrace{30.0 \ \text{V}}_{\text{Potential rises}}$$

Solving for the current gives $\boxed{I = 0.38 \ \text{A}}$.

b. The voltage between points A and B is

$$V_{AB} = 30.0 \ \text{V} - (0.38 \ \text{A})(27 \ \Omega) = \boxed{2.0 \times 10^1 \ \text{V}}$$

c. $\boxed{\text{Point B}}$ is at the higher potential.

74. **REASONING AND SOLUTION**
a. The voltage between A and B is just the voltage across the resistor. Ohm's law gives

$$V_{AB} = IR = (2.0 \text{ A})(6.0 \text{ }\Omega) = \boxed{12 \text{ V}}$$

b. The voltage between A and C is the sum of the drop across the resistor and the rise across the battery:

$$V_{AC} = -12 \text{ V} + 36 \text{ V} = \boxed{24 \text{ V}}$$

---

75. **REASONING AND SOLUTION** Apply the loop rule to get (with the units suppressed)

$$9.0 \text{ V} = I(0.015 \text{ }\Omega) + I(0.015 \text{ }\Omega) + 8.0 \text{ V}$$

Solving this equation for the current gives $I = \boxed{33 \text{ A}}$.

---

76. **REASONING** First, we draw a current I_1 in the 6.00-Ω resistor. We can express I_1 in terms of the other currents in the circuit, I and 3.00 A, by applying the junction rule to junction f; the sum of the currents into the junction must equal the sum of the currents out of the junction.

$$\underbrace{I}_{\text{Current into junction } f} = \underbrace{I_1 + 3.00 \text{ A}}_{\text{Current out of junction } f} \quad \text{or} \quad I_1 = I - 3.00 \text{ A}$$

In order to obtain values for I and V we apply the loop rule to the top and bottom loops of the circuit.

SOLUTION Applying the loop rule to the top loop (*abcf*) we have

$$\underbrace{(3.00 \text{ A})(4.00 \text{ }\Omega) + (3.00 \text{ A})(8.00 \text{ }\Omega)}_{\text{Potential drops}} = \underbrace{24.0 \text{ V} + (I - 3.00 \text{ A})(6.00 \text{ }\Omega)}_{\text{Potential rises}}$$

This equation can be solved directly for the current; $I = \boxed{5.00 \text{ A}}$.

Applying the loop rule to the bottom loop (*fcde*) we have

$$\underbrace{(I - 3.00 \text{ A})(6.00 \text{ }\Omega) + 24.0 \text{ V} + I(2.00 \text{ }\Omega)}_{\text{Potential drops}} = \underbrace{V}_{\text{Potential rises}}$$

Substituting $I = 5.00$ A into this equation and solving for V gives $V = \boxed{46.0 \text{ V}}$.

774 ELECTRIC CIRCUITS

77. **SSM** *REASONING* We begin by labeling the currents in the three resistors. The drawing below shows the directions chosen for these currents. The directions are arbitrary, and if any of them is incorrect, then the analysis will show that the corresponding value for the current is negative.

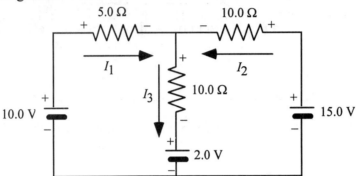

We then mark the resistors with the plus and minus signs that serve as an aid in identifying the potential drops and rises for the loop rule, recalling that conventional current is always directed from a higher potential (+) toward a lower potential (−). Thus, given the directions chosen for I_1, I_2, and I_3, the plus and minus signs *must* be those shown in the drawing. We can now use Kirchhoff's rules to find the voltage across the 5.0-Ω resistor.

SOLUTION Applying the loop rule to the left loop (and suppressing units for onvenience) gives

$$5.0 I_1 + 10.0 I_3 + 2.0 = 10.0 \qquad (1)$$

Similarly, for the right loop,

$$10.0 I_2 + 10.0 I_3 + 2.0 = 15.0 \qquad (2)$$

If we apply the junction rule to the upper junction, we obtain

$$I_1 + I_2 = I_3 \qquad (3)$$

Subtracting Equation (2) from Equation (1) gives

$$5.0 I_1 - 10.0 I_2 = -5.0 \qquad (4)$$

We now multiply Equation (3) by 10 and add the result to Equation (2); the result is

$$10.0 I_1 + 20.0 I_2 = 13.0 \qquad (5)$$

If we then multiply Equation (4) by 2 and add the result to Equation (5), we obtain $20.0 I_1 = 3.0$, or solving for I_1, we obtain $I_1 = 0.15$ A. The fact that I_1 is positive means that the current in the drawing has the correct direction. The voltage across the 5.0-Ω resistor can be found from Ohm's law:

$$V = (0.15 \text{ A})(5.0 \text{ }\Omega) = \boxed{0.75 \text{ V}}$$

Current flows from the higher potential to the lower potential, and the current through the 5.0-Ω flows from left to right, so the $\boxed{\text{left end of the resistor}}$ is at the higher potential.

---

78. **REASONING AND SOLUTION**
    Let the current through the 20.0 Ω be I_1 and flow to the right.
    Let the current through the 10.0 Ω be I_2 and flow up.
    Let the current through the 5.0 Ω be I_3 and flow to the right.

    Applying the loop rule to the left loop gives

    $$20.0 \, I_1 - 10.0 \, I_2 = 0$$

    and to the right loop

    $$10.0 \, I_2 + 5.0 \, I_3 = 30.0$$

    The junction rule applied to the upper junction gives

    $$I_1 + I_2 - I_3 = 0$$

    A simultaneous solution of the above gives $I_2 = \boxed{1.71 \text{ A}}$.

    Since the answer is positive, the current flows in the assumed direction. That is, $\boxed{\text{from the bottom to the top}}$.

776   ELECTRIC CIRCUITS

79. **REASONING** To find the voltage between points B and D, we will find the current in the 60.0-Ω resistor and then use Ohm's law to find the voltage as $V = IR$. To find the current we will use Kirchhoff's laws, the set-up for which is shown in the circuit diagram at the right. In this diagram we have marked the current in each resistor. It is I_{60} that we seek. Note that we have marked each resistor with plus and minus signs, to denote which end of the resistor is at the higher and which end is at the lower potential. Given our choice for the current directions, the plus and minus signs must be those shown, and they will help us apply Kirchhoff's loop rule correctly. If our value for I_{60} turns out to be negative, it will mean that the actual direction for this current is opposite to that in the diagram.

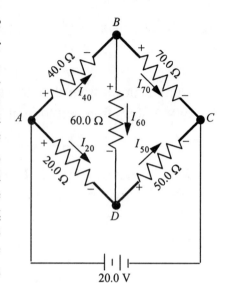

SOLUTION Applying the junction rule to junctions B and D gives

$$\underbrace{I_{40} = I_{60} + I_{70}}_{\text{At junction } B} \quad (1) \qquad \underbrace{I_{60} + I_{20} = I_{50}}_{\text{At junction } D} \quad (2)$$

Applying the loop rule to loops ABD, BCD, and ADC (including the battery) gives

$$\underbrace{I_{40}(40.0\,\Omega) + I_{60}(60.0\,\Omega)}_{\text{Potential drops, loop } ABD} = \underbrace{I_{20}(20.0\,\Omega)}_{\substack{\text{Potential rises,} \\ \text{loop } ABD}} \quad \text{or} \quad I_{40}(2.00) + I_{60}(3.00) = I_{20} \quad (3)$$

$$\underbrace{I_{70}(70.0\,\Omega)}_{\substack{\text{Potential drops,} \\ \text{loop } BCD}} = \underbrace{I_{50}(50.0\,\Omega) + I_{60}(60.0\,\Omega)}_{\text{Potential rises, loop } BCD} \quad \text{or} \quad I_{70}(7.00) = I_{50}(5.00) + I_{60}(6.00) \quad (4)$$

$$\underbrace{I_{20}(20.0\,\Omega) + I_{50}(50.0\,\Omega)}_{\text{Potential drops, loop } ADC} = \underbrace{20.0\text{ V}}_{\substack{\text{Potential rises,} \\ \text{loop } ADC}} \quad \text{or} \quad I_{20}(2.00) + I_{50}(5.00) = 2.00 \quad (5)$$

Equations (1)–(5) are five equations in five unknowns and must be solved simultaneously. Remember that it is I_{60} we seek, so our approach will be to eliminate the other four unknowns. Substituting I_{50} from Equation (2) into Equation (5) gives

$$I_{20}(7.00) + I_{60}(5.00) = 2.00 \quad (6)$$

Substituting I_{40} from Equation (1) into Equation (3) gives

$$I_{60}(5.00) + I_{70}(2.00) = I_{20} \quad (7)$$

Solving Equation (5) for I_{50} and substituting the result into Equation (4) gives

$$I_{70}(7.00) = 2.00 - I_{20}(2.00) + I_{60}(6.00) \qquad (8)$$

Solving Equation (8) for I_{70} and substituting the result into Equation (7) gives

$$I_{60}(47.0) - I_{20}(11.0) = -4.00 \qquad (9)$$

Solving Equation (6) for I_{20} and substituting the result into Equation (9) gives

$$I_{60}(47.0) - \left[\frac{2.00 - I_{60}(5.00)}{7.00}\right](11.0) = -4.00 \qquad \text{or} \qquad I_{60} = -0.0156 \text{ A}$$

Since this result is negative, the current in the 60.0-Ω resistor is opposite to that shown in the diagram given in the reasoning step, that is, from point D up toward point B. Thus, point D is at a higher potential than point B, because conventional current is always directed from the high toward the low potential. Using Ohm's law, we find that the voltage between D and B is

$$V = I_{60} R = (0.0156 \text{ A})(60.0 \text{ Ω}) = \boxed{0.94 \text{ V, with point } D \text{ at the higher potential}}$$

---

80. **REASONING** Since only 0.100 mA out of the available 60.0 mA is needed to cause a full-scale deflection of the galvanometer, the shunt resistor must allow the excess current of 59.9 mA to detour around the meter coil, as the drawing at the right indicates. The value for the shunt resistance can be obtained by recognizing that the 50.0-Ω coil resistance and the shunt resistance are in parallel, both being connected between points A and B in the drawing. Thus, the voltage across each resistance is the same.

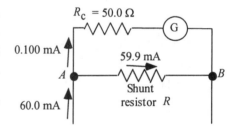

SOLUTION Expressing voltage as the product of current and resistance, we find that

$$\underbrace{(59.9 \times 10^{-3} \text{ A})(R)}_{\text{Voltage across shunt resistance}} = \underbrace{(0.100 \times 10^{-3} \text{ A})(50.0 \text{ Ω})}_{\text{Voltage across coil resistance}}$$

$$R = \frac{(0.100 \times 10^{-3} \text{ A})(50.0 \text{ Ω})}{59.9 \times 10^{-3} \text{ A}} = \boxed{0.0835 \text{ Ω}}$$

778 ELECTRIC CIRCUITS

81. **SSM** *REASONING AND SOLUTION* According to Ohm's law, the voltage across the series combination of the galvanometer and the resistor, under full-scale conditions is $V = I(R + R_g)$ where R_g is the resistance of the galvanometer. Therefore, solving for R, we have

$$R = \frac{V}{I} - R_g = \frac{30.0 \text{ V}}{8.30 \times 10^{-3} \text{ A}} - 180 \, \Omega = \boxed{3.43 \times 10^3 \, \Omega}$$

82. *REASONING* As discussed in Section 20.11, some of the current (6.20 mA) goes directly through the galvanometer and the remainder I goes through the shunt resistor. Since the resistance of the coil R_C and the shunt resistor R are in parallel, the voltage across each is the same. We will use this fact to determine how much current goes through the shunt resistor. This value, plus the 6.20 mA that goes through the galvanometer, is the maximum current that this ammeter can read.

SOLUTION The voltage across the coil resistance is equal to the voltage across the shunt resistor, so

$$\underbrace{(6.20 \times 10^{-3} \text{ A})(20.0 \, \Omega)}_{\text{Voltage across coil resistance}} = \underbrace{(I)(24.8 \times 10^{-3} \, \Omega)}_{\text{Voltage across shunt resistor}}$$

So $I = 5.00$ A. The maximum current is 5.00 A + 6.20 mA = $\boxed{5.01 \text{ A}}$.

83. *REASONING AND SOLUTION* The maximum voltage which can appear across the voltmeter is

$$V = IR_v = (180 \times 10^{-6} \text{ A})(140\,000 \, \Omega) = \boxed{25 \text{ V}}$$

84. *REASONING AND SOLUTION* For the 20.0 V scale

$$V_1 = I(R_1 + R_c)$$

For the 30.0 V scale

$$V_2 = I(R_2 + R_c)$$

Subtracting and rearranging yields

$$I = \frac{V_2 - V_1}{R_2 - R_1} = \frac{30.0 \text{ V} - 20.0 \text{ V}}{2930 \, \Omega - 1680 \, \Omega} = \boxed{8.00 \times 10^{-3} \text{ A}}$$

Substituting this value into either of the equations for V_1 or V_2 gives $R_c = \boxed{820 \, \Omega}$.

---

85. **SSM** *REASONING AND SOLUTION*

a. According to Ohm's law (Equation 20.2, $V = IR$) the current in the circuit is

$$I = \frac{V}{R+R} = \frac{V}{2R}$$

The voltage across either resistor is IR, so that we find

$$IR = \left(\frac{V}{2R}\right)R = \frac{V}{2} = \frac{60.0 \text{ V}}{2} = \boxed{30.0 \text{ V}}$$

b. The voltmeter's resistance is $R_v = V/I = (60.0 \text{ V})/(5.00 \times 10^{-3} \text{ A}) = 12.0 \times 10^3 \, \Omega$, and this resistance is in parallel with the resistance $R = 1550 \, \Omega$. The equivalent resistance of this parallel combination can be obtained as follows:

$$\frac{1}{R_p} = \frac{1}{R_v} + \frac{1}{R} \quad \text{or} \quad R_p = \frac{RR_v}{R+R_v} = \frac{(1550 \, \Omega)(12.0 \times 10^3 \, \Omega)}{1550 \, \Omega + 12.0 \times 10^3 \, \Omega} = 1370 \, \Omega$$

The voltage registered by the voltmeter is IR_p, where I is the current supplied by the battery to the series combination of the other 1550-Ω resistor and R_p. According to Ohm's law

$$I = \frac{60.0 \text{ V}}{1550 \, \Omega + 1370 \, \Omega} = 0.0205 \text{ A}$$

Thus, the voltage registered by the voltmeter is

$$IR_p = (0.0205 \text{ A})(1370 \, \Omega) = \boxed{28.1 \text{ V}}$$

---

86. *REASONING AND SOLUTION*

a. Add the parallel capacitances to get

$$C_p = 4.0 \, \mu\text{F} + 8.0 \, \mu\text{F} = \boxed{12.0 \, \mu\text{F}}$$

b. The total charge stored on the two capacitors is

$$Q = C_p V = (12.0 \times 10^{-6} \text{ F})(25 \text{ V}) = \boxed{3.0 \times 10^{-4} \text{ C}}$$

---

87. **REASONING** Our approach to this problem is to deal with the arrangement in parts. We will combine separately those parts that involve a series connection and those that involve a parallel connection.

    SOLUTION The 24, 12, and 8.0-μF capacitors are in series. Using Equation 20.19, we can find the equivalent capacitance for the three capacitors:

    $$\frac{1}{C_s} = \frac{1}{24 \ \mu\text{F}} + \frac{1}{12 \ \mu\text{F}} + \frac{1}{8.0 \ \mu\text{F}} \qquad \text{or} \qquad C_s = 4.0 \ \mu\text{F}$$

    This 4.0-μF capacitance is in parallel with the 4.0-μF capacitance already shown in the text diagram. Using Equation 20.18, we find that the equivalent capacitance for the parallel group is

    $$C_p = 4.0 \ \mu\text{F} + 4.0 \ \mu\text{F} = 8.0 \ \mu\text{F}$$

    This 8.0-μF capacitance is between the 5.0 and the 6.0-μF capacitances and in series with them. Equation 20.19 can be used, then, to determine the equivalent capacitance between A and B in the text diagram:

    $$\frac{1}{C_s} = \frac{1}{5.0 \ \mu\text{F}} + \frac{1}{8.0 \ \mu\text{F}} + \frac{1}{6.0 \ \mu\text{F}} \qquad \text{or} \qquad C_s = \boxed{2.0 \ \mu\text{F}}$$

---

88. **REASONING AND SOLUTION** The equivalent capacitance of the circuit is

    $$1/C = 1/(4.0 \ \mu\text{F}) + 1/(6.0 \ \mu\text{F}) + 1/(12.0 \ \mu\text{F}) \qquad \text{or} \qquad C = 2.0 \ \mu\text{F}$$

    The total charge provided by the battery is, then,

    $$Q = CV = (2.0 \times 10^{-6} \text{ F})(50.0 \text{ V}) = 1.0 \times 10^{-4} \text{ C}$$

    This charge appears on each capacitor in a series circuit, so the voltage across the 4.0-μF capacitor is

    $$V_1 = Q/C_1 = (1.0 \times 10^{-4} \text{ C})/(4.0 \times 10^{-6} \text{ F}) = \boxed{25 \text{ V}}$$

89. **SSM** **_REASONING_** The magnitude q of the charge on each plate of a capacitor is related to the voltage V across the capacitor plates and the capacitance C by Equation 19.8: $q = CV$. When two or more capacitors are in series, the equivalent capacitance of the combination can be obtained from Equation 20.19: $\dfrac{1}{C_s} = \dfrac{1}{C_1} + \dfrac{1}{C_2} + \dfrac{1}{C_3} \ldots$. Equation 20.18 gives the equivalent capacitance for two or more capacitors in parallel: $C_p = C_1 + C_2 + C_3 + \ldots$.

In this problem two of the capacitors are in series, while the third capacitor is in parallel with the series combination. We proceed by first finding the equivalent capacitance of the circuit, and then we can use Equation 19.8 to find the total charge that the battery delivers to all the capacitors.

SOLUTION The equivalent capacitance of the series combination can be obtained as follows:

$$\dfrac{1}{C_s} = \dfrac{1}{3.0\ \mu F} + \dfrac{1}{4.0\ \mu F} = 0.583\ (\mu F)^{-1} \quad \text{or} \quad C_s = \dfrac{1}{0.583\ (\mu F)^{-1}} = 1.72\ \mu F$$

The equivalent capacitance of the entire circuit is

$$C_{\text{equivalent}} = C_s = 1.72\ \mu F + 10.0\ \mu F = 11.72\ \mu F$$

Therefore the total charge delivered by the battery is

$$q = C_{\text{equivalent}} V = (11.72 \times 10^{-6}\ F)(40.0\ V) = \boxed{4.69 \times 10^{-4}\ C}$$

---

90. **_REASONING AND SOLUTION_** Let C_0 be the capacitance of an empty capacitor. Then the capacitances are

$$C_1 = 3.00\ C_0 \quad \text{and} \quad C_2 = 5.00\ C_0$$

The series capacitance of the two is

$$1/C_s = 1/(2.50\ C_0) + 1/(4.00\ C_0) \quad \text{or} \quad C_s = 1.54\ C_0$$

Now

$$\kappa C_0 = 1.54\ C_0 \quad \text{or} \quad \kappa = \boxed{1.54}$$

782   ELECTRIC CIRCUITS

91. **REASONING AND SOLUTION** The charges stored on capacitors in series are equal and equal to the charge separated by the battery. The total energy stored in the capacitors is

$$\text{Energy} = Q^2/(2C_1) + Q^2/(2C_2)$$

$$\text{Energy} = (1/C_1 + 1/C_2)Q^2/2$$

According to Equation 20.19, the quantity in the parentheses is just the reciprocal of the equivalent capacitance C of the circuit, so

$$\boxed{\text{Energy} = Q^2/(2C)}$$

92. **REASONING AND SOLUTION** The 7.00 and 3.00-μF capacitors are in parallel. According to Equation 20.18, the equivalent capacitance of the two is 7.00 μF + 3.00 μF = 10.0 μF. This 10.0-μF capacitance is in series with the 5.00-μF capacitance. According to Equation 20.19, the equivalent capacitance of the complete arrangement can be obtained as follows:

$$\frac{1}{C} = \frac{1}{10.0\ \mu F} + \frac{1}{5.00\ \mu F} = 0.300\ (\mu F)^{-1} \quad \text{or} \quad C = \frac{1}{0.300\ (\mu F)^{-1}} = 3.33\ \mu F$$

The battery separates an amount of charge

$$Q = CV = (3.33 \times 10^{-6}\ F)(30.0\ V) = 99.9 \times 10^{-6}\ C$$

This amount of charge resides on the 5.00 μF capacitor, so its voltage is

$$V_5 = (99.9 \times 10^{-6}\ C)/(5.00 \times 10^{-6}\ F) = 20.0\ V$$

The loop rule gives the voltage across the 3.00 μF capacitor to be

$$V_3 = 30.0\ V - 20.0\ V = 10.0\ V$$

This is also the voltage across the 7.00 μF capacitor, since it is in parallel, so $V_7 = \boxed{10.0\ V}$.

93. **SSM REASONING** When two or more capacitors are in series, the equivalent capacitance of the combination can be obtained from Equation 20.19, $\frac{1}{C_s} = \frac{1}{C_1} + \frac{1}{C_2} + \frac{1}{C_3} \ldots$. Equation 20.18 gives the equivalent capacitance for two or more capacitors in parallel:

$C_p = C_1 + C_2 + C_3 + \ldots$. The energy stored in a capacitor is given by $\frac{1}{2}CV^2$, according to Equation 19.11. Thus, the energy stored in the series combination is $\frac{1}{2}C_s V_s^2$, where

$$\frac{1}{C_s} = \frac{1}{7.0\ \mu F} + \frac{1}{3.0\ \mu F} = 0.476\ (\mu F)^{-1} \quad \text{or} \quad C_s = \frac{1}{0.476\ (\mu F)^{-1}} = 2.10\ \mu F$$

Similarly, the energy stored in the parallel combination is $\frac{1}{2}C_p V_p^2$ where

$$C_p = 7.0\ \mu F + 3.0\ \mu F = 10.0\ \mu F$$

The voltage required to charge the parallel combination of the two capacitors to the same total energy as the series combination can be found by equating the two energy expressions and solving for V_p.

SOLUTION Equating the two expressions for the energy, we have

$$\tfrac{1}{2} C_s V_s^2 = \tfrac{1}{2} C_p V_p^2$$

Solving for V_p, we obtain the result

$$V_p = V_s \sqrt{\frac{C_s}{C_p}} = (24\ \text{V}) \sqrt{\frac{2.10\ \mu F}{10.0\ \mu F}} = \boxed{11\ \text{V}}$$

---

94. **REASONING AND SOLUTION** Charge is conserved during the re-equilibrium. Therefore, using q_0 and q_f to denote the initial and final charges, respectively, we have

$$q_{10} + q_{20} = 18.0\ \mu C = q_{1f} + q_{2f} \tag{1}$$

After equilibrium has been established the capacitors will have equal voltages across them, since they are connected in parallel. Thus, $V_f = q_{1f}/C_1 = q_{2f}/C_2$, which leads to

$$q_{1f} = q_{2f}(C_1/C_2) = q_{2f}\,(2.00\ \mu F)/(8.00\ \mu F) = 0.250\ q_{2f}$$

Substituting this result into Equation (1) gives

$$18.0\ \mu C = 0.250\ q_{2f} + q_{2f} \quad \text{or} \quad q_{2f} = 14.4\ \mu C$$

Hence,

$$V_f = q_{2f}/C_2 = (14.4 \times 10^{-6}\ \text{C})/(8.00 \times 10^{-6}\ \text{F}) = \boxed{1.80\ \text{V}}$$

95. **SSM REASONING** The charge q on a discharging capacitor in a RC circuit is given by Equation 20.22: $q = q_0 e^{-t/RC}$, where q_0 is the original charge at time $t = 0$ s. Once t (time for one pulse) and the ratio q/q_0 are known, this expression can be solved for C.

SOLUTION Since the pacemaker delivers 81 pulses per minute, the time for one pulse is $[1 \text{ min}/(81 \text{ pulses})] \times [60.0 \text{ s}/(1.00 \text{ min})] = 0.74$ s. Since one pulse is delivered every time the fully-charged capacitor loses 63.2% of its original charge, the charge remaining is 36.8% of the original charge. Thus, we have $q = (0.368)q_0$, or $q/q_0 = 0.368$.

From Equation 20.22, we have

$$\frac{q}{q_0} = e^{-t/RC}$$

Taking the natural logarithm of both sides, we have,

$$\ln\left(\frac{q}{q_0}\right) = -\frac{t}{RC}$$

Solving for C, we find

$$C = \frac{-t}{R \ln(q/q_0)} = \frac{-(0.74 \text{ s})}{(1.8 \times 10^6 \text{ }\Omega) \ln(0.368)} = \boxed{4.1 \times 10^{-7} \text{ F}}$$

___

96. **REASONING AND SOLUTION** According to Equation 20.21, we find that

$$R = \frac{\tau}{C} = \frac{3.0 \text{ s}}{750 \times 10^{-6} \text{ F}} = \boxed{4.0 \times 10^3 \text{ }\Omega}$$

___

97. **REASONING** The time constant of an RC circuit is given by Equation 20.21 as $\tau = RC$, where R is the resistance and C is the capacitance in the circuit. The two resistors are wired in parallel, so we can obtain the equivalent resistance by using Equation 20.17. The two capacitors are also wired in parallel, and their equivalent capacitance is given by Equation 20.18. The time constant is the product of the equivalent resistance and equivalent capacitance.

SOLUTION The equivalent resistance of the two resistors in parallel is

$$\frac{1}{R_P} = \frac{1}{2.0 \text{ k}\Omega} + \frac{1}{4.0 \text{ k}\Omega} = \frac{3}{4.0 \text{ k}\Omega} \quad \text{or} \quad R_P = 1.3 \text{ k}\Omega \qquad (20.17)$$

The equivalent capacitance is

$$C_\text{p} = 3.0 \ \mu\text{F} + 6.0 \ \mu\text{F} = 9.0 \ \mu\text{F} \tag{20.18}$$

The time constant for the charge to build up is

$$\tau = R_\text{p} C_\text{p} = \left(1.3 \times 10^3 \ \Omega\right)\left(9.0 \times 10^{-6} \ \text{F}\right) = \boxed{1.2 \times 10^{-2} \ \text{s}}$$

---

98. **REASONING AND SOLUTION** In drawing (a) the equivalent capacitance is of the series capacitors is $C_\text{s} = C/2$. These are in parallel with a capacitance, C, so that the equivalent capacitance is

$$C_\text{A} = C/2 + C = (3/2) \ C$$

The time constant is $\tau_\text{A} = (3/2) \ C \ R$ so that

$$C = (2/3) \ (\tau_\text{A}/R) \tag{1}$$

In drawing (b) the equivalent capacitance is $C_\text{B} = (2/3)C$. Therefore, with the aid of Equation (1), we have

$$\tau_\text{B} = RC_\text{B} = (2/3) \ C \ R = (4/9) \ \tau_\text{A} = (4/9)(0.34 \ \text{s}) = \boxed{0.15 \ \text{s}}$$

---

99. **REASONING AND SOLUTION** For charging a capacitor, Equations 20.20 and 20.21 indicate that

$$q = q_0(1 - e^{-t/\tau})$$

or

$$t/\tau = -\ln(1 - q/q_0) = -\ln(0.0010) = \boxed{6.9}$$

---

100. **REASONING AND SOLUTION** According to Equation 20.6c, the power dissipated by the iron is

$$P = \frac{V^2}{R} = \frac{(120 \ \text{V})^2}{24 \ \Omega} = \boxed{6.0 \times 10^2 \ \text{W}}$$

786 ELECTRIC CIRCUITS

101. **SSM** *REASONING AND SOLUTION* The equivalent capacitance of the combination can be obtained immediately from Equation 20.19, $\frac{1}{C_s} = \frac{1}{C_1} + \frac{1}{C_2} + \frac{1}{C_3} \ldots$

$$\frac{1}{C_s} = \frac{1}{3.0\ \mu F} + \frac{1}{7.0\ \mu F} + \frac{1}{9.0\ \mu F} = 0.587\ (\mu F)^{-1} \quad \text{or} \quad C_s = \frac{1}{0.587\ (\mu F)^{-1}} = \boxed{1.7\ \mu F}$$

102. *REASONING AND SOLUTION* The definition of current in Equation 20.1 gives

$$I = \frac{\Delta q}{\Delta t} = \frac{35\ C}{1.00 \times 10^{-3}\ s} = \boxed{3.5 \times 10^4\ A}$$

103. **SSM** *REASONING AND SOLUTION* Ohm's law (Equation 20.2), $V = IR$, gives the result directly:

$$R = \frac{V}{I} = \frac{9.0\ V}{0.11\ A} = \boxed{82\ \Omega}$$

104. *REASONING AND SOLUTION* Label the currents with the resistor values. Take I_3 to the right, I_2 to the left and I_1 to the right. Applying the loop rule to the top loop (suppressing the units) gives

$$I_1 + 2.0\ I_2 = 1.0 \tag{1}$$

and to the bottom loop gives

$$2.0\ I_2 + 3.0\ I_3 = 5.0 \tag{2}$$

Applying the junction rule to the left junction gives

$$I_2 = I_1 + I_3 \tag{3}$$

Solving Equations (1), (2) and (3) simultaneously, we find $I_2 = \boxed{0.73\ A}$.

The positive sign shows that the assumed direction is correct. That is, to the $\boxed{\text{left}}$.

105. ***REASONING AND SOLUTION*** From Equation 20.5 we have that $R = R_0[1 + \alpha(T - T_0)]$. Solving for T gives

$$T = T_0 + \frac{(R/R_0) - 1}{\alpha} = 20.0\ °C + \frac{[(99.6\ \Omega)/(125\ \Omega)] - 1}{3.72 \times 10^{-3}\ (C°)^{-1}} = \boxed{-34.6\ °C}$$

---

106. ***REASONING*** The magnitude q of the charge on one plate of a capacitor is given by Equation 19.8 as $q = CV_1$, where $C = 9.0\ \mu F$ and V_1 is the voltage across the capacitor. Since the capacitor and the resistor R_1 are in parallel, the voltage across the capacitor is equal to the voltage across R_1. From Equation 20.2 we know that the voltage across the 4.0-Ω resistor is given by $V_1 = IR_1$, where I is the current in the circuit. Thus, the charge can be expressed as

$$q = CV_1 = C(IR_1)$$

The current is equal to the battery voltage V divided by the equivalent resistance of the two resistors in series, so that

$$I = \frac{V}{R_S} = \frac{V}{R_1 + R_2}$$

Substituting this result for I into the equation for q yields

$$q = C(IR_1) = C\left(\frac{V}{R_1 + R_2}\right)R_1$$

SOLUTION The magnitude of the charge on one of the plates is

$$q = C\left(\frac{V}{R_1 + R_2}\right)R_1 = (9.0 \times 10^{-6}\ F)\left(\frac{12\ V}{4.0\ \Omega + 2.0\ \Omega}\right)(4.0\ \Omega) = \boxed{7.2 \times 10^{-5}\ C}$$

---

107. ***REASONING AND SOLUTION*** The voltages across the galvanometer and shunt resistances are equal since they are in parallel, so

$$I_g R_g = I_s R_s \qquad \text{or} \qquad R_s = (I_g/I_s)R_g$$

where $I_g = 0.150 \times 10^{-3}$ A and $I_s = 4.00 \times 10^{-3}$ A $- 0.150 \times 10^{-3}$ A. Then

$$R_s = (0.150 \times 10^{-3}\ A)(12.0\ \Omega)/(3.85 \times 10^{-3}\ A) = 0.468\ \Omega$$

788   ELECTRIC CIRCUITS

The equivalent parallel resistance can be obtained as follows:

$$\frac{1}{R_p} = \frac{1}{0.468\ \Omega} + \frac{1}{12.0\ \Omega} \quad \text{or} \quad R_p = \boxed{0.450\ \Omega}$$

---

108. **REASONING AND SOLUTION** The voltage V_{Cu} between the ends of the copper rod is given by Ohm's law as $V_{cu} = IR_{Cu}$, where R_{Cu} is the resistance of the copper rod. The current I in the circuit is equal to the voltage V of the battery that is connected across the free ends of the copper-iron rod divided by the equivalent resistance of the rod. The copper and iron rods are joined end-to-end, so the same current passes through each. Thus, they are connected in series, so the equivalent resistance R_S is $R_S = R_{Cu} + R_{Fe}$. Thus, the current is

$$I = \frac{V}{R_S} = \frac{V}{R_{Cu} + R_{Fe}}$$

The voltage across the copper rod is

$$V_{Cu} = IR_{Cu} = \frac{V}{R_{Cu} + R_{Fe}} R_{Cu}$$

The resistance of the copper and iron rods is given by $R_{Cu} = \rho_{Cu} L/A$ and $R_{Fe} = \rho_{Fe} L/A$, where the length L and cross-sectional area A is the same for both rods. Substituting these expressions for the resistances into the equation above and using resistivities from Table 20.1 yield

$$V_{Cu} = \left(\frac{V}{\rho_{Cu} + \rho_{Fe}}\right)\rho_{Cu}$$

$$V_{Cu} = \left(\frac{12\ \text{V}}{1.72 \times 10^{-8}\ \Omega\cdot\text{m} + 9.7 \times 10^{-8}\ \Omega\cdot\text{m}}\right)(1.72 \times 10^{-8}\ \Omega\cdot\text{m}) = \boxed{1.8\ \text{V}}$$

---

109. **SSM** **REASONING** When two or more resistors are in series, the equivalent resistance is given by Equation 20.16: $R_s = R_1 + R_2 + R_3 + \ldots$. When resistors are in parallel, the expression to be solved to find the equivalent resistance is Equation 20.17: $\frac{1}{R_p} = \frac{1}{R_1} + \frac{1}{R_2} + \frac{1}{R_3} + \ldots$. We will use these relations to determine the eight different values of resistance that can be obtained by connecting together the three resistors: 1.00, 2.00, and 3.00 Ω.

SOLUTION When all the resistors are connected in series, the equivalent resistance is the sum of all three resistors and the equivalent resistance is $\boxed{6.00 \, \Omega}$. When all three are in parallel, we have from Equation 20.17, the equivalent resistance is $\boxed{0.545 \, \Omega}$.

We can also connect two of the resistors in parallel and connect the parallel combination in series with the third resistor. When the 1.00 and 2.00-Ω resistors are connected in parallel and the 3.00-Ω resistor is connected in series with the parallel combination, the equivalent resistance is $\boxed{3.67 \, \Omega}$. When the 1.00 and 3.00-Ω resistors are connected in parallel and the 2.00-Ω resistor is connected in series with the parallel combination, the equivalent resistance is $\boxed{2.75 \, \Omega}$. When the 2.00 and 3.00-Ω resistors are connected in parallel and the 1.00-Ω resistor is connected in series with the parallel combination, the equivalent resistance is $\boxed{2.20 \, \Omega}$.

We can also connect two of the resistors in series and put the third resistor in parallel with the series combination. When the 1.00 and 2.00-Ω resistors are connected in series and the 3.00-Ω resistor is connected in parallel with the series combination, the equivalent resistance is $\boxed{1.50 \, \Omega}$. When the 1.00 and 3.00-Ω resistors are connected in series and the 2.00-Ω resistor is connected in parallel with the series combination, the equivalent resistance is $\boxed{1.33 \, \Omega}$. Finally, when the 2.00 and 3.00-Ω resistors are connected in series and the 1.00-Ω resistor is connected in parallel with the series combination, the equivalent resistance is $\boxed{0.833 \, \Omega}$.

---

110. **REASONING AND SOLUTION** The aluminum and copper portions may be viewed as being connected in parallel since the same voltage appears across them. Using a and b to denote the inner and outer radii, respectively, we have for the equivalent resistance that

$$\frac{1}{R_p} = \frac{1}{R_{Al}} + \frac{1}{R_{Cu}} = \frac{A_{Cu}}{\rho_{Cu} L} + \frac{A_{Al}}{\rho_{Al} L}$$

$$\frac{1}{R_p} = \frac{\pi a^2}{\rho_{Cu} L} + \frac{\pi (b^2 - a^2)}{\rho_{Al} L}$$

Using resistivity values from Table 20.1, we find that $R_p = \boxed{0.00116 \, \Omega}$.

---

111. **SSM** **REASONING** The resistance of one of the wires in the extension cord is given by Equation 20.3: $R = \rho L / A$, where the resistivity of copper is $\rho = 1.72 \times 10^{-8} \, \Omega \cdot m$, according to Table 20.1. Since the two wires in the cord are in series with each other, their

total resistance is $R_{cord} = R_{wire\ 1} + R_{wire\ 2} = 2\rho L/A$. Once we find the equivalent resistance of the entire circuit (extension cord + trimmer), Ohm's law can be used to find the voltage applied to the trimmer.

SOLUTION
a. The resistance of the extension cord is

$$R_{cord} = \frac{2\rho L}{A} = \frac{2(1.72 \times 10^{-8}\ \Omega \cdot m)(46\ m)}{1.3 \times 10^{-6}\ m^2} = \boxed{1.2\ \Omega}$$

b. The total resistance of the circuit (cord + trimmer) is, since the two are in series,

$$R_s = 1.2\ \Omega + 15.0\ \Omega = 16.2\ \Omega$$

Therefore from Ohm's law (Equation 20.2: $V = IR$), the current in the circuit is

$$I = \frac{V}{R_s} = \frac{120\ V}{16.2\ \Omega} = 7.4\ A$$

Finally, the voltage applied to the trimmer alone is (again using Ohm's law),

$$V_{trimmer} = (7.4\ A)(15.0\ \Omega) = \boxed{110\ V}$$

---

112. **REASONING AND SOLUTION** The power dissipated in the resistor if the battery has no internal resistance is

$$P_0 = I_0^2 R$$

where $I_0 = V/R$, so

$$P_0 = V^2/R \tag{1}$$

The power dissipated if the battery has an internal resistance is $P = I^2 R$, where $I = V/(R + r)$ and $P = 0.900\ P_0$. Thus,

$$0.900\ P_0 = V^2 R/(R + r)^2 \tag{2}$$

Division of Equation (2) by Equation (1) gives

$$R^2/(R + r)^2 = 0.900 \quad \text{or} \quad R/(R + r) = 0.949 \quad \text{or} \quad r/R = \boxed{0.054}$$

113. **REASONING AND SOLUTION** Using Equation 20.5, we can equate resistances

$$\underbrace{R_{0i}\left[1+\alpha_i(T-T_0)\right]}_{\text{Resistance for iron wire}} = \underbrace{R_{0g}\left[1+\alpha_g(T-T_0)\right]}_{\text{Resistance for gold wire}}$$

We can solve this equation for T to obtain

$$T = T_0 + \frac{R_{0i} - R_{0g}}{R_{0g}\alpha_g - R_{0i}\alpha_i}$$

$$= 20.0\,°C + \frac{5.90\,\Omega - 6.70\,\Omega}{(6.70\,\Omega)\left[0.0034\,(C°)^{-1}\right] - (5.90\,\Omega)\left[0.0050\,(C°)^{-1}\right]} = \boxed{140\,°C}$$

---

114. **REASONING AND SOLUTION** Each resistor can tolerate a current of no more than

$$I = \sqrt{\frac{P}{R}} = \sqrt{\frac{0.25\,W}{47\,\Omega}} = 0.073\,A$$

Ohm's law applied to a series circuit containing N such resistors gives $V = IR_S = INR$, so

$$N = \frac{V}{IR} = \frac{9.0\,V}{(0.073\,A)(47\,\Omega)} = 2.6$$

Only $\boxed{\text{three}}$ resistors can be used.

---

115. **SSM** **REASONING** The foil effectively converts the capacitor into two capacitors in series. Equation 19.10 gives the expression for the capacitance of a capacitor of plate area A and plate separation d (no dielectric): $C_0 = \varepsilon_0 A / d$. We can use this expression to determine the capacitance of the individual capacitors created by the presence of the foil. Then using the fact that the "two capacitors" are in series, we can use Equation 20.19 to find the equivalent capacitance of the system.

SOLUTION Since the foil is placed one-third of the way from one plate of the original capacitor to the other, we have $d_1 = (2/3)d$, and $d_2 = (1/3)d$. Then

$$C_1 = \frac{\varepsilon_0 A}{(2/3)d} = \frac{3\varepsilon_0 A}{2d}$$

and

$$C_2 = \frac{\varepsilon_0 A}{(1/3)d} = \frac{3\varepsilon_0 A}{d}$$

Since these two capacitors are effectively in series, it follows that

$$\frac{1}{C_s} = \frac{1}{C_1} + \frac{1}{C_2} = \frac{1}{3\varepsilon_0 A/(2d)} + \frac{1}{3\varepsilon_0 A/d} = \frac{3d}{3\varepsilon_0 A} = \frac{d}{\varepsilon_0 A}$$

But $C_0 = \varepsilon_0 A/d$, so that $d/(\varepsilon_0 A) = 1/C_0$, and we have

$$\frac{1}{C_s} = \frac{1}{C_0} \quad \text{or} \quad \boxed{C_s = C_0}$$

---

116. **REASONING AND SOLUTION** The resistance of the thermistor decreases by 15% relative to its normal value of 37.0 °C. That is,

$$\Delta R/R_0 = (R - R_0)/R_0 = -0.15$$

According to Equation 20.5, we have

$$R = R_0[1 + \alpha(T - T_0)] \quad \text{or} \quad (R - R_0) = \alpha R_0(T - T_0) \quad \text{or} \quad (R - R_0)/R_0 = \alpha(T - T_0) = -0.15$$

Rearranging this result gives

$$T = T_0 + (-0.15)/\alpha = 37.0\ °C + (-0.15)/(-0.060/C°) = \boxed{39.5\ °C}$$

---

117. **CONCEPT QUESTIONS**
a. The resistance R of a piece of material is related to its length L and cross-sectional area A by Equation 20.3, $R = \rho L/A$, where ρ is the resistivity of the material. In order to rank the resistances, we need to evaluate L and A for each configuration in terms of L_0, the unit of length.

	Resistance	Rank
A	$R = \rho \dfrac{4L_0}{L_0 \times 2L_0} = \rho\left(\dfrac{2}{L_0}\right)$	1
B	$R = \rho \dfrac{L_0}{2L_0 \times 4L_0} = \rho\left(\dfrac{1}{8L_0}\right)$	3
C	$R = \rho \dfrac{2L_0}{L_0 \times 4L_0} = \rho\left(\dfrac{1}{2L_0}\right)$	2

Therefore, the A has the largest resistance, followed by C, and then by B.

b. Equation 20.2 states that the current I is equal to the voltage V divided by the resistance, $I = V/R$. Since the current is inversely proportional to the resistance, the largest current arises when the resistance is smallest, and vice versa. Thus, B has the largest current, followed by C, and then by A.

SOLUTION
a. The resistances can be found by using the results of the Concept Questions:

A $\qquad R = \rho\left(\dfrac{2}{L_0}\right) = (1.50 \times 10^{-2}\ \Omega \cdot \text{m})\left(\dfrac{2}{5.00 \times 10^{-2}\ \text{m}}\right) = \boxed{0.600\ \Omega}$

B $\qquad R = \rho\left(\dfrac{1}{8L_0}\right) = (1.50 \times 10^{-2}\ \Omega \cdot \text{m})\left(\dfrac{1}{8 \times 5.00 \times 10^{-2}\ \text{m}}\right) = \boxed{0.0375\ \Omega}$

C $\qquad R = \rho\left(\dfrac{1}{2L_0}\right) = (1.50 \times 10^{-2}\ \Omega \cdot \text{m})\left(\dfrac{1}{2 \times 5.00 \times 10^{-2}\ \text{m}}\right) = \boxed{0.150\ \Omega}$

b. The current in each case is given by Equation 20.2, where the value of the resistance is obtained from part (a):

A $\qquad I = \dfrac{V}{R} = \dfrac{3.00\ \text{V}}{0.600\ \Omega} = \boxed{5.00\ \text{A}}$

B $\qquad I = \dfrac{V}{R} = \dfrac{3.00\ \text{V}}{0.0375\ \Omega} = \boxed{80.0\ \text{A}}$

C $\qquad I = \dfrac{V}{R} = \dfrac{3.00\ \text{V}}{0.150\ \Omega} = \boxed{20.0\ \text{A}}$

We see that the largest current is in B, with progressively smaller currents in C and A.

---

118. CONCEPT QUESTIONS

a. The power delivered to a resistor is given by Equation 20.6c as $P = V^2/R$, where V is the voltage and R is the resistance. Because of the dependence of the power on V^2, doubling the voltage has a greater effect in increasing the power than halving the resistance. The table shows the power for each circuit, given in terms of these variables:

ELECTRIC CIRCUITS

	Power	Rank
(a)	$P = \dfrac{V^2}{R}$	3
(b)	$P = \dfrac{V^2}{2R}$	4
(c)	$P = \dfrac{(2V)^2}{R} = \dfrac{4V^2}{R}$	1
(d)	$P = \dfrac{(2V)^2}{2R} = \dfrac{2V^2}{R}$	2

b. The current is given by Equation 20.2 as $I = V/R$. Note that the current, unlike the power, depends linearly on the voltage. Therefore, either doubling the voltage or halving the resistance has the same effect on the current. The table shows the current for the four circuits:

	Current	Rank
(a)	$I = \dfrac{V}{R}$	2
(b)	$I = \dfrac{V}{2R}$	3
(c)	$I = \dfrac{2V}{R}$	1
(d)	$I = \dfrac{2V}{2R} = \dfrac{V}{R}$	2

SOLUTION

a. Using the results from part (a) and the values of $V = 12.0$ V and $R = 6.00$ Ω, the power dissipated in each resistor is

	Power	Rank
(a)	$P = \dfrac{V^2}{R} = \dfrac{(12.0\text{ V})^2}{6.00\text{ }\Omega} = \boxed{24.0\text{ W}}$	3
(b)	$P = \dfrac{V^2}{2R} = \dfrac{(12.0\text{ V})^2}{2(6.00\text{ }\Omega)} = \boxed{12.0\text{ W}}$	4
(c)	$P = \dfrac{4V^2}{R} = \dfrac{4(12.0\text{ V})^2}{(6.00\text{ }\Omega)} = \boxed{96.0\text{ W}}$	1
(d)	$P = \dfrac{2V^2}{R} = \dfrac{2(12.0\text{ V})^2}{(6.00\text{ }\Omega)} = \boxed{48.0\text{ W}}$	2

b. Using the results from part (b) and the values of $V = 12.0$ V and $R = 6.00$ Ω, the current in each circuit is

	Current	Rank
(a)	$I = \dfrac{V}{R} = \dfrac{12.0\text{ V}}{6.00\text{ }\Omega} = \boxed{2.00\text{ A}}$	2
(b)	$I = \dfrac{V}{2R} = \dfrac{12.0\text{ V}}{2(6.00\text{ }\Omega)} = \boxed{1.00\text{ A}}$	3
(c)	$I = \dfrac{2V}{R} = \dfrac{2(12.0\text{ V})}{6.00\text{ }\Omega} = \boxed{4.00\text{ A}}$	1
(d)	$I = \dfrac{2V}{2R} = \dfrac{2(12.0\text{ V})}{2(6.00\text{ }\Omega)} = \boxed{2.00\text{ A}}$	2

119. *CONCEPT QUESTIONS*

a. The three resistors are in series, so the same current goes through each resistor: $I_1 = I_2 = I_3$. The voltage across each resistor is given by Equation 20.2 as $V = IR$. Because the current through each resistor is the same, the voltage across each is proportional to the resistance. Since $R_1 > R_2 > R_3$, the ranking of the voltages is $V_1 > V_2 > V_3$.

b. The three resistors are in parallel, so the same voltage exists across each: $V_1 = V_2 = V_3$. The current through each resistor is given by Equation 20.2 as $I = V/R$. Because the voltage across each resistor is the same, the current through each is inversely proportional to the resistance. Since $R_1 > R_2 > R_3$, the ranking of the currents is $I_3 > I_2 > I_1$.

796 ELECTRIC CIRCUITS

SOLUTION

a. The current through the three resistors is given by $I = V/R_S$, where R_S is the equivalent resistance of the series circuit. From Equation 20.16, the equivalent resistance is $R_S = 50.0\ \Omega + 25.0\ \Omega + 10.0\ \Omega = 85.0\ \Omega$. The current through each resistor is

$$I_1 = I_2 = I_3 = \frac{V}{R_S} = \frac{24.0\ \text{V}}{85.0\ \Omega} = \boxed{0.282\ \text{A}}$$

The voltage across each resistor is

$$V_1 = IR_1 = (0.282\ \text{A})(50.0\ \Omega) = \boxed{14.1\ \text{V}}$$

$$V_2 = IR_2 = (0.282\ \text{A})(25.0\ \Omega) = \boxed{7.05\ \text{V}}$$

$$V_3 = IR_3 = (0.282\ \text{A})(10.0\ \Omega) = \boxed{2.82\ \text{V}}$$

b. The resistors are in parallel, so the voltage across each is the same as the voltage of the battery:

$$V_1 = V_2 = V_3 = \boxed{24.0\ \text{V}}$$

The current through each resistor is equal to the voltage across each divided by the resistance:

$$I_1 = \frac{V}{R_1} = \frac{24.0\ \text{V}}{50.0\ \Omega} = \boxed{0.480\ \text{A}}$$

$$I_2 = \frac{V}{R_2} = \frac{24.0\ \text{V}}{25.0\ \Omega} = \boxed{0.960\ \text{A}}$$

$$I_3 = \frac{V}{R_3} = \frac{24.0\ \text{V}}{10.0\ \Omega} = \boxed{2.40\ \text{A}}$$

---

120. **CONCEPT QUESTION** Between points a and b there is only one resistor, so the equivalent resistance is $R_{ab} = R$. Between points b and c the two resistors are in parallel. The equivalent resistance can be found from Equation 20.17:

$$\frac{1}{R_{bc}} = \frac{1}{R} + \frac{1}{R} = \frac{2}{R} \quad \text{so} \quad R_{bc} = \tfrac{1}{2} R$$

The equivalent resistance between a and b is in series with the equivalent resistance between b and c, so the equivalent resistance between a and c is

$$R_{ac} = R_{ab} + R_{bc} = R + \tfrac{1}{2}R = \tfrac{3}{2}R$$

Thus, we see that $R_{ac} > R_{ab} > R_{bc}$.

SOLUTION Since the resistance is $R = 10.0\ \Omega$, the equivalent resistances are:

$$R_{ab} = R = \boxed{10.0\ \Omega}$$

$$R_{bc} = \tfrac{1}{2}R = \boxed{5.00\ \Omega}$$

$$R_{ac} = \tfrac{3}{2}R = \boxed{15.0\ \Omega}$$

---

121. **CONCEPT QUESTION** In part a, the current goes from left-to-right through the resistor. Since the current always goes from a higher to a lower potential, the left end of the resistor is + and the right end is −. In part b, the current goes from right-to-left through the resistor. The right end of the resistor is + and the left end is −. The potential drops and rises for the two cases are:

	Potential drops	Potential rises
Part a	IR	V
Part b	V	IR

SOLUTION Since the current I goes from left-to-right through the 3.0- and 4.0-Ω resistors, the left end of each resistor is + and the right end is −. The current goes through the 5.0-Ω resistor from right-to-left, so the right end is + and the left end is −. Starting at the upper left corner of the circuit, and proceeding clockwise around it, Kirchhoff's loop rule is written as

$$\underbrace{(3.0\ \Omega)I + 12\ \text{V} + (4.0\ \Omega)I + (5.0\ \Omega)I}_{\text{Potential drops}} = \underbrace{36\ \text{V}}_{\substack{\text{Potential}\\ \text{rises}}}$$

Solving this equation for the current gives $I = \boxed{2.0\ \text{A}}$

---

122. **CONCEPT QUESTION** When capacitors are connected in parallel, each receives the entire voltage V of the battery. Thus, the total energy stored in the two capacitors is $\tfrac{1}{2}C_1V^2 + \tfrac{1}{2}C_2V^2$. When the capacitors are connected in series, the sum of the voltages across each capacitor equals the battery voltage: $V_1 + V_2 = V$. Thus, the voltage across each capacitor is series is *less than* the battery voltage, so the total energy, $\tfrac{1}{2}C_1V_1^2 + \tfrac{1}{2}C_2V_2^2$, is *less than* when the capacitors are wired in parallel.

SOLUTION

a. The voltage across each capacitor is the battery voltage, or 60.0 V. The energy stored in both capacitors is

$$\text{Total energy} = \tfrac{1}{2}C_1V^2 + \tfrac{1}{2}C_2V^2 = \tfrac{1}{2}(C_1 + C_2)V^2$$

$$= \tfrac{1}{2}(2.00 \times 10^{-6}\text{ F} + 4.00 \times 10^{-6}\text{ F})(60.0\text{ V})^2 = \boxed{1.08 \times 10^{-2}\text{ J}}$$

b. According to the discussion in Section 20.12, the total energy stored by capacitors in series is Total energy $= \tfrac{1}{2}C_S V^2$, where C_S is the equivalent capacitance of the series combination:

$$\frac{1}{C_S} = \frac{1}{C_1} + \frac{1}{C_2} = \frac{1}{2.00 \times 10^{-6}\text{ F}} + \frac{1}{4.00 \times 10^{-6}\text{ F}} \qquad (20.19)$$

Solving this equation yields $C_S = 1.33 \times 10^{-6}$ F. The total energy is

$$\text{Total energy} = \tfrac{1}{2}(1.33 \times 10^{-6}\text{ F})(60.0\text{ V})^2 = \boxed{2.39 \times 10^{-3}\text{ J}}$$

CHAPTER 21 | MAGNETIC FORCES AND MAGNETIC FIELDS

PROBLEMS

1. **SSM** *REASONING AND SOLUTION* According to Equation 21.1, the magnitude B of the magnetic field is

$$B = \frac{F}{q_0 v \sin\theta} = \frac{8.7 \times 10^{-3}\,\text{N}}{(12 \times 10^{-6}\,\text{C})(9.0 \times 10^6\,\text{m/s})\sin 90.0°} = \boxed{8.1 \times 10^{-5}\,\text{T}}$$

2. *REASONING AND SOLUTION* The magnitude of the force can be determined using Equation 21.1, $F = qvB \sin\theta$, where θ is the angle between the velocity and the magnetic field. The direction of the force is determined by using Right-Hand Rule No. 1.

   a. $F = qvB \sin 30.0° = (8.4 \times 10^{-6}\,\text{C})(45\,\text{m/s})(0.30\,\text{T})\sin 30.0° = \boxed{5.7 \times 10^{-5}\,\text{N}}$, directed $\boxed{\text{into the paper}}$.

   b. $F = qvB \sin 90.0° = (8.4 \times 10^{-6}\,\text{C})(45\,\text{m/s})(0.30\,\text{T})\sin 90.0° = \boxed{1.1 \times 10^{-4}\,\text{N}}$, directed $\boxed{\text{into the paper}}$.

   c. $F = qvB \sin 150° = (8.4 \times 10^{-6}\,\text{C})(45\,\text{m/s})(0.30\,\text{T})\sin 150° = \boxed{5.7 \times 10^{-5}\,\text{N}}$, directed $\boxed{\text{into the paper}}$.

3. *REASONING AND SOLUTION*

   a. The magnitude of the magnetic force is given by $F = qvB \sin\theta$. We have, therefore,

   $$B = \frac{F}{qv\sin\theta} = \frac{7.31 \times 10^{-3}\,\text{N}}{(25.0 \times 10^{-6}\,\text{C})(4.50 \times 10^3\,\text{m/s})\sin 90.0°} = \boxed{6.50 \times 10^{-2}\,\text{T}}$$

   b. For the second charge,

   $$v_2 = \frac{F}{qB\sin\theta} = \frac{1.90 \times 10^{-3}\,\text{N}}{(5.00 \times 10^{-6}\,\text{C})(6.50 \times 10^{-2}\,\text{T})\sin 40.0°} = \boxed{9.10 \times 10^3\,\text{m/s}}$$

800 MAGNETIC FORCES AND MAGNETIC FIELDS

4. **REASONING** According to Equation 21.1, the magnetic force has a magnitude of $F = qvB \sin\theta$. The field B and the directional angle θ are the same for each particle. Particle 1, however, travels faster than particle 2. By itself, a faster speed v would lead to a greater force magnitude F. But the force on each particle is the same. Therefore, particle 1 must have a smaller charge to counteract the effect of its greater speed.

SOLUTION Applying Equation 21.1 to each particle, we have

$$\underbrace{F = q_1 v_1 B \sin\theta}_{\text{Particle 1}} \quad \text{and} \quad \underbrace{F = q_2 v_2 B \sin\theta}_{\text{Particle 2}}$$

Dividing the equation for particle 1 by the equation for particle 2 and remembering that $v_1 = 3v_2$ gives

$$\frac{F}{F} = \frac{q_1 v_1 B \sin\theta}{q_2 v_2 B \sin\theta} \quad \text{or} \quad 1 = \frac{q_1 v_1}{q_2 v_2} \quad \text{or} \quad \frac{q_1}{q_2} = \frac{v_2}{v_1} = \frac{v_2}{3v_2} = \boxed{\tfrac{1}{3}}$$

---

5. **SSM REASONING** According to Equation 21.1, the magnitude of the magnetic force on a moving charge is $F = q_0 v B \sin\theta$. Since the magnetic field points due north and the proton moves eastward, $\theta = 90.0°$. Furthermore, since the magnetic force on the moving proton balances its weight, we have $mg = q_0 v B \sin\theta$, where m is the mass of the proton. This expression can be solved for the speed v.

SOLUTION Solving for the speed v, we have

$$v = \frac{mg}{q_0 B \sin\theta} = \frac{(1.67 \times 10^{-27} \text{ kg})(9.80 \text{ m/s}^2)}{(1.6 \times 10^{-19} \text{ C})(2.5 \times 10^{-5} \text{ T}) \sin 90.0°} = \boxed{4.1 \times 10^{-3} \text{ m/s}}$$

---

6. **REASONING** According to Equation 21.1, the magnetic force has a magnitude of $F = qvB \sin\theta$, where q is the magnitude of the charge, B is the magnitude of the magnetic field, v is the speed, and θ is the angle of the velocity with respect to the field. As θ increases from 0° to 90°, the force increases. Therefore, the angle we seek must lie between 25° and 90°.

SOLUTION Letting $\theta_1 = 25°$ and θ_2 be the desired angle, we can apply Equation 21.1 to both situations as follows:

$$\underbrace{F = qvB \sin\theta_1}_{\text{Situation 1}} \quad \text{and} \quad \underbrace{2F = qvB \sin\theta_2}_{\text{Situation 2}}$$

Dividing the equation for situation 2 by the equation for situation 1 gives

$$\frac{2F}{F} = \frac{qvB\sin\theta_2}{qvB\sin\theta_1} \quad \text{or} \quad \sin\theta_2 = 2\sin\theta_1 = 2\sin 25° = 0.85$$

$$\theta_2 = \sin^{-1}(0.85) = \boxed{58°}$$

7. **REASONING** The angle θ between the electron's velocity and the magnetic field can be found from Equation 21.1,

$$\sin\theta = \frac{F}{qvB}$$

According to Newton's second law, the magnitude F of the force is equal to the product of the electron's mass m and the magnitude a of its acceleration, $F = ma$.

SOLUTION The angle θ is

$$\theta = \sin^{-1}\left(\frac{ma}{qvB}\right) = \sin^{-1}\left[\frac{(9.11\times 10^{-31}\text{ kg})(3.50\times 10^{14}\text{ m/s}^2)}{(1.60\times 10^{-19}\text{ C})(6.80\times 10^6\text{ m/s})(8.70\times 10^{-4}\text{ T})}\right] = \boxed{19.7°}$$

8. **REASONING AND SOLUTION** Equation 21.1 gives the magnitude F of the maximum force on the particle as $F = q_0 vB \sin 90.0° = q_0 vB$, where B is the magnitude of the net magnetic field. Since the two field components B_x and B_y are perpendicular, the Pythagorean theorem indicates that $B = \sqrt{B_x^2 + B_y^2}$. Therefore, we find that $B = \sqrt{B_x^2 + B_y^2} = \frac{F}{q_0 v}$. Squaring this result and solving for the y component of the magnetic field gives

$$B_y = \sqrt{\frac{F^2}{(q_0 v)^2} - B_x^2} = \sqrt{\frac{(0.455\text{ N})^2}{\left[(6.50\times 10^{-5}\text{ C})(2.00\times 10^4\text{ m/s})\right]^2} - (0.200\text{ T})^2} = \boxed{0.287\text{ T}}$$

9. **SSM** **WWW** **REASONING** The direction in which the electrons are deflected can be determined using Right-Hand Rule No. 1 and reversing the direction of the force (RHR-1 applies to positive charges, and electrons are negatively charged).

Each electron experiences an acceleration a given by Newton's second law of motion, $a = F/m$, where F is the net force and m is the mass of the electron. The only force acting on the electron is the magnetic force, $F = q_0 vB \sin\theta$, so it is the net force. The speed v of the

802 MAGNETIC FORCES AND MAGNETIC FIELDS

electron is related to its kinetic energy KE by the relation $KE = \frac{1}{2}mv^2$. Thus, we have enough information to find the acceleration.

SOLUTION

a. According to RHR-1, if you extend your right hand so that your fingers point along the direction of the magnetic field **B** and your thumb points in the direction of the velocity **v** of a *positive* charge, your palm will face in the direction of the force **F** on the positive charge.

For the electron in question, the fingers of the right hand should be oriented downward (direction of **B**) with the thumb pointing to the east (direction of **v**). The palm of the right hand points due north (the direction of **F** on a positive charge). Since the electron is negatively charged, it will be deflected $\boxed{\text{due south}}$.

b. The acceleration of an electron is given by Newton's second law, where the net force is the magnetic force. Thus,

$$a = \frac{F}{m} = \frac{q_0 v B \sin\theta}{m}$$

Since the kinetic energy is $KE = \frac{1}{2}mv^2$, the speed of the electron is $v = \sqrt{2(KE)/m}$. Thus, the acceleration of the electron is

$$a = \frac{q_0 v B \sin\theta}{m} = \frac{q_0 \sqrt{\frac{2(KE)}{m}} B \sin\theta}{m}$$

$$= \frac{(1.60 \times 10^{-19} \text{ C})\sqrt{\frac{2(2.40 \times 10^{-15} \text{ J})}{9.11 \times 10^{-31} \text{ kg}}}(2.00 \times 10^{-5} \text{ T})\sin 90.0°}{9.11 \times 10^{-31} \text{ kg}} = \boxed{2.55 \times 10^{14} \text{ m/s}^2}$$

---

10. **REASONING AND SOLUTION**

b. The magnitude of the magnetic field is $B = mv/qr$ (see Equation 21.2). Therefore,

$$B = \frac{(9.11 \times 10^{-31} \text{ kg})(6.0 \times 10^6 \text{ m/s})}{(1.60 \times 10^{-19} \text{ C})(1.3 \times 10^{-3} \text{ m})} = \boxed{2.6 \times 10^{-2} \text{ T}}$$

c. From Newton's second law, the electron's acceleration is $a = F/m$, where the force can be obtained using Equation 21.1:

$F = qvB \sin \theta$

$= (1.60 \times 10^{-19} \text{ C})(6.0 \times 10^6 \text{ m/s})(2.6 \times 10^{-2} \text{ T}) \sin 90.0° = 2.5 \times 10^{-14} \text{ N}$

The acceleration is, therefore,

$$a = \frac{F}{m} = \frac{2.5 \times 10^{-14} \text{ N}}{9.11 \times 10^{-31} \text{ kg}} = \boxed{2.7 \times 10^{16} \text{ m/s}^2}$$

---

11. **REASONING AND SOLUTION** The charge can be found from Equation 21.2 as

$$q = \frac{mv}{Br} = \frac{(6.6 \times 10^{-27} \text{ kg})(4.4 \times 10^5 \text{ m/s})}{(0.75 \text{ T})(0.012 \text{ m})} = 3.2 \times 10^{-19} \text{ C}$$

Since $e = 1.6 \times 10^{-19}$ C, we see that the charge of the ionized helium is $\boxed{+2e}$.

---

12. **REASONING** The magnetic field applies the maximum magnetic force to the moving charge, because the motion is perpendicular to the field. This force is perpendicular to both the field and the velocity. The electric field applies an electric force to the charge that is in the same direction as the field, since the charge is positive. These two forces are shown in the drawing, and they are perpendicular to one another. Therefore, the magnitude of the net field can be obtained using the Pythagorean theorem.

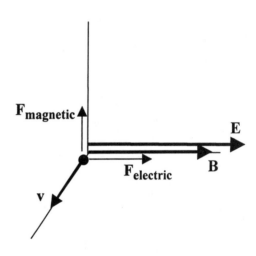

SOLUTION According to Equation 21.1, the magnetic force has a magnitude of $F_{\text{magnetic}} = qvB \sin \theta$, where q is the magnitude of the charge, B is the magnitude of the magnetic field, v is the speed, and $\theta = 90°$ is the angle of the velocity with respect to the field. Thus, $F_{\text{magnetic}} = qvB$. According to Equation 18.2, the electric force has a magnitude of $F_{\text{electric}} = qE$. Using the Pythagorean theorem, we find the magnitude of the net force to be

$$F = \sqrt{F_{\text{magnetic}}^2 + F_{\text{electric}}^2} = \sqrt{(qvB)^2 + (qE)^2} = q\sqrt{(vB)^2 + E^2}$$

$$= (1.8 \times 10^{-6} \text{ C})\sqrt{[(3.1 \times 10^6 \text{ m/s})(1.2 \times 10^{-3} \text{ T})]^2 + (4.6 \times 10^3 \text{ N/C})^2} = \boxed{1.1 \times 10^{-2} \text{ N}}$$

13. **SSM** *REASONING AND SOLUTION*

    a. The speed of a proton can be found from Equation 21.2 ($r = mv/qB$),

    $$v = \frac{qBr}{m} = \frac{(1.6\times 10^{-19}\text{ C})(0.30\text{ T})(0.25\text{ m})}{1.67\times 10^{-27}\text{ kg}} = \boxed{7.2\times 10^6 \text{ m/s}}$$

    b. The magnitude F_c of the centripetal force is given by Equation 5.3,

    $$F_c = \frac{mv^2}{r} = \frac{(1.67\times 10^{-27}\text{ kg})(7.2\times 10^6\text{ m/s})^2}{0.25\text{ m}} = \boxed{3.5\times 10^{-13}\text{ N}}$$

14. *REASONING AND SOLUTION* The radius of the circular path of a charged particle in a magnetic field is given by $r = mv/qB$.

    a. For an electron

    $$r = \frac{mv}{qB} = \frac{(9.11\times 10^{-31}\text{ kg})(9.0\times 10^6\text{ m/s})}{(1.6\times 10^{-19}\text{ C})(1.2\times 10^{-7}\text{ T})} = \boxed{4.3\times 10^2 \text{ m}}$$

    b. For a proton, only the mass changes in the calculation above. Using $m = 1.67\times 10^{-27}$ kg, we obtain $r = \boxed{7.8\times 10^5 \text{ m}}$.

15. **SSM** **WWW** *REASONING AND SOLUTION* In one revolution, the particle moves once around the circumference of the circle. Therefore, it travels a distance of $d = 2\pi r$, where r is the radius of the circle. Since the particle moves at constant speed, $v = d/t$, and the time required for one revolution is $t = d/v$. According to Equation 21.2, $r = mv/qB$, so $v = qBr/m$. Thus, the time required for the particle to complete one revolution is

    $$t = \frac{d}{v} = \frac{2\pi r}{qBr/m} = \frac{2\pi}{B(q/m)} = \frac{2\pi}{(0.72\text{ T})(5.7\times 10^8\text{ C/kg})} = \boxed{1.5\times 10^{-8}\text{ s}}$$

16. *REASONING AND SOLUTION* The magnitudes of the magnetic and electric forces are equal. Therefore, $F_B = F_E$, or $qvB = qE$. Solving for v yields,

    $$v = \boxed{E/B}$$

17. **REASONING AND SOLUTION** The radius of curvature for a particle in a mass spectrometer is given by (see Section 21.4)

$$r = \sqrt{\frac{2mV}{qB^2}} = \sqrt{\frac{2(3.27 \times 10^{-25} \text{ kg})(1.00 \times 10^3 \text{ V})}{(3.2 \times 10^{-19} \text{ C})(0.500 \text{ T})^2}} = \boxed{0.0904 \text{ m}}$$

18. **REASONING** The radius of the circular path is given by Equation 21.2 as $r = mv/(qB)$, where m is the mass of the species, v is the speed, q is the magnitude of the charge, and B is the magnitude of the magnetic field. To use this expression, we must know something about the speed. Information about the speed can be obtained by applying the conservation of energy principle. The electric potential energy lost as a charged particle "falls" from a higher to a lower electric potential is gained by the particle as kinetic energy.

SOLUTION For an electric potential difference V and a charge q, the electric potential energy lost is qV, according to Equation 19.4. The kinetic energy gained is $\tfrac{1}{2}mv^2$. Thus, energy conservation dictates that

$$qV = \tfrac{1}{2}mv^2 \quad \text{or} \quad v = \sqrt{\frac{2qV}{m}}$$

Substituting this result into Equation 21.2 for the radius gives

$$r = \frac{mv}{qB} = \frac{m}{qB}\sqrt{\frac{2qV}{m}} = \frac{1}{B}\sqrt{\frac{2mV}{q}}$$

Using e to denote the magnitude of the charge on an electron, we note that the charge for species X^+ is $+e$, while the charge for species X^{2+} is $+2e$. With this in mind, we find for the ratio of the radii that

$$\frac{r_1}{r_2} = \frac{\dfrac{1}{B}\sqrt{\dfrac{2mV}{e}}}{\dfrac{1}{B}\sqrt{\dfrac{2mV}{2e}}} = \sqrt{2} = \boxed{1.41}$$

19. **SSM** **REASONING** From the discussion in Section 21.3, we know that when a charged particle moves perpendicular to a magnetic field, the trajectory of the particle is a circle. The drawing shows a particle moving in the plane of the paper (the magnetic field is perpendicular to the paper). If the particle is moving initially through the coordinate origin and to the right (along the $+x$ axis), the subsequent circular path of the particle will intersect the y axis at the greatest possible value, which is equal to twice the radius r of the circle.

806 MAGNETIC FORCES AND MAGNETIC FIELDS

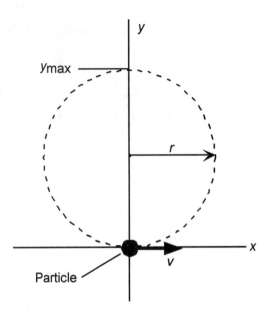

SOLUTION

a. From the drawing above, it can be seen that the largest value of y is equal to the diameter ($2r$) of the circle. When the particle passes through the coordinate origin its velocity must be parallel to the $+x$ axis. Thus, the angle is $\boxed{\theta = 0°}$.

b. The maximum value of y is twice the radius r of the circle. According to Equation 21.2, the radius of the circular path is $r = mv/qB$. The maximum value y_{max} is, therefore,

$$y_{max} = 2r = 2\left(\frac{mv}{qB}\right) = 2\left[\frac{(3.8 \times 10^{-8} \text{ kg})(44 \text{ m/s})}{(7.3 \times 10^{-6} \text{ C})(1.6 \text{ T})}\right] = \boxed{0.29 \text{ m}}$$

---

20. **REASONING AND SOLUTION** The magnitudes of the magnetic and electric forces must be equal. Therefore,

$$F_B = F_E \quad \text{or} \quad qvB = qE$$

This relation can be solved to give the speed of the particle, $v = E/B$. We also know that when the electric field is turned off, the particle travels in a circular path of radius $r = mv/qB$. Substituting $v = E/B$ into this equation and solving for q/m gives

$$\frac{q}{m} = \frac{E}{rB^2} = \frac{3.80 \times 10^3 \text{ N/C}}{(4.30 \times 10^{-2} \text{ m})(0.360 \text{ T})^2} = \boxed{6.8 \times 10^5 \text{ C/kg}}$$

21. **REASONING AND SOLUTION** The proton will miss the opposite plate if the distance between the plates is equal to the radius of the circular orbit of the proton. Therefore,

$$B = \frac{mv}{qr} = \frac{(1.67 \times 10^{-27} \text{ kg})(2.2 \times 10^6 \text{ m/s})}{(1.6 \times 10^{-19} \text{ C})(0.18 \text{ m})} = \boxed{0.13 \text{ T}}$$

22. **REASONING AND SOLUTION** Equation 21.2 gives the speed as $v = rqB/m$, or $mv^2 = r^2q^2B^2/m$. Since the kinetic energy is $KE = \frac{1}{2}mv^2$, it follows that $KE = \frac{r^2q^2B^2}{2m}$. Recognizing that q, B, and m are the same for both particles, we find that

$$\frac{KE_3}{KE_1} = \frac{\frac{r_3^2 q^2 B^2}{2m}}{\frac{r_1^2 q^2 B^2}{2m}} = \left(\frac{r_3}{r_1}\right)^2 = (22)^2 = \boxed{484}$$

23. **SSM REASONING AND SOLUTION** According to Right-Hand Rule No. 1, the magnetic force on the positively charged particle is toward the bottom of the page in the drawing in the text. If the presence of the electric field is to double the magnitude of the net force on the charge, the electric field must also be $\boxed{\text{directed toward the bottom of the page}}$.

Note that this results in the electric field being perpendicular to the magnetic field, even though the electric force and the magnetic force are in the same direction.

Furthermore, if the magnitude of the net force on the particle is twice the magnetic force, the electric force must be equal in magnitude to the magnetic force. In other words, combining Equations 18.2 and 21.1, we find $qE = qvB\sin\theta$, with $\sin\theta = \sin 90.0° = 1.0$. Then, solving for E

$$E = vB\sin\theta = (270 \text{ m/s})(0.52 \text{ T})(1.0) = \boxed{140 \text{ V/m}}$$

24. **REASONING AND SOLUTION** The angular speed is $\omega = v/r = qB/m$. Therefore,

$$\omega = \frac{qB}{m} = \frac{(3.2 \times 10^{-19} \text{ C})(0.0210 \text{ T})}{6.64 \times 10^{-27} \text{ kg}} = \boxed{1.01 \times 10^6 \text{ rad/s}}$$

25. **REASONING AND SOLUTION** The drawings show the two circular paths leading to the target T when the proton is projected from the origin O. In each case, the center of the circle is at C. Since the target is located at $\dot{x} = -0.10$ m and $y = -0.10$ m, the radius of each circle is $r = 0.10$ m. The speed with which the proton is projected can be obtained from

Equation 21.2, if we remember that the charge and mass of a proton are $q = 1.6 \times 10^{-19}$ C and $m = 1.67 \times 10^{-27}$ kg, respectively:

$$v = \frac{rqB}{m} = \frac{(0.10 \text{ m})(1.6 \times 10^{-19} \text{ C})(0.010 \text{ T})}{1.67 \times 10^{-27} \text{ kg}} = \boxed{9.6 \times 10^4 \text{ m/s}}$$

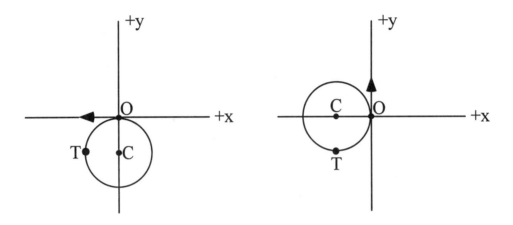

26. **REASONING** According to Equation 21.3, the magnetic force has a magnitude of $F = ILB \sin \theta$, where I is the current, B is the magnitude of the magnetic field, L is the length of the wire, and $\theta = 90°$ is the angle of the wire with respect to the field.

    SOLUTION Using Equation 21.3, we find that

    $$L = \frac{F}{IB \sin \theta} = \frac{7.1 \times 10^{-5} \text{ N}}{(0.66 \text{ A})(4.7 \times 10^{-5} \text{ T}) \sin 58°} = \boxed{2.7 \text{ m}}$$

27. [SSM] **REASONING AND SOLUTION** The magnetic force exerted on the power line is given by Equation 21.3,

    $$F = ILB \sin \theta = (1400 \text{ A})(120 \text{ m})(5.0 \times 10^{-5} \text{ T})(\sin 75°) = \boxed{8.1 \text{ N}}$$

28. **REASONING AND SOLUTION** The magnitude F of the force on a current I is given by Equation 21.3 as $F = ILB \sin \theta$, where L is the length of the wire and θ is the angle between the wire and a magnetic field that has a magnitude B. Applying this equation in the two situations described in this problem and recognizing that L, B, and θ are the same in each, we find that

    $$\frac{F_1}{F_2} = \frac{I_1 LB \sin \theta}{I_2 LB \sin \theta} = \frac{I_1}{I_2} \quad \text{so} \quad \frac{0.030 \text{ N}}{0.047 \text{ N}} = \frac{2.7 \text{ A}}{I_2} \quad \text{or} \quad I_2 = (2.7 \text{ A})\left(\frac{0.047 \text{ N}}{0.030 \text{ N}}\right) = \boxed{4.2 \text{ A}}$$

29. ***REASONING AND SOLUTION*** The force on each side can be found from $F = ILB \sin \theta$. For the top side, $\theta = 90.0°$, so

$$F = (12 \text{ A})(0.32 \text{ m})(0.25 \text{ T}) \sin 90.0° = \boxed{0.96 \text{ N}}$$

The force on the bottom side ($\theta = 90.0°$) is the same as that on the top side, $F = \boxed{0.96 \text{ N}}$.

For each of the other two sides $\theta = 0°$, so that the force is $F = \boxed{0 \text{ N}}$.

30. ***REASONING*** The magnitude of the magnetic force exerted on a long straight wire is given by Equation 21.3 as $F = ILB \sin \theta$. The direction of the magnetic force is predicted by Right-Hand Rule No. 1. The net force on the triangular loop is the vector sum of the forces on the three sides.

SOLUTION
a. The direction of the current in side AB is opposite to the direction of the magnetic field, so the angle θ between them is $\theta = 180°$. The magnitude of the magnetic force is

$$F_{AB} = ILB \sin \theta = ILB \sin 180° = \boxed{0 \text{ N}}$$

For the side BC, the angle is $\theta = 55.0°$, and the length of the side is $L = \dfrac{2.00 \text{ m}}{\cos 55.0°} = 3.49 \text{ m}$. The magnetic force is

$$F_{BC} = ILB \sin \theta = (4.70 \text{ A})(3.49 \text{ m})(1.80 \text{ T}) \sin 55.0° = \boxed{24.2 \text{ N}}$$

An application of Right-Hand No. 1 shows that the magnetic force on side BC is directed $\boxed{\text{perpendicularly out of the paper}}$, toward the reader.

For the side AC, the angle is $\theta = 90.0°$, and the length of the side is $L = (2.00 \text{ m}) \tan 55.0° = 2.86 \text{ m}$. The magnetic force is

$$F_{AC} = ILB \sin \theta = (4.70 \text{ A})(2.86 \text{ m})(1.80 \text{ T}) \sin 90.0° = \boxed{24.2 \text{ N}}$$

An application of Right-Hand No. 1 shows that the magnetic force on side AC is directed $\boxed{\text{perpendicularly into the paper}}$, away from the reader.

b. The net force is the vector sum of the forces on the three sides. Taking the positive direction as being out of the paper, the net force is

$$\sum F = 0\,\text{N} + 24.2\,\text{N} + (-24.2\,\text{N}) = \boxed{0\,\text{N}}$$

---

31. **SSM** **REASONING AND SOLUTION** The force on a current-carrying wire is given by Equation 21.3: $F = ILB\sin\theta$. Solving for the angle θ, we find that the angle between the wire and the magnetic field is

$$\theta = \sin^{-1}\left(\frac{F}{ILB}\right) = \sin^{-1}\left[\frac{5.46\,\text{N}}{(21.0\,\text{A})(0.655\,\text{m})(0.470\,\text{T})}\right] = \boxed{57.6°}$$

---

32. **REASONING** A maximum magnetic force is exerted on the wire by the field components that are perpendicular to the wire, and no magnetic force is exerted by field components that are parallel to the wire. Thus, the wire experiences a force only from the x- and y-components of the field. The z-component of the field may be ignored, since it is parallel to the wire. We can use the Pythagorean theorem to find the net field in the x, y plane. This net field, then, is perpendicular to the wire and makes an angle of $\theta = 90°$ with respect to the wire. Equation 21.3 can be used to calculate the magnitude of the magnetic force that this net field applies to the wire.

SOLUTION According to Equation 21.3, the magnetic force has a magnitude of $F = ILB \sin\theta$, where I is the current, B is the magnitude of the magnetic field, L is the length of the wire, and θ is the angle of the wire with respect to the field. Using the Pythagorean theorem, we find that the net field in the x, y plane is

$$B = \sqrt{B_x^2 + B_y^2}$$

Using this field in Equation 21.3, we calculate the magnitude of the magnetic force to be

$$F = ILB \sin\theta = IL\sqrt{B_x^2 + B_y^2}\,\sin\theta$$

$$= (4.3\,\text{A})(0.25\,\text{m})\sqrt{(0.10\,\text{T})^2 + (0.15\,\text{T})^2}\,\sin 90° = \boxed{0.19\,\text{N}}$$

---

33. **SSM** **REASONING AND SOLUTION**
a. From Right-Hand Rule No. 1, if we extend the right hand so that the fingers point in the direction of the magnetic field, and the thumb points in the direction of the current, the palm of the hand faces the direction of the magnetic force on the current.

The springs will stretch when the magnetic force exerted on the copper rod is downward, toward the bottom of the page. Therefore, if you extend your right hand with your fingers pointing out of the page and the palm of your hand facing the bottom of the page, your thumb points left-to-right along the copper rod. Thus, the current flows $\boxed{\text{left-to-right}}$ in the copper rod.

b. The downward magnetic force exerted on the copper rod is, according to Equation 21.3

$$F = ILB\sin\theta = (12\text{ A})(0.85\text{ m})(0.16\text{ T})\sin 90.0° = 1.6\text{ N}$$

From Hooke's law, we know that the net force exerted on the rod by the two springs is $F = kx + kx = 2kx$, where k is the spring constant. Solving for x, we find that

$$x = \frac{F}{2k} = \frac{1.6\text{ N}}{2(75\text{ N/m})} = \boxed{1.1\times 10^{-2}\text{ m}}$$

---

34. **REASONING AND SOLUTION** Since the rod rotates neither clockwise nor counterclockwise about the axis at P, the torques relative to that axis must balance to zero. There are two torques that must be considered, that due to the magnetic force and that due to the weight of the rod. We consider both of these to act at the rod's center of gravity, which is at the center of the rod (length = L), because the rod is uniform. According to RHR-1, the magnetic force acts perpendicular to the rod and is directed up and to the left in the drawing. Equation 21.3 gives the magnitude F of the magnetic force as $F = ILB\sin 90.0°$, since the current is perpendicular to the field. Therefore, the magnetic torque is counterclockwise. The weight is mg and acts downward, producing a clockwise torque. Thus, we have for the torques:

$$\tau_{\text{magnetic}} = \underbrace{(ILB)}_{\text{force}}\underbrace{(L/2)}_{\text{lever arm}} \quad \text{and} \quad \tau_{\text{weight}} = \underbrace{(mg)}_{\text{force}}\underbrace{\left[(L/2)\cos\theta\right]}_{\text{lever arm}}$$

Equating these torques, we find that

$$\underbrace{(ILB)(L/2)}_{\text{magnetic torque}} = \underbrace{(mg)\left[(L/2)\cos\theta\right]}_{\text{weight torque}} \quad \text{or} \quad \cos\theta = \frac{ILB}{mg}$$

$$\theta = \cos^{-1}\left[\frac{(3.8\text{ A})(0.40\text{ m})(0.31\text{ T})}{(0.080\text{ kg})(9.80\text{ m/s}^2)}\right] = \boxed{53°}$$

---

35. **REASONING** The drawing below shows a side view of the conducting rails and the aluminum rod. Three forces act on the rod: (1) its weight mg, (2) the magnetic force F, and the normal force F_N. An application of Right-Hand Rule No. 1 shows that the magnetic force is directed to the left, as shown in the drawing. Since the rod slides down the rails at a

constant velocity, its acceleration is zero. If we choose the x-axis to be along the rails, Newton's second law states that the net force along the x-direction is zero: $\Sigma F_x = ma_x = 0$. Using the components of F and mg that are along the x-axis, Newton's second law becomes

$$\underbrace{-F \cos 30.0° + mg \sin 30.0°}_{\Sigma F_x} = 0$$

The magnetic force is given by Equation 21.3 as $F = ILB \sin \theta$, where $\theta = 90.0°$ is the angle between the magnetic field and the current. We can use these two relations to find the current in the rod.

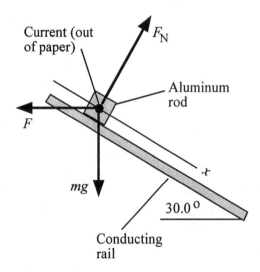

SOLUTION Substituting the expression $F = ILB \sin 90.0°$ into Newton's second law and solving for the current I, we obtain

$$I = \frac{mg \sin 30.0°}{(LB \sin 90.0°) \cos 30.0°} = \frac{(0.20 \text{ kg})(9.80 \text{ m/s}^2) \sin 30.0°}{[(1.6 \text{ m})(0.050 \text{ T}) \sin 90.0°] \cos 30.0°} = \boxed{14 \text{ A}}$$

36. **REASONING AND SOLUTION** The maximum torque occurs when $\phi = 90.0°$ so that $\tau = NIAB$. For a square loop, $A = (L/4)^2 = (0.50 \text{ m}/4)^2 = 1.6 \times 10^{-2} \text{ m}^2$. So,

$$\tau = NIAB = (1)(12 \text{ A})(1.6 \times 10^{-2} \text{ m}^2)(0.12 \text{ T}) = \boxed{0.023 \text{ N} \cdot \text{m}}$$

37. **SSM REASONING AND SOLUTION**
a. The magnetic moment of the coil is

$$\text{Magnetic moment} = NIA = (50)(15 \text{ A})\left[\pi(0.10 \text{ m})^2\right] = \boxed{24 \text{ A} \cdot \text{m}^2}$$

where we have used the fact that the area of the circular loop is $A = \pi r^2 = \pi(0.10 \text{ m})^2$.

b. According to Equation 21.4, the torque is the product of the magnetic moment NIA and $B \sin \phi$. However, the maximum torque occurs when $\phi = 90.0°$, so we have

$$\tau = (\text{Magnetic moment})(B \sin 90.0°) = (24 \text{ A} \cdot \text{m}^2)(0.20 \text{ T}) = \boxed{4.8 \text{ N} \cdot \text{m}}$$

---

38. **REASONING** According to Equation 21.4, the maximum torque is $\tau_{max} = NIAB$, where N is the number of turns in the coil, I is the current, $A = \pi r^2$ is the area of the circular coil, and B is the magnitude of the magnetic field. We can apply the maximum-torque expression to each coil, noting that τ_{max}, N, and I are the same for each.

SOLUTION Applying Equation 21.4 to each coil, we have

$$\underbrace{\tau_{max} = NI\pi r_1^2 B_1}_{\text{Coil 1}} \quad \text{and} \quad \underbrace{\tau_{max} = NI\pi r_2^2 B_2}_{\text{Coil 2}}$$

Dividing the expression for coil 2 by the expression for coil 1 gives

$$\frac{\tau_{max}}{\tau_{max}} = \frac{NI\pi r_2^2 B_2}{NI\pi r_1^2 B_1} \quad \text{or} \quad 1 = \frac{r_2^2 B_2}{r_1^2 B_1}$$

Solving for r_2, we obtain

$$r_2 = r_1 \sqrt{\frac{B_1}{B_2}} = (5.0 \text{ cm})\sqrt{\frac{0.18 \text{ T}}{0.42 \text{ T}}} = \boxed{3.3 \text{ cm}}$$

---

39. **REASONING** The magnetic moment of the current-carrying triangle is the product of the current I and the area A of the triangle. The magnitude τ of the torque exerted on the triangle by the magnetic field is equal to the product of the magnetic moment, the magnitude B of the magnetic field, and the sine of the angle ϕ between the magnetic field and the normal to the plane of the triangle, $\tau = (\text{Magnetic moment}) B \sin \phi$.

SOLUTION
a. The area of a triangle is equal to one-half the base times the height,

$$A = \tfrac{1}{2}(2.00 \text{ m})\left[(2.00 \text{ m}) \tan 55.0°\right] = 2.86 \text{ m}^2$$

The magnetic moment is

$$\text{Magnetic moment} = IA = (4.70 \text{ A})(2.86 \text{ m}^2) = \boxed{13.4 \text{ A} \cdot \text{m}^2}$$

814   MAGNETIC FORCES AND MAGNETIC FIELDS

b. The magnitude of the net torque exerted on the triangle is

$$\tau = (\text{Magnetic moment})B \sin\phi \qquad (21.4)$$

$$= (13.4 \text{ A}\cdot\text{m}^2)(1.80 \text{ T})\sin 90.0° = \boxed{24.1 \text{ N}\cdot\text{m}}$$

---

40. **REASONING AND SOLUTION** The torque is given by $\tau = NIAB \sin\phi$.

a. The maximum torque occurs when $\phi = 90.0°$ ($\sin\phi = 1$). In this case we want the torque to be 80.0% of the maximum value, so

$$NIAB\sin\phi = 0.800(NIAB\sin 90.0°) \quad \text{so that} \quad \phi = \sin^{-1}(0.800) = \boxed{53.1°}$$

---

41. **SSM WWW REASONING** The torque on the loop is given by Equation 21.4, $\tau = NIAB\sin\phi$. From the drawing in the text, we see that the angle ϕ between the normal to the plane of the loop and the magnetic field is $90° - 35° = 55°$. The area of the loop is $0.70 \text{ m} \times 0.50 \text{ m} = 0.35 \text{ m}^2$.

SOLUTION
a. The magnitude of the net torque exerted on the loop is

$$\tau = NIAB\sin\phi = (75)(4.4 \text{ A})(0.35 \text{ m}^2)(1.8 \text{ T})\sin 55° = \boxed{170 \text{ N}\cdot\text{m}}$$

b. As discussed in the text, when a current-carrying loop is placed in a magnetic field, the loop tends to rotate such that its normal becomes aligned with the magnetic field. The normal to the loop makes an angle of 55° with respect to the magnetic field. Since this angle decreases as the loop rotates, the $\boxed{35° \text{ angle increases}}$.

---

42. **REASONING** According to Equation 21.4, the maximum torque is $\tau_{max} = NIAB$, where N is the number of turns in the coil, I is the current, $A = \pi r^2$ is the area of the circular coil, and B is the magnitude of the magnetic field. Since the coil contains only one turn, the length L of the wire is the circumference of the circle, so that $L = 2\pi r$ or $r = L/(2\pi)$. Since N, I, and B are known we can solve for L.

SOLUTION According to Equation 21.4 and the fact that $r = L/(2\pi)$, we have

$$\tau_{max} = NI\pi r^2 B = NI\pi \left(\frac{L}{2\pi}\right)^2 B$$

Solving this result for L gives

$$L = \sqrt{\frac{4\pi \tau_{max}}{NIB}} = \sqrt{\frac{4\pi(8.4 \times 10^{-4} \text{ N} \cdot \text{m})}{(1)(3.7 \text{ A})(0.75 \text{ T})}} = \boxed{0.062 \text{ m}}$$

---

43. **REASONING AND SOLUTION** In Figure 21.22a the magnetic torque is a maximum. Equation 21.4 gives this maximum torque as $\tau_{max} = NIAB$. The torque from the brake balances this magnetic torque. The brake torque is $\tau_{brake} = F_{brake} r$, where F_{brake} is the brake force, and r is the radius of the shaft and also the lever arm. The maximum value for the brake force available from static friction is $F_{brake} = \mu_s F_N$, where F_N is the normal force pressing the brake shoe against the shaft. The brake torque, then, is $\tau_{brake} = \mu_s F_N r$. Therefore, we can obtain the magnitude of the minimum normal force as follows:

$$\tau_{brake} = \tau_{max} \quad \text{or} \quad \mu_s F_N r = NIAB$$

$$F_N = \frac{NIAB}{\mu_s r} = \frac{(380)(0.16 \text{ A})(2.5 \times 10^{-3} \text{ m}^2)(0.12 \text{ T})}{(0.70)(0.010 \text{ m})} = \boxed{2.6 \text{ N}}$$

---

44. **REASONING AND SOLUTION** According to Equation 21.4, the maximum torque for a single turn is $\tau_{max} = IAB$. When the length L of the wire is used to make the square, each side of the square has a length $L/4$. The area of the square is $A_{square} = (L/4)^2$. For the rectangle, since two sides have a length d, while the other two sides have a length $2d$, it follows that $L = 6d$, or $d = L/6$. The area is $A_{rectangle} = 2d^2 = 2(L/6)^2$. Using Equation 21.4 for the square and the rectangle, we obtain

$$\frac{\tau_{square}}{\tau_{rectangle}} = \frac{IA_{square}B}{IA_{rectangle}B} = \frac{A_{square}}{A_{rectangle}} = \frac{(L/4)^2}{2(L/6)^2} = \boxed{1.13}$$

---

45. [SSM] **REASONING** The magnetic moment of the orbiting electron can be found from the expression *Magnetic moment* $= NIA$. For this situation, $N = 1$. Thus, we need to find the current and the area for the orbiting charge.

SOLUTION The current for the orbiting charge is, by definition (see Equation 20.1), $I = \Delta q/\Delta t$, where Δq is the amount of charge that passes a given point during a time interval Δt. Since the charge ($\Delta q = e$) passes by a given point once per revolution, we can find the current by dividing the total orbiting charge by the period T of revolution.

$$I = \frac{\Delta q}{T} = \frac{\Delta q}{2\pi r/v} = \frac{(1.6\times 10^{-19}\text{ C})(2.2\times 10^{6}\text{ m/s})}{2\pi(5.3\times 10^{-11}\text{ m})} = 1.06\times 10^{-3}\text{ A}$$

The area of the orbiting charge is
$$A = \pi r^2 = \pi(5.3\times 10^{-11}\text{ m})^2 = 8.82\times 10^{-21}\text{ m}^2$$

Therefore, the magnetic moment is

$$\text{Magnetic moment} = NIA = (1)(1.06\times 10^{-3}\text{ A})(8.82\times 10^{-21}\text{ m}^2) = \boxed{9.3\times 10^{-24}\text{ A}\cdot\text{m}^2}$$

---

46. **REASONING AND SOLUTION** The magnitude B of the magnetic field at a distance r from a long straight wire carrying a current I is $B = \mu_0 I/(2\pi r)$. Thus, the distance is

$$r = \frac{\mu_0 I}{2\pi B} = \frac{(4\pi\times 10^{-7}\text{ T}\cdot\text{m/A})(48\text{ A})}{2\pi(8.0\times 10^{-5}\text{ T})} = \boxed{0.12\text{ m}} \qquad (21.5)$$

---

47. **REASONING AND SOLUTION** The current associated with the lightning bolt is

$$I = \frac{\Delta q}{\Delta t} = \frac{15\text{ C}}{1.5\times 10^{-3}\text{ s}} = 1.0\times 10^{4}\text{ A}$$

The magnetic field near this current is given by

$$B = \frac{\mu_0 I}{2\pi r} = \frac{(4\pi\times 10^{-7}\text{ T}\cdot\text{m/A})(1.0\times 10^{4}\text{ A})}{2\pi(25\text{ m})} = \boxed{8.0\times 10^{-5}\text{ T}}$$

---

48. **REASONING AND SOLUTION** The magnetic field at the center of a current loop of radius R is given by $B = \mu_0 I/(2R)$, so that

$$R = \frac{\mu_0 I}{2B} = \frac{(4\pi\times 10^{-7}\text{ T}\cdot\text{m/A})(12\text{ A})}{2(1.8\times 10^{-4}\text{ T})} = \boxed{4.2\times 10^{-2}\text{ m}}$$

---

49. **SSM** **REASONING AND SOLUTION** The magnetic field inside a long solenoid is given by Equation 21.7:

$$B = \mu_0 nI = (4\pi\times 10^{-7}\text{ T}\cdot\text{m/A})\left(\frac{1400\text{ turns}}{0.65\text{ m}}\right)(4.7\text{ A}) = \boxed{1.3\times 10^{-2}\text{ T}}$$

50. **REASONING AND SOLUTION**

a. In Figure 21.28a the magnetic field that exists at the location of each wire points upward. Since the current in each wire is the same, the fields at the locations of the wires also have the same magnitudes. Therefore, a single external field that points ⟨downward⟩ will cancel the mutual repulsion of the wires, if this external field has a magnitude that equals that of the field produced by either wire.

b. Equation 21.5 gives the magnitude of the field produced by a long straight wire. The external field must have this magnitude:

$$B = \frac{\mu_0 I}{2\pi r} = \frac{(4\pi \times 10^{-7} \text{ T·m/A})(25 \text{ A})}{2\pi(0.016 \text{ m})} = \boxed{3.1 \times 10^{-4} \text{ T}}$$

---

51. **SSM** **REASONING** The two rods attract each other because they each carry a current I in the same direction. The bottom rod floats because it is in equilibrium. The two forces that act on the bottom rod are the downward force of gravity $m\mathbf{g}$ and the upward magnetic force of attraction to the upper rod. If the two rods are a distance s apart, the magnetic field generated by the top rod at the location of the bottom rod is (see Equation 21.5) $B = \mu_0 I/(2\pi s)$. According to Equation 21.3, the magnetic force exerted on the bottom rod is $F = ILB\sin\theta = \mu_0 I^2 L \sin\theta/(2\pi s)$, where θ is the angle between the magnetic field at the location of the bottom rod and the direction of the current in the bottom rod. Since the rods are parallel, the magnetic field is perpendicular to the direction of the current (RHR-2), and $\theta = 90.0°$, so that $\sin\theta = 1.0$.

SOLUTION Taking upward as the positive direction, the net force on the bottom rod is

$$\frac{\mu_0 I^2 L \sin\theta}{2\pi s} - mg = 0$$

Solving for I, we find

$$I = \sqrt{\frac{2\pi mgs}{\mu_0 L}} = \sqrt{\frac{2\pi(0.073 \text{ kg})(9.80 \text{ m/s}^2)(8.2 \times 10^{-3} \text{ m})}{(4\pi \times 10^{-7} \text{ T·m/A})(0.85 \text{ m})}} = \boxed{190 \text{ A}}$$

---

52. **REASONING** The magnitude of the magnetic field at the center of a circular loop of current is given by Equation 21.6 as $B = N\mu_0 I/(2R)$, where N is the number of turns, μ_0 is the permeability of free space, I is the current, and R is the radius of the loop. The field is perpendicular to the plane of the loop. Magnetic fields are vectors, and here we have two fields, each perpendicular to the plane of the loop producing it. Therefore, the two field vectors are perpendicular, and we must add them as vectors to get the net field. Since they are perpendicular, we can use the Pythagorean theorem to calculate the magnitude of the net field.

SOLUTION Using Equation 21.6 and the Pythagorean theorem, we find that the magnitude of the net magnetic field at the common center of the two loops is

$$B_{net} = \sqrt{\left(\frac{N\mu_0 I}{2R}\right)^2 + \left(\frac{N\mu_0 I}{2R}\right)^2} = \sqrt{2}\left(\frac{N\mu_0 I}{2R}\right)$$

$$= \frac{\sqrt{2}(1)(4\pi \times 10^{-7} \text{ T·m/A})(1.7 \text{ A})}{2(0.040 \text{ m})} = \boxed{3.8 \times 10^{-5} \text{ T}}$$

---

53. **REASONING AND SOLUTION** Let the current in the left-hand wire be labeled I_1 and that in the right-hand wire I_2.

   a. At point A: B_1 is *up* and B_2 is *down*, so we subtract them to get the net field. We have

   $$B_1 = \mu_0 I_1/(2\pi d_1) = \mu_0(8.0 \text{ A})/[2\pi(0.030 \text{ m})]$$
   $$B_2 = \mu_0 I_2/(2\pi d_2) = \mu_0(8.0 \text{ A})/[2\pi(0.150 \text{ m})]$$

   So the net field at point A is

   $$B_A = B_1 - B_2 = \boxed{4.3 \times 10^{-5} \text{ T}}$$

   b. At point B: B_1 and B_2 are both *down* so we add the two. We have

   $$B_1 = \mu_0(8.0 \text{ A})/[2\pi(0.060 \text{ m})]$$
   $$B_2 = \mu_0(8.0 \text{ A})/[2\pi(0.060 \text{ m})]$$

   So the net field at point B is

   $$B_B = B_1 + B_2 = \boxed{5.3 \times 10^{-5} \text{ T}}$$

---

54. **REASONING AND SOLUTION** The net force on the wire loop is a sum of the forces on each segment of the loop. The forces on the two segments perpendicular to the long straight wire cancel each other out. The net force on the loop is therefore the sum of the forces on the parallel segments (near and far). These are

$$F_n = \mu_0 I_1 I_2 L/(2\pi d_n) = \mu_0(12 \text{ A})(25 \text{ A})(0.50 \text{ m})/[2\pi(0.11 \text{ m})] = 2.7 \times 10^{-4} \text{ N}$$

$$F_f = \mu_0 I_1 I_2 L/(2\pi d_f) = \mu_0(12 \text{ A})(25 \text{ A})(0.50 \text{ m})/[2\pi(0.26 \text{ m})] = 1.2 \times 10^{-4} \text{ N}$$

Note: F_n is a force of attraction, while F_f is a repulsive one. The magnitude of the net force is, therefore,

$$F = F_n - F_f = 2.7 \times 10^{-4} \text{ N} - 1.2 \times 10^{-4} \text{ N} = \boxed{1.5 \times 10^{-4} \text{ N}}$$

---

55. **REASONING AND SOLUTION** The magnitude B_i of the magnetic field due to the inner coil is

$$B_i = \mu_0 I_i N_i / (2 r_i)$$

The magnetic field B_0 due to the outer coil must cancel that due to the inner coil. We have

$$B_0 = \mu_0 I_0 N_0 / (2 r_0)$$

and this must equal B_i. Therefore,

$$\mu_0 I_i N_i / (2 r_i) = \mu_0 I_0 N_0 / (2 r_0)$$

so that

$$I_0 = I_i \left(\frac{N_i}{N_0}\right)\left(\frac{r_0}{r_i}\right) = (6.0 \text{ A})\left(\frac{120 \text{ turns}}{150 \text{ turns}}\right)\left(\frac{0.017 \text{ m}}{0.012 \text{ m}}\right) = \boxed{6.8 \text{ A}}$$

In order for the two fields to be equal in magnitude but opposite in direction, the current in the outer coil must flow in the $\boxed{\text{opposite direction}}$ as the inner coil current.

---

56. **REASONING** Two wires that are parallel and carry current in the same direction exert attractive magnetic forces on one another, as Section 21.7 discusses. This attraction between the wires causes the spring to compress. When compressed, the spring exerts an elastic restoring force on each wire, as Section 10.1 discusses. For each wire, this restoring force acts to push the wires apart and balances the magnetic force, thus keeping the separation between the wires from decreasing to zero. Equation 10.2 (without the minus sign) gives the magnitude of the restoring force as F = kx, where k is the spring constant and x is the magnitude of the displacement of the spring from its unstrained length. By setting the magnitude of the magnetic force equal to the magnitude of the restoring force, we will be able to find the separation between the rods when the current is present.

SOLUTION According to Equation 21.3, the magnetic force has a magnitude of $F = ILB \sin \theta$, where I is the current, B is the magnitude of the magnetic field, L is the length of the wire, and θ is the angle of the wire with respect to the field. Using RHR-2 reveals that the magnetic field produced by either wire is perpendicular to the other wire, so that $\theta = 90°$ in Equation 21.3, which becomes $F = ILB$. According to Equation 21.5 the magnitude of the magnetic field produced by a long straight wire is $B = \mu_0 I/(2\pi r)$. Substituting this expression into Equation 21.3 gives the magnitude of the magnetic force as

820 MAGNETIC FORCES AND MAGNETIC FIELDS

$$F = IL\left(\frac{\mu_0 I}{2\pi r}\right) = \frac{\mu_0 I^2 L}{2\pi r}$$

Equating this expression to the magnitude of the restoring force from the spring gives

$$\frac{\mu_0 I^2 L}{2\pi r} = kx$$

Solving for the separation r, we find

$$r = \frac{\mu_0 I^2 L}{2\pi kx} = \frac{(4\pi \times 10^{-7} \text{ T}\cdot\text{m/A})(950 \text{ A})^2 (0.50 \text{ m})}{2\pi(150 \text{ N/m})(0.020 \text{ m})} = \boxed{0.030 \text{ m}}$$

---

57. **SSM** **REASONING** According to Equation 21.6 the magnetic field at the center of a circular, current-carrying loop of N turns and radius r is $B = N\mu_0 I/(2r)$. The number of turns N in the coil can be found by dividing the total length L of the wire by the circumference after it has been wound into a circle. The current in the wire can be found by using Ohm's law, $I = V/R$.

SOLUTION The number of turns in the wire is

$$N = \frac{L}{2\pi r}$$

The current in the wire is

$$I = \frac{V}{R} = \frac{12.0 \text{ V}}{(5.90 \times 10^{-3} \text{ }\Omega/\text{m}) L} = \frac{2.03 \times 10^3}{L} \text{ A}$$

Therefore, the magnetic field at the center of the coil is

$$B = N\left(\frac{\mu_0 I}{2r}\right) = \left(\frac{L}{2\pi r}\right)\left(\frac{\mu_0 I}{2r}\right) = \frac{\mu_0 LI}{4\pi r^2}$$

$$= \frac{(4\pi \times 10^{-7} \text{ T}\cdot\text{m/A}) L \left(\frac{2.03 \times 10^3}{L} \text{ A}\right)}{4\pi(0.140 \text{ m})^2} = \boxed{1.04 \times 10^{-2} \text{ T}}$$

58. **REASONING AND SOLUTION** The forces acting on each wire are the magnetic force F, the gravitational force mg, and the tension T in the strings. Each string makes an angle of 7.5° with respect to the vertical. From the drawing at the right we can relate the magnetic force to the gravitational force. Since the wire is in equilibrium, Newton's second law requires that $\Sigma F_x = 0$ and $\Sigma F_y = 0$. These equations become

$$\underbrace{-T \sin 7.5° + F = 0}_{\Sigma F_x} \quad \text{and} \quad \underbrace{T \cos 7.5° - mg = 0}_{\Sigma F_y}$$

Solving the first equation for T, and then substituting the result into the second equation gives (after some simplification)

$$\tan 7.5° = \frac{F}{mg} \quad (1)$$

The magnetic force F exerted on one wire by the other is $F = \dfrac{\mu_0 I^2 L}{2\pi d}$, where d is the distance between the wires [$d/2 = (1.2 \text{ m}) \sin 7.5°$, so that $d = 0.31$ m], I is the current (which is the same for each wire), and L is the length of each wire. Substituting this relation for F into Equation (1) and then solving for the current, gives

$$I = \sqrt{\left(\frac{m}{L}\right) g \tan 7.5° \left(\frac{2\pi d}{\mu_0}\right)}$$

$$= \sqrt{(0.050 \text{ kg/m})(9.80 \text{ m/s}^2) \tan 7.5° \left[\frac{2\pi(0.31 \text{ m})}{\mu_0}\right]} = \boxed{320 \text{ A}}$$

---

59. **REASONING AND SOLUTION** The currents in wires 1 and 2 produce the magnetic fields B_1 and B_2 at the empty corner, as shown in the drawing. The directions of these fields can be obtained using RHR-2. Since there are equal currents in wires 1 and 2 and since these wires are each the same distance r from the empty corner, B_1 and B_2 have equal magnitudes. Using Equation 21.5, we can write the field magnitude as $B_1 = B_2 = \mu_0 I/(2\pi r)$. Since B_1 and B_2 are perpendicular, it follows from the Pythagorean theorem that they combine to produce a net magnetic field that has the direction shown in the following drawing and has a magnitude B_{1+2} given by

$$B_{1+2} = \sqrt{B_1^2 + B_2^2} = \sqrt{\left(\frac{\mu_0 I}{2\pi r}\right)^2 + \left(\frac{\mu_0 I}{2\pi r}\right)^2} = \frac{\sqrt{2}\,\mu_0 I}{2\pi r}$$

822 **MAGNETIC FORCES AND MAGNETIC FIELDS**

The current in wire 3 produces a field B_3 at the empty corner. Since B_3 and B_{1+2} combine to give a zero net field, B_3 must have a direction opposite to that of B_{1+2}. Thus, B_3 must point upward and to the left, and RHR-2 indicates that

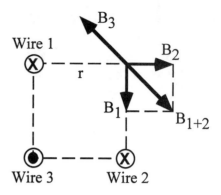

| the current in wire 3 must be directed out of the plane of the paper |

Moreover, the magnitudes of B_3 and B_{1+2} must be the same. Recognizing that wire 3 is a distance of $d = \sqrt{r^2 + r^2} = \sqrt{2}\,r$ from the empty corner, we have

$$B_3 = B_{1+2} \quad \text{or} \quad \frac{\mu_0 I_3}{2\pi(\sqrt{2}\,r)} = \frac{\sqrt{2}\,\mu_0 I}{2\pi r} \quad \text{so that} \quad \boxed{\frac{I_3}{I} = 2}$$

---

60. **REASONING AND SOLUTION** Ampere's law in the form of Equation 21.8 indicates that $\Sigma B_\parallel \Delta \ell = \mu_0 I$. Since the magnetic field is everywhere perpendicular to the plane of the paper, it is everywhere perpendicular to the circular path and has no component $B_\parallel$ that is parallel to the circular path. Therefore, Ampere's law reduces to $\Sigma B_\parallel \Delta \ell = 0 = \mu_0 I$, so that the net current passing through the circular surface is zero .

---

61. **SSM** **REASONING** Since the two wires are next to each other, the net magnetic field is everywhere parallel to $\Delta \ell$ in Figure 21.40. Moreover, the net magnetic field **B** has the same magnitude B at each point along the circular path, because each point is at the same distance from the wires. Thus, in Ampere's law (Equation 21.8), $B_\parallel = B$, $I = I_1 + I_2$, and we have

$$\Sigma B_\parallel \Delta \ell = B(\Sigma \Delta \ell) = \mu_0 (I_1 + I_2)$$

But $\Sigma \Delta \ell$ is just the circumference ($2\pi r$) of the circle, so Ampere's law becomes

$$B(2\pi r) = \mu_0 (I_1 + I_2)$$

This expression can be solved for B.

SOLUTION
a. When the currents are in the same direction, we find that

$$B = \frac{\mu_0(I_1+I_2)}{2\pi r} = \frac{(4\pi \times 10^{-7}\text{ T·m/A})(28\text{ A}+12\text{ A})}{2\pi(0.72\text{ m})} = \boxed{1.1 \times 10^{-5}\text{ T}}$$

b. When the currents have opposite directions, a similar calculation shows that

$$B = \frac{\mu_0(I_1-I_2)}{2\pi r} = \frac{(4\pi \times 10^{-7}\text{ T·m/A})(28\text{ A}-12\text{ A})}{2\pi(0.72\text{ m})} = \boxed{4.4 \times 10^{-6}\text{ T}}$$

---

62. **REASONING AND SOLUTION**

a. An end-on view of the copper cylinder is a circle, as the drawing shows. The dots around the circle represent the current coming out of the paper toward you. The larger dashed circle of radius r is the closed path used in Ampere's law and is centered on the axis of the cylinder. Equation 21.8 gives Ampere's law as $\Sigma B_\parallel \Delta \ell = \mu_0 I$. Because of the symmetry of the arrangement in the drawing, we have $B_\parallel = B$ for all $\Delta \ell$ on the circular path, so that Ampere's law becomes

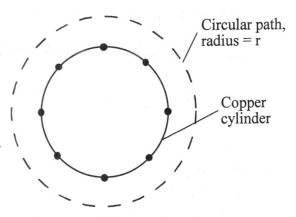

$$\Sigma B_\parallel \Delta \ell = B(\Sigma \Delta \ell) = \mu_0 I$$

In this result, I is the net current through the circular surface bounded by the dashed path. In other words, it is the current I in the copper tube. Furthermore, $\Sigma \Delta \ell$ is the circumference of the circle, so we find that

$$B(\Sigma \Delta \ell) = B(2\pi r) = \mu_0 I \quad \text{or} \quad \boxed{B = \frac{\mu_0 I}{2\pi r}}$$

b. The setup here is similar to that in part a, except that the smaller dashed circle of radius r is now the closed path used in Ampere's law (see the drawing at the right). With this change, the derivation then proceeds exactly as in part a. Now, however, there is no current through the circular surface bounded by the dashed path, because all of the current is outside the path. Therefore, $I = 0$, and

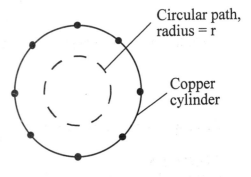

$$B = \frac{\mu_0 I}{2\pi r} = \boxed{0\text{ T}}.$$

63. **REASONING AND SOLUTION** The drawing shows an end-on view of the solid cylinder. The dots represent the current in the cylinder coming out of the paper toward you. The dashed circle of radius r is the closed path used in Ampere's law and is centered on the axis of the cylinder. Equation 21.8 gives Ampere's law as $\Sigma B_{\parallel} \Delta \ell = \mu_0 I$. Because of the symmetry of the arrangement in the drawing, we have $B_{\parallel} = B$ for all $\Delta \ell$ on the circular path, so that Ampere's law becomes

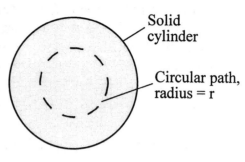

$$\Sigma B_{\parallel} \Delta \ell = B(\Sigma \Delta \ell) = \mu_0 I$$

In this result, $\Sigma \Delta \ell = 2\pi r$, the circumference of the circle. The current I is the part of the total current that comes through the area πr^2 bounded by the dashed path. We can calculate this current by using the current per unit cross-sectional area of the solid cylinder. This current per unit area is called the current density. The current I is the current density times the area πr^2:

$$I = \underbrace{\left(\frac{I_0}{\pi R^2}\right)}_{\text{current density}} (\pi r^2) = \frac{I_0 r^2}{R^2}$$

Thus, Ampere's law becomes

$$B(\Sigma \Delta \ell) = \mu_0 I \quad \text{or} \quad B(2\pi r) = \mu_0 \left(\frac{I_0 r^2}{R^2}\right) \quad \text{or} \quad \boxed{B = \frac{\mu_0 I_0 r}{2\pi R^2}}$$

---

64. **REASONING AND SOLUTION** The speed of the electron can be determined using $eV = (1/2)mv^2$ so that

$$v = \sqrt{\frac{2eV}{m}} = \sqrt{\frac{2(1.60 \times 10^{-19} \text{ C})(19\,000 \text{ V})}{9.11 \times 10^{-31} \text{ kg}}} = 8.17 \times 10^7 \text{ m/s}$$

The magnetic force is given by

$$F = qvB \sin\theta = (1.60 \times 10^{-19} \text{ C})(8.17 \times 10^7 \text{ m/s})(0.28 \text{ T}) \sin 90.0° = \boxed{3.7 \times 10^{-12} \text{ N}}$$

65. **SSM** *REASONING* The coil carries a current and experiences a torque when it is placed in an external magnetic field. Thus, when the coil is placed in the magnetic field due to the solenoid, it will experience a torque given by Equation 21.4: $\tau = NIAB\sin\phi$, where N is the number of turns in the coil, A is the area of the coil, B is the magnetic field inside the solenoid, and ϕ is the angle between the normal to the plane of the coil and the magnetic field. The magnetic field in the solenoid can be found from Equation 21.7: $B = \mu_0 nI$, where n is the number of turns per unit length of the solenoid and I is the current.

SOLUTION The magnetic field inside the solenoid is

$$B = \mu_0 nI = (4\pi \times 10^{-7} \text{ T} \cdot \text{m/A})(1400 \text{ turns/m})(3.5 \text{ A}) = 6.2 \times 10^{-3} \text{ T}$$

The torque exerted on the coil is

$$\tau = NIAB\sin\phi = (50)(0.50 \text{ A})(1.2 \times 10^{-3} \text{ m}^2)(6.2 \times 10^{-3} \text{ T})(\sin 90.0°) = \boxed{1.9 \times 10^{-4} \text{ N} \cdot \text{m}}$$

66. *REASONING AND SOLUTION* Use $B = F/(qv \sin\theta)$. For both the electron and the proton the magnitude of B is the same,

$$B = (8.0 \times 10^{-14} \text{ N})/[(1.60 \times 10^{-19} \text{ C})(4.5 \times 10^6 \text{ m/s}) \sin 90.0°] = \boxed{0.11 \text{ T}}$$

a. For a proton, the direction of the magnetic field would be $\boxed{\text{upward}}$, away from the surface of the earth, perpendicular to both v and F.

b. For an electron, which has a negative charge, the magnetic field would be directed $\boxed{\text{downward}}$, toward the surface of the earth, perpendicular to both v and F.

67. *REASONING AND SOLUTION* The torque is given by

$$\tau = (\text{Magnetic Moment}) B \sin\phi = (1.4 \times 10^{-26} \text{ A} \cdot \text{m}^2)(0.65 \text{ T}) \sin 64° = \boxed{8.2 \times 10^{-27} \text{ N} \cdot \text{m}}$$

68. *REASONING AND SOLUTION* We require the magnetic field of the wire to be equal in magnitude but opposite in direction to the external magnetic field. We know that $B = \mu_0 I/(2\pi r)$, so we can solve for r to get

$$r = \frac{\mu_0 I}{2\pi B} = \frac{(4\pi \times 10^{-7} \text{ T} \cdot \text{m/A})(305 \text{ A})}{2\pi(7.00 \times 10^{-3} \text{ T})} = \boxed{8.71 \times 10^{-3} \text{ m}}$$

826   MAGNETIC FORCES AND MAGNETIC FIELDS

69. **SSM** *REASONING AND SOLUTION* According to Equation 21.1, $B = \dfrac{F}{q_0 (v \sin\theta)}$.

Solving this for the angle θ, we find

$$\theta = \sin^{-1}\left(\dfrac{F}{q_0 vB}\right) = \sin^{-1}\left[\dfrac{2.30 \times 10^{-7}\text{ N}}{(1.70 \times 10^{-5}\text{ C})(2.80 \times 10^{2}\text{ m/s})(5.00 \times 10^{-5}\text{ T})}\right] = \boxed{75.1° \text{ and } 105°}$$

---

70. *REASONING AND SOLUTION* The direction of the magnetic force on the current is given by Right-Hand Rule No. 1. Using this rule, it can be shown that the current is directed $\boxed{\text{east-to-west}}$. The magnitude of the force is given by $F = ILB \sin\theta$. Solving for I we get

$$I = \dfrac{F}{LB \sin\theta} = \dfrac{0.058\text{ N}}{(46\text{ m})(3.2 \times 10^{-5}\text{ T}) \sin 90.0°} = \boxed{39\text{ A}}$$

---

71. *REASONING AND SOLUTION* The radius of the circular path of the ion in a mass spectrometer can be obtained using the relation $r = \sqrt{\dfrac{2mV}{eB^2}}$ (see Section 21.4). For the deuteron, we have

$$r = \sqrt{\dfrac{2mV}{eB^2}} = \sqrt{\dfrac{2(3.34 \times 10^{-27}\text{ kg})(2.00 \times 10^{3}\text{ V})}{(1.60 \times 10^{-19}\text{ C})(0.600\text{ T})^2}} = \boxed{0.0152\text{ m}}$$

---

72. *REASONING AND SOLUTION* The magnetic field due to the circular loop alone is $B_1 = \dfrac{\mu_0 I_1}{2R}$. The field due to the straight wire is $B_2 = \dfrac{\mu_0 I_2}{2\pi H}$. These two fields cancel at the center of the loop, so that their magnitudes must be equal:

$$\dfrac{\mu_0 I_1}{2R} = \dfrac{\mu_0 (6.6 I_1)}{2\pi H} \quad\text{or}\quad \boxed{H = 2.1\,R}$$

---

73. **SSM** **WWW** *REASONING* The particle travels in a semicircular path of radius r, where r is given by Equation 21.2 ($r = mv/qB$). The time spent by the particle in the magnetic field is given by $t = s/v$, where s is the distance traveled by the particle and v is its speed. The distance s is equal to one-half the circumference of a circle ($s = \pi r$).

SOLUTION We find that

$$t = \frac{s}{v} = \frac{\pi r}{v} = \frac{\pi}{v}\left(\frac{mv}{qB}\right) = \frac{\pi m}{qB} = \frac{\pi(6.0\times 10^{-8}\text{ kg})}{(7.2\times 10^{-6}\text{ C})(3.0\text{ T})} = \boxed{8.7\times 10^{-3}\text{ s}}$$

---

74. **REASONING AND SOLUTION** Using Right-Hand Rule No. 2, the magnetic fields due to wires A and B will cancel out on the side of wire A away from wire B. If x is the distance (in meters) from this point to wire A, the distance to wire B is $(1.0 + x)$ meters. At this point then,

$$B_A = B_B \quad \text{so that} \quad \mu_0 I_A/(2\pi x) = \mu_0 I_B/[2\pi(1.0 + x)]$$

Using $I_A = I_B/3$ and simplifying, we obtain

$$(I_B/3)/x = I_B/(1.0 + x), \text{ which reduces to } 1.0 + x = 3x$$

Solving this equation yields $x = \boxed{0.50\text{ m}}$, on the side of wire A that is away from wire B.

---

75. **REASONING** The positive plate has a charge q and is moving downward with a speed v at right angles to a magnetic field of magnitude B. The magnitude F of the magnetic force exerted on the positive plate is $F = qvB \sin 90.0°$. The charge on the positive plate is related to the magnitude E of the electric field that exists between the plates by (see Equation 18.4) $q = \varepsilon_0 AE$, where A is the area of the positive plate. Substituting this expression for q into the expression for the magnetic force gives the answer in terms of known quantities.

SOLUTION

$$F = (\varepsilon_0 AE)vB$$

$$= [8.85\times 10^{-12}\text{ C}^2/(\text{N}\cdot\text{m}^2)](7.5\times 10^{-4}\text{ m}^2)(170\text{ N}/\text{C})(32\text{ m}/\text{s})(3.6\text{ T}) = \boxed{1.3\times 10^{-10}\text{ N}}$$

An application of Right-Hand Rule No. 1 shows that the magnetic force is perpendicular to the plane of the page and $\boxed{\text{directed out of the page}}$, toward the reader.

---

76. **REASONING AND SOLUTION** Find the magnitude of the magnetic field due to each wire, with the left-hand wire being #1 and the right-hand wire being #2. We have, for the magnitudes of B_1 and B_2

$$B_1 = B_2 = \mu_0 I/(2\pi r) = \mu_0(85.0\text{ A})/[2\pi(0.150\text{ m})] = 1.13\times 10^{-4}\text{ T}$$

The direction of B_1 is 30.0° below the horizontal to the right. The direction of B_2 is 30.0° below the horizontal to the left. The components of B_1 and B_2 are

$$B_{1x} = +B_1 \cos 30.0° = +9.79 \times 10^{-5} \text{ T}$$
$$B_{1y} = -B_1 \sin 30.0° = -5.65 \times 10^{-5} \text{ T}$$
$$B_{2x} = -B_2 \cos 30.0° = -9.79 \times 10^{-5} \text{ T}$$
$$B_{2y} = -B_2 \sin 30.0° = -5.65 \times 10^{-5} \text{ T}$$

The components of the magnetic field at P, then, are

$$B_x = B_{1x} + B_{2x} = 0, \quad \text{and} \quad B_y = B_{1y} + B_{2y} = -2(5.65 \times 10^{-5} \text{ T}) = -1.13 \times 10^{-4} \text{ T}$$

The net field is

$$B = \sqrt{B_x^2 + B_y^2} = \boxed{1.13 \times 10^{-4} \text{ T, down (toward the bottom of the page)}}$$

---

77. **SSM REASONING** The magnetic moment of the rotating charge can be found from the expression *Magnetic moment* $= NIA$. For this situation, $N = 1$. Thus, we need to find the current and the area for the rotating charge. This can be done by resorting to first principles.

SOLUTION The current for the rotating charge is, by definition (see Equation 20.1), $I = \Delta q / \Delta t$, where Δq is the amount of charge that passes by a given point during a time interval Δt. Since the charge passes by once per revolution, we can find the current by dividing the total rotating charge by the period T of revolution.

$$I = \frac{\Delta q}{T} = \frac{\Delta q}{2\pi/\omega} = \frac{\omega \Delta q}{2\pi} = \frac{(150 \text{ rad/s})(4.0 \times 10^{-6} \text{ C})}{2\pi} = 9.5 \times 10^{-5} \text{ A}$$

The area of the rotating charge is $A = \pi r^2 = \pi (0.20 \text{ m})^2 = 0.13 \text{ m}^2$

Therefore, the magnetic moment is

$$\text{Magnetic moment} = NIA = (1)(9.5 \times 10^{-5} \text{ A})(0.13 \text{ m}^2) = \boxed{1.2 \times 10^{-5} \text{ A} \cdot \text{m}^2}$$

---

78. **CONCEPT QUESTIONS** a. A moving charge experiences no magnetic force when its velocity points in the direction of the magnetic field or in the direction opposite to the magnetic field. Thus, the magnetic field must point either in the direction of the $+x$ axis or in the direction of the $-x$ axis.

b. If a moving charge experiences the maximum possible magnetic force when moving in a magnetic field, then the velocity must be perpendicular to the field. In other words, the angle θ that the charge's velocity makes with respect to the magnetic field is $\theta = 90°$.

SOLUTION The magnitude of the magnetic field can be determined using Equation 21.1 as follows:

$$B = \frac{F}{qv \sin\theta} = \frac{0.48 \text{ N}}{(8.2 \times 10^{-6} \text{ C})(5.0 \times 10^5 \text{ m/s})\sin 90°} = \boxed{0.12 \text{ T}}$$

In this calculation we use $\theta = 90°$, because the 0.48-N force is the maximum possible force. Since the particle experiences no magnetic force when it moves along the +x axis, we can conclude that the magnetic field points

$\boxed{\text{either in the direction of the +x axis or in the direction of the -x axis}}$.

---

79. **CONCEPT QUESTIONS** a. Figure 21.12 shows an example of the circular path followed by the proton (positive charge). If the charge in that drawing were an electron (negative charge), the force on it at point 1 would be downward rather than upward. As a result, the electron would move downward in a clockwise direction around its circular path. Thus, the electron does not follow the exact same circular path as the proton.

b. To make the electron follow the exact same circular path as the proton, it is necessary to make the force on it at point 1 in Figure 21.12 point upward just as it does for the proton. Therefore, the field direction in that figure must be reversed for the electron. Then, RHR-1 would predict an upward force for the electron.

c. Equation 21.2 gives the radius of the circular path as $r = mv/(qB)$. We wish the radius to be the same for both the proton and the electron. The speed v and the charge magnitude q are the same for the proton and the electron, but the mass of the electron is 9.11×10^{-31} kg, while the mass of the proton is 1.67×10^{-27} kg. Therefore, to offset the effect of the smaller electron mass m in Equation 21.2, the magnitude B of the field must be reduced for the electron.

SOLUTION Applying Equation 21.2 to the proton and the electron, both of which carry charges of the same magnitude e, we obtain

$$\underbrace{r = \frac{m_p v}{eB_p}}_{\text{Proton}} \quad \text{and} \quad \underbrace{r = \frac{m_e v}{eB_e}}_{\text{Electron}}$$

Dividing the proton-equation by the electron-equation gives

830   MAGNETIC FORCES AND MAGNETIC FIELDS

$$\frac{r}{r} = \frac{\frac{m_p v}{eB_p}}{\frac{m_e v}{eB_e}} \quad \text{or} \quad 1 = \frac{m_p B_e}{m_e B_p}$$

Solving for B_e, we obtain

$$B_e = \frac{m_e B_p}{m_p} = \frac{(9.11 \times 10^{-31} \text{ kg})(0.50 \text{ T})}{1.67 \times 10^{-27} \text{ kg}} = \boxed{2.7 \times 10^{-4} \text{ T}}$$

---

80. **CONCEPT QUESTIONS** a. According to the principle of conservation of energy, the electric potential energy lost as the particles accelerate is converted into kinetic energy. Equation 19.4 indicates that the electric potential energy lost is qV. Since the charge q and the electric potential difference V are the same for each particle, each loses the same amount of potential energy. Energy conservation, then, dictates that each gains the same amount of kinetic energy. Since each particle starts from rest, each enters the field region with the same amount of kinetic energy.

b. Kinetic energy is $\frac{1}{2}mv^2$. Since each particle has the same kinetic energy, and since particle 1 has the smaller mass, it must have the greater speed v.

SOLUTION According to Equation 21.2, $r = mv/(qB)$. To determine the speed v with which each particle enters the field, we use Equation 19.4 and the energy-conservation principle as follows:

$$\underbrace{qV}_{\text{Electric potential energy lost}} = \underbrace{\tfrac{1}{2}mv^2}_{\text{Kinetic energy gained}} \quad \text{or} \quad v = \sqrt{\frac{2qV}{m}}$$

Substituting this result into Equation 21.2 gives the radius:

$$r = \frac{mv}{qB} = \frac{m}{qB}\sqrt{\frac{2qV}{m}} = \frac{1}{B}\sqrt{\frac{2mV}{q}}$$

Applying this result to each particle, we obtain

$$\underbrace{r_1 = \frac{1}{B}\sqrt{\frac{2m_1 V}{q}}}_{\text{Particle 1}} \quad \text{and} \quad \underbrace{r_2 = \frac{1}{B}\sqrt{\frac{2m_2 V}{q}}}_{\text{Particle 2}}$$

Dividing the equation for particle 2 by the equation for particle 1 gives

$$\frac{r_2}{r_1} = \frac{\frac{1}{B}\sqrt{\frac{2m_2 V}{q}}}{\frac{1}{B}\sqrt{\frac{2m_1 V}{q}}} = \sqrt{\frac{m_2}{m_1}}$$

$$r_2 = r_1 \sqrt{\frac{m_2}{m_1}} = (12 \text{ cm})\sqrt{\frac{5.9 \times 10^{-8} \text{ kg}}{2.3 \times 10^{-8} \text{ kg}}} = \boxed{19 \text{ cm}}$$

---

81. **CONCEPT QUESTION** We begin by noting that segments AB and BC are both perpendicular to the magnetic field. Therefore, they experience magnetic forces. However, segment CD is parallel to the field. As a result no magnetic force acts on it. According to Equation 21.3, the magnitude of the magnetic force on a current I is $F = ILB \sin \theta$, where L is the length of the wire segment and θ is the angle that the current makes with respect to the field. For both segments AB and BC the value of the current is the same and the value of θ is 90°. The length of segment AB is greater, however. Because of its greater length, segment AB experiences the greater force. Thus, segment AB experiences the greatest force, followed by segment BC, and finally segment CD.

SOLUTION Using Equation 21.3, we find that the magnitudes of the forces acting on the segments are

Segment AB $F = ILB \sin \theta = (2.8 \text{ A})(1.1 \text{ m})(0.26 \text{ T})\sin 90° = \boxed{0.80 \text{ N}}$

Segment BC $F = ILB \sin \theta = (2.8 \text{ A})(0.55 \text{ m})(0.26 \text{ T})\sin 90° = \boxed{0.40 \text{ N}}$

Segment CD $F = ILB \sin \theta = (2.8 \text{ A})(0.55 \text{ m})(0.26 \text{ T})\sin 0° = \boxed{0 \text{ N}}$

---

82. **CONCEPT QUESTION** When the wire is used to make a single-turn square coil, each side of the square has a length of $\frac{1}{4}L$. When the wire is used to make a two-turn coil, each side of the square has a length of $\frac{1}{8}L$. The drawing shows these two options and indicates that the total effective area of NA is greater for the single-turn option. Hence, more torque is obtained by using the single-turn option.

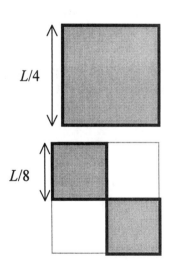

SOLUTION According to Equation 21.4 the maximum torque experienced by the coil is $\tau_{\max} = NIAB$, where N is the number of turns, I is the current, A is the area of each turn, and B is the magnitude of the magnetic field. Applying this expression to each option gives

Single - turn $\tau_{max} = NIAB = NI\left(\frac{1}{4}L\right)^2 B$

$= (1)(1.7 \text{ A})\left[\frac{1}{4}(1.00 \text{ m})\right]^2 (0.34 \text{ T}) = \boxed{0.036 \text{ N}\cdot\text{m}}$

Two - turn $\tau_{max} = NIAB = NI\left(\frac{1}{8}L\right)^2 B$

$= (2)(1.7 \text{ A})\left[\frac{1}{8}(1.00 \text{ m})\right]^2 (0.34 \text{ T}) = \boxed{0.018 \text{ N}\cdot\text{m}}$

---

83. **CONCEPT QUESTIONS** a. Using RHR-2, we can see that at point A the magnetic field due to the horizontal current points perpendicularly out of the plane of the paper. Similarly, the magnetic field due to the vertical current points perpendicularly into the plane of the paper. Point A is closer to the horizontal wire, so that the effect of the horizontal current dominates and the net field is directed out of the plane of the paper.

b. Using RHR-2, we can see that at point B the magnetic field due to the horizontal current points perpendicularly into the plane of the paper. Similarly, the magnetic field due to the vertical current points perpendicularly out of the plane of the paper. Point B is closer to the horizontal wire, so that the effect of the horizontal current dominates and the net field is directed into the plane of the paper.

c. Points A and B are the same distance of 0.20 m from the horizontal wire. They are also the same distance of 0.40 m from the vertical wire. Therefore, the magnitude of the field contribution from the horizontal current is the same at both points, although the directions of the field contributions are opposite. Likewise, the magnitude of the field contribution from the vertical current is the same at both points, although the directions of the field contributions are opposite. At either point the magnitude of the net field is the magnitude of the difference between the two contributions, and this is the same at points A and B.

SOLUTION According to Equation 21.5, the magnitude of the magnetic field from the current in a long straight wire is $B = \mu_0 I/(2\pi r)$, where μ_0 is the permeability of free space, I is the current, and r is the distance from the wire. Applying this equation to each wire at each point, we see that the magnitude of the net field B_{net} is

Point A $\quad B_{net} = \underbrace{\dfrac{\mu_0 I}{2\pi r_{A,H}}}_{\text{Horizontal wire}} - \underbrace{\dfrac{\mu_0 I}{2\pi r_{A,V}}}_{\text{Vertical wire}} = \dfrac{\mu_0 I}{2\pi}\left(\dfrac{1}{r_{A,H}} - \dfrac{1}{r_{A,V}}\right)$

$\quad\quad\quad\quad = \dfrac{(4\pi \times 10^{-7}\ \text{T·m/A})(5.6\ \text{A})}{2\pi}\left(\dfrac{1}{0.20\ \text{m}} - \dfrac{1}{0.40\ \text{m}}\right) = \boxed{2.8 \times 10^{-6}\ \text{T}}$

Point B $\quad B_{net} = \underbrace{\dfrac{\mu_0 I}{2\pi r_{B,H}}}_{\text{Horizontal wire}} - \underbrace{\dfrac{\mu_0 I}{2\pi r_{B,V}}}_{\text{Vertical wire}} = \dfrac{\mu_0 I}{2\pi}\left(\dfrac{1}{r_{B,H}} - \dfrac{1}{r_{B,V}}\right)$

$\quad\quad\quad\quad = \dfrac{(4\pi \times 10^{-7}\ \text{T·m/A})(5.6\ \text{A})}{2\pi}\left(\dfrac{1}{0.20\ \text{m}} - \dfrac{1}{0.40\ \text{m}}\right) = \boxed{2.8 \times 10^{-6}\ \text{T}}$

CHAPTER 22 | ELECTROMAGNETIC INDUCTION

PROBLEMS

1. **SSM REASONING AND SOLUTION** According to Equation 22.1, the motional emf when the velocity **v**, the magnetic field **B**, and the length **L** are mutually perpendicular is given by Emf = vBL. Therefore in order for a spark to jump across the gap, the rod would have to be moving in the magnetic field with speed

$$v = \frac{\text{Emf}}{BL} = \frac{940 \text{ V}}{(4.8 \text{ T})(1.3 \text{ m})} = \boxed{150 \text{ m/s}}$$

2. **REASONING AND SOLUTION**

   a. The right-hand rule shows that positive charge would be forced to the $\boxed{\text{driver's side}}$.

   b. According to Equation 22.1, the width of the car is

$$L = \frac{\text{Emf}}{vB} = \frac{2.4 \times 10^{-3} \text{ V}}{(25 \text{ m/s})(4.8 \times 10^{-5} \text{ T})} = \boxed{2.0 \text{ m}}$$

3. **REASONING AND SOLUTION** Using Equation 22.1, we find

$$\text{Emf} = vBL = (220 \text{ m/s})(5.0 \times 10^{-6} \text{ T})(59 \text{ m}) = \boxed{0.065 \text{ V}}$$

4. **REASONING AND SOLUTION** The motional emf generated by a conductor moving perpendicular to a magnetic field is given by Equation 22.1 as Emf = vBL, where v and L are the speed and length, respectively, of the conductor, and B is the magnitude of the magnetic field. The emf would have been

$$\text{Emf} = vBL = (7.6 \times 10^{3} \text{ m/s})(5.1 \times 10^{-5} \text{ T})(2.0 \times 10^{4} \text{ m}) = \boxed{7800 \text{ V}}$$

5. **WWW** **REASONING AND SOLUTION** For each of the three rods in the drawing, we have the following:

Rod A: The motional emf is $\boxed{\text{zero}}$, because the velocity of the rod is parallel to the direction of the magnetic field, and the charges do not experience a magnetic force.

Rod B: The motional emf is, according to Equation 22.1,

$$\text{Emf} = vBL = (2.7 \text{ m/s})(0.45 \text{ T})(1.3 \text{ m}) = \boxed{1.6 \text{ V}}$$

The positive end of Rod 2 is $\boxed{\text{end 2}}$.

Rod C: The motional emf is $\boxed{\text{zero}}$, because the magnetic force F on each charge is directed perpendicular to the length of the rod. For the ends of the rod to become charged, the magnetic force must be directed parallel to the length of the rod.

---

6. **REASONING AND SOLUTION** Ohm's law requires $I = \text{Emf}/R$, where $\text{Emf} = vBL$ and $R = \rho L/A$. Hence,

$$I = \frac{BvA}{\rho} = \frac{(0.050 \text{ T})(2.0 \text{ m/s})(3.1 \times 10^{-6} \text{ m}^2)}{9.7 \times 10^{-8} \text{ }\Omega\cdot\text{m}} = \boxed{3.2 \text{ A}}$$

---

7. **SSM** **REASONING** The moving rod acts as a battery does in supplying an emf or voltage in the circuit. The speed of the rod is related to the emf by Equation 22.1, $\text{Emf} = vBL$. From Equation 20.6c, we have $P = (\text{Emf})^2/R$, where P is the power consumed by the circuit and R is the resistance in the circuit.

SOLUTION
a. Solving Equation 20.6c for the emf, we have

$$\text{Emf} = \sqrt{PR} = \sqrt{(15 \text{ W})(6.0 \text{ }\Omega)} = 9.5 \text{ V}$$

The speed of the rod is

$$v = \frac{\text{Emf}}{LB} = \frac{9.5 \text{ V}}{(2.4 \text{ T})(1.2 \text{ m})} = \boxed{3.3 \text{ m/s}}$$

836 ELECTROMAGNETIC INDUCTION

b. The force exerted on the rod by the magnetic field is given by Equation 21.3 with $\theta = 90°$, and $I = \text{Emf}/R$. Therefore,

$$F = ILB(\sin 90°) = \left(\frac{\text{Emf}}{R}\right)LB = \left(\frac{9.5 \text{ V}}{6.0 \text{ }\Omega}\right)(1.2 \text{ m})(2.4 \text{ T}) = \boxed{4.6 \text{ N}}$$

---

8. **REASONING** Once the switch is closed, there is a current in the rod. The magnetic field applies a force to this current and accelerates the rod to the right. As the rod begins to move, however, a motional emf appears between the ends of the rod. This motional emf depends on the speed of the rod and increases as the speed increases. Equally important is the fact that the motional emf opposes the emf of the battery. The net emf causing the current in the rod is the algebraic sum of the two emf contributions. Thus, as the speed of the rod increases and the motional emf increases, the net emf decreases. As the net emf decreases, the current in the rod decreases and so does the force that field applies to the current. Eventually, the speed reaches the point when the motional emf has the same magnitude as the battery emf, and the net emf becomes zero. At this point, there is no longer a net force acting on the rod and the speed remains constant from this point onward, according to Newton's second law. This maximum speed can be determined by using Equation 22.1 for the motional emf, with a value of the motional emf that equals the battery emf.

SOLUTION Using Equation 22.1 (Emf = vBL) and a value of 3.0 V for the emf, we find that the maximum speed of the rod is

$$v = \frac{\text{Emf}}{BL} = \frac{3.0 \text{ V}}{(0.60 \text{ T})(0.20 \text{ m})} = \boxed{25 \text{ m/s}}$$

---

9. **REASONING AND SOLUTION**
a. Newton's second law gives the magnetic retarding force to be

$$F = mg = IBL$$

Now the current, I, is

$$I = (\text{Emf})/R = vBL/R$$

so

$$m = \frac{v(BL)^2}{Rg} = \frac{(4.0 \text{ m/s})(0.50 \text{ T})^2(1.3 \text{ m})^2}{(0.75 \text{ }\Omega)(9.80 \text{ m/s}^2)} = \boxed{0.23 \text{ kg}}$$

b. The change in height in a time Δt is $\Delta h = -v\Delta t$. The change in gravitational potential energy is

$$\Delta PE = mg\Delta h = -mgv\Delta t = -(0.23 \text{ kg})(9.80 \text{ m/s}^2)(4.0 \text{ m/s})(0.20 \text{ s}) = \boxed{-1.8 \text{ J}}$$

c. The energy dissipated in the resistor is the amount by which the gravitational potential energy decreases or $\boxed{1.8 \text{ J}}$.

---

10. **REASONING AND SOLUTION**
    a. The magnetic flux is $\Phi = BA \cos \phi$. Since $\phi = 0.0°$, we have

    $$\Phi = (0.35 \text{ T})(0.0160 \text{ m}^2) \cos 0.0° = \boxed{5.6 \times 10^{-3} \text{ Wb}}$$

    b. Since $\phi = 90°$, we have

    $$\Phi = (0.35 \text{ T})(0.0160 \text{ m}^2) \cos 90° = \boxed{0 \text{ Wb}}$$

---

11. **SSM** **REASONING** The general expression for the magnetic flux through an area A is given by Equation 22.2: $\Phi = BA \cos\phi$ where B is the magnitude of the magnetic field and ϕ is the angle of inclination of the magnetic field **B** with respect to the normal to the area.

    The magnetic flux through the door is a maximum when the magnetic field lines are perpendicular to the door and $\phi_1 = 0.0°$ so that $\Phi_1 = \Phi_{\max} = BA(\cos 0.0°) = BA$.

    SOLUTION When the door rotates through an angle ϕ_2, the magnetic flux that passes through the door decreases from its maximum value to one-third of its maximum value. Therefore $\Phi_2 = \frac{1}{3}\Phi_{\max}$, and we have

    $$\Phi_2 = BA\cos\phi_2 = \tfrac{1}{3}BA \quad \text{or} \quad \cos\phi_2 = \tfrac{1}{3} \quad \text{or} \quad \phi_2 = \cos^{-1}\left(\tfrac{1}{3}\right) = \boxed{70.5°}$$

---

12. **REASONING** The magnetic flux is defined in Equation 22.2 as $\Phi = BA\cos\phi$, where B is the magnitude of the magnetic field, A is the area of the loop, and ϕ is the angle between the normal to the surface of the loop and the magnetic field. The change $\Delta\Phi$ in flux is the final flux Φ minus the initial flux Φ_0; $\Delta\Phi = \Phi - \Phi_0$.

    SOLUTION Let L_1 and L_2 be the lengths of the left and bottom edges of the loop. The initial area is the area of this rectangular loop plus the area of the semicircle: $A_0 = L_1 \times L_2 + \tfrac{1}{2}\pi r^2$. The initial magnetic flux is $\Phi_0 = B\left(L_1 \times L_2 + \tfrac{1}{2}\pi r^2\right)\cos 0°$. The final area is the area of the rectangular loop minus the area of the semicircle: $A_0 = L_1 \times L_2 - \tfrac{1}{2}\pi r^2$. The final magnetic flux is $\Phi = B\left(L_1 \times L_2 - \tfrac{1}{2}\pi r^2\right)\cos 0°$. The change in flux is

$$\Delta \Phi = \Phi - \Phi_0 = B\left(L_1 \times L_2 - \tfrac{1}{2}\pi r^2\right)\cos 0° - B\left(L_1 \times L_2 + \tfrac{1}{2}\pi r^2\right)\cos 0°$$

$$= -B\pi r^2 \cos 0° = -(0.75\text{ T})\pi(0.20\text{ m})^2 \cos 0° = \boxed{-9.4 \times 10^{-2}\text{ Wb}}$$

Note that the change in flux does not depend on L_1 or L_2.

---

13. **REASONING AND SOLUTION** When computing the magnetic flux, we multiply the component of the magnetic field that is parallel to the normal to the surface by the area of the surface.

    a. For the surface that lies in the xy plane, the normal is parallel to the z axis, so

    $$\Phi = B_z A = (0.30\text{ T})(2.0 \times 10^{-2}\text{ m})^2 = \boxed{1.2 \times 10^{-4}\text{ Wb}}$$

    b. For the surface that lies in the xz plane, the normal is parallel to the y axis, so

    $$\Phi = B_y A = (0.80\text{ T})(2.0 \times 10^{-2}\text{ m})^2 = \boxed{3.2 \times 10^{-4}\text{ Wb}}$$

    c. For the surface that lies in the yz plane, the normal is parallel to the x axis, so

    $$\Phi = B_x A = (0.50\text{ T})(2.0 \times 10^{-2}\text{ m})^2 = \boxed{2.0 \times 10^{-4}\text{ Wb}}$$

---

14. **REASONING AND SOLUTION**
    a. According to Equation 22.2, we have $\Phi = BA\cos\phi$. If the wall faces north, then

    $$\Phi = B_H A \cos 0.0° + B_V A \cos 90° = B_H A = (2.6 \times 10^{-5}\text{ T})(28\text{ m}^2) = \boxed{7.3 \times 10^{-4}\text{ Wb}}$$

    b. If the wall faces east, then

    $$\Phi = B_H A \cos 90° + B_V A \cos 90° = \boxed{0\text{ Wb}}$$

    c. The normal to the floor is vertical, so

    $$\Phi = B_H A \cos 90° + B_V A \cos 0.0° = B_V A = (4.2 \times 10^{-5}\text{ T})(112\text{ m}^2) = \boxed{4.7 \times 10^{-3}\text{ Wb}}$$

15. **SSM WWW** *REASONING* The general expression for the magnetic flux through an area A is given by Equation 22.2: $\Phi = BA\cos\phi$, where B is the magnitude of the magnetic field and ϕ is the angle of inclination of the magnetic field **B** with respect to the normal to the surface.

SOLUTION Since the magnetic field **B** is parallel to the surface for the triangular ends and the bottom surface, the flux through each of these three surfaces is $\boxed{0 \text{ Wb}}$.

The flux through the 1.2 m by 0.30 m face is

$$\Phi = (0.25 \text{ T})(1.2 \text{ m})(0.30 \text{ m}) \cos 0.0° = \boxed{0.090 \text{ Wb}}$$

For the 1.2 m by 0.50 m side, the area makes an angle ϕ with the magnetic field **B**, where

$$\phi = 90° - \tan^{-1}\left(\frac{0.30 \text{ m}}{0.40 \text{ m}}\right) = 53°$$

Therefore,

$$\Phi = (0.25 \text{ T})(1.2 \text{ m})(0.50 \text{ m}) \cos 53° = \boxed{0.090 \text{ Wb}}$$

---

16. *REASONING AND SOLUTION* The change in flux $\Delta\Phi$ is given by $\Delta\Phi = B \Delta A \cos\phi$, where ΔA is the area of the loop that leaves the region of the magnetic field in a time Δt. This area is the product of the width of the rectangle (0.080 m) and the length $v \Delta t$ of the side that leaves the magnetic field, $\Delta A = (0.080 \text{ m}) v \Delta t$.

$$\Delta\Phi = B \Delta A \cos\phi = (2.4 \text{ T})(0.080 \text{ m})(0.020 \text{ m/s})(2.0 \text{ s}) \cos 0.0° = \boxed{7.7 \times 10^{-3} \text{ Wb}}$$

---

17. *REASONING AND SOLUTION*
a. The change in flux through the loop is $\Delta\Phi = (\Delta B)A \cos\phi$, so the magnitude for the emf given by Faraday's law is

$$\text{Emf} = N[(\Delta B)/(\Delta t)]A \cos\phi$$

$$\text{Emf} = 300[(0.40 \text{ T})/(0.80 \text{ s})](5.0 \times 10^{-3} \text{ m}^2) \cos 30.0° = \boxed{0.65 \text{ V}}$$

b. The current in the resistor is, according to Ohm's law,

$$I = (\text{Emf})/R = (0.65 \text{ V})/(6.0 \text{ }\Omega) = \boxed{0.11 \text{ A}}$$

840 ELECTROMAGNETIC INDUCTION

18. **REASONING AND SOLUTION** The average emf induced in the circular wire is given by

$$\text{Emf} = -B\left(\frac{A - A_0}{t - t_0}\right)\cos\phi$$

The change in the area is equal to the final area of the circle ($A = 0$ m^2) minus the initial area ($A_0 = \pi r^2$).

$$\text{Emf} = -B\left(\frac{0 - \pi r^2}{t - t_0}\right)\cos\phi$$

$$\text{Emf} = -(0.55\text{ T})\left[\frac{0 - \pi(2.0 \times 10^{-2}\text{ m})^2}{0.25\text{ s} - 0}\right]\cos 0.0° = \boxed{2.8 \times 10^{-3}\text{ V}}$$

---

19. **SSM REASONING** According to Equation 22.3, the average emf induced in a coil of N loops is $\text{Emf} = -N\Delta\Phi/\Delta t$.

SOLUTION For the circular coil in question, the flux through a single turn changes by

$$\Delta\Phi = BA\cos 45° - BA\cos 90° = BA\cos 45°$$

during the interval of $\Delta t = 0.010$ s. Therefore, for N turns, Faraday's law gives the magnitude of the emf (without the minus sign) as

$$\text{Emf} = \frac{NBA\cos 45°}{\Delta t}$$

Since the loops are circular, the area A of each loop is equal to πr^2. Solving for B, we have

$$B = \frac{(\text{Emf})\Delta t}{N\pi r^2 \cos 45°} = \frac{(0.065\text{ V})(0.010\text{ s})}{(950)\pi(0.060\text{ m})^2 \cos 45°} = \boxed{8.6 \times 10^{-5}\text{ T}}$$

---

20. **REASONING** According to Faraday's law as given in Equation 22.3 (without the minus sign), the magnitude of the emf is $\text{Emf} = \Delta\Phi/\Delta t$ for a single turn ($N = 1$). Since the normal is parallel to the magnetic field, the angle ϕ between the normal and the field is $\phi = 0°$ when calculating the flux Φ from Equation 22.2: $\Phi = BA\cos 0° = BA$. We will use this expression for the flux in Faraday's law.

SOLUTION Representing the flux as $\Phi = BA$, we find that Faraday's law (without the minus sign) becomes

$$\text{Emf} = \frac{\Delta \Phi}{\Delta t} = \frac{\Delta(BA)}{\Delta t} = \frac{B\Delta A}{\Delta t}$$

In this result we have used the fact that the field magnitude B is constant. Rearranging this equation gives

$$\frac{\Delta A}{\Delta t} = \frac{\text{Emf}}{B} = \frac{2.6 \text{ V}}{1.7 \text{ T}} = \boxed{1.5 \text{ m}^2/\text{s}}$$

21. **REASONING AND SOLUTION** Faraday's law gives (using only the magnitude of the induced emf)

$$\Delta t = BA \cos 0.0°/(\text{Emf}) = (1.5 \text{ T})(0.032 \text{ m}^2)/(0.010 \text{ V}) = \boxed{4.8 \text{ s}}$$

22. **REASONING** We will use Faraday's law of electromagnetic induction, Equation 22.3, to find the emf induced in the loop. Once this value has been determined, we can employ Equation 22.3 again to find the rate at which the area changes.

SOLUTION
a. The magnitude of the induced emf is given by Equation 22.3 without the minus sign:

$$\text{Emf} = N\left(\frac{\Phi - \Phi_0}{t - t_0}\right) = N\left(\frac{BA\cos\phi - B_0 A\cos\phi}{t - t_0}\right) = NA\cos\phi\left(\frac{B - B_0}{t - t_0}\right)$$

$$= (1)(0.35 \text{ m} \times 0.55 \text{ m})\cos 65°\left(\frac{2.1 \text{ T} - 0 \text{ T}}{0.45 \text{ s} - 0 \text{ s}}\right) = \boxed{0.38 \text{ V}}$$

b. When the magnetic field is constant and the area is changing in time, Faraday's law can be written as (again, without the minus sign)

$$\text{Emf} = N\left(\frac{\Phi - \Phi_0}{t - t_0}\right) = N\left(\frac{BA\cos\phi - BA_0\cos\phi}{t - t_0}\right)$$

$$= NB\cos\phi\left(\frac{A - A_0}{t - t_0}\right) = NB\cos\phi\left(\frac{\Delta A}{\Delta t}\right)$$

Solving this equation for $\Delta A/\Delta t$ and substituting in the value of 0.38 V for the emf, we find that

$$\frac{\Delta A}{\Delta t} = \frac{\text{Emf}}{NB\cos\phi} = \frac{0.38 \text{ V}}{(1)(2.1 \text{ T})\cos 65°} = \boxed{0.43 \text{ m}^2/\text{s}}$$

842 ELECTROMAGNETIC INDUCTION

23. **REASONING AND SOLUTION** The rod may be viewed as sweeping out an area whose initial value is 0.0 m² and final value is $\pi r^2/4$. The flux change due to this change in area is

$$\Delta\Phi = B\pi r^2/4$$

Faraday's law gives the magnitude of the emf to be

$$\text{Emf} = \Delta\Phi/\Delta t = (1/4)(0.16\ \text{T})(\pi)(1.5\ \text{m})^2/(0.66\ \text{s}) = \boxed{0.43\ \text{V}}$$

---

24. **REASONING** The energy dissipated in the resistance is given by Equation 6.10b as the power dissipated multiplied by the time, $E = Pt$. The power, according to Equation 20.6c, is the square of the induced emf divided by the resistance, $P = (\text{Emf})^2/R$. The induced emf can be determined from Faraday's law of electromagnetic induction, Equation 22.3.

SOLUTION Expressing the energy consumed as $E = Pt$, and substituting in $P = (\text{Emf})^2/R$, we find

$$E = Pt = \frac{(\text{Emf})^2 t}{R}$$

The induced emf is given by Faraday's law as $\text{Emf} = -N(\Delta\Phi/\Delta t)$, and the resistance R is equal to the resistance per unit length (3.3×10^{-2} Ω/m) times the length of the circumference of the loop, $2\pi r$. Thus, the energy consumed is

$$E = \frac{\left(-N\dfrac{\Delta\Phi}{\Delta t}\right)^2 t}{(3.3 \times 10^{-2}\ \Omega/\text{m})2\pi r} = \frac{\left[-N\left(\dfrac{\Phi - \Phi_0}{t - t_0}\right)\right]^2 t}{(3.3 \times 10^{-2}\ \Omega/\text{m})2\pi r}$$

$$= \frac{\left[-N\left(\dfrac{BA\cos\phi - B_0 A\cos\phi}{t - t_0}\right)\right]^2 t}{(3.3 \times 10^{-2}\ \Omega/\text{m})2\pi r} = \frac{\left[-NA\cos\phi\left(\dfrac{B - B_0}{t - t_0}\right)\right]^2 t}{(3.3 \times 10^{-2}\ \Omega/\text{m})2\pi r}$$

$$= \frac{\left[-(1)\pi(0.12\ \text{m})^2(\cos 0°)\left(\dfrac{0.60\ \text{T} - 0\ \text{T}}{0.45\ \text{s} - 0\ \text{s}}\right)\right]^2 (0.45\ \text{s})}{(3.3 \times 10^{-2}\ \Omega/\text{m})2\pi(0.12\ \text{m})} = \boxed{6.6 \times 10^{-2}\ \text{J}}$$

25. **SSM WWW** **REASONING AND SOLUTION** According to Ohm's law, the emf developed by the coil is Emf = IR, which, when combined with the definition of electric current becomes Emf = $(\Delta q)R/(\Delta t)$. The magnitude of the emf is also given by Equation 22.3 (Faraday's law) without the minus sign:

$$\text{Emf} = N\frac{\Delta\Phi}{\Delta t} = N\left(\frac{BA\cos 0° - 0}{\Delta t}\right) = \frac{NBA}{\Delta t}$$

Combining the two expressions for the emf, we have

$$\frac{(\Delta q)R}{\Delta t} = \frac{NBA}{\Delta t} \quad \text{or} \quad B = \frac{(\Delta q)R}{NA} = \frac{(8.87\times 10^{-3}\text{ C})(45.0\text{ }\Omega)}{(1850)(4.70\times 10^{-4}\text{ m}^2)} = \boxed{0.459\text{ T}}$$

26. **REASONING AND SOLUTION** In time, Δt, the flux change through the loop ABC is $\Delta\Phi = B\Delta A$, where

$$\Delta A = [\omega\Delta t/(2\pi)](\pi R^2)$$

Faraday's law gives the magnitude of the induced emf to be

$$\text{Emf} = \Delta\Phi/\Delta t = B\omega R^2/2 = (3.8\times 10^{-3}\text{ T})(15\text{ rad/s})(0.50\text{ m})^2/2 = 7.1\times 10^{-3}\text{ V}$$

The current through the resistor is

$$I = (\text{Emf})/R = (7.1\times 10^{-3}\text{ V})/(3.0\text{ }\Omega) = \boxed{2.4\times 10^{-3}\text{ A}}$$

27. **REASONING AND SOLUTION** Consider one revolution of either rod. The magnitude of the emf induced across the rod is

$$\text{Emf} = B\Delta A/\Delta t = B(\pi L^2)/\Delta t$$

The angular speed of the rods is $\omega = 2\pi/\Delta t$, so Emf = $BL^2\omega/2$. The rod tips have opposite polarity since they are rotating in opposite directions. Hence, the difference in potentials of the tips is

$$\Delta V = BL^2\omega$$

so

$$\omega = \frac{\Delta V}{BL^2} = \frac{4.5\times 10^3\text{ V}}{(4.7\text{ T})(0.68\text{ m})^2} = \boxed{2100\text{ rad/s}}$$

844 ELECTROMAGNETIC INDUCTION

28. **REASONING AND SOLUTION**
a. When the triangle is crossing the + y axis, the magnetic flux through it begins to increase. The induced field must be into the paper by Lenz's law. The induced current must, therefore, be CW .

b. When the triangle crosses the –x axis, the flux through it is not changing at all. No induced field will be produced and there will be NO induced current .

c. Upon crossing the –y axis the coil will experience a decrease in flux so the induced field will be out of the paper. The induced current must then be CCW .

d. As the triangle crosses the +x axis, the flux through it is not changing at all. No induced field will be produced and there will be NO induced current .

29. SSM **REASONING AND SOLUTION** If the north and south poles of the magnet are interchanged, the currents in the ammeter would simply be reversed in direction.

Therefore, in Figure 22.1b, the current will flow right to left through the ammeter.

Similarly, in Figure 22.1c, the current will flow left to right through the ammeter.

30. **REASONING AND SOLUTION** According to Lenz's law, the induced emf has a polarity that leads to an induced current whose direction is such that the induced magnetic field opposes the original flux change. As the loop rotates through one-half a revolution, the area through which the magnetic field passes decreases. The magnetic flux, being proportional to the area, also decreases. The induced magnetic field must oppose this decrease in flux, so the induced magnetic field must strengthen the original magnetic field. Thus, the induced magnetic field points in the same direction as the original magnetic field, or into the plane of the page. According to RHR-2, the induced current flows clockwise around the loop (and right-to-left through the resistor) in order that its magnetic field be directed into the page. Since conventional current flows through a resistor from higher potential toward lower potential, the right end of the resistor must be the positive end .

31. SSM **REASONING** In solving this problem, we apply Lenz's law, which essentially says that the change in magnetic flux must be opposed by the induced magnetic field.

SOLUTION
a. The magnetic field due to the wire in the vicinity of position 1 is directed out of the paper. The coil is moving closer to the wire into a region of higher magnetic field, so the flux through

the coil is increasing. Lenz's law demands that the induced field counteract this increase. The direction of the induced field, therefore, must be into the paper. The current in the coil must be clockwise.

b. At position 2 the magnetic field is directed into the paper and is decreasing as the coil moves away from the wire. The induced magnetic field, therefore, must be directed into the paper, so the current in the coil must be clockwise.

---

32. **REASONING AND SOLUTION**
a. When the magnet is above the ring its magnetic field points down through the ring and is increasing as the magnet falls. The induced magnetic field attempts to reduce the increasing field and points up. The induced current in the ring is as shown in the drawing at the right, and then the ring looks like a magnet with its north pole at the top, repelling the north pole of the falling magnet and retarding its motion.

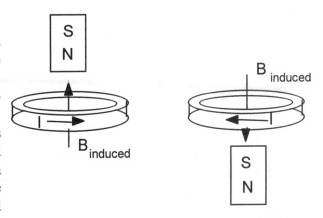

When the magnet is below the ring, its magnetic field still points down through the ring but is decreasing as the magnet falls. The induced magnetic field attempts to bolster the decreasing field and points down. The induced current in the ring is as shown in the drawing, and the ring then looks like a magnet with its north pole at the bottom, attracting the south pole of the falling magnet and retarding its motion.

b. The motion of the magnet is unaffected, since no induced current can flow in the cut ring. No induced current means that no induced magnetic field can be produced to repel or attract the falling magnet.

---

33. **REASONING** The current I in the straight wire produces a circular pattern of magnetic field lines around the wire. The magnetic field at any point is tangent to one of these circular field lines. Thus, the field points perpendicular to the plane of the table. Furthermore, according to Right-Hand Rule No. 2, the field is directed down into the table surface in region 1 above the wire and is directed up out of the table surface in region 2 below the wire (see the drawing at the right). To deduce

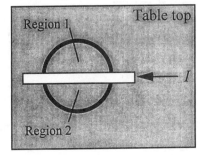

the direction of any induced current in the circular loop, we consider Faraday's law and the change that occurs in the magnetic flux through the loop due to the field of the straight wire.

SOLUTION As the current I increases, the magnitude of the field that it produces also increases. However, the directions of the fields in regions 1 and 2 do not change and remain as discussed in the reasoning. Since the fields in these two regions always have opposite directions and equal magnitudes at any given radial distance from the straight wire, the flux through the regions add up to give zero for any value of the current. With the flux remaining constant as time passes, Faraday's law indicates that there is no induced emf in the coil. Since there is no induced emf in the coil, there is no induced current .

---

34. REASONING AND SOLUTION
a. Location I
As the loop swings downward, the normal to the loop makes a smaller angle with the applied field. Hence, the flux through the loop is increasing. The induced magnetic field must point generally to the left to counteract this increase. The induced current flows

$$x \rightarrow y \rightarrow z$$

Location II
The angle between the normal to the loop and the applied field is now increasing, so the flux through the loop is decreasing. The induced field must now be generally to the right, and the current flows

$$z \rightarrow y \rightarrow x$$

b. Location I
The argument is the same as for location II in part a.

$$z \rightarrow y \rightarrow x$$

Location II
The argument is the same as for location I in part a.

$$x \rightarrow y \rightarrow z$$

---

35. SSM WWW REASONING AND SOLUTION
a. From the drawing, we see that the period of the generator (the time for one full cycle) is 0.42 s; therefore, the frequency of the generator is

$$f = \frac{1}{T} = \frac{1}{0.42 \text{ s}} = \boxed{2.4 \text{ Hz}}$$

b. The angular speed of the generator is related to its frequency by $\omega = 2\pi f$, so the angular speed is

$$\omega = 2\pi(2.4 \text{ Hz}) = \boxed{15 \text{ rad/s}}$$

c. The magnitude of the maximum emf induced in a rotating planar coil is given by $(\text{Emf})_0 = NAB\omega$ (see Equation 22.4). The magnitude of the magnetic field can be found by solving this expression for B and noting from the drawing that $(\text{Emf})_0 = 28$ V:

$$B = \frac{(\text{Emf})_0}{NA\omega} = \frac{28 \text{ V}}{(150)(0.020 \text{ m}^2)(15 \text{ rad/s})} = \boxed{0.62 \text{ T}}$$

---

36. **REASONING AND SOLUTION** If the generators have the same peak emf, number of turns and angular frequency, then $NB_1A_1 \sin \omega t = NB_2A_2 \sin \omega t$, so that $B_1A_1 = B_2A_2$. Thus,

$$B_1 = (0.10 \text{ T})(0.045 \text{ m}^2)/(0.015 \text{ m}^2) = \boxed{0.30 \text{ T}}$$

---

37. **REASONING AND SOLUTION** The maximum emf produced by the generator is $NAB\omega$, so

$$B = \frac{5500 \text{ V}}{150(0.85 \text{ m}^2) 2\pi (60.0 \text{ Hz})} = \boxed{0.11 \text{ T}}$$

---

38. **REASONING AND SOLUTION** Using Equation 22.5 to take the back emf into account, we find

$$R = \frac{V - \text{Emf}}{I} = \frac{(120.0 \text{ V}) - (72.0 \text{ V})}{3.0 \text{ A}} = \boxed{16 \text{ }\Omega}$$

---

39. **SSM REASONING** We can use the information given in the problem statement to determine the area of the coil A. Since it is square, the length of one side is $\ell = \sqrt{A}$.

SOLUTION According to Equation 22.4, the maximum emf induced in the coil is $(\text{Emf})_0 = NAB\omega$. Therefore, the length of one side of the coil is

$$\ell = \sqrt{A} = \sqrt{\frac{(\text{Emf})_0}{NB\omega}} = \sqrt{\frac{75.0 \text{ V}}{(248)(0.170 \text{ T})(79.1 \text{ rad/s})}} = \boxed{0.150 \text{ m}}$$

---

40. **REASONING** When the motor is running at normal speed, the current is the net emf divided by the resistance R of the armature wire. The net emf is the applied voltage V minus the back emf developed by the rotating coil. We can use this relation to find the back emf. When the motor is just turned on, there is no back emf, so the current is just the applied voltage divided by the resistance of the wire. To limit the starting current to 15.0 A, a resistor R_1 is placed in series with the resistance of the wire, so the equivalent resistance is

848 ELECTROMAGNETIC INDUCTION

$R + R_1$. The current to the motor is equal to the applied voltage divided by the equivalent resistance.

SOLUTION

a. According to Equation 22.5, the back emf generated by the motor is

$$\text{Emf} = V - IR = 120.0 \text{ V} - (7.00 \text{ A})(0.720 \text{ }\Omega) = \boxed{115 \text{ V}}$$

b. When the motor has been just turned on, the back emf is zero, so the current is

$$I = \frac{V - \text{Emf}}{R} = \frac{120.0 \text{ V} - 0 \text{ V}}{0.720 \text{ }\Omega} = \boxed{167 \text{ A}}$$

c. When a resistance R_1 is placed in series with the resistance R of the wire, the equivalent resistance is $R_1 + R$. The current to the motor is

$$I = \frac{V - \text{Emf}}{R_1 + R}$$

Solving this expression for R_1 gives

$$R_1 = \frac{V - \text{Emf}}{I} - R = \frac{120.0 \text{ V} - 0 \text{ V}}{15.0 \text{ A}} - 0.720 \text{ }\Omega = \boxed{7.28 \text{ }\Omega}$$

---

41. **SSM** **REASONING AND SOLUTION** When the motor just begins to turn the fan blade, there is no back emf. Therefore, Ohm's law indicates that $120.0 \text{ V} = I_0 R$, where I_0 is the current in the motor and R is the resistance of the motor coils. Solving for I_0 gives

$$I_0 = \frac{120.0 \text{ V}}{R} \qquad (1)$$

At normal operating speed, there is a back emf, and Equation 22.5 gives the current in the motor as

$$I = \frac{120.0 \text{ V} - \text{Emf}}{R} \qquad (2)$$

But we know that $I = 0.150 I_0$. Therefore, using Equations (1) and (2), we find that

$$\frac{120.0 \text{ V} - \text{Emf}}{R} = 0.150 \left(\frac{120.0 \text{ V}}{R} \right)$$

Eliminating R algebraically, we have

$$120.0 \text{ V} - \text{Emf} = 0.150(120.0 \text{ V}) \quad \text{or} \quad \text{Emf} = 0.850(120.0 \text{ V}) = \boxed{102 \text{ V}}$$

---

42. **REASONING AND SOLUTION** The peak emf is $NAB\omega$, where N is the length of the wire divided by the circumference of a turn.

$$\text{Peak emf} = [L/(2\pi r)](\pi r^2)\omega B$$

$$\text{Peak emf} = Lr\omega B/2 = (5.7 \text{ m})(0.14 \text{ m})(25 \text{ rad/s})(0.20 \text{ T})/2 = \boxed{2.0 \text{ V}}$$

---

43. **REASONING AND SOLUTION**
   a. On startup, the back emf of the generator is zero. Then,

   $$R = V/I = (117 \text{ V})/(12.2 \text{ A}) = \boxed{9.59 \text{ }\Omega}$$

   b. At normal speed
   $$\text{Emf} = V - IR = 117 \text{ V} - (2.30 \text{ A})(9.59 \text{ }\Omega) = \boxed{95 \text{ V}}$$

   c. The back emf of the motor is proportional to the rotational speed, so at 1/3 the normal speed, the back emf is

   $$(1/3)(95 \text{ V}) = 32 \text{ V}$$

   The voltage applied to the resistor is then $V = 117 \text{ V} - 32 \text{ V} = 85 \text{ V}$, so the current is

   $$I = V/R = (85 \text{ V})/(9.59 \text{ }\Omega) = \boxed{8.9 \text{ A}}$$

---

44. **REASONING AND SOLUTION** Using Equation 22.7, we find

$$(\text{Emf})_2 = -M\frac{\Delta I_1}{\Delta t} \quad \text{or} \quad M = \frac{-(\text{Emf})_2 \Delta t}{\Delta I_1} = \frac{-(0.12 \text{ V})(0.14 \text{ s})}{1.6 \text{ A} - 3.4 \text{ A}} = \boxed{9.3 \times 10^{-3} \text{ H}}$$

45. **SSM REASONING AND SOLUTION** From the results of Example 13, the self-inductance L of a long solenoid is given by $L = \mu_0 n^2 A \ell$. Solving for the number of turns n per unit length gives

$$n = \sqrt{\frac{L}{\mu_0 A \ell}} = \sqrt{\frac{1.4 \times 10^{-3} \text{ H}}{(4\pi \times 10^{-7} \text{ T} \cdot \text{m/A})(1.2 \times 10^{-3} \text{ m}^2)(0.052 \text{ m})}} = 4.2 \times 10^3 \text{ turns/m}$$

Therefore, the total number of turns N is the product of n and the length ℓ of the solenoid:

$$N = n\ell = (4.2 \times 10^3 \text{ turns/m})(0.052 \text{ m}) = \boxed{220 \text{ turns}}$$

46. **REASONING** When the current through an inductor changes, the induced emf is given by Equation 22.9 as

$$\text{Emf} = -L\frac{\Delta I}{\Delta t}$$

where L is the inductance, ΔI is the change in the current, and Δt is the time interval during which the current changes. For each interval, we can determine ΔI and Δt from the graph.

SOLUTION

a. $\text{Emf} = -L\dfrac{\Delta I}{\Delta t} = -(3.2 \times 10^{-3} \text{ H})\left(\dfrac{4.0 \text{ A} - 0 \text{ A}}{2.0 \times 10^{-3} \text{ s} - 0 \text{ s}}\right) = \boxed{-6.4 \text{ V}}$

b. $\text{Emf} = -L\dfrac{\Delta I}{\Delta t} = -(3.2 \times 10^{-3} \text{ H})\left(\dfrac{4.0 \text{ A} - 4.0 \text{ A}}{5.0 \times 10^{-3} \text{ s} - 2.0 \times 10^{-3} \text{ s}}\right) = \boxed{0 \text{ V}}$

c. $\text{Emf} = -L\dfrac{\Delta I}{\Delta t} = -(3.2 \times 10^{-3} \text{ H})\left(\dfrac{0 \text{ A} - 4.0 \text{ A}}{9.0 \times 10^{-3} \text{ s} - 5.0 \times 10^{-3} \text{ s}}\right) = \boxed{+3.2 \text{ V}}$

47. **REASONING AND SOLUTION** The induced emf in the secondary coil is proportional to the mutual inductance. If the primary coil is assumed to be unaffected by the metal, that is $\Delta I_1/\Delta t$ is the same for both cases, then

$$\text{New emf} = 3(0.46 \text{ V}) = \boxed{1.4 \text{ V}}$$

48. **REASONING AND SOLUTION** The magnitude of the emf induced in the secondary coil is

$$(\text{Emf})_2 = M\Delta I_1/\Delta t = (2.5 \times 10^{-3} \text{ H})(3.0 \text{ A})/(0.040 \text{ s}) = 0.188 \text{ V}$$

This emf drives a current in the secondary coil according to Ohm's law:

$$I_2 = (\text{Emf})_2/R_2 = (0.188 \text{ V})/(2.0 \text{ }\Omega) = \boxed{0.094 \text{ A}}$$

---

49. **SSM REASONING** The energy density is given by Equation 22.11 as

$$\text{Energy density} = \frac{\text{Energy}}{\text{Volume}} = \frac{1}{2\mu_0} B^2$$

The energy stored is the energy density times the volume.

SOLUTION The volume is the area A times the height h. Therefore, the energy stored is

$$\text{Energy} = \frac{B^2 Ah}{2\mu_0} = \frac{(7.0 \times 10^{-5} \text{ T})^2 (5.0 \times 10^8 \text{ m}^2)(1500 \text{ m})}{2(4\pi \times 10^{-7} \text{ T}\cdot\text{m/A})} = \boxed{1.5 \times 10^9 \text{ J}}$$

---

50. **REASONING AND SOLUTION** The energy density in an electric field is Energy density = $\varepsilon_0 E^2/2$, and the energy density in a magnetic field is Energy density = $B^2/(2\mu_0)$. Equating and rearranging yields

$$E = B\sqrt{\frac{1}{\varepsilon_0 \mu_0}} = Bc = (12 \text{ T})(3.0 \times 10^8 \text{ m/s}) = \boxed{3.6 \times 10^9 \text{ N/C}}$$

---

51. **REASONING AND SOLUTION** The mutual inductance is $M = N_2\Phi_2/I_1$. The flux through the 125 turn coil is $\Phi_2 = B_1 A_2 = \mu_0 n_1 I_1 A_2$, so

$$M = N_2\mu_0 n_1 A_2 = (125)(4\pi \times 10^{-7} \text{ T}\cdot\text{m/A})(1750/\text{m})(\pi)(0.0180 \text{ m})^2 = \boxed{2.80 \times 10^{-4} \text{ H}}$$

---

52. **REASONING AND SOLUTION** If the rectangle is made $\Delta I = I_f$ wide, then the top of the rectangle will intersect the line at LI_f. The work is one-half the area of the rectangle, so

$$\boxed{W = \tfrac{1}{2}(LI_f)I_f = \tfrac{1}{2}LI_f^2}$$

852 ELECTROMAGNETIC INDUCTION

53. **SSM** *REASONING* According to Ohm's law, the average current I induced in the coil is given by $I = (\text{Emf})/R$, where R is the resistance of the coil. To find the induced emf, we use Faraday's law of electromagnetic induction

SOLUTION The magnitude of the induced emf can be found from Faraday's law of electromagnetic induction and is given by Equation 22.3 without the minus sign:

$$\text{Emf} = N\frac{\Delta \Phi}{\Delta t} = N\frac{\Delta(BA\cos 0°)}{\Delta t} = NA\frac{\Delta B}{\Delta t}$$

We have used the fact that the field within a long solenoid is perpendicular to the cross-sectional area A of the solenoid and makes an angle of 0° with respect to the normal to the area. The field is given by Equation 21.7 as $B = \mu_0 n I$, so the change ΔB in the field is $\Delta B = \mu_0 n \Delta I$, where ΔI is the change in the current. The induced current is, then,

$$I = \frac{NA\dfrac{\Delta B}{\Delta t}}{R} = \frac{NA\dfrac{\mu_0 n \Delta I}{\Delta t}}{R}$$

$$= \frac{(10)(6.0 \times 10^{-4} \text{ m}^2)\dfrac{(4\pi \times 10^{-7} \text{ T}\cdot\text{m/A})(400 \text{ turns/m})(0.40 \text{ A})}{(0.050 \text{ s})}}{1.5 \text{ }\Omega} = \boxed{1.6 \times 10^{-5} \text{ A}}$$

54. *REASONING AND SOLUTION*
    a. The transformer is a $\boxed{\text{step-up}}$ transformer.

    b. The turns ratio is
    $$\frac{N_s}{N_p} = \frac{V_s}{V_p} = \frac{12\,000 \text{ V}}{220 \text{ V}} = \boxed{55:1}$$

55. **SSM** *REASONING AND SOLUTION* Since the secondary voltage is less than the primary voltage, we can conclude that the transformer used in the doorbell described in the problem is a $\boxed{\text{step-down transformer}}$. The turns ratio is given by the transformer equation, Equation 22.12:

$$\frac{N_s}{N_p} = \frac{V_s}{V_p} = \frac{10.0 \text{ V}}{120 \text{ V}} = \frac{1}{12}$$

so the turns ratio is $\boxed{1:12}$.

56. **REASONING AND SOLUTION** The step down transformer reduces the voltage by a factor of 13.

$$V_2 = (120 \text{ V})/13 = \boxed{9.2 \text{ V}}$$

57. **REASONING AND SOLUTION** The power consumed by the air filter is $P_s = I_s V_s$. But according to the transformer equation, we have $V_s = (N_s/N_p)V_p$. Therefore,

$$P_s = I_s V_p (N_s/N_p) = (1.5 \times 10^{-3} \text{ A})(120 \text{ V})(43) = \boxed{7.7 \text{ W}}$$

58. **REASONING AND SOLUTION** According to the transformer equation (Equation 22.12), we have

$$N_s = (V_s/V_p)N_p = [(4320 \text{ V})/(120.0 \text{ V})](21) = \boxed{756}$$

59. **SSM REASONING** The power lost in heating the wires is given by Equation 20.6b: $P = I^2 R$. Before we can use this equation, however, we must determine the total resistance R of the wire and the current that flows through the wire.

SOLUTION
a. The total resistance of one of the wires is equal to the resistance per unit length multiplied by the length of the wire. Thus, we have

$$(5.0 \times 10^{-2} \text{ }\Omega/\text{km})(7.0 \text{ km}) = 0.35 \text{ }\Omega$$

and the total resistance of the transmission line is twice this value or 0.70 Ω. According to Equation 20.6a ($P = IV$), the current flowing into the town is

$$I = \frac{P}{V} = \frac{1.2 \times 10^6 \text{ W}}{1200 \text{ V}} = 1.0 \times 10^3 \text{ A}$$

Thus, the power lost in heating the wire is

$$P = I^2 R = (1.0 \times 10^3 \text{ A})^2 (0.70 \text{ }\Omega) = \boxed{7.0 \times 10^5 \text{ W}}$$

b. According to the transformer equation (Equation 22.12), the stepped-up voltage is

854 ELECTROMAGNETIC INDUCTION

$$V_s = V_p\left(\frac{N_s}{N_p}\right) = (1200 \text{ V})\left(\frac{100}{1}\right) = 1.2 \times 10^5 \text{ V}$$

According to Equation 20.6a ($P = IV$), the current in the wires is

$$I = \frac{P}{V} = \frac{1.2 \times 10^6 \text{ W}}{1.2 \times 10^5 \text{ V}} = 1.0 \times 10^1 \text{ A}$$

The power lost in heating the wires is now

$$P = I^2 R = (1.0 \times 10^1 \text{ A})^2 (0.70 \text{ }\Omega) = \boxed{7.0 \times 10^1 \text{ W}}$$

---

60. **REASONING** The starting point here is to realize that the power used to operate the tube comes from the secondary coil and, therefore, is $P_s = I_s V_s$ according to Equation 20.15a. The current I_s in the secondary is given, as is the power P_s. We seek the turns ratio, and the voltage V_s across the secondary is related to the turns ratio by the transformer equation (Equation 22.12).

SOLUTION Using Equation 20.15a for the power delivered by the secondary, we have $P_s = I_s V_s$. According to the transformer equation, the voltage V_s across the secondary is

$$V_s = \left(\frac{N_s}{N_p}\right) V_p$$

Substituting this result into the power expression, we find

$$P_s = I_s V_s = I_s \left(\frac{N_s}{N_p}\right) V_p$$

Rearranging this result shows that the turns ratio is

$$\frac{N_s}{N_p} = \frac{P_s}{I_s V_p} = \frac{95 \text{ W}}{(5.3 \times 10^{-3} \text{ A})(120 \text{ V})} = \boxed{150}$$

---

61. **REASONING AND SOLUTION** Ohm's law written for the secondary is $V_s = I_s R_2$. We know that

$$V_s = (N_s/N_p)V_p \qquad \text{and} \qquad I_s = (N_p/N_s)I_p$$

Substituting these expressions for V_s and I_s into $V_s = I_s R_2$ and recognizing that $R_1 = V_p/I_p$, we find that

$$R_1 = \left(\frac{N_p}{N_s}\right)^2 R_2$$

---

62. **REASONING AND SOLUTION** Faraday's law gives for the magnitude of the change in the magnetic field

$$\Delta B = \frac{(\text{Emf})\Delta t}{NA} = \frac{(1.5 \text{ V})(0.050 \text{ s})}{(100)(0.040 \text{ m})(0.060 \text{ m})} = \boxed{0.31 \text{ T}}$$

---

63. **REASONING AND SOLUTION** An rms voltage of 120 V corresponds to a maximum emf of $(120 \text{ V})\sqrt{2}$ or 170 V. The maximum voltage produced by the generator is $NAB\omega$, so

$$N = \frac{170 \text{ V}}{(0.016 \text{ m}^2)(7.0 \times 10^{-5} \text{ T})2\pi(60.0 \text{ Hz})} = \boxed{4.0 \times 10^5}$$

---

64. **REASONING** Using Equation 22.3 (Faraday's law) without the minus sign and recognizing that $N = 1$, we can write the magnitudes of the emf for parts a and b as follows:

$$(\text{Emf})_a = \left(\frac{\Delta\Phi}{\Delta t}\right)_a \quad (1) \quad \text{and} \quad (\text{Emf})_b = \left(\frac{\Delta\Phi}{\Delta t}\right)_b \quad (2)$$

To solve this problem, we need to consider the change in flux $\Delta\Phi$ and the time interval Δt for both parts of the drawing.

SOLUTION The change in flux is the same for both parts of the drawing and is given by

$$(\Delta\Phi)_a = (\Delta\Phi)_b = \Phi_{\text{inside}} - \Phi_{\text{outside}} = \Phi_{\text{inside}} = BA \quad (3)$$

In Equation (3) we have used the fact that initially the coil is outside the field region, so that $\Phi_{\text{outside}} = 0$ Wb for both cases. Moreover, the field is perpendicular to the plane of the coil and has the same magnitude B over the entire area A of the coil, once it has completely entered the field region. Thus, $\Phi_{\text{inside}} = BA$ in both cases, according to Equation 22.2.

The time interval required for the coil to entire the field region completely can be expressed as the distance the coil travels divided by the speed at which it is pushed. In part a of the drawing the distance traveled is W, while in part b it is L. Thus, we have

$$(\Delta t)_a = W/v \quad (4) \qquad\qquad (\Delta t)_b = L/v \quad (5)$$

Substituting Equations (3), (4), and (5) into Equations (1) and (2), we find

$$(\text{Emf})_a = \left(\frac{\Delta \Phi}{\Delta t}\right)_a = \frac{BA}{W/v} \quad (6) \qquad (\text{Emf})_b = \left(\frac{\Delta \Phi}{\Delta t}\right)_b = \frac{BA}{L/v} \quad (7)$$

Dividing Equation (6) by Equation (7) gives

$$\frac{(\text{Emf})_a}{(\text{Emf})_b} = \frac{\frac{BA}{W/v}}{\frac{BA}{L/v}} = \frac{L}{W} = 3.0 \qquad \text{or} \qquad (\text{Emf})_b = \frac{(\text{Emf})_a}{3.0} = \frac{0.15\text{ V}}{3.0} = \boxed{0.050\text{ V}}$$

---

65. **SSM** **REASONING AND SOLUTION** According to Faraday's law, only the relative motion between the coil and magnet is significant.. It makes no difference if the coil is held fixed and the magnet is moved or the magnet is held fixed and the coil is moved. The results are the same. Therefore,

a. The current will flow from $\boxed{\text{left to right}}$ as shown in Figure 22.1b.

b. The current will flow from $\boxed{\text{right to left}}$ as shown in Figure 22.1c.

---

66. **REASONING AND SOLUTION** The resistance of the primary is

$$R_p = \rho L_p / A \quad (1)$$

The resistance of the secondary is

$$R_s = \rho L_s / A \quad (2)$$

In writing Equations (1) and (2) we have assumed that both coils are made of the same wire, so that the resistivity ρ and the cross-sectional area A is the same for each. Division of the equations gives

$$R_s/R_p = L_s/L_p = (14\ \Omega)/(56\ \Omega) = 1/4$$

The lengths of the wires are proportional to the number of turns so

$$N_s/N_p = L_s/L_p = \boxed{1/4}$$

67. **SSM** *REASONING AND SOLUTION* The emf induced in the coil of an ac generator is given by Equation 22.4 as

$$\text{Emf} = NAB\omega \sin\omega t = (500)(1.2 \times 10^{-2} \text{ m}^2)(0.13 \text{ T})(34 \text{ rad/s}) \sin 27° = \boxed{12 \text{ V}}$$

68. *REASONING AND SOLUTION* The energy stored in a capacitor is given by Equation 19.11 as $\frac{1}{2}CV^2$. The energy stored in an inductor is given by Equation 22.10 as $\frac{1}{2}LI^2$. Setting these two equations equal to each other and solving for the current I, we get

$$I = \sqrt{\frac{C}{L}} V = \sqrt{\frac{3.0 \times 10^{-6} \text{ F}}{5.0 \times 10^{-3} \text{ H}}} (35 \text{ V}) = \boxed{0.86 \text{ A}}$$

69. *REASONING AND SOLUTION* Faraday's law can be used if we can find the flux change through the triangle as the bar slides.

$$\Delta\Phi = B\Delta A = B(bh/2 - b_0 h_0/2)$$

where the b's and h's represent the bases and heights of the triangle. At time zero, the height and base of the triangle are both zero. At time, $\Delta t = t = 5.0$ s, $h = b \tan 15°$ and $b = vt$. Now

$$\Delta\Phi = (1/2) B(vt)^2 \tan 15°$$

Faraday's law gives the magnitude of the emf to be

$$\text{Emf} = \Delta\Phi/\Delta t = (1/2) (Bv^2 t) \tan 15°$$

$$\text{Emf} = (1/2) (0.42 \text{ T})(0.40 \text{ m/s})^2 (5.0 \text{ s}) \tan 15° = \boxed{0.045 \text{ V}}$$

70. *REASONING AND SOLUTION* If the applied magnetic field is decreasing in time, then the flux through the circuit is decreasing. Lenz's law requires that an induced magnetic field be produced which attempts to counteract this decrease; hence its direction is out of the paper. The sense of the induced current in the circuit must be CCW. Therefore, the lower plate of the capacitor is positive while the upper plate is negative. The electric field between the plates of the capacitor points from positive to negative so the electric field points $\boxed{\text{upward}}$.

858 ELECTROMAGNETIC INDUCTION

71. **SSM** *REASONING* According to Equation 22.3, the average emf induced in a single loop is emf $= -\Delta\Phi/\Delta t$. Since the magnitude of the magnetic field is changing, the area of the loop remains constant, and the direction of the field is parallel to the normal to the loop, the change in flux through the loop is given by $\Delta\Phi = (\Delta B)A$. Thus the magnitude of the induced emf in the loop is given by Emf $= (\Delta B)A/\Delta t$.

Similarly, when the area of the loop is changed and the field B has a given value, we find the magnitude of the induced emf to be Emf $= B\Delta A/\Delta t$.

SOLUTION
a. The magnitude of the induced emf when the field changes in magnitude is

$$\text{Emf} = \frac{(\Delta B)A}{\Delta t} = (0.20 \text{ T/s})(0.018 \text{ m}^2) = \boxed{3.6\times 10^{-3} \text{ V}}$$

b. At a particular value of B (when B is changing), the rate at which the area must change can be obtained from

$$\text{Emf} = \frac{B\Delta A}{\Delta t} \quad \text{or} \quad \frac{\Delta A}{\Delta t} = \frac{\text{Emf}}{B} = \frac{3.6\times 10^{-3} \text{ V}}{1.8 \text{ T}} = \boxed{2.0\times 10^{-3} \text{ m}^2/\text{s}}$$

In order for the induced emf to be zero, the magnitude of the magnetic field and the area of the loop must change in such a way that the flux remains constant. Since the magnitude of the magnetic field is increasing, the area of the loop must decrease, if the flux is to remain constant. Therefore, $\boxed{\text{the area of the loop must be shrunk}}$.

---

72. *REASONING* From the results of Example 13, the self-inductance L of a solenoid is given by

$$L = \mu_0 n^2 A \ell \qquad (1)$$

where ℓ is the length of the solenoid. A toroid is a solenoid that is bent to form a circle of radius R. The length ℓ of the toroid is the circumference of the circle, $\ell = 2\pi R$. Substituting this value for ℓ into the equation above, we can obtain and expression for the self-inductance of the toroid.

SOLUTION Substituting $\ell = 2\pi R$ into Equation (1) gives

$$L = 2\pi\mu_0 n^2 AR$$

$$= 2\pi\left(4\pi\times 10^{-7} \text{ T}\cdot\text{m/A}\right)\left(2400 \text{ m}^{-1}\right)^2 \left(1.0\times 10^{-6} \text{ m}^2\right)(0.050 \text{ m}) = \boxed{2.3\times 10^{-6} \text{ H}}$$

Chapter 22 Problems 859

73. **REASONING AND SOLUTION**
a. Using Equation 22.5, we find

$$I = (V - \text{Emf})/R = (120.0 \text{ V} - 108 \text{ V})/(15.0 \text{ }\Omega) = \boxed{0.80 \text{ A}}$$

b. If the armature is not rotating, then the back emf is zero and

$$I = V/R = (120.0 \text{ V})/(15.0 \text{ }\Omega) = \boxed{8.00 \text{ A}}$$

c. Since the back emf is proportional to the angular speed of the armature, the half speed emf is Emf = (108 V)/2 = 54.0 V. The current is then

$$I = (120.0 \text{ V} - 54.0 \text{ V})/(15.0 \text{ }\Omega) = \boxed{4.40 \text{ A}}$$

---

74. **REASONING AND SOLUTION** The mutual inductance is

$$M = \frac{N_2 \Phi_2}{I_1}$$

The flux through loop 2 is

$$\Phi_2 = B_1 A_2 = \left(\frac{\mu_0 N_1 I_1}{2 R_1} \right) \left(\pi R_2^2 \right)$$

Then

$$M = \frac{N_2 \Phi_2}{I_1} = \frac{N_2}{I_1} \left(\frac{\mu_0 N_1 I_1}{2 R_1} \right) \left(\pi R_2^2 \right) = \boxed{\frac{\mu_0 \pi N_1 N_2 R_2^2}{2 R_1}}$$

---

75. **CONCEPT QUESTIONS**
a. The motional emf generated by the moving metal rod depends only on its speed, its length, and the magnitude of the magnetic field (see Equation 22.1). The motional emf does not depend on the resistance in the circuit. Therefore, the emfs are the same.

b. According to Equation 20.2, the current I is equal to the emf divided by the resistance R of the circuit. Since the emfs in the two circuits are the same, the circuit with the smaller resistance has the larger current. Since circuit 1 has one-half the resistance of circuit 2, the current in circuit 1 is twice as large.

c. According to Equation 20.6a, the power is equal to the product of the current and the emf. We have already seen that the current in circuit 1 is twice as large as that in circuit 2. Thus, even when the rods have the same speeds, circuit 1 delivers twice as much power as circuit 2. The emf produced by the moving rod is directly proportional to its speed (see Equation 22.1),

860 ELECTROMAGNETIC INDUCTION

so that circuit 1 produces twice the emf when its rod is moving twice as fast. Thus, circuit 1 would deliver 2×2 = 4 times more power to the light bulb than circuit 2 would.

SOLUTION
a. The ratio of the emfs is, according to Equation 22.1

$$\frac{(\text{Emf})_1}{(\text{Emf})_2} = \frac{vBL}{vBL} = \boxed{1}$$

b. Equation 20.2 states that the current is equal to the emf divided by the resistance. The ratio of the currents is

$$\frac{I_1}{I_2} = \frac{(\text{Emf})_1/R_1}{(\text{Emf})_2/R_2} = \frac{R_2}{R_1} = \frac{110\,\Omega}{55\,\Omega} = \boxed{2}$$

c. The power, according to Equation 20.6a, is the product of the current and the emf. The motional emf is given by Equation 22.1 as Emf = vBL. The ratio of the powers is

$$\frac{P_1}{P_2} = \frac{I_1\,(\text{Emf})_1}{I_2\,(\text{Emf})_2} = \left(\frac{I_1}{I_2}\right)\left(\frac{v_1 BL}{v_2 BL}\right) = \left(\frac{I_1}{I_2}\right)\left(\frac{2v_2}{v_2}\right) = (2)(2) = \boxed{4}$$

---

76. **CONCEPT QUESTIONS**
a. The induced emf is zero during the second interval, 3.0 – 6.0 s. According to Faraday's law of electromagnetic induction, Equation 22.3, an induced emf arises only when the magnetic flux changes. During this interval, the magnetic field, the area of the loop, and the orientation of the field relative to the loop are constant. Thus, the magnetic flux does not change, so there is no induced emf.

b. An emf is induced during the first and third intervals, because the magnetic field is changing in time. The time interval is the same (3.0 s) for the two cases. However, the magnitude of the field changes more during the first interval. Therefore, the magnetic flux is changing at a greater rate in that interval, which means that the magnitude of the induced emf is greatest during the first interval.

c. During the first interval the magnetic field in increasing with time. During the third interval, the field is decreasing with time. As a result, the induced emfs will have opposite polarities during these intervals. If the direction of the induced current is clockwise during the first interval, it will be counterclockwise during the third interval.

SOLUTION
a. The induced emf is given by Equations 22.3 and 22.3:

0–3.0 s:

$$\text{Emf} = -N\frac{\Delta\Phi}{\Delta t} = -N\left(\frac{BA\cos\phi - B_0 A\cos\phi}{t - t_0}\right)$$

$$= -NA\cos\phi\left(\frac{B - B_0}{t - t_0}\right) = -(50)(0.15\text{ m}^2)(\cos 0°)\left(\frac{0.40\text{ T} - 0\text{ T}}{3.0\text{ s} - 0\text{ s}}\right) = \boxed{-1.0\text{ V}}$$

3.0–6.0 s:

$$\text{Emf} = -NA\cos\phi\left(\frac{B - B_0}{t - t_0}\right) = -(50)(0.15\text{ m}^2)(\cos 0°)\left(\frac{0.40\text{ T} - 0.40\text{ T}}{6.0\text{ s} - 3.0\text{ s}}\right) = \boxed{0\text{ V}}$$

6.0–9.0 s:

$$\text{Emf} = -NA\cos\phi\left(\frac{B - B_0}{t - t_0}\right) = -(50)(0.15\text{ m}^2)(\cos 0°)\left(\frac{0.20\text{ T} - 0.40\text{ T}}{9.0\text{ s} - 6.0\text{ s}}\right) = \boxed{+0.50\text{ V}}$$

b. The induced current is given by Equation 20.2 as $I = (\text{Emf})/R$.

0–3.0 s:
$$I = \frac{\text{Emf}}{R} = \frac{-1.0\text{ V}}{0.50\text{ }\Omega} = \boxed{-2.0\text{ A}}$$

6.0–9.0 s:
$$I = \frac{\text{Emf}}{R} = \frac{+0.50\text{ V}}{0.50\text{ }\Omega} = \boxed{+1.0\text{ A}}$$

As expected, the currents are in opposite directions.

---

77. **CONCEPT QUESTIONS**
a. The magnetic field produced by I extends throughout the space surrounding the loop. Using RHR-2, it can be shown that the magnetic field is parallel to the normal to the loop. Thus, the magnetic field pentrates the loop and generates a magnetic flux.

b. According to Faraday's law of electromagnetic induction, an emf is induced when the magnetic flux through the loop is changing in time. If the current I is constant, the magnetic flux is constant, and no emf is induced in the loop. However, if the current is decreasing in time, the magnetic flux is decreasing and an induced current exists in the loop.

c. No. Lenz's law states that the induced magnetic field opposes the *change* in the magnetic field produced by the current I. The induced magnetic field does not necessarily oppose the magnetic field itself. Thus, the induced magnetic field does not always have a direction that is opposite to the direction of the field produced by I.

SOLUTION At the location of the loop, the magnetic field produced by the current I is directed into the page (this can be verified by using RHR-2). The current is decreasing, so the magnetic field is decreasing. Therefore, the magnetic flux that penetrates the loop is decreasing. According to Lenz's law, the induced emf has a polarity that leads to an induced

862 ELECTROMAGNETIC INDUCTION

current whose direction is such that the induced magnetic field opposes this flux change. The induced magnetic field will oppose this decrease in flux by pointing into the page, in the same direction as the field produced by I. According to RHR-2, the induced current must flow clockwise around the loop in order to produce such an induced field. The current then flows from right-to-left through the resistor.

---

78. **CONCEPT QUESTIONS**
    a. An emf is induced in the coil because the magnetic flux through the coil is changing in time. The flux is changing because the angle ϕ between the normal to the coil and the magnetic field is changing.

    b. The amount of induced current is equal to the induced emf divided by the resistance of the coil (see Equation 20.2).

    c. According to Equation 20.1, the amount of charge Δq that flows is equal to the induced current I multiplied by the time interval Δt during which the coil rotates or $\Delta q = I \Delta t$.

SOLUTION According to Equation 20.1, the amount of charge that flows is $\Delta q = I \Delta t$. The current is related to the emf in the coil and the resistance by Equation 20.2 as $I = (\text{Emf})/R$. The amount of charge that flows can, therefore, be written as

$$\Delta q = I \Delta t = \left(\frac{\text{Emf}}{R}\right) \Delta t$$

The emf is given by Faraday's law of electromagnetic induction as

$$\text{Emf} = -N\left(\frac{\Delta \Phi}{\Delta t}\right) = -N\left(\frac{BA \cos \phi - BA \cos \phi_0}{\Delta t}\right)$$

where we have also used Equation 22.2, which gives the definition of magnetic flux as $\Phi = BA \cos \phi$. With this emf, the expression for the amount of charge becomes

$$\Delta q = I \Delta t = \left[\frac{-N\left(\dfrac{BA \cos \phi - BA \cos \phi_0}{\Delta t}\right)}{R}\right] \Delta t = \frac{-NBA(\cos \phi - \cos \phi_0)}{R}$$

Solving for the magnitude of the magnetic field yields

$$B = \frac{-R\Delta q}{NA(\cos\phi - \cos\phi_0)}$$

$$= \frac{-(140\,\Omega)(8.5\times 10^{-5}\,\text{C})}{(50)(1.5\times 10^{-3}\,\text{m}^2)(\cos 90° - \cos 0°)} = \boxed{0.16\,\text{T}}$$

---

79. **CONCEPT QUESTIONS**
    a. If the wire has no resistance, there is no voltage across the solenoid. This conclusion follows from Equation 20.2, $V = IR$. If $R = 0\,\Omega$, then $V = 0$ V.

    b. Yes. According to Faraday's law of electromagnetic induction, expressed as Emf $= -L(\Delta I/\Delta t)$, an emf is induced in the solenoid as long as the current is changing in time.

    c. Yes. The amount of electrical energy E stored by an inductor is given by Equation 22.10 as $E = \tfrac{1}{2}LI^2$, where L is the inductance and I is the current.

    d. The power is equal to the energy dissipated divided by the time (Equation 6.10b), $P = \dfrac{E}{t} = \dfrac{\tfrac{1}{2}LI^2}{t}$.

    SOLUTION
    a. The emf induced in the solenoid is

    $$\text{Emf} = -L\left(\frac{\Delta I}{\Delta t}\right) = -(3.1\,\text{H})\left(\frac{0\,\text{A} - 15\,\text{A}}{75\times 10^{-3}\,\text{s}}\right) = \boxed{620\,\text{V}} \tag{22.9}$$

    b. The energy stored in the solenoid is

    $$E = \tfrac{1}{2}LI^2 = \tfrac{1}{2}(3.1\,\text{H})(15\,\text{A})^2 = \boxed{350\,\text{J}} \tag{22.10}$$

    c. The power dissipated is

    $$P = \frac{E}{t} = \frac{\tfrac{1}{2}LI^2}{t} = \frac{\tfrac{1}{2}(3.1\,\text{H})(15\,\text{A})^2}{75\times 10^{-3}\,\text{s}} = \boxed{4700\,\text{W}} \tag{6.10b}$$

864 ELECTROMAGNETIC INDUCTION

80. CONCEPT QUESTIONS
a. Since the secondary voltage (the voltage to charge the batteries) is less than the primary voltage (the voltage at the wall socket), the transformer is a step-down transformer.

b. In a step-down transformer, the voltage across the secondary coil is less than the voltage across the primary coil. However, the current in the secondary coil is greater than the current in the primary coil. Thus, the current that goes through the batteries is greater than the current from the wall socket.

c. If the transformer has negligible resistance, the power delivered to the batteries is equal to the power coming from the wall socket.

SOLUTION
a. The turns ratio N_s / N_p is equal to the ratio of secondary voltage to the primary voltage:

$$\frac{N_s}{N_p} = \frac{V_s}{V_p} = \frac{9.0 \text{ V}}{120 \text{ V}} = \boxed{1:13} \tag{22.12}$$

b. The current from the wall socket is given by Equation 22.13:

$$I_p = I_s \left(\frac{N_s}{N_p} \right) = \left(225 \times 10^{-3} \text{ A} \right) \left(\frac{1}{13} \right) = \boxed{17 \times 10^{-3} \text{ A}} \tag{22.13}$$

c. The power delivered by the wall socket is the product of the primary current and voltage:

$$P_p = I_p V_p = \left(17 \times 10^{-3} \text{ A} \right) (120 \text{ V}) = \boxed{2.0 \text{ W}} \tag{20.15a}$$

The power delivered to the batteries is the same as that coming from the wall socket, so $P_s = \boxed{2.0 \text{ W}}$.

CHAPTER 23 ALTERNATING CURRENT CIRCUITS

PROBLEMS

1. **SSM** **REASONING AND SOLUTION** The capacitive reactance is given by Equation 23.2: $X_C = \dfrac{1}{2\pi f C}$. Solving for the frequency f, we find that

$$f = \dfrac{1}{2\pi X_C C} = \dfrac{1}{2\pi (168\ \Omega)(7.50 \times 10^{-6}\ \text{F})} = \boxed{126\ \text{Hz}}$$

2. **REASONING AND SOLUTION** The rms voltage can be calculated using $V = IX_C$, where the capacitive reactance X_C can be found using

$$X_C = \dfrac{1}{2\pi f C} = \dfrac{1}{2\pi (3.4 \times 10^3\ \text{Hz})(0.86 \times 10^{-6}\ \text{F})} = \boxed{54\ \Omega}$$

The voltage is, therefore,

$$V = IX_C = (35 \times 10^{-3}\ \text{A})(54\ \Omega) = \boxed{1.9\ \text{V}}$$

3. **REASONING** We will first find the capacitance C using the relation $C = 1/(2\pi f X_C)$, where f is the frequency and X_C is the capacitive reactance. The value of the capacitive reactance at a frequency of 510 Hz can then be determined by using this value of the capacitance.

SOLUTION The capacitance is

$$C = \dfrac{1}{2\pi f X_C} = \dfrac{1}{2\pi (170\ \text{Hz})(36\ \Omega)} = 2.6 \times 10^{-5}\ \text{F} \tag{23.2}$$

When the frequency is 510 Hz, the capacitive reactance is

$$X_C = \dfrac{1}{2\pi f C} = \dfrac{1}{2\pi (510\ \text{Hz})(2.6 \times 10^{-5}\ \text{F})} = \boxed{12\ \Omega}$$

866 **ALTERNATING CURRENT CIRCUITS**

4. **REASONING** The rms current in a capacitor is $I_{rms} = V_{rms}/X_C$, according to Equation 23.1. The capacitive reactance is $X_C = 1/(2\pi f C)$, according to Equation 23.2. For the first capacitor, we use $C = C_1$ in these expressions. For the two capacitors in parallel, we use $C = C_P$, where C_P is the equivalent capacitance from Equation 20.18 ($C_P = C_1 + C_2$). Taking the difference between the currents and using the given data, we can obtain the desired value for C_2. The capacitance C_1 is unknown, but it will be eliminated algebraically from the calculation.

SOLUTION Using Equations 23.1 and 23.2, we find that the current in a capacitor is

$$I_{rms} = \frac{V_{rms}}{X_C} = \frac{V_{rms}}{1/(2\pi f C)} = V_{rms} 2\pi f C$$

Applying this result to the first capacitor and the parallel combination of the two capacitors, we obtain

$$\underbrace{I_1 = V_{rms} 2\pi f C_1}_{\text{Single capacitor}} \quad \text{and} \quad \underbrace{I_{\text{Combination}} = V_{rms} 2\pi f (C_1 + C_2)}_{\text{Parallel combination}}$$

Subtracting I_1 from $I_{\text{Combination}}$, reveals that

$$I_{\text{Combination}} - I_1 = V_{rms} 2\pi f (C_1 + C_2) - V_{rms} 2\pi f C_1 = V_{rms} 2\pi f C_2$$

Solving for C_2 gives

$$C_2 = \frac{I_{\text{Combination}} - I_1}{V_{rms} 2\pi f} = \frac{0.18 \text{ A}}{(24 \text{ V}) 2\pi (440 \text{ Hz})} = \boxed{2.7 \times 10^{-6} \text{ F}}$$

---

5. **SSM** **REASONING AND SOLUTION**
a. The equivalent capacitance C_s of two capacitors in series is given by Equation 20.19 as

$$\frac{1}{C_s} = \frac{1}{C_1} + \frac{1}{C_2} = \frac{1}{3.00 \times 10^{-6} \text{ F}} + \frac{1}{6.00 \times 10^{-6} \text{ F}} \quad \text{or} \quad \boxed{C_s = 2.00 \text{ }\mu\text{F}}$$

b. The current in the circuit can be found by solving Equation 23.1, $V_{rms} = I_{rms} X_C$, for I_{rms}. However, we must first find the capacitive reactance X_C. From Equation 23.2, we have

$$X_C = \frac{1}{2\pi f C_s} = \frac{1}{2\pi (510 \text{ Hz})(2.00 \times 10^{-6} \text{ F})} = 156 \text{ }\Omega$$

The current in the circuit is given by

$$I_{rms} = \frac{V_{rms}}{X_C} = \frac{120 \text{ V}}{156 \text{ }\Omega} = \boxed{0.77 \text{ A}}$$

6. **REASONING** The capacitance C is related to the capacitive reactance X_C and the frequency f via Equation 23.2 as $C = 1/(2\pi f X_C)$. The capacitive reactance, in turn is related to the rms-voltage V_{rms} and the rms-current I_{rms} by $X_C = V_{rms}/I_{rms}$ (see Equation 23.1). Thus, the capacitance can be written as $C = I_{rms}/(2\pi f V_{rms})$. The magnitude of the maximum charge q on one plate of the capacitor is, from Equation 19.8, the product of the capacitance C and the peak voltage V.

SOLUTION

a. Recall that the rms-voltage V_{rms} is related to the peak voltage V by $V_{rms} = \frac{V}{\sqrt{2}}$. The capacitance is, then,

$$C = \frac{I_{rms}}{2\pi f V_{rms}} = \frac{3.0 \text{ A}}{2\pi (750 \text{ Hz})\left(\frac{140 \text{ V}}{\sqrt{2}}\right)} = \boxed{6.4 \times 10^{-6} \text{ F}}$$

b. The maximum charge on one plate of the capacitor is

$$q = CV = (6.4 \times 10^{-6} \text{ F})(140 \text{ V}) = \boxed{9.0 \times 10^{-4} \text{ C}}$$

7. **REASONING AND SOLUTION** Equations 23.1 and 23.2 indicate that the rms current in a capacitor is $I = V/X_C$, where V is the rms voltage and $X_C = 1/(2\pi f C)$. Therefore, the current is $I = V 2\pi f C$. For a single capacitor $C = C_1$, and we have

$$I = V 2\pi f C_1$$

For two capacitors in series, Equation 20.19 indicates that the equivalent capacitance can be obtained from $C^{-1} = C_1^{-1} + C_2^{-1}$, which can be solved to show that $C = C_1 C_2 / (C_1 + C_2)$. The total series current is, then,

$$I_{series} = V 2\pi f C = V 2\pi f \left(\frac{C_1 C_2}{C_1 + C_2}\right)$$

The series current is one-third of the current I. It follows, therefore, that

868 ALTERNATING CURRENT CIRCUITS

$$\frac{I_{series}}{I} = \frac{V2\pi f\left(\frac{C_1 C_2}{C_1+C_2}\right)}{V2\pi f C_1} = \frac{C_2}{C_1+C_2} = \frac{1}{3} \quad \text{or} \quad \frac{C_1}{C_2} = 2$$

For two capacitors in parallel, Equation 20.18 indicates that the equivalent capacitance is $C = C_1 + C_2$. The total current in this case is

$$I_{parallel} = V2\pi f C = V2\pi f(C_1+C_2)$$

The ratio of $I_{parallel}$ to the current I in the single capacitor is

$$\frac{I_{parallel}}{I} = \frac{V2\pi f(C_1+C_2)}{V2\pi f C_1} = \frac{C_1+C_2}{C_1} = 1+\frac{C_2}{C_1} = 1+\frac{1}{2} = \boxed{\frac{3}{2}}$$

---

8. **REASONING AND SOLUTION** We know that $V = IX_L = I(2\pi f L)$, so the frequency is

$$f = \frac{V}{2\pi I L} = \frac{2.1\ \text{V}}{2\pi(0.023\ \text{A})(0.047\ \text{H})} = \boxed{310\ \text{Hz}}$$

---

9. **SSM REASONING** The individual reactances are given by Equations 23.2 and 23.4, respectively,

[Capacitive reactance] $\quad X_C = \dfrac{1}{2\pi f C}$

[Inductive reactance] $\quad X_L = 2\pi f L$

When the reactances are equal, we have $X_C = X_L$, from which we find

$$\frac{1}{2\pi f C} = 2\pi f L \quad \text{or} \quad 4\pi^2 f^2 LC = 1$$

The last expression may be solved for the frequency f.

SOLUTION Solving for f with $L = 52 \times 10^{-3}$ H and $C = 76 \times 10^{-6}$ F, we obtain

$$f = \frac{1}{2\pi\sqrt{LC}} = \frac{1}{2\pi\sqrt{(52\times 10^{-3}\ \text{H})(76\times 10^{-6}\ \text{F})}} = \boxed{8.0 \times 10^1\ \text{Hz}}$$

10. **REASONING AND SOLUTION** The reactance of an inductor is given by Equation 23.4 as $X_L = 2\pi f L$. At the frequencies $f_1 = 1350$ Hz and $f_2 = 450$ Hz, we can use this equation as follows:

$$\frac{X_{L2}}{X_{L1}} = \frac{2\pi f_2 L}{2\pi f_1 L} = \frac{f_2}{f_1} \quad \text{or} \quad X_{L2} = X_{L1}\left(\frac{f_2}{f_1}\right) = (480\,\Omega)\left(\frac{450\text{ Hz}}{1350\text{ Hz}}\right) = \boxed{160\,\Omega}$$

---

11. **SSM WWW** **REASONING** The rms current can be calculated from Equation 23.3, $I_{rms} = V_{rms}/X_L$, provided that the inductive reactance is obtained first. Then the peak value of the current I_0 supplied by the generator can be calculated from the rms current I_{rms} by using Equation 20.12, $I_0 = \sqrt{2}\, I_{rms}$.

SOLUTION At the frequency of $f = 620$ Hz, we find, using Equations 23.4 and 23.3, that

$$X_L = 2\pi f L = 2\pi(620\text{ Hz})(8.2 \times 10^{-3}\text{ H}) = 32\,\Omega$$

$$I_{rms} = \frac{V_{rms}}{X_L} = \frac{10.0\text{ V}}{32\,\Omega} = 0.31\text{ A}$$

Therefore, from Equation 20.12, we find that the *peak value* I_0 of the current supplied by the generator must be

$$I_0 = \sqrt{2}\, I_{rms} = \sqrt{2}\,(0.31\text{ A}) = \boxed{0.44\text{ A}}$$

---

12. **REASONING** The rms current in an inductor is $I_{rms} = V_{rms}/X_L$, according to Equation 23.3. The inductive reactance is $X_L = 2\pi f L$, according to Equation 23.4. Applying these expressions to both generators will allow us to obtain the desired current.

SOLUTION Using Equations 23.3 and 23.4, we find that the current in an inductor is

$$I_{rms} = \frac{V_{rms}}{X_L} = \frac{V_{rms}}{2\pi f L}$$

Applying this result to the two generators gives

$$\underbrace{I_1 = \frac{V_{rms}}{2\pi f_1 L}}_{\text{Generator 1}} \quad \text{and} \quad \underbrace{I_2 = \frac{V_{rms}}{2\pi f_2 L}}_{\text{Generator 2}}$$

Dividing the equation for generator 2 by the equation for generator 1, we obtain

870 ALTERNATING CURRENT CIRCUITS

$$\frac{I_2}{I_1} = \frac{\dfrac{V_{rms}}{2\pi f_2 L}}{\dfrac{V_{rms}}{2\pi f_1 L}} = \frac{f_1}{f_2} \quad \text{or} \quad I_2 = I_1 \frac{f_1}{f_2} = (0.30 \text{ A})\frac{1.5 \text{ kHz}}{6.0 \text{ kHz}} = \boxed{0.075 \text{ A}}$$

---

13. ***REASONING AND SOLUTION*** The inductance of the solenoid is given by (see Section 22.8)

$$L = \mu_0 N^2 A/l = (4\pi \times 10^{-7} \text{ T·m/A})(135)^2 (3.1 \times 10^{-5} \text{ m}^2)/(0.025 \text{ m}) = 2.8 \times 10^{-5} \text{ H}$$

Thus,

$$X_L = 2\pi fL = 2\pi(18 \times 10^3 \text{ Hz})(2.8 \times 10^{-5} \text{ H}) = 3.2 \text{ }\Omega$$

The rms voltage is

$$V = IX_L = (0.036 \text{ A})(3.2 \text{ }\Omega) = 0.12 \text{ V}$$

The *peak* voltage V_0 can be obtained from the rms-voltage V as

$$V_0 = V\sqrt{2} = (0.12 \text{ V})\sqrt{2} = \boxed{0.17 \text{ V}}$$

---

14. ***REASONING AND SOLUTION*** Equations 23.3 and 23.4 indicate that the rms current in a single inductance L_1 is $I_1 = V/X_{L1}$, where V is the rms voltage and $X_{L1} = 2\pi f L_1$. Therefore, the current is $I_1 = V/(2\pi f L_1)$. Similarly, the current in the second inductor connected across the terminals of the generator is $I_2 = V/(2\pi f L_2)$. The total current delivered by the generator is the sum of these two values:

$$I_{total} = I_1 + I_2 = \frac{V}{2\pi f L_1} + \frac{V}{2\pi f L_2}$$

But this same total current is delivered to the single inductance L, so it also follows that $I_{total} = V/(2\pi f L)$. Equating the two expressions for I_{total} shows that

$$\frac{V}{2\pi f L} = \frac{V}{2\pi f L_1} + \frac{V}{2\pi f L_2} \quad \text{or} \quad \frac{1}{L} = \frac{1}{L_1} + \frac{1}{L_2}$$

Using this result, we determine the value of L as follows:

$$\frac{1}{L} = \frac{1}{L_1} + \frac{1}{L_2} \quad \text{or} \quad L = \frac{L_1 L_2}{L_1 + L_2} = \frac{(0.030 \text{ H})(0.060 \text{ H})}{0.030 \text{ H} + 0.060 \text{ H}} = \boxed{0.020 \text{ H}}$$

15. **SSM REASONING** The voltage supplied by the generator can be found from Equation 23.6, $V_{rms} = I_{rms} Z$. The value of I_{rms} is given in the problem statement, so we must obtain the impedance of the circuit.

SOLUTION The impedance of the circuit is, according to Equation 23.7,

$$Z = \sqrt{R^2 + (X_L - X_C)^2} = \sqrt{(275\ \Omega)^2 + (648\ \Omega - 415\ \Omega)^2} = 3.60 \times 10^2\ \Omega$$

The rms voltage of the generator is

$$V_{rms} = I_{rms} Z = (0.233\ \text{A})(3.60 \times 10^2\ \Omega) = \boxed{83.9\ \text{V}}$$

16. **REASONING AND SOLUTION**
a. At a frequency of 609 Hz:

$$X_L = 2\pi f L = 2\pi(609\ \text{Hz})(0.0310\ \text{H}) = 119\ \Omega$$

$$X_C = \frac{1}{2\pi f C} = \frac{1}{2\pi(609\ \text{Hz})(3.30 \times 10^{-6}\ \text{F})} = 79.2\ \Omega$$

so,

$$Z = \sqrt{R^2 + (X_L - X_C)^2} = \sqrt{(106\ \Omega)^2 + (119\ \Omega - 79.2\ \Omega)^2} = \boxed{113\ \Omega}$$

b. The phase angle between the current and the voltage is

$$\phi = \tan^{-1}\left(\frac{X_L - X_C}{R}\right) = \tan^{-1}\left(\frac{119\ \Omega - 79.2\ \Omega}{106\ \Omega}\right) = \boxed{+21°}$$

17. **REASONING AND SOLUTION** The power factor is $\cos \phi$, where ϕ can be obtained from Equation 23.8 $\left[\tan \phi = (X_L - X_C)/R\right]$. To obtain ϕ, we need values for the capacitive and inductive reactances, X_C and X_L, which are obtained from Equations 23.2 and 23.4, respectively:

$$X_C = \frac{1}{2\pi f C} = \frac{1}{2\pi(2550\ \text{Hz})(2.00 \times 10^{-6}\ \text{F})} = 31.2\ \Omega$$

$$X_L = 2\pi f L = 2\pi(2550\ \text{Hz})(4.00 \times 10^{-3}\ \text{H}) = 64.1\ \Omega$$

872 ALTERNATING CURRENT CIRCUITS

Using these values and the given value for R, we can obtain the power factor:

$$\tan\phi = \frac{X_L - X_C}{R} \quad \text{or} \quad \phi = \tan^{-1}\left(\frac{X_L - X_C}{R}\right) = \tan^{-1}\left(\frac{64.1\,\Omega - 31.2\,\Omega}{47.0\,\Omega}\right) = 35.0°$$

Thus, the power factor is $\cos\phi = \cos 35.0° = \boxed{0.819}$.

---

18. **REASONING AND SOLUTION** In order to find the voltage and phase angle we need to know the impedance of the circuit. We know that

$$X_C = \frac{1}{2\pi f C} = \frac{1}{2\pi(4.80 \times 10^3\text{ Hz})(0.250 \times 10^{-6}\text{ F})} = 133\,\Omega$$

The impedance of the circuit can be found using

$$Z = \sqrt{R^2 + (X_L - X_C)^2} = \sqrt{(232\,\Omega)^2 + (0 - 133\,\Omega)^2} = 267\,\Omega$$

a. The voltage of the generator is

$$V = IZ = (0.0400\text{ A})(267\,\Omega) = \boxed{10.7\text{ V}}$$

b. The phase angle is

$$\phi = \tan^{-1}\left(\frac{X_L - X_C}{R}\right) = \tan^{-1}\left(\frac{0 - 133\,\Omega}{232\,\Omega}\right) = \boxed{-29.8°}$$

---

19. **SSM** **REASONING** We can use the equations for a series RCL circuit to solve this problem provided that we set $X_C = 0$ since there is no capacitor in the circuit. The current in the circuit can be found from Equation 23.6, $V_{\text{rms}} = I_{\text{rms}} Z$, once the impedance of the circuit has been obtained. Equation 23.8, $\tan\phi = (X_L - X_C)/R$, can then be used (with $X_C = 0$) to find the phase angle between the current and the voltage.

SOLUTION The inductive reactance is (Equation 23.4)

$$X_L = 2\pi f L = 2\pi(106\text{ Hz})(0.200\text{ H}) = 133\,\Omega$$

The impedance of the circuit is

$$Z = \sqrt{R^2 + (X_L - X_C)^2} = \sqrt{R^2 + X_L^2} = \sqrt{(215\,\Omega)^2 + (133\,\Omega)^2} = 253\,\Omega$$

a. The current through each circuit element is, using Equation 23.6,

$$I_{rms} = \frac{V_{rms}}{Z} = \frac{234 \text{ V}}{253 \text{ }\Omega} = \boxed{0.925 \text{ A}}$$

b. The phase angle between the current and the voltage is, according to Equation 23.8 (with $X_C = 0$),

$$\tan\phi = \frac{X_L - X_C}{R} = \frac{X_L}{R} = \frac{133 \text{ }\Omega}{215 \text{ }\Omega} = 0.619$$

$$\phi = \tan^{-1}(0.619) = \boxed{31.8°}$$

20. **REASONING** The average power dissipated is that dissipated in the resistor and is $\overline{P} = I_{rms}^2 R$, according to Equation 20.15b. We are given the current I_{rms} but need to find the resistance R. Since the inductive reactance X_L is known, we can find the resistance from the impedance, which is $Z = \sqrt{R^2 + X_L^2}$, according to Equation 23.7. Since the voltage and the current are known, we can obtain the impedance from Equation 23.6 as $Z = V_{rms}/I_{rms}$.

SOLUTION From Equation 23.7, we can determine the resistance as $R = \sqrt{Z^2 - X_L^2}$. With this expression for the resistance, Equation 20.15b for the power becomes

$$\overline{P} = I_{rms}^2 R = I_{rms}^2 \sqrt{Z^2 - X_L^2}$$

Using Equation 23.6 to express the impedance, we obtain the following value for the dissipated power

$$\overline{P} = I_{rms}^2 \sqrt{Z^2 - X_L^2} = I_{rms}^2 \sqrt{\left(\frac{V_{rms}}{I_{rms}}\right)^2 - X_L^2}$$

$$= (1.75 \text{ A})^2 \sqrt{\left(\frac{115 \text{ V}}{1.75 \text{ A}}\right)^2 - (52.0 \text{ }\Omega)^2} = \boxed{123 \text{ W}}$$

21. [SSM] [WWW] **REASONING** We can use the equations for a series RCL circuit to solve this problem, provided that we set $X_L = 0$ since there is no inductance in the circuit. Thus, according to Equations 23.6 and 23.7, the current in the circuit is $I_{rms} = V_{rms}/\sqrt{R^2 + X_C^2}$. When the frequency f is very large, the capacitive reactance is zero, or $X_C = 0$, in which case the current becomes $I_{rms}(\text{large } f) = V_{rms}/R$. When the current I_{rms} in the circuit is one-half the value of $I_{rms}(\text{large } f)$ that exists when the frequency is very large, we have

874 ALTERNATING CURRENT CIRCUITS

$$\frac{I_{rms}}{I_{rms}(\text{large }f)} = \frac{1}{2}$$

We can use these expressions to write the ratio above in terms of the resistance and the capacitive reactance. Once the capacitive reactance is known, the frequency can be determined.

SOLUTION The ratio of the currents is

$$\frac{I_{rms}}{I_{rms}(\text{large }f)} = \frac{V_{rms}/\sqrt{R^2+X_C^2}}{V_{rms}/R} = \frac{R}{\sqrt{R^2+X_C^2}} = \frac{1}{2} \quad \text{or} \quad \frac{R^2}{R^2+X_C^2} = \frac{1}{4}$$

Taking the reciprocal of this result gives

$$\frac{R^2+X_C^2}{R^2} = 4 \quad \text{or} \quad 1+\frac{X_C^2}{R^2} = 4$$

Therefore,

$$\frac{X_C}{R} = \sqrt{3}$$

According to Equation 23.2, $X_C = 1/(2\pi f C)$, so it follows that

$$\frac{X_C}{R} = \frac{1/(2\pi f C)}{R} = \sqrt{3}$$

Thus,

$$f = \frac{1}{2\pi RC\sqrt{3}} = \frac{1}{2\pi(85\,\Omega)(4.0\times 10^{-6}\,\text{F})\sqrt{3}} = \boxed{270\text{ Hz}}$$

---

22. **REASONING AND SOLUTION**
a. At very large frequencies the impedance of the inductor becomes much larger than that of the capacitor or the resistances of the two resistors. Therefore, current flows only through the top branch of the circuit, i.e., through the capacitor and 290 Ω resistor. For a very large frequency, $X_C = 0$, so $Z = R = 290\,\Omega$. Therefore,

$$I = V/Z = (75\text{ V})/(290\,\Omega) = \boxed{0.26\text{ A}}$$

b. At very small frequencies the impedance of the capacitor becomes much larger than that of the inductor or the resistances of the two resistors. Therefore, current flows only through the middle branch of the circuit, i.e., through the inductor and 710 Ω resistor. At very small frequencies, $X_L = 0$, so $Z = R = 710$ Ω. Therefore,

$$I = V/Z = (75 \text{ V})/(710 \text{ Ω}) = \boxed{0.11 \text{ A}}$$

---

23. **REASONING** The instantaneous value of the generator voltage is given by $V(t) = V_0 \sin 2\pi f t$, where V_0 is the peak voltage and f is the frequency. We will see that the inductive reactance is greater than the capacitive reactance, $X_L > X_C$, so that the current in the circuit *lags* the voltage by $\pi/2$ radians, or 90°. Thus, the current in the circuit obeys the relation $I(t) = I_0 \sin(2\pi f t - \pi/2)$, where I_0 is the peak current.

SOLUTION

a. The instantaneous value of the voltage at a time of 1.5×10^{-4} s is

$$V(t) = V_0 \sin 2\pi f t = (64.0 \text{ V}) \sin\left[2\pi(1.00 \times 10^3 \text{ Hz})(1.50 \times 10^{-4} \text{ s})\right] = \boxed{51.8 \text{ V}}$$

Note: When evaluating the sine function in the expression above, be sure to set your calculator to the *radian* mode.

b. The inductive and capacitive reactances are

$$X_L = 2\pi f L = 2\pi(1.00 \times 10^3 \text{ Hz})(4.30 \times 10^{-3} \text{ H}) = 27.02 \text{ Ω}$$

$$X_C = \frac{1}{2\pi f C} = \frac{1}{2\pi(1.00 \times 10^3 \text{ Hz})(8.80 \times 10^{-6} \text{ F})} = 18.08 \text{ Ω}$$

Since X_L is greater than X_C, the current lags the voltage by $\pi/2$ radians. Thus, the instantaneous current in the circuit is $I(t) = I_0 \sin(2\pi f t - \pi/2)$, where $I_0 = V_0/Z$. The impedance Z of the circuit is

$$Z = \sqrt{R^2 + (X_L - X_C)^2} = \sqrt{0 + (27.0 \text{ Ω} - 18.1 \text{ Ω})^2} = 8.94 \text{ Ω}$$

The instantaneous current is

$$I = \frac{V_0}{Z} \sin\left(2\pi f t - \tfrac{1}{2}\pi\right) = \frac{64.0 \text{ V}}{8.94 \text{ Ω}} \sin\left[2\pi(1.00 \times 10^3 \text{ Hz})(1.50 \times 10^{-4} \text{ s}) - \tfrac{1}{2}\pi\right] = \boxed{-4.21 \text{ A}}$$

876 ALTERNATING CURRENT CIRCUITS

24. **REASONING AND SOLUTION** At very low frequencies the capacitors behave as if they were cut out of the circuit, while the inductors behave as if they were replaced with wires that have zero resistance. The circuit behaves as shown in drawing A, and the current delivered by the generator in the limit of very low frequency is $I_{\text{low frequency}} = V/R_1$.

A very high frequencies the capacitors behave as if they were replaced with wires that have zero resistance, while the inductors behave as if they were cut out of the circuit. The circuit now behaves as in drawing B, and the two resistors are in series. The current at very high frequencies is $I_{\text{high frequency}} = V/(R_1 + R_2)$.

The ratio of the currents is known, so that we can obtain the ratio of the resistances:

$$\frac{I_{\text{low frequency}}}{I_{\text{high frequency}}} = \frac{V/R_1}{V/(R_1 + R_2)} = \frac{R_1 + R_2}{R_1} = 1 + \frac{R_2}{R_1} = 4 \quad \text{or} \quad \boxed{\frac{R_2}{R_1} = 3}$$

A: Low frequency   B: High frequency

---

25. **REASONING AND SOLUTION** With only the resistor in the circuit, the power dissipated is $P_1 = V_0^2/R = 1.000$ W. Therefore, $V_0^2 = 1.000\,R$. When the capacitor is added in series with the resistor, the power dissipated is given by $P_2 = I_2 V_0 \cos\phi = 0.500$ W, where $\cos\phi$ is the power factor, with $\cos\phi = R/Z_2$, and $I_2 = V_0/Z_2$. Substituting yields,

$$P_2 = V_0^2 R/Z_2^2 = R^2/(R^2 + X_C^2) = 0.500 \text{ W}$$

Solving for X_C gives, $X_C = R$. When the inductor is added in series with the resistor, we have $P_3 = V_0 I_3 \cos\phi = 0.250$ W, where $I_3 = V_0/Z_3$ and $\cos\phi = R/Z_3$. Thus,

$$P_3 = R^2/(R^2 + X_L^2) = 0.250 \text{ W}$$

Solving for X_L, we find that $X_L = R\sqrt{3}$. Finally, when both the inductor and capacitor are added in series with the resistor we have

$$P_4 = \frac{V_0^2 R}{R^2 + (X_L - X_C)^2} = \frac{R^2}{R^2 + (R\sqrt{3} - R)^2} = \boxed{0.651 \text{ W}}$$

26. **REASONING AND SOLUTION** The capacitance can be obtained from Equation 23.10:

$$C = \frac{1}{4\pi^2 L f_0^2} = \frac{1}{4\pi^2 (15\times 10^{-6}\text{ H})(740\times 10^3\text{ Hz})^2} = \boxed{3.1\times 10^{-9}\text{ F}}$$

27. [SSM] **REASONING** The current in an RCL circuit is given by Equation 23.6, $I_{rms} = V_{rms}/Z$, where the impedance Z of the circuit is given by Equation 23.7 as $Z = \sqrt{R^2 + (X_L - X_C)^2}$. The current is a maximum when the impedance is a minimum for a given generator voltage. The minimum impedance occurs when the frequency is f_0, corresponding to the condition that $X_L = X_C$, or $2\pi f_0 L = 1/(2\pi f_0 C)$. Solving for the frequency f_0, called the resonant frequency, we find that

$$f_0 = \frac{1}{2\pi\sqrt{LC}}$$

Note that the resonant frequency depends on the inductance and the capacitance, but does not depend on the resistance.

SOLUTION
a. The frequency at which the current is a maximum is

$$f_0 = \frac{1}{2\pi\sqrt{LC}} = \frac{1}{2\pi\sqrt{(17.0\times 10^{-3}\text{ H})(12.0\times 10^{-6}\text{ F})}} = \boxed{352\text{ Hz}}$$

b. The maximum value of the current occurs when $f = f_0$. This occurs when $X_L = X_C$, so that $Z = R$. Therefore, according to Equation 23.6, we have

$$I_{rms} = \frac{V_{rms}}{Z} = \frac{V_{rms}}{R} = \frac{155\text{ V}}{10.0\ \Omega} = \boxed{15.5\text{ A}}$$

28. **REASONING AND SOLUTION** At resonance, $P = IV$. Therefore,

$$V = P/I = (65.0\text{ W})/(0.530\text{ A}) = \boxed{123\text{ V}}$$

29. [SSM] **REASONING AND SOLUTION** The resonant frequency is given by Equation 23.10 as

$$f_0 = \frac{1}{2\pi\sqrt{LC}} = 9.3\text{ kHz}$$

878 **ALTERNATING CURRENT CIRCUITS**

If the inductance and capacitance of the circuit are each tripled, the new resonant frequency is

$$f_0' = \frac{1}{2\pi\sqrt{(3L)(3C)}} = \frac{1}{2\pi\sqrt{9LC}} = \frac{1}{3}\left(\frac{1}{2\pi\sqrt{LC}}\right) = \frac{1}{3}f_0 = \frac{1}{3}(9.3 \text{ kHz}) = \boxed{3.1 \text{ kHz}}$$

---

30. **REASONING** The resonant frequency is given by Equation 23.10 as $f_0 = 1/(2\pi\sqrt{LC})$, where L is the inductance and C is the capacitance. In this problem the capacitance remains unchanged, and with this in mind, we will apply Equation 23.10 twice, once for each value of the inductance. The unknown capacitance can then be eliminated from the resulting equations.

SOLUTION Applying Equation 23.10 for each value of the inductance gives

$$f_{01} = \frac{1}{2\pi\sqrt{L_1 C}} \quad \text{and} \quad f_{02} = \frac{1}{2\pi\sqrt{L_2 C}}$$

Dividing the expression for f_{02} by the expression for f_{01}, we obtain

$$\frac{f_{02}}{f_{01}} = \frac{1/(2\pi\sqrt{L_2 C})}{1/(2\pi\sqrt{L_1 C})} = \sqrt{\frac{L_1}{L_2}}$$

Solving for f_{02} shows that

$$f_{02} = f_{01}\sqrt{\frac{L_1}{L_2}} = (1.3 \text{ kHz})\sqrt{\frac{7.0 \text{ mH}}{1.5 \text{ mH}}} = \boxed{2.8 \text{ kHz}}$$

---

31. **REASONING** Since the resonant frequency f_0 is known, we may use Equation 23.10, $f_0 = \frac{1}{2\pi\sqrt{LC}}$ to find the inductance L, provided the capacitance C can be determined. The capacitance can be found by using the definitions of capacitive and inductive reactances.

SOLUTION
a. Solving Equation 23.10 for the inductance, we have

$$L = \frac{1}{4\pi^2 f_0^2 C} \tag{1}$$

where f_0 is the resonant frequency. From Equations 23.2 and 23.4, the capacitive and inductive reactances are

$$X_C = \frac{1}{2\pi f C} \quad \text{and} \quad X_L = 2\pi f L$$

where f is any frequency. Solving the first of these equations for f, substituting the result into the second equation, and solving for C yields $C = \dfrac{L}{X_L X_C}$. Substituting this result into Equation (1) above and solving for L gives

$$L = \frac{1}{2\pi f_0}\sqrt{X_L X_C} = \frac{1}{2\pi(1500\text{ Hz})}\sqrt{(30.0\,\Omega)(5.0\,\Omega)} = \boxed{1.3\times 10^{-3}\text{ H}}$$

b. The capacitance is

$$C = \frac{L}{X_L X_C} = \frac{1.3\times 10^{-3}\text{ H}}{(30.0\,\Omega)(5.0\,\Omega)} = \boxed{8.7\times 10^{-6}\text{ F}}$$

---

32. **REASONING AND SOLUTION** The resonant frequency of the circuit is 225 Hz. Solving for LC at resonance, we get $LC = 1/(4\pi^2 f_0^2)$. When the circuit is *not* at resonance, $X_L/X_C = 5.36$. We have

$$\frac{X_L}{X_C} = \frac{2\pi f L}{\dfrac{1}{2\pi f C}} = 4\pi^2 f^2 LC = 4\pi^2 f^2\left(\frac{1}{4\pi^2 f_0^2}\right) = \frac{f^2}{f_0^2} = 5.36$$

Solving for the frequency f yields,

$$f = f_0\sqrt{5.36} = (225\text{ Hz})\sqrt{5.36} = \boxed{521\text{ Hz}}$$

---

33. **SSM REASONING** At the resonant frequency f_0, we have $C = 1/(4\pi^2 f_0^2 L)$. We want to determine some series combination of capacitors whose equivalent capacitance C'_s is such that $f'_0 = 3 f_0$. Thus,

$$C'_s = \frac{1}{4\pi^2 f_0'^2 L} = \frac{1}{4\pi^2 (3f_0)^2 L} = \frac{1}{9}\left(\frac{1}{4\pi^2 f_0^2 L}\right) = \frac{1}{9}C$$

880 ALTERNATING CURRENT CIRCUITS

The equivalent capacitance of a series combination of capacitors is $1/C'_s = 1/C_1 + 1/C_2 + \ldots$ If we require that all the capacitors have the same capacitance C, the equivalent capacitance is

$$\frac{1}{C'_s} = \frac{1}{C} + \frac{1}{C} + \ldots = \frac{n}{C}$$

where n is the total number of identical capacitors. Using the result above, we find that

$$\frac{1}{C'_s} = \frac{1}{\frac{1}{9}C} = \frac{n}{C} \quad \text{or} \quad n = 9$$

Therefore, the number of *additional* capacitors that must be inserted in series in the circuit so that the frequency triples is $n' = n - 1 = \boxed{8}$.

---

34. **REASONING AND SOLUTION** At resonance, the power dissipated by the circuit is $P = IV$. At a *non-resonant* frequency, the power dissipated is

$$P' = I'V \cos \phi = \tfrac{1}{2} P = \tfrac{1}{2} IV$$

so that

$$I' \cos \phi = \tfrac{1}{2} I \quad \text{and} \quad \cos \phi = \frac{I}{2I'} = \frac{R}{Z'}$$

We also know that

$$I = \frac{V}{R} \quad \text{and} \quad I' = \frac{V}{Z'} \quad \text{so that} \quad \frac{I}{I'} = \frac{Z'}{R}$$

Combining these results yields

$$\underbrace{\frac{1}{2}\left(\frac{I}{I'}\right)}_{\frac{R}{Z}} = \frac{1}{2}\left(\frac{Z'}{R}\right) \quad \text{so that} \quad Z' = R\sqrt{2}$$

Finally, the power factor, $\cos \phi$, is

$$\cos \phi = \frac{R}{Z'} = \frac{R}{R\sqrt{2}} = \boxed{0.707}$$

---

35. **REASONING AND SOLUTION**

   a. We are given that when $f = 2f_0$ that $Z = 2R = \sqrt{R^2 + (X_L - X_C)^2}$. We also know by rearranging the resonant frequency equation that

$$L = 1/(4\pi^2 f_0^2 C) \quad \text{and} \quad C = 1/(4\pi^2 f_0^2 L)$$

Squaring the equation for Z we get

$$Z^2 = 4R^2 = R^2 + (X_L - X_L)^2 \quad \text{so that} \quad X_L - X_C = R\sqrt{3}$$

Now,

$$X_L = 2\pi f L = 4\pi f_0 L \quad \text{and} \quad X_C = \frac{1}{2\pi f C} = \frac{1}{4\pi f_0 C}$$

so that

$$X_L - X_C = 4\pi f_0 L - \frac{1}{4\pi f_0 C} = 4\pi f_0 L - \frac{1}{4\pi f_0 \left(\frac{1}{4\pi^2 f_0^2 L}\right)} = 3\pi f_0 L$$

Combining yields, $\sqrt{3}\,\pi f_0 L = R$. Now look at X_L/R.

$$\frac{X_L}{R} = \frac{4\pi f_0 L}{\sqrt{3}\pi f_0 L} = \boxed{\frac{4}{\sqrt{3}}}$$

b. A similar treatment yields X_C/R:

$$\frac{X_C}{R} = \frac{\frac{1}{4\pi f_0 C}}{\frac{3}{4\sqrt{3}\pi f_0 C}} = \boxed{\frac{1}{\sqrt{3}}}$$

---

36. **REASONING** From Figure 23.11 we see that $V_0^2 = (V_L - V_C)^2 + V_R^2$. Since $V_L = 0$ ($L = 0$), and we know V_0 and V_R, we can use this equation to find V_C.

    SOLUTION Solving the equation above for V_C gives

    $$V_C = \sqrt{V_0^2 - V_R^2} = \sqrt{(45\text{ V})^2 - (24\text{ V})^2} = \boxed{38\text{ V}}$$

---

37. [SSM] **REASONING** The voltage across the capacitor reaches its maximum instantaneous value when the generator voltage reaches its maximum instantaneous value. The maximum value of the capacitor voltage first occurs one-fourth of the way, or one-quarter of a period, through a complete cycle (see the voltage curve in Figure 23.4).

882 ALTERNATING CURRENT CIRCUITS

SOLUTION The period of the generator is $T = 1/f = 1/(5.00 \text{ Hz}) = 0.200 \text{ s}$. Therefore, the least amount of time that passes before the instantaneous voltage across the capacitor reaches its maximum value is $\frac{1}{4}T = \frac{1}{4}(0.200 \text{ s}) = \boxed{5.00 \times 10^{-2} \text{ s}}$.

---

38. **REASONING AND SOLUTION** At very high frequencies the capacitors behave as if they were replaced with wires that have zero resistance, while the inductors behave as if they were cut out of the circuit. The drawings below show the circuits under this condition. Circuit I behaves as if the two resistors are in parallel, and the equivalent resistance can be obtained from $R_P^{-1} = R^{-1} + R^{-1}$ as $R_P = R/2$. Circuit II behaves as if the two resistors are in series, and the equivalent resistance is $R_S = R + R = 2R$. In either case, the current is the voltage divided by the resistance. Therefore, the ratio of the currents in the two circuits is

$$\frac{I_{\text{circuit I}}}{I_{\text{circuit II}}} = \frac{V/(R/2)}{V/(2R)} = \boxed{4}$$

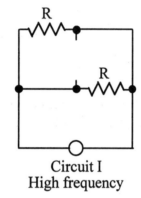
Circuit I
High frequency

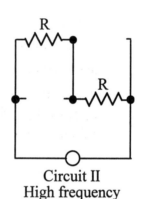
Circuit II
High frequency

---

39. **REASONING AND SOLUTION**
a. Equation 20.14 gives

$$I = \frac{V}{R} = \frac{120 \text{ V}}{240 \text{ }\Omega} = \boxed{0.50 \text{ A}}$$

b. Since there is no inductor, Equations 23.6 and 23.7 apply with $X_L = 0$. Therefore, the current is $I = V/Z = V/\sqrt{R^2 + X_C^2}$. Using Equation 23.2 for X_C, we find

$$I = \frac{V}{\sqrt{R^2 + \left(\frac{1}{2\pi f C}\right)^2}} = \frac{120 \text{ V}}{\sqrt{(240 \text{ }\Omega)^2 + \left[\frac{1}{2\pi(60.0 \text{ Hz})(10.0 \times 10^{-6} \text{ F})}\right]^2}} = \boxed{0.34 \text{ A}}$$

c. When an inductor is present, Equations 23.6 and 23.7 give the current as $I = V/Z = V/\sqrt{R^2 + (X_L - X_C)^2}$. This expression reduces to $I = V/R$ when $X_L = X_C$. Using Equations 23.2 and 23.4 for the reactances, we find that

$$2\pi f L = \frac{1}{2\pi f C} \quad \text{or} \quad L = \frac{1}{4\pi^2 f^2 C} = \frac{1}{4\pi^2 (60.0 \text{ Hz})^2 (10.0 \times 10^{-6} \text{ F})} = \boxed{0.704 \text{ H}}$$

$\boxed{\text{Yes,}}$ it is possible to return the current to the value calculated in part (a).

---

40. **REASONING AND SOLUTION** We know that

$$V = IX_L = I(2\pi f L) = (0.20 \text{ A})(2\pi)(750 \text{ Hz})(0.080 \text{ H}) = \boxed{75 \text{ V}}$$

---

41. **SSM WWW** **REASONING** Since the capacitor and the inductor are connected in parallel, the voltage across each of these elements is the same or $V_L = V_C$. Using Equations 23.3 and 23.1, respectively, this becomes $I_{rms} X_L = I_{rms} X_C$. Since the currents in the inductor and capacitor are equal, this relation simplifies to $X_L = X_C$. Therefore, we can find the value of the inductance by equating the expressions (Equations 23.4 and 23.2) for the inductive reactance and the capacitive reactance, and solving for L.

SOLUTION Since $X_L = X_C$, we have

$$2\pi f L = \frac{1}{2\pi f C}$$

Therefore, the value of the inductance is

$$L = \frac{1}{4\pi^2 f^2 C} = \frac{1}{4\pi^2 (60.0 \text{ Hz})^2 (40.0 \times 10^{-6} \text{ F})} = 0.176 \text{ H} = \boxed{176 \text{ mH}}$$

---

42. **REASONING AND SOLUTION** We begin by calculating the impedance of the circuit using $Z = \sqrt{R^2 + (X_L - X_C)^2}$. We have

$$X_C = \frac{1}{2\pi f C} = \frac{1}{2\pi (1350 \text{ Hz})(4.10 \times 10^{-6} \text{ F})} = 28.8 \; \Omega$$

$$X_L = 2\pi f L = 2\pi (1350 \text{ Hz})(5.30 \times 10^{-3} \text{ H}) = 45.0 \; \Omega$$

884 ALTERNATING CURRENT CIRCUITS

$$Z = \sqrt{(16.0\,\Omega)^2 + (45.0\,\Omega - 28.8\,\Omega)^2} = 22.8\,\Omega$$

The current is therefore,

$$I = V/Z = (15.0\text{ V})/(22.8\,\Omega) = 0.658\text{ A}$$

Since the circuit elements are in series, the current through each element is the same. The voltage across each element is

$$V_R = IR = (0.658\text{A})(16.0\,\Omega) = \boxed{10.5\text{ V}}$$

$$V_C = IX_C = (0.658\text{ A})(28.8\,\Omega) = \boxed{19.0\text{ V}}$$

$$V_L = IX_L = (0.658\text{ A})(45.0\,\Omega) = \boxed{29.6\text{ V}}$$

---

43. **SSM** **WWW** *REASONING* Since we know the values of the resonant frequency of the circuit, the capacitance, and the generator voltage, we can find the value of the inductance from Equation 23.10, the expression for the resonant frequency. The resistance can be found from energy considerations at resonance; the power factor is given by $\cos\phi$, where the phase angle ϕ is given by Equation 23.8, $\tan\phi = (X_L - X_C)/R$.

SOLUTION
a. Solving Equation 23.10 for the inductance L, we find that

$$L = \frac{1}{4\pi^2 f_0^2 C} = \frac{1}{4\pi^2 (1.30\times 10^3\text{ Hz})^2 (5.10\times 10^{-6}\text{ F})} = \boxed{2.94\times 10^{-3}\text{ H}}$$

b. At resonance, $f = f_0$, and the current is a maximum. This occurs when $X_L = X_C$, so that $Z = R$. Thus, the average power $\overline{P}$ provided by the generator is $\overline{P} = V_{rms}^2 / R$, and solving for R we find

$$R = \frac{V_{rms}^2}{\overline{P}} = \frac{(11.0\text{ V})^2}{25.0\text{ W}} = \boxed{4.84\,\Omega}$$

c. When the generator frequency is 2.31 kHz, the individual reactances are

$$X_C = \frac{1}{2\pi f C} = \frac{1}{2\pi(2.31\times 10^3\text{ Hz})(5.10\times 10^{-6}\text{ F})} = 13.5\,\Omega$$

$$X_L = 2\pi f L = 2\pi(2.31\times 10^3\text{ Hz})(2.94\times 10^{-3}\text{ H}) = 42.7\,\Omega$$

The phase angle ϕ is, from Equation 23.8,

$$\phi = \tan^{-1}\left(\frac{X_L - X_C}{R}\right) = \tan^{-1}\left(\frac{42.7\ \Omega - 13.5\ \Omega}{4.84\ \Omega}\right) = 80.6°$$

The power factor is then given by

$$\cos\phi = \cos 80.6° = \boxed{0.163}$$

---

44. **REASONING AND SOLUTION** With the inductor hooked up to the battery we have $V = I_0 R$. When the inductor is hooked up to the generator we have $V = (I_0/4)Z$. Since the voltage is the same in each case, we can equate these two expressions: $I_0 R = (I_0/4)Z$. Therefore,

$$R = \frac{1}{4}Z = \frac{1}{4}\sqrt{R^2 + X_L^2}$$

Squaring gives

$$R^2 = \frac{1}{16}(R^2 + X_L^2) \quad \text{or} \quad R = \frac{X_L}{\sqrt{15}}$$

Thus,

$$R = \frac{2\pi f L}{\sqrt{15}} = \frac{2\pi(1200\ \text{Hz})(4.5 \times 10^{-3}\ \text{H})}{\sqrt{15}} = \boxed{8.8\ \Omega}$$

---

45. **REASONING AND SOLUTION** The current in an RCL-circuit is

$$I = \frac{V}{\sqrt{R^2 + (X_L - X_C)^2}}$$

Rearranging terms

$$(X_L - X_C)^2 = (V/I)^2 - R^2$$

Using Equations 23.2 and 23.4 for X_C and X_L, respectively, we obtain

$$2\pi f L - \frac{1}{2\pi f C} = \sqrt{\left(\frac{V}{I}\right)^2 - R^2} = \sqrt{\left(\frac{26.0\ \text{V}}{0.141\ \text{A}}\right)^2 - (108\ \Omega)^2} = 149\ \Omega$$

Multiplying by f leads to

$$2\pi f^2 L - (149\ \Omega)f - 1/(2\pi C) = 0$$

or

886 ALTERNATING CURRENT CIRCUITS

$$2\pi f^2 (5.42 \times 10^{-3} \text{ H}) - (149 \text{ }\Omega)f - 1/[(2\pi(0.200 \times 10^{-6} \text{ F})] = 0$$

We can solve this quadratic equation for the frequencies. We obtain

$$f_1 = \boxed{3.11 \times 10^3 \text{ Hz}} \quad \text{and} \quad f_2 = \boxed{7.50 \times 10^3 \text{ Hz}}$$

---

46. **CONCEPT QUESTIONS** a. The capacitance of a parallel plate capacitor is given by Equation 19.10 as $C = \kappa \varepsilon_0 A/d$, where κ is the dielectric constant of the material between the plates, ε_0 is the permittivity of free space, A is the area of each plate, and d is the separation between the plates. When the capacitor is empty, $\kappa = 1$, so that $C = \kappa C_{\text{empty}}$. Thus, the capacitance increases when the dielectric material is inserted.

b. According to Equation 20.18, the equivalent capacitance of two capacitors in parallel is $C_P = C_1 + C_2$, where C_1 and C_2 are the individual capacitances. Therefore, C_P is greater than either C_1 or C_2.

c. According to Equation 23.1, the current is given by $I_{\text{rms}} = V_{\text{rms}}/X_C$, where V_{rms} is the rms voltage of the generator, and $X_C = 1/(2\pi f C)$ is the capacitive reactance. For a given voltage, smaller reactances lead to greater currents. Thus, when the capacitors are connected in parallel, the greater capacitance leads to a smaller reactance, which in turn leads to a greater current. As a result, the current delivered by the generator increases when the second capacitor is connected.

SOLUTION Using Equation 23.1 to express the current as $I_{\text{rms}} = V_{\text{rms}}/X_C$ and Equation 23.2 to express the reactance as $X_C = 1/(2\pi f C)$, we have for the current that

$$I_{\text{rms}} = \frac{V_{\text{rms}}}{X_C} = \frac{V_{\text{rms}}}{1/(2\pi f C)} = V_{\text{rms}} 2\pi f C$$

Applying this result to the case where the empty capacitor C_1 is connected alone to the generator and to the case where the full capacitor C_2 is connected in parallel with C_1, we obtain

$$\underbrace{I_{1,\text{rms}} = V_{\text{rms}} 2\pi f C_1}_{C_1 \text{ alone}} \quad \text{and} \quad \underbrace{I_{P,\text{rms}} = V_{\text{rms}} 2\pi f C_P}_{C_1 \text{ and } C_2 \text{ in parallel}}$$

Dividing the two expressions gives

$$\frac{I_{P,\text{rms}}}{I_{1,\text{rms}}} = \frac{V_{\text{rms}} 2\pi f C_P}{V_{\text{rms}} 2\pi f C_1} = \frac{C_P}{C_1}$$

According to Equation 20.18, the equivalent capacitance of the two capacitors in parallel is $C_P = C_1 + C_2$, so that the result for the current ratio becomes

$$\frac{I_{P,\text{rms}}}{I_{1,\text{rms}}} = \frac{C_1 + C_2}{C_1} = 1 + \frac{C_2}{C_1}$$

Since the capacitance of a filled capacitor is given by Equation 19.10 as $C = \kappa \varepsilon_0 A/d$, we find that

$$\frac{I_{P,\text{rms}}}{I_{1,\text{rms}}} = 1 + \frac{\kappa \varepsilon_0 A/d}{\varepsilon_0 A/d} = 1 + \kappa$$

Solving for $I_{P,\text{rms}}$ gives

$$I_{P,\text{rms}} = I_{1,\text{rms}}(1+\kappa) = (0.22 \text{ A})(1+4.2) = \boxed{1.1 \text{ A}}$$

---

47. **CONCEPT QUESTIONS** a. The current in L_1 is given by Equation 23.3 as $I_{\text{rms}} = V_{\text{rms}}/X_L$, where $X_L = 2\pi f L_1$ is the inductive reactance of L_1 according to Equation 23.4. This current does not depend in any way on L_2 and exists whether or not L_2 is present.

b. The current in L_2 is given by Equation 23.3 as $I_{\text{rms}} = V_{\text{rms}}/X_L$, where $X_L = 2\pi f L_2$ is the inductive reactance of L_2 according to Equation 23.4. This current does not depend in any way on L_1 and exists whether or not L_1 is present.

c. The current delivered to the parallel combination is the sum of the currents delivered to each inductance and is, therefore, greater than either individual current.

SOLUTION Using Equation 23.3 to express the current as $I_{\text{rms}} = V_{\text{rms}}/X_L$ and Equation 23.4 to express the reactance as $X_L = 2\pi f L$, we have for the current that

$$I_{\text{rms}} = \frac{V_{\text{rms}}}{X_L} = \frac{V_{\text{rms}}}{2\pi f L}$$

Applying this result to the case where L_1 or L_2 is connected alone to the generator, we obtain

$$\underbrace{I_{1,\text{rms}} = \frac{V_{\text{rms}}}{X_L} = \frac{V_{\text{rms}}}{2\pi f L_1}}_{L_1 \text{ alone}} \quad \text{and} \quad \underbrace{I_{2,\text{rms}} = \frac{V_{\text{rms}}}{X_L} = \frac{V_{\text{rms}}}{2\pi f L_2}}_{L_2 \text{ alone}}$$

The current delivered to L_1 alone is

888 ALTERNATING CURRENT CIRCUITS

$$I_{1,\text{rms}} = \frac{V_{\text{rms}}}{2\pi f L_1} = \frac{240 \text{ V}}{2\pi (2200 \text{ Hz})(6.0 \times 10^{-3} \text{ H})} = \boxed{2.9 \text{ A}}$$

The current delivered to the parallel combination of L_1 and L_2 is the sum of that delivered individually to each inductor and is

$$I_{P,\text{rms}} = I_{1,\text{rms}} + I_{2,\text{rms}} = \frac{V_{\text{rms}}}{2\pi f L_1} + \frac{V_{\text{rms}}}{2\pi f L_2} = \frac{V_{\text{rms}}}{2\pi f}\left(\frac{1}{L_1} + \frac{1}{L_2}\right)$$

$$= \frac{240 \text{ V}}{2\pi (2200 \text{ Hz})}\left(\frac{1}{6.0 \times 10^{-3} \text{ H}} + \frac{1}{9.0 \times 10^{-3} \text{ H}}\right) = \boxed{4.8 \text{ A}}$$

As expected, this value is greater than the current in L_1 alone.

---

48. **CONCEPT QUESTIONS** a. The phase angle is given by Equation 23.8 as $\tan \phi = (X_L - X_C)/R$. When a series circuit contains only a resistor and a capacitor, the inductive reactance X_L is zero, and the phase angle is negative, signifying that the current leads the voltage of the generator. The impedance is given by Equation 23.7 with $X_L = 0 \text{ }\Omega$, or $Z = \sqrt{R^2 + X_C^2}$.

b. The phase angle is given by Equation 23.8 as $\tan \phi = (X_L - X_C)/R$. When a series circuit contains only a resistor and an inductor, the capacitive reactance X_C is zero, and the phase angle is positive, signifying that the current lags behind the voltage of the generator. The impedance is given by Equation 23.7 with $X_C = 0 \text{ }\Omega$, or $Z = \sqrt{R^2 + X_L^2}$.

SOLUTION Since the phase angle is negative, we can conclude that only a resistor and a capacitor are present. Using Equations 23.8, then, we have

$$\tan \phi = \frac{-X_C}{R} \quad \text{or} \quad X_C = -R\tan(-75.0°) = 3.73 R$$

According to Equation 23.7, the impedance is

$$Z = 192 \text{ }\Omega = \sqrt{R^2 + X_C^2}$$

Substituting $X_C = 3.73R$ into this expression for Z gives

$$192 \, \Omega = \sqrt{R^2 + (3.73 \, R)^2} = \sqrt{14.9 \, R^2} \quad \text{or} \quad (192 \, \Omega)^2 = 14.9 \, R^2$$

$$R = \sqrt{\frac{(192 \, \Omega)^2}{14.9}} = \boxed{49.7 \, \Omega}$$

Using the fact that $X_C = 3.73 \, R$, we obtain

$$X_C = 3.73 \, (49.7 \, \Omega) = \boxed{185 \, \Omega}$$

---

49. **CONCEPT QUESTIONS** a. Only the resistor, on average, dissipates power.

b. According to Equation 23.6, the current is given by $I_{rms} = V_{rms}/Z$, where V_{rms} is the voltage of the generator and Z is the circuit impedance.

c. The impedance is given by Equation 23.7 as $Z = \sqrt{R^2 + (X_L - X_C)^2}$. At resonance the inductive reactance X_L and the capacitive reactance X_C are equal, so that $Z = R$.

d. Since average power is dissipated only by the resistor, we have $\overline{P} = I_{rms}^2 R$, according to Equation 20.15b. But the current at resonance is $I_{rms} = V_{rms}/R$. Therefore, the power is

$$\overline{P} = I_{rms}^2 R = \left(\frac{V_{rms}}{R}\right)^2 R = \frac{V_{rms}^2}{R}$$

SOLUTION The average power dissipated in the circuit at resonance is

$$\overline{P} = \frac{V_{rms}^2}{R} = \frac{(3.0 \, \text{V})^2}{92 \, \Omega} = \boxed{0.098 \, \text{W}}$$

---

50. **CONCEPT QUESTIONS** a. The resonant frequency is given by Equation 23.10 as $f_0 = 1/(2\pi\sqrt{LC})$ and is inversely proportional to the square root of the circuit capacitance C. Therefore, to reduce the resonant frequency, it is necessary to increase the circuit capacitance.

b. The equivalent capacitance of two capacitors in parallel is $C_P = C_1 + C_2$, according to Equation 20.18 The equivalent capacitance is greater than either capacitance individually. Therefore, to increase the circuit capacitance, C_2 should be added in parallel with C_1.

890 ALTERNATING CURRENT CIRCUITS

SOLUTION The initial resonant frequency is f_{01}. The resonant frequency that results after C_2 is added in parallel with C_1 is f_{0P}. Using Equation 23.10, we can express both of these frequencies as follows:

$$f_{01} = \frac{1}{2\pi\sqrt{LC_1}} \quad \text{and} \quad f_{0P} = \frac{1}{2\pi\sqrt{LC_P}}$$

Here, C_P is the equivalent parallel capacitance. Dividing the expression for f_{01} by the expression for f_{0P} shows that

$$\frac{f_{01}}{f_{0P}} = \frac{1/\left(2\pi\sqrt{LC_1}\right)}{1/\left(2\pi\sqrt{LC_P}\right)} = \sqrt{\frac{C_P}{C_1}}$$

According to Equation 20.18, the equivalent capacitance is $C_P = C_1 + C_2$, so that this frequency ratio becomes

$$\frac{f_{01}}{f_{0P}} = \sqrt{\frac{C_1 + C_2}{C_1}} = \sqrt{1 + \frac{C_2}{C_1}}$$

Squaring both sides of this result and solving for C_2, we find

$$\left(\frac{f_{01}}{f_{0P}}\right)^2 = 1 + \frac{C_2}{C_1}$$

$$C_2 = C_1\left[\left(\frac{f_{01}}{f_{0P}}\right)^2 - 1\right] = (2.60\ \mu\text{F})\left[\left(\frac{7.30\ \text{kHz}}{5.60\ \text{kHz}}\right)^2 - 1\right] = \boxed{1.8\ \mu\text{F}}$$

CHAPTER 24 | ELECTROMAGNETIC WAVES

PROBLEMS

1. **SSM** *REASONING AND SOLUTION* This is a standard exercise in units conversion. We first determine the number of meters in one light-year. The distance that light travels in one year is

$$d = ct = (3.00 \times 10^8 \text{ m/s})(1.00 \text{ year})\left(\frac{365.25 \text{ days}}{1 \text{ year}}\right)\left(\frac{24 \text{ hours}}{1.0 \text{ day}}\right)\left(\frac{3600 \text{ s}}{1 \text{ hour}}\right) = 9.47 \times 10^{15} \text{ m}$$

Thus, 1 light year = 9.47×10^{15} m. Then, the distance to Alpha Centauri is

$$(4.3 \text{ light-years})\left(\frac{9.47 \times 10^{15} \text{ m}}{1 \text{ light-year}}\right) = \boxed{4.1 \times 10^{16} \text{ m}}$$

2. *REASONING AND SOLUTION* The time is the distance divided by the speed of the waves.

   a. $t = s/c = (3.85 \times 10^8 \text{ m})/(3.00 \times 10^8 \text{ m/s}) = \boxed{1.28 \text{ s}}$

   b. A calculation similar to that in part a yields $\boxed{t = 1.9 \times 10^2 \text{ s}}$.

3. *REASONING AND SOLUTION* Let $f_1 = 88.0 \times 10^6$ Hz then

$$f_1 = \frac{1}{2\pi\sqrt{LC_1}}$$

Let $f_2 = 108.0 \times 10^6$ Hz then

$$f_2 = \frac{1}{2\pi\sqrt{LC_2}}$$

Division and rearrangement of the above expressions give

$$C_2 = (f_1/f_2)^2 C_1 = [(88.0 \times 10^6 \text{ Hz})/(108.0 \times 10^6 \text{ Hz})]^2 (33.0 \times 10^{-12} \text{ F}) = \boxed{2.19 \times 10^{-11} \text{ F}}$$

892 ELECTROMAGNETIC WAVES

4. **REASONING** The electromagnetic wave will be picked up by the radio when the resonant frequency f_0 of the circuit in Figure 24.5 is equal to the frequency of the broadcast wave, or $f_0 = 1400$ kHz. This frequency, in turn, is related to the capacitance C and inductance L of the circuit through Equation 23.10, $f_0 = 1/(2\pi\sqrt{LC})$. Since C is known, we can use this relation to find the inductance.

SOLUTION Solving the relation $f_0 = 1/(2\pi\sqrt{LC})$ for the inductance L, we find that

$$L = \frac{1}{4\pi^2 f_0^2 C} = \frac{1}{4\pi^2 (1400 \times 10^3 \text{ Hz})^2 (8.4 \times 10^{-11} \text{ F})} = \boxed{1.5 \times 10^{-4} \text{ H}}$$

5. **SSM REASONING** The equation that represents the wave mathematically is $y = A\sin(2\pi ft - 2\pi x/\lambda)$. In this expression the amplitude is $A = 156$ N/C. The wavelength λ can be calculated using Equation 16.1, and we obtain

$$\lambda = \frac{c}{f} = \frac{3.00 \times 10^8 \text{ m/s}}{1.50 \times 10^8 \text{ Hz}} = 2.00 \text{ m}$$

SOLUTION
a. For $t = 0$ s, the wave expression becomes

$$y = A\sin(2\pi ft - 2\pi x/\lambda) = 156 \sin\left[2\pi f(0) - \frac{2\pi x}{2.00}\right] = -156 \sin\left(\frac{2\pi x}{2.00}\right) = -156 \sin(\pi x)$$

In this result, the units are suppressed for convenience. The following table gives the values of the electric field obtained using this version of the wave expression with the given values of the position x. The term πx is in radians when x is in meters, and conversion from radians to degrees is accomplished using the fact that 2π rad $= 360°$.

x	$y = -156 \sin(\pi x)$
0 m	$-156 \sin(0) = -156 \sin(0°) = 0$
0.50 m	$-156 \sin(0.50\pi) = -156 \sin(90°) = -156$
1.00 m	$-156 \sin(1.00\pi) = -156 \sin(180°) = 0$
1.50 m	$-156 \sin(1.50\pi) = -156 \sin(270°) = +156$
2.00 m	$-156 \sin(2.00\pi) = -156 \sin(360°) = 0$

These values for the electric field are plotted in the graph shown at the right.

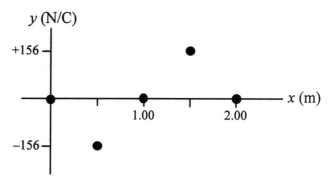

b. For $t = T/4$, we use the fact that $f = 1/T$, and the wave expression becomes

$$y = A\sin(2\pi f t - 2\pi x/\lambda) = 156\sin\left[2\pi\left(\frac{1}{T}\right)\left(\frac{T}{4}\right) - \frac{2\pi x}{2.00}\right] = 156\sin\left(\frac{\pi}{2} - \pi x\right) = 156\cos(\pi x)$$

In this result, the units are suppressed for convenience. The following table gives the values of the electric field obtained using this version of the wave expression with the given values of the position x. The term πx is in radians when x is in meters, and conversion from radians to degrees is accomplished using the fact that 2π rad $= 360°$.

x	$y = 156 \cos(\pi x)$
0 m	$156 \cos(0) = 156 \cos(0°) = +156$
0.50 m	$156 \cos(0.50\,\pi) = 156 \cos(90°) = 0$
1.00 m	$156 \cos(1.00\,\pi) = 156 \cos(180°) = -156$
1.50 m	$156 \cos(1.50\,\pi) = 156 \cos(270°) = 0$
2.00 m	$156 \cos(2.00\,\pi) = 156 \cos(360°) = +156$

These values for the electric field are plotted in the graph shown at the right.

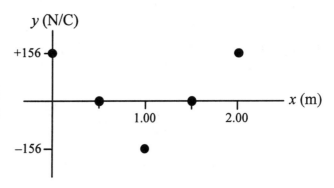

894 ELECTROMAGNETIC WAVES

6. **REASONING AND SOLUTION** The average flux change through the coil in one fourth of the wave period is, according to Faraday's law, $\Delta\Phi = NAB = NB_0A$. The magnitude of the average emf is then emf $= \Delta\Phi/\Delta t = NB_0A/\Delta t$. Now $\Delta t = T/4 = 1/(4f)$, so

$$\text{Emf} = 4NfB_0A = 4(450)(1.2 \times 10^6 \text{ Hz})(2.0 \times 10^{-13} \text{ T})(\pi)(0.25 \text{ m})^2 = \boxed{8.5 \times 10^{-5} \text{ V}}$$

---

7. **REASONING AND SOLUTION** Using Equation 16.1, we obtain

$$\lambda = c/f = (2.9979 \times 10^8 \text{ m/s})/(26.965 \times 10^6 \text{ Hz}) = \boxed{11.118 \text{ m}}$$

---

8. **REASONING AND SOLUTION** The longest FM radio wavelength is

$$\lambda = c/f = (3.00 \times 10^8 \text{ m/s})/(88.0 \times 10^6 \text{ Hz}) = \boxed{3.41 \text{ m}}$$

The shortest wavelength is

$$\lambda = c/f = (3.00 \times 10^8 \text{ m/s})/(108.0 \times 10^6 \text{ Hz}) = \boxed{2.78 \text{ m}}$$

---

9. **SSM WWW REASONING AND SOLUTION** According to Equation 16.1, the wavelength is $\lambda = c/f$. Since the rods are adjusted so that each one has a length of $\lambda/4$, the length of each rod is

$$L = \frac{\lambda}{4} = \frac{(c/f)}{4} = \frac{(3.00 \times 10^8 \text{ m/s})/(60.0 \times 10^6 \text{ Hz})}{4} = \boxed{1.25 \text{ m}}$$

---

10. **REASONING AND SOLUTION** According to Equation 16.1, the wavelength of these waves is $\lambda = c/f$. Therefore,

$$\frac{\lambda_{MRI}}{\lambda_{PET}} = \frac{c/f_{MRI}}{c/f_{PET}} = \frac{f_{PET}}{f_{MRI}} = \frac{1.23 \times 10^{20} \text{ Hz}}{6.38 \times 10^7 \text{ Hz}} = \boxed{1.93 \times 10^{12}}$$

---

11. **SSM REASONING AND SOLUTION** The number of wavelengths that can fit across the width W of your thumb is W/λ. From Equation 16.1, we know that $\lambda = c/f$, so

$$\text{No. of wavelengths} = \frac{W}{\lambda} = \frac{Wf}{c} = \frac{(2.0 \times 10^{-2} \text{ m})(5.5 \times 10^{14} \text{ Hz})}{3.0 \times 10^8 \text{ m/s}} = \boxed{3.7 \times 10^4}$$

12. ***REASONING AND SOLUTION*** The VHF wavelength is

$$\lambda = c/f = (3.00 \times 10^8 \text{ m/s})/(63.0 \times 10^6 \text{ Hz}) = 4.76 \text{ m}$$

The UHF wavelength is

$$\lambda = (3.00 \times 10^8 \text{ m/s})/(527 \times 10^6 \text{ Hz}) = 0.569 \text{ m}$$

The ratio of the wavelengths is $(4.76 \text{ m})/(0.569 \text{ m}) = \boxed{8.37}$

---

13. ***REASONING AND SOLUTION*** The wavelength of the microwaves is $\lambda = 4L$, so the frequency is

$$f = c/\lambda = c/(4L) = (3.00 \times 10^8 \text{ m/s})/(4 \times 0.50 \times 10^{-2} \text{ m}) = \boxed{1.5 \times 10^{10} \text{ Hz}}$$

---

14. ***REASONING AND SOLUTION*** The distance traveled by each wave is $s = ct$. The difference in distance traveled is then

$$\Delta s = c\Delta t = (3.00 \times 10^8 \text{ m/s})(5.0 \times 10^{-7} \text{ s}) = \boxed{150 \text{ m}}$$

---

15. **SSM** ***REASONING*** We proceed by first finding the time t for sound waves to travel between the astronauts. Since this is the same time it takes for the electromagnetic waves to travel to earth, the distance between earth and the spaceship is $d_{\text{earth-ship}} = ct$.

SOLUTION The time it takes for sound waves to travel at 343 m/s through the air between the astronauts is

$$t = \frac{d_{\text{astronaut}}}{v_{\text{sound}}} = \frac{1.5 \text{ m}}{343 \text{ m/s}} = 4.4 \times 10^{-3} \text{ s}$$

Therefore, the distance between the earth and the spaceship is

$$d_{\text{earth-ship}} = ct = (3.0 \times 10^8 \text{ m/s})(4.4 \times 10^{-3} \text{ s}) = \boxed{1.3 \times 10^6 \text{ m}}$$

896 ELECTROMAGNETIC WAVES

16. **REASONING AND SOLUTION** The time it takes for light from Sirius to reach earth is the distance s divided by the speed of light,

$$t = \frac{s}{c} = \frac{8.3 \times 10^{16} \text{ m}}{3.0 \times 10^8 \text{ m/s}} \left(\frac{1 \text{ h}}{3600 \text{ s}}\right)\left(\frac{1 \text{ d}}{24 \text{ h}}\right)\left(\frac{1 \text{ y}}{365.25 \text{ d}}\right) = \boxed{8.8 \text{ y}}$$

17. **REASONING** We see from Figure 24.11, that the rotating mirror is eight-sided. Therefore, the minimum angular speed is that angular speed at which one side of the mirror rotates one-eighth of a revolution during the time it takes for the light to make the round-trip between Mt. San Antonio and Mt. Wilson. The angular speed of the mirror can be determined by dividing the angular displacement of one-eighth of a revolution by the round-trip time, where the round-trip time is the round-trip distance divided by the speed of light.

SOLUTION The round-trip travel time is

$$t = \frac{2(35 \times 10^3 \text{ m})}{3.00 \times 10^8 \text{ m/s}} = 2.3 \times 10^{-4} \text{ s}$$

Therefore, the minimum angular speed for the rotating mirror is

$$\omega = \frac{\frac{1}{8} \text{ revolution}}{2.3 \times 10^{-4} \text{ s}} = \boxed{540 \text{ rev/s}}$$

18. **REASONING AND SOLUTION** The time t that it takes for the telephone call to go from one city to the other is equal to the distance s traveled by the electromagnetic wave divided by the speed of light, $t = s/c$. The distance $s/2$ from one city to the satellite is given by the Pythagorean theorem as (see the drawing)

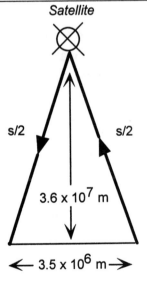

$$\frac{s}{2} = \sqrt{(3.6 \times 10^7 \text{ m})^2 + \left(\frac{3.5}{2} \times 10^6 \text{ m}\right)^2} = 3.6 \times 10^7 \text{ m}$$

The time for the wave to travel from one city up to the satellite and back to the other is

$$t = \frac{s}{c} = \frac{2(3.6 \times 10^7 \text{ m})}{3.0 \times 10^8 \text{ m/s}} = \boxed{0.24 \text{ s}}$$

19. **SSM** ***REASONING AND SOLUTION*** If d is the distance between the mirrors, the time required for the echo to be heard is $t = 2d/v_{sound}$. In this time, the light will travel a total distance D where $D = ct = c(2d/v_{sound})$. Therefore, the number of times that the flash travels the round-trip distance between the mirrors in the time t is

$$N = \frac{D}{2d} = \frac{2cd/v_{sound}}{2d} = \frac{c}{v_{sound}} = \frac{3.00 \times 10^8 \text{ m/s}}{343 \text{ m/s}} = \boxed{8.75 \times 10^5}$$

20. ***REASONING AND SOLUTION*** The time required for the sound to travel from the TV to the viewer is $t_1 = s_1/v$. The time required for the EM waves to travel to the viewer's set is $t = s/c$. The time required for the sound to travel from the celebrity to the reporter is $t_2 = s_2/v$. Now $t_1 + t = t_2$, so

$$s = (s_2 - s_1)(c/v) = (4.1 \text{ m} - 2.3 \text{ m})(3.00 \times 10^8 \text{ m/s})/(343 \text{ m/s}) = \boxed{1.6 \times 10^6 \text{ m}}$$

21. ***REASONING AND SOLUTION*** Using Equations 24.5b and 24.3, we find the following results.

a. $$E_{rms} = \sqrt{\frac{\overline{S}}{c\varepsilon_0}} = \sqrt{\frac{1.23 \times 10^9 \text{ W/m}^2}{(3.00 \times 10^8 \text{ m/s})[8.85 \times 10^{-12} \text{ C}^2/(\text{N} \cdot \text{m}^2)]}} = \boxed{6.81 \times 10^5 \text{ N/C}}$$

b. $B_{rms} = E_{rms}/c = \boxed{2.27 \times 10^{-3} \text{ T}}$

22. ***REASONING AND SOLUTION*** According to Equation 24.3, we obtain

$$B = E/c = (1470 \text{ N/C})/(3.00 \times 10^8 \text{ m/s}) = \boxed{4.90 \times 10^{-6} \text{ T}}$$

23. **SSM** ***REASONING*** The rms value E_{rms} of the electric field is related to the average energy density $\overline{u}$ of the microwave radiation according to Equation 24.2b: $\overline{u} = \varepsilon_0 E_{rms}^2$.

SOLUTION Solving for E_{rms} gives

$$E_{rms} = \sqrt{\frac{\overline{u}}{\varepsilon_0}} = \sqrt{\frac{4 \times 10^{-14} \text{ J/m}^3}{8.85 \times 10^{-12} \text{ C}^2/(\text{N} \cdot \text{m}^2)}} = \boxed{0.07 \text{ N/C}}$$

24. **REASONING AND SOLUTION** The intensity S of a wave is the power passing perpendicularly through a surface divided by the area A of the surface. But power is the total energy U per unit time t, so the intensity can be written as

$$S = \frac{\text{Total energy}}{\text{Time} \cdot \text{Area}} = \frac{U}{tA}$$

Equation 24.5c relates the intensity S of the electromagnetic wave to the magnitude B of its magnetic field; namely $S = (c/\mu_0)B^2$. Combining these two results, we have

$$\frac{U}{tA} = \frac{c}{\mu_0} B^2$$

If the rms value for the magnetic field is used, the energy becomes the average energy $\bar{U}$. Thus, the average energy that this wave carries through the window in a 45 s phone call is

$$\bar{U} = \frac{c}{\mu_0} B_{\text{rms}}^2 tA = \left(\frac{3.0 \times 10^8 \text{ m/s}}{4\pi \times 10^{-7} \text{ T} \cdot \text{m/A}}\right)(1.5 \times 10^{-10} \text{ T})^2 (45 \text{ s})(0.20 \text{ m}^2) = \boxed{4.8 \times 10^{-5} \text{ J}}$$

---

25. **SSM REASONING AND SOLUTION**

   a. According to Equation 24.5b, the average intensity is $\bar{S} = c\varepsilon_0 E_{\text{rms}}^2$. In addition, the average intensity is the average power $\bar{P}$ divided by the area A. Therefore,

$$E_{\text{rms}} = \sqrt{\frac{\bar{S}}{c\varepsilon_0}} = \sqrt{\frac{\bar{P}}{c\varepsilon_0 A}}$$

$$= \sqrt{\frac{1.20 \times 10^4 \text{ W}}{(3.00 \times 10^8 \text{ m/s})[8.85 \times 10^{-12} \text{C}^2/(\text{N} \cdot \text{m}^2)](135 \text{ m}^2)}} = \boxed{183 \text{ N/C}}$$

   b. Then, from Equation 24.3 ($E_{\text{rms}} = cB_{\text{rms}}$), we have

$$B_{\text{rms}} = \frac{E_{\text{rms}}}{c} = \frac{183 \text{ N/C}}{3.00 \times 10^8 \text{ m/s}} = \boxed{6.10 \times 10^{-7} \text{ T}}$$

---

26. **REASONING AND SOLUTION** The energy is equal to the power P multiplied by the time t. The power, on the other hand, is equal to product of the the intensity S of the wave and the area A through which the wave passes.

$$\text{Energy} = Pt = (SA)t = (1390 \text{ W/m}^2)(25 \text{ m} \times 45 \text{ m})(3600 \text{ s}) = \boxed{5.6 \times 10^9 \text{ J}}$$

27. **REASONING** The electromagnetic solar power that strikes an area $A_\perp$ oriented perpendicular to the direction in which the sunlight is radiated is $P = SA_\perp$, where S is the intensity of the sunlight. In the problem, the solar panels are not oriented perpendicular to the direction of the sunlight, and sunlight strikes the panels at an angle θ with respect to the normal. We wish to find the solar power that impinges on the solar panels when $\theta = 35°$, given that the incident power is 3200 W when $\theta = 55°$.

SOLUTION When the angle that the sunlight makes with the normal to the solar panel is θ, the power that strikes the solar panel is given by $P = SA\cos\theta$, where the area perpendicular to the sunlight is $A_\perp = A\cos\theta$ (see the drawing). Therefore we can write

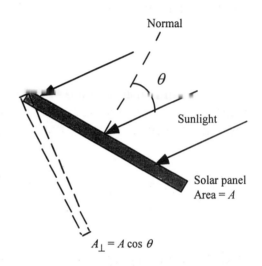

$$\frac{P_2}{P_1} = \frac{SA\cos\theta_2}{SA\cos\theta_1}$$

where the intensity S of the sunlight that reaches the panel, as well as the area A, are the same in both cases. Therefore, we have

$$\frac{P_2}{P_1} = \frac{\cos\theta_2}{\cos\theta_1}$$

Solving for P_2, we find that the when $\theta_2 = 35°$, the solar power impinging on the panel is

$$P_2 = P_1\left(\frac{\cos\theta_2}{\cos\theta_1}\right) = (3200\text{ W})\left(\frac{\cos 35°}{\cos 55°}\right) = \boxed{4600\text{ W}}$$

28. **REASONING** According to Equation 24.5b, the average intensity $\bar{S}$ of the infrared radiation is related to the rms value of the electric field E_{rms} by $\bar{S} = c\varepsilon_0 E_{\text{rms}}^2$. According to Equation 16.8, the average power $\bar{P}$ is equal to the average intensity times the area A to which the power is being delivered, the area being that of a circle or $A = \pi r^2$. Thus, $\bar{P} = \bar{S}A = \bar{S}(\pi r^2)$. The average power is given by Equation 6.10b as the energy Q absorbed by the leg divided by the time t, so that $t = Q/\bar{P}$. The energy absorbed by the leg is related to the rise in temperature ΔT by Equation 12.4 as $Q = cm\Delta T$, where c is the specific heat capacity and m is the mass.

SOLUTION

a. The average intensity of the infrared radiation is

$$\overline{S} = c\varepsilon_0 E_{rms}^2 \tag{24.5b}$$

$$= (3.0 \times 10^8 \text{ m/s})\left[8.85 \times 10^{-12} \text{ C}^2/(\text{N} \cdot \text{m}^2)\right](2800 \text{ N/C})^2 = \boxed{2.1 \times 10^4 \text{ W/m}^2}$$

b. The average power delivered to the leg is

$$\overline{P} = \overline{S}A = \overline{S}(\pi r^2) = (2.1 \times 10^4 \text{ W/m}^2)\pi(4.0 \times 10^{-2} \text{ m})^2 = \boxed{1.1 \times 10^2 \text{ W}} \tag{16.8}$$

c. Combining the relations $t = Q/\overline{P}$ and $Q = cm\Delta T$, the time required to raise the temperature by 2.0 C° is

$$t = \frac{Q}{\overline{P}} = \frac{cm\Delta T}{\overline{P}} = \frac{\left[3500 \text{ J}/(\text{kg} \cdot \text{C}°)(0.28 \text{ kg})\right](2.0 \text{ C}°)}{1.1 \times 10^2 \text{ W}} = \boxed{18 \text{ s}}$$

---

29. **SSM** **WWW** **REASONING AND SOLUTION** The sun radiates sunlight (electromagnetic waves) uniformly in all directions, so the intensity at a distance r from the sun is given by Equation 16.9 as $S = P/(4\pi r^2)$, where P is the power radiated by the sun. The power that strikes an area $A_\perp$ oriented perpendicular to the direction in which the sunlight is radiated is $P' = SA_\perp$, according to Equation 16.8. The 0.75-m² patch of flat land on the equator at point Q is not perpendicular to the direction of the sunlight, however.

The figure at the right shows that

$$A_\perp = (0.75 \text{ m}^2) \cos 27°$$

$A_\perp = (0.75 \text{ m}^2) \cos 27°$

Sunlight

27°

0.75-m² patch of land (edge-on view)

Therefore, the power striking the patch of land is

$$P' = SA_\perp = \left(\frac{P}{4\pi r^2}\right)(0.75 \text{ m}^2) \cos 27°$$

$$= \left[\frac{3.9 \times 10^{26} \text{ W}}{4\pi(1.5 \times 10^{11} \text{ m})^2}\right](0.75 \text{ m}^2) \cos 27° = \boxed{920 \text{ W}}$$

30. **REASONING AND SOLUTION**
a. The maximum electrical force on the charge is $F = qE_0$. The rms electric field generated by the wave is

$$E_{rms} = \sqrt{\frac{\overline{S}}{c\varepsilon_0}} = \sqrt{\frac{1390 \text{ W/m}^2}{(3.00 \times 10^8 \text{ m/s})\left[8.85 \times 10^{-12} \text{ C}^2/(\text{N} \cdot \text{m}^2)\right]}} = 724 \text{ N/C}$$

The maximum electric field is

$$E_0 = \sqrt{2} \, E_{rms} = \sqrt{2} \, (724 \text{ N/C}) = 1020 \text{ N/C}$$

The force is then

$$F = (2.6 \times 10^{-8} \text{ C})(1020 \text{ N/C}) = \boxed{2.7 \times 10^{-5} \text{ N}}$$

b. The maximum magnetic force on the charge is $F = qvB_0$. The rms magnetic field is $B_{rms} = E_{rms}/c$, and the maximum magnetic field is

$$B_0 = \sqrt{2} \, B_{rms} = \sqrt{2} \, E_{rms}/c = \sqrt{2} \, (724 \text{ N/C})/(3.00 \times 10^8 \text{ m/s}) = 3.41 \times 10^{-6} \text{ T}$$

so

$$F = (2.6 \times 10^{-8} \text{ C})(3.7 \times 10^4 \text{ m/s})(3.41 \times 10^{-6} \text{ T}) = \boxed{3.3 \times 10^{-9} \text{ N}}$$

---

31. **REASONING** The Doppler effect for electromagnetic radiation is given by Equation 24.6;

$$f_o = f_s\left(1 \pm \frac{v_{rel}}{c}\right) \qquad \text{if } v_{rel} \ll c$$

where f_o is the observed frequency, f_s is the frequency emitted by the source, and v_{rel} is the speed of the source relative to the observer. As discussed in the text, the plus sign applies when the source and the observer are moving toward one another, while the minus sign applies when they are moving apart. According to Equation 16.1, the wavelength of these waves is $\lambda = c/f$. Therefore, the Doppler shift can be written in terms of wavelengths:

$$\frac{1}{\lambda_o} = \frac{1}{\lambda_s}\left(1 \pm \frac{v_{rel}}{c}\right) \qquad \text{if } v_{rel} \ll c$$

902 ELECTROMAGNETIC WAVES

SOLUTION

a. The wavelength λ_o of the light observed on earth is greater than the wavelength λ_s of the light when it is emitted from the distant galaxy (the source). Therefore, the frequency of the light observed on earth is less than the frequency of the light when it is emitted from the distant galaxy. Thus, the quantity in the brackets in Equation 24.6 must be less than one. It is equal to $[1-(v_{rel}/c)]$, and since the minus sign applies, we can conclude that

$$\boxed{\text{the galaxy must be receding from the earth}}.$$

b. We can find the speed of the galaxy relative to the earth by solving the wavelength version of Equation 24.6 for v_{rel}:

$$v_{rel} = c\left(1-\frac{\lambda_s}{\lambda_o}\right) = (3.0\times 10^8 \text{ m/s})\left(1-\frac{500.7 \text{ nm}}{503.7 \text{ nm}}\right) = \boxed{1.8\times 10^6 \text{ m/s}}$$

---

32. **REASONING** Using the Doppler effect, we will find the relative speed between the speeding car and the police car. Since we know the speed of the police car relative to the ground, we can determine the speed of the car relative to the ground, once the relative speed v_{rel} is found.

There are two Doppler frequency changes in this situation. First, the speeder's car observes the wave frequency coming from the radar gun to have a frequency f_o that is different from the emitted frequency f_s. The second Doppler shift occurs after the wave reflects from the speeder's car and returns to the police car.

The Doppler frequency for electromagnetic radiation is given by Equation 24.6, $f_o = f_s[1\pm(v_{rel}/c)]$, where v_{rel} is the relative speed between the source and the observer of the radiation, and the plus sign applies when the source and the observer are moving toward one another, while the minus sign applies when they are moving apart. Since the distance between the police car and the speeder's car is increasing, they are moving apart, and according to Equation 24.6, the first Doppler frequency change is given by $f_o - f_s = -f_s(v_{rel}/c)$. After the wave reflects from the speeder's car it returns to the police car where it is observed to have a frequency f_o' that is different from its frequency f_o at the instant of reflection. Equation 24.6 may again be used, this time to determine the second Doppler frequency shift: $f_o' - f_o = -f_o(v_{rel}/c)$. We can use these two equations for the frequency shifts to determine an expression for the total Doppler change in frequency. Adding the two equations, we have

$$(f_o' - f_o) + (f_o - f_s) = -f_o\left(\frac{v_{rel}}{c}\right) - f_s\left(\frac{v_{rel}}{c}\right)$$

$$f_o' - f_s = -\left[f_o\left(\frac{v_{rel}}{c}\right) + f_s\left(\frac{v_{rel}}{c}\right)\right] \approx -2f_s\left(\frac{v_{rel}}{c}\right)$$

where we have assumed that f_s and f_o differ only by a negligibly small amount, so that $f_o \approx f_s$. Rearranging, we have

$$f_s - f_o' \approx 2f_s\left(\frac{v_{rel}}{c}\right)$$

SOLUTION Solving for the relative speed v_{rel} gives

$$v_{rel} \approx \left(\frac{f_s - f_o'}{2f_s}\right)c = \left[\frac{320 \text{ Hz}}{2(7.0\times10^9 \text{ Hz})}\right](3.0\times10^8 \text{ m/s}) = 6.9 \text{ m/s}$$

The relative speed v_{rel} is related to the speeds of the vehicles with respect to the ground by $v_{rel} = v_{speeder} - v_{police}$. Therefore, the speeder's speed with respect to the ground is

$$v_{speeder} = v_{rel} + v_{police} = 6.9 \text{ m/s} + 25 \text{ m/s} = \boxed{32 \text{ m/s}}$$

---

33. **SSM** **REASONING** The Doppler effect for electromagnetic radiation is given by Equation 24.6, $f_o = f_s(1 \pm v_{rel}/c)$, where f_o is the observed frequency, f_s is the frequency emitted by the source, and v_{rel} is the speed of the source relative to the observer. As discussed in the text, the plus sign applies when the source and the observer are moving toward one another, while the minus sign applies when they are moving apart.

SOLUTION
a. At location A, the galaxy is moving away from the earth with a relative speed of

$$v_{rel} = (1.6\times10^6 \text{ m/s}) - (0.4\times10^6 \text{ m/s}) = 1.2\times10^6 \text{ m/s}$$

Therefore, the minus sign in Equation 24.6 applies and the observed frequency for the light from region A is

$$f_o = f_s\left(1 - \frac{v_{rel}}{c}\right) = (6.200\times10^{14} \text{ Hz})\left(1 - \frac{1.2\times10^6 \text{ m/s}}{3.0\times10^8 \text{ m/s}}\right) = \boxed{6.175\times10^{14} \text{ Hz}}$$

b. Similarly, at location B, the galaxy is moving away from the earth with a relative speed of

$$v_{rel} = (1.6 \times 10^6 \text{ m/s}) + (0.4 \times 10^6 \text{ m/s}) = 2.0 \times 10^6 \text{ m/s}$$

The observed frequency for the light from region B is

$$f_o = f_s \left(1 - \frac{v_{rel}}{c}\right) = (6.200 \times 10^{14} \text{ Hz})\left(1 - \frac{2.0 \times 10^6 \text{ m/s}}{3.0 \times 10^8 \text{ m/s}}\right) = \boxed{6.159 \times 10^{14} \text{ Hz}}$$

---

34. **REASONING AND SOLUTION**

   a. The polarizer reduces the intensity of the light by a factor of two or to $\boxed{0.55 \text{ W/m}^2}$.

   b. The intensity of the light leaving the analyzer is given by Malus' law.

   $$S = (0.55 \text{ W/m}^2) \cos^2 75° = \boxed{3.7 \times 10^{-2} \text{ W/m}^2}$$

---

35. **SSM REASONING AND SOLUTION** The average intensity of light leaving the polarizing material is given by Malus' Law (Equation 24.7). Therefore, using Malus' law, we obtain the following results.

   a. $\dfrac{\overline{S}}{\overline{S}_0} = \cos^2 25° = \boxed{0.82}$

   b. $\dfrac{\overline{S}}{\overline{S}_0} = \cos^2 65° = \boxed{0.18}$

---

36. **REASONING** Since the incident beam is unpolarized, the intensity of the light transmitted by the first sheet of polarizing material is one-half the intensity of the incident beam. The beams striking the second and third sheets of polarizing material are polarized, so the average intensity $\overline{S}$ of the light transmitted by each sheet is given by Malus' law, $\overline{S} = \overline{S}_0 \cos^2 \theta$, where $\overline{S}_0$ is the average intensity of the light incident on each sheet.

   SOLUTION The average intensity $\overline{S}_1$ of the light leaving the first sheet is one-half the intensity of the incident beam, so $\overline{S}_1 = \frac{1}{2}(1260.0 \text{ W/m}^2) = 630.0 \text{ W/m}^2$. The intensity $\overline{S}_2$ of the light leaving the second sheet of polarizing material is given by Malus' law, Equation 24.7, $\overline{S}_2 = \overline{S}_1 \cos^2 \theta$, where θ is the angle between the polarization of the incident beam and the transmission axis of the second sheet:

   $$\overline{S}_2 = (630.0 \text{ W/m}^2) \cos^2 (55.0° - 19.0°) = 412 \text{ W/m}^2$$

The intensity $\overline{S}_3$ of the light leaving the third sheet of polarizing material is $\overline{S}_3 = \overline{S}_2 \cos^2 \theta$, where θ is the angle between the polarization of the incident beam and the transmission axis of the third sheet:

$$\overline{S}_3 = (412 \text{ W/m}^2) \cos^2 (100.0° - 55.0°) = \boxed{206 \text{ W/m}^2}$$

---

37. **REASONING AND SOLUTION** Malus' law gives

$$S/S_0 = \cos^2 \theta = \cos^2 38° = 0.62$$

so that $\boxed{62\%}$ of the light is transmitted.

---

38. **REASONING AND SOLUTION** If the intensity of the unpolarized light is I_0, the intensity of the polarized light leaving the polarizer is $\frac{1}{2} I_0$. By Malus' law, the intensity of the light leaving the insert is $\frac{1}{2} I_0 \cos^2 \theta$. From the results of Conceptual Example 8, the intensity of light leaving the analyzer is $\frac{1}{2} I_0 \cos^2 \theta \sin^2 \theta$. Thus, the intensity I of light that reaches the photocell is

$$I = \tfrac{1}{2} I_0 \cos^2 \theta \sin^2 \theta = \tfrac{1}{2}(150 \text{ W/m}^2) \cos^2 30.0° \sin^2 30.0° = \boxed{14 \text{ W/m}^2}$$

---

39. **SSM** **WWW** **REASONING** The average intensity of light leaving each analyzer is given by Malus' Law (Equation 24.7). Thus, intensity of the light transmitted through the first analyzer is

$$\overline{S}_1 = \overline{S}_0 \cos^2 27°$$

Similarly, the intensity of the light transmitted through the second analyzer is

$$\overline{S}_2 = \overline{S}_1 \cos^2 27° = \overline{S}_0 \cos^4 27°$$

And the intensity of the light transmitted through the third analyzer is

$$\overline{S}_3 = \overline{S}_2 \cos^2 27° = \overline{S}_0 \cos^6 27°$$

If we generalize for the Nth analyzer, we deduce that

$$\overline{S}_N = \overline{S}_{N-1} \cos^2 27° = \overline{S}_0 \cos^{2N} 27°$$

Since we want the light reaching the photocell to have an intensity that is reduced by a least a factor of one hundred relative to the first analyzer, we want $\overline{S}_N/\overline{S}_0 = 0.010$. Therefore, we need to find N such that $\cos^{2N} 27° = 0.010$. This expression can be solved for N.

SOLUTION Taking the common logarithm of both sides of the last expression gives

$$2N \log(\cos 27°) = \log 0.010 \quad \text{or} \quad N = \frac{\log 0.010}{2 \log (\cos 27°)} = \boxed{20}$$

---

40. **REASONING** The polarizer, the insert and the analyzer in the set-up in Figure 24.22a all reduce the intensity of the light that reaches the photocell. The polarizer reduces the intensity by a factor of one-half, as described in Section 24.6 of the text; if the average intensity of the incident unpolarized light is $\overline{I}$, the average intensity of the polarized light that leaves the polarizer and strikes the insert is $\overline{S}_0 = \overline{I}/2$. According to Malus' law (see Equation 24.7), the average intensity $\overline{S}_{insert}$ of the light leaving the insert is $\overline{S}_{insert} = \overline{S}_0 \cos^2 \theta$, where θ is the relative angle between the transmission axes of the polarizer and the insert. The intensity of the light is further reduced as the polarized light passes through the analyzer. Malus' law can be used in succession at each piece of polarizing material to determine the intensity that reaches the photocell, both with and without the presence of the analyzer.

SOLUTION When the analyzer is present, the average intensity reaching the photocell is equal to the average intensity that leaves the analyzer. The average intensity leaving the analyzer is, according to Malus' law, $\overline{S}_{insert} \cos^2 \phi$ where $\overline{S}_{insert}$ is the average intensity that leaves the insert and reaches the analyzer, and ϕ is the relative angle between the transmission axes of the analyzer and insert. From Figure 24.22a we see that $\phi = 90° - \theta$. The average intensity of the light leaving the insert is $\overline{S}_{insert} = \overline{S}_0 \cos^2 \theta$, according to Malus' law. Therefore, when the analyzer is present as shown in Figure 24.22b, the average intensity leaving the analyzer and reaching the photocell is

$$\overline{S}_{photocell} = \overline{S}_{insert} \cos^2 \phi = \left(\overline{S}_0 \cos^2 \theta\right) \cos^2 (90° - \theta)$$

This expression can be solved for $\overline{S}_0$ to determine the average intensity leaving the polarizer:

$$\overline{S}_0 = \frac{\overline{S}_{photocell}}{(\cos^2 \theta) \cos^2 (90° - \theta)} = \frac{110 \text{ W/m}^2}{(\cos^2 23°) \cos^2 (90° - 23°)} = 850 \text{ W/m}^2$$

If the analyzer were removed from the setup, everything else remaining the same, the intensity reaching the photocell would be equal to the intensity that leaves the insert. Therefore, if the analyzer were removed, the intensity reaching the photocell would be

$$\overline{S}_{photocell} = \overline{S}_{insert} = \overline{S}_0 \cos^2 \theta = (850 \text{ W/m}^2)\cos^2 23° = \boxed{720 \text{ W/m}^2}$$

---

41. **REASONING AND SOLUTION** Using Equation 16.1, we find

$$f = c/\lambda = (3.00 \times 10^8 \text{ m/s})/(2.1 \times 10^{-9} \text{ m}) = \boxed{1.4 \times 10^{17} \text{ Hz}}$$

---

42. **REASONING AND SOLUTION** According to Equation 16.8, we have

$$S = \frac{P}{A} = \frac{P}{\pi r^2} = \frac{1.2 \times 10^{-3} \text{ W}}{\pi (1.0 \times 10^{-3} \text{ m})^2} = \boxed{3.8 \times 10^2 \text{ W/m}^2}$$

---

43. **SSM WWW REASONING AND SOLUTION** The actual time t and round-trip distance s are related by $s = ct$. An uncertainty of Δt in the time introduces an uncertainty in the round trip distance Δs, so that $s + \Delta s = c(t + \Delta t)$. Thus, subtraction of these two expression leads to

$$\Delta s = c\Delta t = (3.0 \times 10^8 \text{ m/s})(0.10 \times 10^{-9} \text{ s}) = 0.030 \text{ m}$$

This is the error in the round-trip distance. Therefore, the error in the earth-moon distance is half this distance or $\boxed{0.015 \text{ m}}$.

---

44. **REASONING AND SOLUTION** The average electromagnetic energy contained in a volume is the product of the average energy density $\overline{u}$ and the volume V. However, the average energy density is related to the average intensity $\overline{S}$ of the electromagnetic wave by Equation 24.4 as $\overline{u} = \overline{S}/c$. Therefore,

$$\text{Average energy} = \overline{u}V = \left(\frac{\overline{S}}{c}\right)V = \left(\frac{1.0 \times 10^3 \text{ W/m}^2}{3.0 \times 10^8 \text{ m/s}}\right)(5.5 \text{ m}^3) = \boxed{1.8 \times 10^{-5} \text{ J}}$$

---

45. **REASONING AND SOLUTION** The sun radiates sunlight uniformly in all directions. The intensity S, or power per unit area, of the sunlight at a distance r from the sun is given by Equation 16.9 as $S = P/(4\pi r^2)$, where P is the power radiated by the sun. The intensity of the sunlight at the surface of Mars and Earth, respectively, is

$$S_{Mars} = \frac{P}{4\pi r_{Mars}^2} \quad \text{and} \quad S_{Earth} = \frac{P}{4\pi r_{Earth}^2}$$

Solving the equation on the right for P and then substituting it into the equation on the left yield, after canceling the factors of 4π,

$$S_{Mars} = S_{Earth}\left(\frac{r_{Earth}}{r_{Mars}}\right)^2 = (1390 \text{ W/m}^2)\left(\frac{1.50 \times 10^{11} \text{ m}}{2.28 \times 10^{11} \text{ m}}\right)^2 = \boxed{602 \text{ W/m}^2}$$

46. **REASONING AND SOLUTION**
    a. The frequency of the wave is

    $$f = c/\lambda = (3.00 \times 10^8 \text{ m/s})/(274 \text{ m}) = \boxed{1.09 \times 10^6 \text{ Hz}}$$

    b. It is an $\boxed{\text{AM radio wave}}$.

47. **SSM REASONING AND SOLUTION** The average intensity of light leaving each polarizer is given by Malus' Law (Equation 24.7): $\bar{S} = \bar{S}_0 \cos^2 \theta$. Solving for the angle θ and noting that $\bar{S} = 0.10\, \bar{S}_0$, we have

    $$\theta = \cos^{-1}\sqrt{\frac{0.100\, \bar{S}_0}{\bar{S}_0}} = \boxed{71.6°}$$

48. **REASONING** There are two Doppler frequency changes in the emitted wave in this case. First, the speeder's car observes the wave frequency coming from the radar gun to have a frequency f_o that is different from the emitted (source) frequency f_s. The wave then reflects and returns to the police car, where it is observed to have a frequency f_o' that is different than its frequency f_o at the instant of reflection. Although the police car is now moving, the relative motion of the two vehicles is one of approach. In Example 6, it is shown that, when the source and the observer of the radar are approaching each other, the magnitude of the difference between frequency of the emitted wave and the wave that returns to the police car after reflecting from the speeder's car is

    $$f_o' - f_s \approx 2f_s\left(\frac{v_{rel}}{c}\right)$$

    where v_{rel} is the relative speed between the speeding car and the police car.

SOLUTION Since the police car is moving to the right at 27 m/s, while the speeder is coming from behind at 39 m/s, the relative speed v_{rel} is 39 m/s − 27 m/s = 12 m/s.

a. The total Doppler change in frequency is, therefore,

$$f'_o - f_s = 2(8.0 \times 10^9 \text{ Hz}) \left(\frac{12 \text{ m/s}}{3.0 \times 10^8 \text{ m/s}} \right) = \boxed{640 \text{ Hz}}$$

b. Since the two cars are approaching each other, the plus sign in Equation 24.6 applies, the quantity in brackets in Equation 24.6 will be greater than one, and the Doppler shifted frequency will be greater than the source frequency.

Therefore, $\boxed{\text{the wave that returns to the police car has the greater frequency}}$.

---

49. **REASONING** According to Equation 16.3, the displacement y of a wave that travels in the +x direction and has amplitude A, frequency f, and wavelength λ is given by

$$y = A \sin\left(2\pi f t - \frac{2\pi x}{\lambda}\right)$$

This equation, with $y = E$, applies to the traveling electromagnetic wave in the problem, which is represented mathematically as

$$E = E_0 \sin\left[(1.5 \times 10^{10} \text{ s}^{-1})t - (5.0 \times 10^1 \text{ m}^{-1})x \right]$$

As E_0 is the maximum field strength, it represents the amplitude A of the wave. We can find the frequency and wavelength of this electromagnetic wave by comparing the mathematical form of the electric field with Equation 16.3.

SOLUTION
a. By inspection, we see that $2\pi f = 1.5 \times 10^{10}$ s^{-1}. Therefore, the frequency of the wave is

$$f = \frac{1.5 \times 10^{10} \text{ s}^{-1}}{2\pi} = \boxed{2.4 \times 10^9 \text{ Hz}}$$

b. As shown in Figure 17.18, the separation between adjacent nodes in any standing wave is one-half of a wavelength. By inspection of the mathematical form of the electric field and comparison with Equation 16.3, we infer that $2\pi/\lambda = 5.0 \times 10^1$ m^{-1}. Therefore,

$$\lambda = \frac{2\pi}{5.0 \times 10^1 \text{ m}^{-1}} = 0.126 \text{ m}$$

Therefore, the nodes in the standing waves formed by this electromagnetic wave are separated by $\lambda/2 = \boxed{0.063 \text{ m}}$.

---

50. **REASONING** The rms value E_{rms} of the electric field is related to the average intensity $\overline{S}$ of the light by Equation 24.5b, $\overline{S} = c\varepsilon_0 E_{rms}^2$. The average intensity $\overline{S}$ of the light transmitted by the polarizer is related to the incident intensity $\overline{S}_0$ by Malus' law, $\overline{S} = \overline{S}_0 \cos^2\theta$, where θ is the angle between the transmission axis and the direction of polarization. These two relations will allow us to determine the rms value of the electric field.

SOLUTION Combining the two equations given above and solving for the rms value of the electric field, we have

$$E_{rms} = \sqrt{\frac{\overline{S}_0 \cos^2\theta}{c\varepsilon_0}} = \sqrt{\frac{(15 \text{ W/m}^2)\cos^2 25°}{(3.0 \times 10^8 \text{ m/s})[8.85 \times 10^{-12} \text{ C}^2/(\text{N}\cdot\text{m}^2)]}} = \boxed{68 \text{ N/C}}$$

---

51. **SSM REASONING AND SOLUTION** Since the sun emits radiation uniformly in all directions, at a distance r from the sun's center, the energy spreads out over a sphere of surface area $4\pi r^2$. Therefore, according to Equation 16.9, $S = P/(4\pi r^2)$, the total power radiated by the sun is

$$P = S(4\pi r^2) = (1390 \text{ W/m}^2)(4\pi)(1.50 \times 10^{11} \text{ m})^2 = \boxed{3.93 \times 10^{26} \text{ W}}$$

---

52. **REASONING AND SOLUTION** The intensity of the laser light is $S = P/A = cu$, where u is the energy density of the light. The energy in a section of length L of the cylindrical beam is $U = uAL$ or

$$U = PL/c = (0.750 \text{ W})(2.50 \text{ m})/(3.00 \times 10^8 \text{ m/s}) = \boxed{6.25 \times 10^{-9} \text{ J}}$$

---

53. **REASONING AND SOLUTION** The light energy spreads out uniformly over the surface of a sphere centered on the source. The average intensity $\overline{S}$ of the light at the point $r = 8.00$ m away from the source is then

$$\overline{S} = \frac{\overline{P}}{A} = \frac{\overline{P}}{4\pi r^2} = \frac{60.0 \text{ W}}{4\pi(8.00 \text{ m})^2} = 7.46 \times 10^{-2} \text{ W/m}^2$$

a. The rms value of the electric field at the point is

$$E_{rms} = \sqrt{\frac{\overline{S}}{c\varepsilon_0}} = \sqrt{\frac{7.46 \times 10^{-2} \text{ W/m}^2}{(3.00 \times 10^8 \text{ m/s})(8.85 \times 10^{-12} \text{ C}^2/\text{N} \cdot \text{m}^2)}} = \boxed{5.30 \text{ N/C}}$$

b. Then

$$B_{rms} = E_{rms}/c = \boxed{1.77 \times 10^{-8} \text{ T}}$$

---

54. **REASONING AND SOLUTION** The polarizer will transmit a maximum intensity of $(1/2)S_u + S_p$, when its axis is parallel to the polarization direction of the polarized component of the incident light. Then the light intensity at the photocell is

$$S_{max} = [(1/2) S_u + S_p] \cos^2 \theta \qquad (1)$$

The polarizer transmits minimum light intensity of $(1/2) S_u$ when its axis is perpendicular to the polarization direction of the polarized incident light, so

$$S_{min} = (1/2) S_u \cos^2 \theta \qquad (2)$$

Solving Equation (2) for S_u gives

$$S_u = 2S_{min}/\cos^2 \theta \qquad (3)$$

Using Equation (3) in Equation (1) and solving give

$$S_p = (S_{max} - S_{min})/\cos^2 \theta \qquad (4)$$

Using Equations (3) and (4), we find that the percent polarization is

$$\frac{100 S_p}{S_p + S_u} = \frac{\dfrac{100 (S_{max} - S_{min})}{\cos^2 \theta}}{\dfrac{S_{max} - S_{min} + 2S_{min}}{\cos^2 \theta}}$$

$$\frac{100 S_p}{S_p + S_u} = \frac{100 (S_{max} - S_{min})}{S_{max} + S_{min}}$$

912 ELECTROMAGNETIC WAVES

55. CONCEPT QUESTIONS

a. According to Equation 16.1, the wavelength λ is related to the frequency f by $\lambda = c/f$, where c is the speed of light in a vacuum. Therefore, the greater the frequency, the smaller the wavelength.

b. The length of each pulse is equal to the product of its speed and the time, or $x = ct_0$.

SOLUTION

a. The number of wavelengths in one pulse is equal to the length of the pulse divided by the wavelength. The length of each pulse is $x = ct_0$ and the wavelength is $\lambda = c/f$, so

$$\text{Number of wavelengths} = \frac{x}{\lambda} = \frac{ct_0}{c/f}$$

$$= ft_0 = (5.2 \times 10^{14} \text{ Hz})(2.7 \times 10^{-11} \text{ s}) = \boxed{1.4 \times 10^4}$$

b. When the light is traveling in water, its speed is v, which is less than the speed of light in a vacuum. The length of each pulse is now $x = vt_0$ and the wavelength is $\lambda = v/f$, so

$$\text{Number of wavelengths} = \frac{x}{\lambda} = \frac{vt_0}{v/f}$$

$$= ft_0 = (5.2 \times 10^{14} \text{ Hz})(2.7 \times 10^{-11} \text{ s}) = \boxed{1.4 \times 10^4}$$

---

56. CONCEPT QUESTIONS

a. The relationship between the intensity S of an electromagnetic wave and its electric field E is given by Equation 24.5b as $S = c\varepsilon_0 E^2$. Therefore, if the magnitude of the electric field triples, the intensity increases by a factor of $3^2 = 9$.

b. Even though the magnitude of the magnetic field is much smaller than that of the electric field, tripling the magnetic field also causes the intensity to increase by a factor of $3^2 = 9$. This can be seen by examining Equation 24.5c, $S = \dfrac{c}{\mu_0} B^2$.

SOLUTION

a. When the magnitude of the electric field is 315 N/C, the intensity of the electromagnetic wave is

$$S = c\varepsilon_0 E^2 \qquad (24.5b)$$

$$= (3.00 \times 10^8 \text{ m/s})[8.85 \times 10^{-12} \text{ C}^2/(\text{N}\cdot\text{m}^2)](315 \text{ N/C})^2 = \boxed{263 \text{ W/m}^2}$$

When the magnitude of the electric field is 945 N/C, the intensity of the electromagnetic wave is

$$S = c\varepsilon_0 E^2 = (3.00 \times 10^8 \text{ m/s})\left[8.85 \times 10^{-12} \text{ C}^2/(\text{N} \cdot \text{m}^2)\right](945 \text{ N/C})^2 = \boxed{2370 \text{ W/m}^2}$$

b. The magnitude of the magnetic field associated with each electric field is

$$B = \frac{E}{c} = \frac{315 \text{ N/C}}{3.00 \times 10^8 \text{ m/s}} = \boxed{1.05 \times 10^{-6} \text{ T}} \qquad (24.3)$$

$$B = \frac{E}{c} = \frac{945 \text{ N/C}}{3.00 \times 10^8 \text{ m/s}} = \boxed{3.15 \times 10^{-6} \text{ T}}$$

c. The intensity of the wave associated with each value of the magnetic field is

$$S = \frac{c}{\mu_0} B^2 = \frac{3.00 \times 10^8 \text{ m/s}}{4\pi \times 10^{-7} \text{ T} \cdot \text{m/A}}(1.05 \times 10^{-6} \text{ T})^2 = \boxed{263 \text{ W/m}^2} \qquad (24.5c)$$

$$S = \frac{c}{\mu_0} B^2 = \frac{3.00 \times 10^8 \text{ m/s}}{4\pi \times 10^{-7} \text{ T} \cdot \text{m/A}}(3.15 \times 10^{-6} \text{ T})^2 = \boxed{2370 \text{ W/m}^2}$$

---

57. **CONCEPT QUESTIONS**

a. The average intensity $\overline{S}$ is related to the distance r from the source by Equation 16.9, $\overline{S} = \overline{P}/(4\pi r^2)$, where $\overline{P}$ is the average power radiated by the source. Thus, as r increases, the average intensity decreases.

b. The intensity of an electromagnetic wave is related to the magnitude E of its electric field by Equation 24.5b, $S = c\varepsilon_0 E^2$. According to the discussion in Section 24.4, if the intensity is an average intensity, then the value for the electric field must be an rms value, not a peak value.

SOLUTION
a. The average intensity of the wave is

$$\overline{S} = \frac{\overline{P}}{4\pi r^2} = \frac{150.0 \text{ W}}{4\pi (5.00 \text{ m})^2} = \boxed{0.477 \text{ W/m}^2} \qquad (16.9)$$

b. The average intensity $\overline{S}$ is related to the rms value E_{rms} of the electric field by Equation 24.5b, $\overline{S} = c\varepsilon_0 E_{rms}^2$. Solving for the electric field gives

$$E_{rms} = \sqrt{\frac{\overline{S}}{c\varepsilon_0}} = \sqrt{\frac{0.477 \text{ W/m}^2}{(3.00\times 10^8 \text{ m/s})[8.85\times 10^{-12} \text{ C}^2/(\text{N}\cdot\text{m}^2)]}} = \boxed{13.4 \text{ N/C}}$$

c. The rms value E_{rms} of the electric field is related to the peak value E_0 by $E_{rms} = E_0/\sqrt{2}$. The peak electric field is, therefore,

$$E_0 = \sqrt{2}E_{rms} = \sqrt{2}(13.4 \text{ N/C}) = \boxed{19.0 \text{ N/C}}$$

---

58. CONCEPT QUESTIONS

a. When a stationary charge is placed in an electric field, it experiences an electric force. The magnitude F of the electric force is given by Equation 18.2 as $F = qE$, where q is the magnitude of the charge and E is the magnitude of the electric field.

b. When a stationary charge is placed in a magnetic field, it does not experience a magnetic force, because the charge is not moving. According to Equation 21.1, the magnitude of the magnetic force is related to the magnitude B of the magnetic field by $F = qvB\sin\theta$, where v is the speed of the charge and θ is the angle between the velocity of the charge and the magnetic field. Since the charge is stationary, $v = 0$ m/s and the magnetic force is zero.

c. When a moving charge is placed in an electric field, it experiences an electric force that is given by Equation 18.2. It does not matter whether the charge is stationary or moving.

d. Since the charge is now moving $(v \neq 0 \text{ m/s})$ and its velocity is perpendicular to the magnetic field $(\theta = 90°)$, it experiences a magnetic force, as specified by Equation 21.1.

SOLUTION

a. The magnitude of the electric force is $F = qE$, where the magnitude of the electric field is related to the intensity S of the laser beam by Equation 24.5b $(S = c\varepsilon_0 E^2)$. Therefore, the magnitude of the electric force is

$$F = qE = q\sqrt{\frac{S}{c\varepsilon_0}}$$

$$= (2.6\times 10^{-8} \text{ C})\sqrt{\frac{2.5\times 10^3 \text{ W/m}^2}{(3.00\times 10^8 \text{ m/s})[8.85\times 10^{-12} \text{ C}^2/(\text{N}\cdot\text{m}^2)]}} = \boxed{2.5\times 10^{-5} \text{ N}}$$

b. Since the particle is not moving, the magnetic force on it is zero, $F = \boxed{0 \text{ N}}$.

c. The electric force on the particle is the same whether it is moving or not, so the answer is the same as in part (a); $F = \boxed{2.5 \times 10^{-5} \text{ N}}$.

d. The magnitude of the magnetic force is given by Equation 21.1 as $F = qvB\sin\theta$. The magnitude B of the magnetic field is related to the intensity S of the laser beam by Equation 24.5c ($S = cB^2/\mu_0$). Thus, the magnetic force is

$$F = qvB\sin\theta = qv\sqrt{\frac{\mu_0 S}{c}}\sin\theta$$

$$= (2.6 \times 10^{-8} \text{ C})(3.7 \times 10^4 \text{ m/s})\sqrt{\frac{(4\pi \times 10^{-7} \text{ T}\cdot\text{m/A})(2.5 \times 10^3 \text{ W/m}^2)}{3.00 \times 10^8 \text{ m/s}}}\sin 90.0°$$

$$= \boxed{3.1 \times 10^{-9} \text{ N}}$$

---

59. **CONCEPT QUESTIONS**
a. If the incident light is unpolarized, the intensity of the transmitted light is one-half the intensity of the incident light, independent of the angle of the transmission axis. Thus, the intensity of the transmitted light remains the same as the polarizing material is rotated.

b. If the incident light is polarized along the z axis, the direction of polarization and the transmission axis are initially parallel to each other, and the maximum amount of light is transmitted. As the polarizing material is rotated, the intensity of the transmitted light decreases in accord with Malus' law.

c. If the incident light is polarized along the y axis, the direction of polarization and the transmission axis are initially perpendicular to each other, and no light is transmitted. As the polarizing material is rotated, the intensity of the transmitted light increases.

SOLUTION
a. Since the incident light is unpolarized, the intensity of the transmitted light is one-half the intensity of the incident light, so $\overline{S} = \tfrac{1}{2}\overline{S}_0 = \tfrac{1}{2}(7.0 \text{ W/m}^2) = \boxed{3.5 \text{ W/m}^2}$ for $\alpha = 0°$ and $35°$.

b. When the incident light is polarized along the z axis, the direction of polarization and the transmission axis are initially parallel to each other. Therefore, the angle α is the same as the angle θ between the transmission axis of the polarizer and the direction of the polarization. According to Malus' law (Equation 24.7), the intensity of the transmitted light is given by

916 ELECTROMAGNETIC WAVES

$$\bar{S} = \bar{S}_0 \cos^2 \theta = (7.0 \text{ W/m}^2)\cos^2 0° = \boxed{7.0 \text{ W/m}^2}$$

$$\bar{S} = \bar{S}_0 \cos^2 \theta = (7.0 \text{ W/m}^2)\cos^2 35° = \boxed{4.7 \text{ W/m}^2}$$

c. When the incident light is polarized along the y axis, the direction of polarization and the transmission axis are initially perpendicular to each other. The angle θ in Malus' law is the angle between the direction of polarization (along the y axis) and the transmission axis (measured relative to the z axis). It is related to the angle α according to $\theta = 90.0° - \alpha$. The intensity of the transmitted light is, therefore,

$$\bar{S} = \bar{S}_0 \cos^2 \theta = (7.0 \text{ W/m}^2)\cos^2 (90.0° - 0°) = \boxed{0 \text{ W/m}^2}$$

$$\bar{S} = \bar{S}_0 \cos^2 \theta = (7.0 \text{ W/m}^2)\cos^2 (90.0° - 35°) = \boxed{2.3 \text{ W/m}^2}$$

The table below summarizes the results:

Incident Light	Intensity of Transmitted Light	
	$\alpha = 0°$	$\alpha = 35°$
Unpolarized	3.5 W/m^2	3.5 W/m^2
Polarized parallel to z axis	7.0 W/m^2	4.7 W/m^2
Polarized parallel to y axis	0 W/m^2	2.3 W/m^2

---

60. **CONCEPT QUESTIONS**
a. The transmission axes of the polarizer and analyzer are parallel to each other, so all the light transmitted by the polarizer is completely transmitted by the analyzer.

b. The transmission axes of the polarizer and analyzer are perpendicular to each other, so no light is transmitted through the analyzer.

c. The transmission axes of the polarizer and analyzer make an angle of 30.0° with respect to each other. Thus, some of the light transmitted by the polarizer, but not all, is transmitted through the analyzer. The ranking, largest transmitted intensity first, is: (a), (c), (b).

SOLUTION Since the incident light is unpolarized, the intensity $\bar{S}_1$ of the light transmitted by the polarizer is one-half the intensity $\bar{S}_0$ of the incident light, or $\bar{S}_1 = \frac{1}{2}\bar{S}_0 = \frac{1}{2}(48 \text{ W/m}^2) = 24 \text{ W/m}^2$. The intensity $\bar{S}_2$ of the light transmitted by the analyzer is given by Malus' law, Equation 24.7, as $\bar{S}_2 = \bar{S}_1 \cos^2 \theta$, where θ is the angle between the direction of polarization and the transmission axis. The intensity of the transmitted beams for each of the three cases is

(a) $\bar{S}_2 = \bar{S}_1 \cos^2 \theta = (24 \text{ W/m}^2) \cos^2 0° = \boxed{24 \text{ W/m}^2}$

(b) $\bar{S}_2 = \bar{S}_1 \cos^2 \theta = (24 \text{ W/m}^2) \cos^2 90° = \boxed{0 \text{ W/m}^2}$

(c) $\bar{S}_2 = \bar{S}_1 \cos^2 \theta = (24 \text{ W/m}^2) \cos^2 (60.0° - 30.0°) = \boxed{18 \text{ W/m}^2}$

CHAPTER 25 | THE REFLECTION OF LIGHT: MIRRORS

PROBLEMS

1. **SSM REASONING** The geometry is shown below. According to the law of reflection, the incident ray, the reflected ray, and the normal to the surface all lie in the same plane, and the angle of reflection θ_r equals the angle of incidence θ_i. We can use the law of reflection and the properties of triangles to determine the angle θ at which the ray leaves M_2.

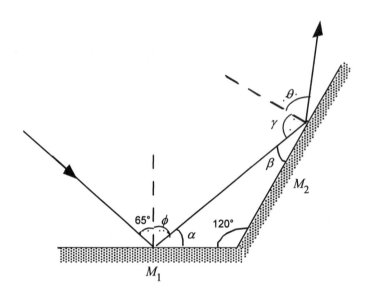

SOLUTION From the law of reflection, we know that $\phi = 65°$. We see from the figure that $\phi + \alpha = 90°$, or $\alpha = 90° - \phi = 90° - 65° = 25°$. From the figure and the fact that the sum of the interior angles in any triangle is 180°, we have $\alpha + \beta + 120° = 180°$. Solving for β, we find that $\beta = 180° - (120° + 25°) = 35°$. Therefore, since $\beta + \gamma = 90°$, we find that the angle γ is given by $\gamma = 90° - \beta = 90° - 35° = 55°$. Since γ is the angle of incidence of the ray on mirror M_2, we know from the law of reflection that $\boxed{\theta = 55°}$.

2. **REASONING AND SOLUTION** The two arrows, A and B are located in front of a plane mirror, and a person at point P is viewing the image of each arrow. As discussed in Conceptual Example 1, light emanating from the arrow is reflected from the mirror and is reflected toward the observer at P. In order for the observer to see the arrow in its entirety, both rays, the one from the top of the arrow and the one from the bottom of the arrow, must pass through the point P.

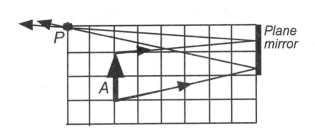
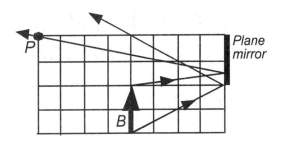

According to the law of reflection, all rays will be reflected so that the angle of reflection is equal to the angle of incidence. The ray from the top of arrow A strikes the mirror and reflects so that it passes through point P. Likewise, the ray from the bottom of the arrow is reflected such that it too passes through point P. Therefore, the observer at P sees the arrow at A in its entirety.

Similar reasoning shows that the ray from the top of arrow B passes through point P. However, as the drawing shows, the ray from the bottom of the arrow does not pass through P. This conclusion is true no matter where the bottom ray strikes the mirror. The observer does *not* see the arrow at B in its entirety.

3. **REASONING AND SOLUTION** As seen in the diagram, the angle of incidence θ is found from $\tan\theta = (0.90 \text{ m})/(3.6 \text{ m}) = 0.25$. Thus, $\theta = \tan^{-1}(0.25) = \boxed{14°}$.

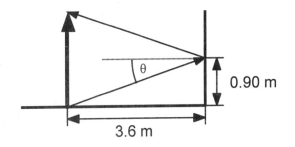

4. **REASONING AND SOLUTION** Referring to Figure 25.9b and Conceptual Example 2, we find the following locations for the three images:

Image 1:	$x = -2.0$ m,	$y = +1.0$ m
Image 2:	$x = +2.0$ m,	$y = -1.0$ m
Image 3:	$x = +2.0$ m,	$y = +1.0$ m

5. [SSM] [WWW] **REASONING AND SOLUTION** The drawing at the right shows a ray diagram in which the reflected rays have been projected behind the mirror. We can see by inspection of this drawing that, after the rays reflect from the plane mirror, the angle α between them is still $\boxed{10°}$.

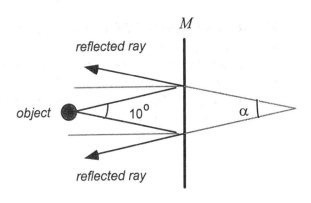

6. **REASONING** The drawing shows the situation described. The law of reflection indicates that the angle of incidence θ_i is equal to the angle of reflection θ_r.

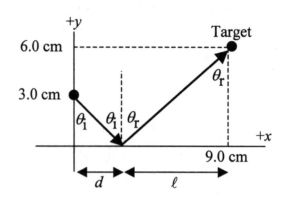

SOLUTION For the incident and reflected light, we have

$$\underbrace{\tan\theta_i = \frac{d}{3.0\text{ cm}}}_{\text{Incident light}} \quad \text{and} \quad \underbrace{\tan\theta_r = \frac{\ell}{6.0\text{ cm}}}_{\text{Reflected light}}$$

But $\theta_i = \theta_r$, according to the law of reflection, so that $\tan\theta_i = \tan\theta_r$, and we have

$$\frac{d}{3.0\text{ cm}} = \frac{\ell}{6.0\text{ cm}} \quad \text{or} \quad \ell = 2d$$

From the drawing we can see that $d + \ell = 9.0$ cm, and using the fact that $\ell = 2.0d$, we obtain

$$d + \ell = d + 2.0d = 9.0\text{ cm} \quad \text{or} \quad d = 3.0\text{ cm}$$

Therefore, the laser should be aimed at the point at $\boxed{x = +3.0\text{ cm}}$.

---

7. **REASONING AND SOLUTION**
   a. After the mirror has been rotated, the new angle of incidence is $\theta_i = 45° + 15° = 60°$. The angle of reflection, then, is also equal to 60°. The reflected ray which was originally 90° (45° + 45°) from the original angle of incidence, is now 120° (60° + 60°) from the incident ray's direction. Therefore, the reflected ray has been rotated through

$$\beta = 120° - 90° = \boxed{30°}$$

   b. The angle through which the reflected ray is rotated depends only on the angle through which the mirror is rotated, and is independent of the angle of incidence. Therefore, $\boxed{\beta' = 30°}$.

8. **REASONING AND SOLUTION** Denote the walls as side A (left-hand mirror opposite the target), side B (top wall), and side C (where the target is). The directions in the which the laser will be fired will be the angles measured counterclockwise with respect to the +x axis (i.e., the line from P to the bottom of wall C). We have

$\boxed{\theta_1 = 11.3°}$, follows a path from C to A to the target.

$\boxed{\theta_2 = 45.0°}$, follows a path directly to the target.

$\boxed{\theta_3 = 71.6°}$, follows a path from B to the target.

$\boxed{\theta_4 = 135°}$, follows a path from A to B to the target.

$\boxed{\theta_5 = 162°}$, follows a path from A to the target.

---

9. **SSM REASONING** The time of travel is proportional to the total distance for each path. Therefore, using the distances identified in the drawing, we have

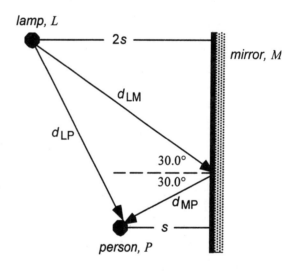

$$\frac{t_{\text{reflected}}}{t_{\text{direct}}} = \frac{d_{LM} + d_{MP}}{d_{LP}} \quad (1)$$

We know that the angle of incidence is equal to the angle of reflection. We also know that the lamp L is a distance $2s$ from the mirror, while the person P is a distance s from the mirror.

Therefore, it follows that if $d_{MP} = d$, then $d_{LM} = 2d$. We may use the law of cosines (see Appendix E) to express the distance d_{LP} as

$$d_{LP} = \sqrt{(2d)^2 + d^2 - 2(2d)(d)\cos 2(30.0°)}$$

SOLUTION Substituting the expressions for d_{MP}, d_{LM}, and d_{LP} into Equation (1), we find that the ratio of the travel times is

$$\frac{t_{\text{reflected}}}{t_{\text{direct}}} = \frac{2d + d}{\sqrt{(2d)^2 + d^2 - 2(2d)(d)\cos 2(30.0°)}} = \frac{3}{\sqrt{5 - 4\cos 60.0°}} = \boxed{1.73}$$

10. **REASONING AND SOLUTION** The ray diagram is shown in the figure (Note: $f = 5.0$ cm and $d_o = 15.0$ cm).

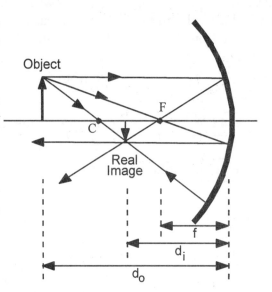

a. The ray diagram indicates that the image distance is 7.5 cm in front of the mirror.

b. The image height is 1.0 cm, and the image is inverted relative to the object.

11. **REASONING AND SOLUTION** An accurately drawn ray diagram, which will look similar, but not identical, to that in Figure 25.20a, reveals that

a. the image distance is 3.0×10^1 cm behind the mirror, and

b. the image height is 5.0 cm, and the image is upright relative to the object.

12. **REASONING AND SOLUTION**

a. A ray diagram, which will look similar, but not identical, to that in Figure 25.22a, reveals that the image distance is 20.0 cm behind the mirror, or $d_i = -20.0$ cm.

b. The ray diagram also shows that the image height is 6.0 cm, and the image is upright relative to the object.

13. **SSM REASONING AND SOLUTION** The ray diagram is shown below. (Note: $f = -50.0$ cm and $d_o = 25.0$ cm)

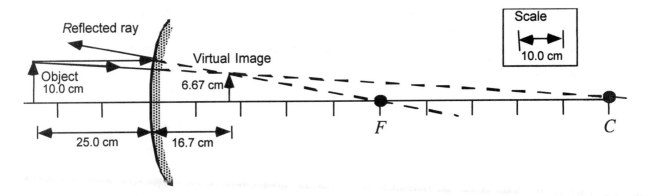

a. The ray diagram indicates that the image is 16.7 cm behind the mirror.
b. The ray diagram indicates that the image height is 6.67 cm.

---

14. **REASONING AND SOLUTION** An accurately drawn ray diagram, which will look similar to that in Figure 25.19a, reveals that

a. the image distance is 23.6 cm in front of the mirror, and

b. the image height is 2.14 cm, and the image is inverted relative to the object.

---

15. **REASONING AND SOLUTION** A plane mirror faces a concave mirror ($f = 8.0$ cm). The following is a ray diagram of an object placed 10.0 cm in front of the plane mirror.

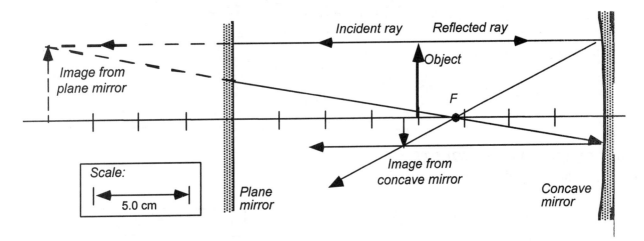

The ray diagram shows the light that is first reflected from the plane mirror and then the concave mirror. The scale is shown in the figure. For the reflection from the plane mirror,

924 THE REFLECTION OF LIGHT: MIRRORS

as discussed in the text, the image is upright, the same size as the object, and is located as far behind the mirror as the object is in front of it. When the reflected ray reaches the concave mirror, the ray that is initially parallel to the principal axis passes through the focal point F after reflection from the concave mirror. The ray that passes directly through the focal point emerges parallel to the principal axis after reflection from the concave surface. The point of intersection of these two rays locates the position of the image. By inspection, we see that the image is located at $\boxed{10.9 \text{ cm}}$ from the concave mirror.

16. **REASONING AND SOLUTION** We know that $d_o = 8.0$ cm and that the image is real and enlarged. A concave mirror produces a real, enlarged image when the object is located between the focal point and the center of curvature, as in Figure 25.19a. As a result, the image is also inverted, so that $m = -d_i/d_o = -4$, according to the magnification equation. Thus,

$$d_i = 4d_o = 4(8.0 \text{ cm}) = \boxed{+32 \text{ cm}}$$

17. **SSM REASONING** This problem can be solved using the mirror equation, Equation 25.3.

    SOLUTION Using the mirror equation with $d_i = +97$ cm and $f = 42$ cm, we find

    $$\frac{1}{d_o} = \frac{1}{f} - \frac{1}{d_i} = \frac{1}{42 \text{ cm}} - \frac{1}{97 \text{ cm}} \quad \text{or} \quad \boxed{d_o = +74 \text{ cm}}$$

18. **REASONING AND SOLUTION**
    a. The focal length is found from

    $$\frac{1}{f} = \frac{1}{d_o} + \frac{1}{d_i} = \frac{1}{27 \text{ cm}} + \frac{1}{-65 \text{ cm}} \quad \text{so} \quad \boxed{f = +46 \text{ cm}}$$

    b. The magnification is

    $$m = \frac{-d_i}{d_o} = \frac{-(-65 \text{ cm})}{27 \text{ cm}} = \boxed{2.4}$$

19. **REASONING AND SOLUTION**
    a. The image distance is

    $$\frac{1}{d_i} = \frac{1}{f} - \frac{1}{d_o} = \frac{1}{32.0 \text{ cm}} - \frac{1}{38.0 \text{ cm}} \quad \text{so} \quad \boxed{d_i = 2.0 \times 10^2 \text{ cm}}$$

b. The image height is

$$h_i = -h_o \frac{d_i}{d_o} = -(1.20 \text{ cm})\left(\frac{2.0 \times 10^2 \text{ cm}}{38.0 \text{ cm}}\right) = \boxed{-6.3 \text{ cm}}$$

c. Since h_i is negative, the image is inverted. Therefore, to make the picture on the wall appear normal, the slide must be oriented $\boxed{\text{upside down}}$.

---

20. **REASONING** The mirror equation relates the object and image distances to the focal length. Thus, we can apply the mirror equation once with the given object and image distances to determine the focal length. Then, we can use the mirror equation again with the focal length and the second object distance to determine the unknown image distance.

SOLUTION According to the mirror equation, we have

$$\underbrace{\frac{1}{d_{o1}} + \frac{1}{d_{i1}} = \frac{1}{f}}_{\text{First position of object}} \quad \text{and} \quad \underbrace{\frac{1}{d_{o2}} + \frac{1}{d_{i2}} = \frac{1}{f}}_{\text{Second position of object}}$$

Since the focal length is the same in both cases, it follows that

$$\frac{1}{d_{o1}} + \frac{1}{d_{i1}} = \frac{1}{d_{o2}} + \frac{1}{d_{i2}}$$

$$\frac{1}{d_{i2}} = \frac{1}{d_{o1}} + \frac{1}{d_{i1}} - \frac{1}{d_{o2}} = \frac{1}{25 \text{ cm}} + \frac{1}{(-17 \text{ cm})} - \frac{1}{19 \text{ cm}} = -0.071 \text{ cm}^{-1}$$

$$d_{i2} = -14 \text{ cm}$$

The negative value for d_{i2} indicates that the image is located $\boxed{14 \text{ cm}}$ behind the mirror.

---

21. **SSM REASONING** We have seen that a convex mirror always forms a *virtual image* as shown in Figure 25.22a of the text, where the image is *upright* and *smaller* than the object. These characteristics should bear out in the results of our calculations.

SOLUTION The radius of curvature of the convex mirror is 68 cm. Therefore, the focal length is, from Equation 25.2, $f = -(1/2)R = -34$ cm. Since the image is virtual, we know that $d_i = -22$ cm.

a. With $d_i = -22$ cm and $f = -34$ cm, the mirror equation gives

$$\frac{1}{d_o} = \frac{1}{f} - \frac{1}{d_i} = \frac{1}{-34 \text{ cm}} - \frac{1}{-22 \text{ cm}} \quad \text{or} \quad \boxed{d_o = +62 \text{ cm}}$$

b. According to the magnification equation, the magnification is

$$m = -\frac{d_i}{d_o} = -\frac{-22 \text{ cm}}{62 \text{ cm}} = \boxed{+0.35}$$

c. Since the magnification m is positive, the image is $\boxed{\text{upright}}$.

d. Since the magnification m is less than one, the image is $\boxed{\text{smaller}}$ than the object.

22. **REASONING AND SOLUTION** We can find the focal length from the mirror equation, provided we have values for both the image and object distances. The image in a convex mirror is upright, so the magnification is a positive quantity. Since the image is half the size of the object, the magnification equation indicates that $m = -d_i/d_o = \frac{1}{2}$, or $d_i = -\frac{1}{2}d_o$. Therefore, the image distance is $d_i = -\frac{1}{2}(13 \text{ cm}) = -6.5 \text{ cm}$. Using this result in the mirror equation, we find that

$$\frac{1}{d_o} + \frac{1}{d_i} = \frac{1}{f} \quad \text{or} \quad \frac{1}{13 \text{ cm}} + \frac{1}{-6.5 \text{ cm}} = \frac{1}{f} \quad \text{or} \quad \boxed{f = -13 \text{ cm}}$$

23. **SSM WWW REASONING** This problem can be solved by using the mirror equation, Equation 25.3, and the magnification equation, Equation 25.4.

    SOLUTION
    a. Using the mirror equation with $d_i = d_o$ and $f = R/2$, we have

    $$\frac{1}{d_o} = \frac{1}{f} - \frac{1}{d_i} = \frac{1}{R/2} - \frac{1}{d_o} \quad \text{or} \quad \frac{2}{d_o} = \frac{2}{R}$$

    Therefore, we find that $\boxed{d_o = R}$.

    b. According to the magnification equation, the magnification is

    $$m = -\frac{d_i}{d_o} = -\frac{d_o}{d_o} = \boxed{-1}$$

    c. Since the magnification m is negative, the image is $\boxed{\text{inverted}}$.

24. **REASONING AND SOLUTION**
    a. Since the image of the tooth is enlarged, it cannot be a plane mirror. Convex mirrors will produce smaller images in all cases. Therefore, the mirror is $\boxed{\text{concave}}$.

b. The focal length is

$$\frac{1}{f} = \frac{1}{d_o} + \frac{1}{d_i} = \frac{1}{2.0\text{ cm}} + \frac{1}{-5.6\text{ cm}} \quad \text{so} \quad \boxed{f = +3.1\text{ cm}}$$

c. The magnification is

$$m = -\frac{d_i}{d_o} = -\frac{-5.6\text{ cm}}{2.0\text{ cm}} = \boxed{+2.8}$$

d. Since m is positive, the image is $\boxed{\text{upright}}$ relative to the object.

---

25. **REASONING** Since the size of the image is one-third that of the object, we know from Equation 25.4 that

$$\underbrace{m}_{1/3} = -\frac{d_i}{d_o} \quad \text{so that} \quad d_i = -\tfrac{1}{3} d_o$$

We will substitute this relation into the mirror equation to find the ratio d_o/f.

SOLUTION Substituting $d_i = -\tfrac{1}{3} d_o$ into the mirror equation gives

$$\frac{1}{d_o} + \frac{1}{d_i} = \frac{1}{f} \quad \text{or} \quad \frac{1}{d_o} + \frac{1}{-\tfrac{1}{3} d_o} = \frac{1}{f}$$

Solving this last equation for the ratio d_o/f yields $d_o/f = \boxed{-2}$.

---

26. **REASONING** The mirror equation relates the object and image distances to the focal length. The magnification equation relates the magnification to the object and image distances. The problem neither gives nor asks for information about the image distance. Therefore, we can solve the magnification equation for the image distance and substitute the result into the mirror equation to obtain an expression relating the object distance, the magnification, and the focal length. This expression can be applied to both mirrors A and B to obtain the ratio of the focal lengths.

SOLUTION The magnification equation gives the magnification as $m = -d_i/d_o$. Solving for d_i, we obtain $d_i = -md_o$. Substituting this result into the mirror equation, we obtain

$$\frac{1}{d_o} + \frac{1}{d_i} = \frac{1}{d_o} + \frac{1}{(-md_o)} = \frac{1}{f} \quad \text{or} \quad \frac{1}{d_o}\left(1 - \frac{1}{m}\right) = \frac{1}{f} \quad \text{or} \quad f = \frac{d_o m}{m-1}$$

928 THE REFLECTION OF LIGHT: MIRRORS

Applying this result for the focal length f to each mirror gives

$$f_A = \frac{d_o m_A}{m_A - 1} \quad \text{and} \quad f_B = \frac{d_o m_B}{m_B - 1}$$

Dividing the expression for f_A by the expression for f_B, we find

$$\frac{f_A}{f_B} = \frac{d_o m_A / (m_A - 1)}{d_o m_B / (m_B - 1)} = \frac{m_A (m_B - 1)}{m_B (m_A - 1)} = \frac{4.0(2.0 - 1)}{2.0(4.0 - 1)} = \boxed{0.67}$$

---

27. **SSM WWW** *REASONING AND SOLUTION* The magnification equation, Equation 25.4, indicates that $d_i = -m d_o$. Substituting this result into the mirror equation, we find that

$$\frac{1}{d_o} + \frac{1}{d_i} = \frac{1}{d_o} + \frac{1}{-m d_o} = \frac{1}{f}$$

Solving for the object distance d_o, we find

$$d_o = \frac{(m-1) f}{m}$$

After the object is moved to its new position, its object distance becomes d_o' and its magnification becomes $m' = 2m$. Therefore, the amount by which the object is moved is

$$d_o - d_o' = \frac{(m-1) f}{m} - \frac{(m'-1) f}{m'} = \frac{(m-1) f}{m} - \frac{(2m-1) f}{2m}$$

$$= -\frac{f}{2m} = -\frac{-24.0 \text{ cm}}{2(0.150)} = \boxed{+80.0 \text{ cm}}$$

The positive answer means that the initial object distance is larger than the final object distance, so that $\boxed{\text{the object is moved toward the mirror}}$.

---

28. *REASONING AND SOLUTION* We know that $d_o - d_i = 45.0$ cm. We also have $1/d_o + 1/d_i = 1/f$. Solving the first equation for d_i and substituting the result into the second equation yields,

$$\frac{1}{d_o} + \frac{1}{d_o - 45.0 \text{ cm}} = \frac{1}{30.0 \text{ cm}}$$

Cross multiplying gives $d_o^2 - 105\, d_o + 1350 = 0$, which we can solve using the quadratic equation to yield two roots, $d_o = (105 \pm 75)/2$.

a. When the object lies beyond the center of curvature we have

$$d_{o+} = (1.80 \times 10^2 \text{ cm})/2 = \boxed{+9.0 \times 10^1 \text{ cm}} \quad \text{and} \quad d_{i+} = \boxed{+45 \text{ cm}}$$

b. When the object lies within the focal point

$$d_{o-} = (3.0 \times 10^1 \text{ cm})/2 = \boxed{+15 \text{ cm}}, \quad \text{and} \quad d_{i-} = \boxed{-3.0 \times 10^1 \text{ cm}}$$

---

29. **REASONING AND SOLUTION**
   a. From the mirror equation we have

$$\frac{1}{d_o} + \frac{1}{d_i} = \frac{1}{f} \quad \text{so that} \quad d_i = \frac{1}{\dfrac{1}{f} - \dfrac{1}{d_o}}$$

Since f is negative for a convex mirror, and since d_o is always positive, the quantity $(1/f - 1/d_o)$ must always be negative. Hence, d_i will always be negative, i.e., the image is $\boxed{\text{virtual}}$.

b. We have

$$d_i = \frac{1}{\dfrac{1}{f} - \dfrac{1}{d_o}} = \frac{f d_o}{d_o - f}$$

Since $m = -d_i/d_o$, we have that

$$m = \frac{-f}{d_o - f}$$

This expression shows that the magnitude of m is less than 1 for the following reason. Because f is always negative for a convex mirror and d_o is always positive, the denominator is always greater than the numerator. So the image is always smaller than the object. In addition, $m = -d_i/d_o > 0$, since d_i is always negative and d_o is positive. Thus, the image is always $\boxed{\text{upright and smaller}}$.

930   THE REFLECTION OF LIGHT: MIRRORS

30. **REASONING AND SOLUTION**

   a. Since the car is "very distant," we can assume that it is infinitely far from the mirror. Therefore, when paraxial rays leave the car and travel parallel to the principal axis, they appear to come from the focal point after reflecting from the mirror. Thus, the focal length is $f = -12$ cm (negative, because the mirror is a convex mirror), and the radius of curvature is $R = -2f = \boxed{24 \text{ cm}}$.

   b. The ray diagram is similar to that shown in Figure 25.16.

---

31. [SSM] [WWW] **REASONING** When paraxial light rays that are parallel to the principal axis strike a convex mirror, the reflected rays diverge after being reflected, and appear to originate from the focal point F behind the mirror (see Figure 25.16). We can treat the sun as being infinitely far from the mirror, so it is reasonable to treat the incident rays as paraxial rays that are parallel to the principal axis.

   SOLUTION

   a. Since the sun is infinitely far from the mirror and its image is a virtual image that lies *behind* the mirror, we can conclude that the mirror is a $\boxed{\text{convex mirror}}$.

   b. With $d_i = -12.0$ cm and $d_o = \infty$, the mirror equation (Equation 25.3) gives

   $$\frac{1}{f} = \frac{1}{d_o} + \frac{1}{d_i} = \frac{1}{\infty} + \frac{1}{d_i} = \frac{1}{d_i}$$

   Therefore, the focal length f lies 12.0 cm behind the mirror (this is consistent with the reasoning above that states that, after being reflected, the rays appear to originate from the focal point behind the mirror). In other words, $f = -12.0$ cm. Then, according to Equation 25.2, $f = -\frac{1}{2}R$, and the radius of curvature is

   $$R = -2f = -2(-12.0 \text{ cm}) = \boxed{24.0 \text{ cm}}$$

---

32. **REASONING AND SOLUTION**

   a. The height of the shortest mirror would be one-half the height of the person. Therefore,

   $$h = H/2 = (1.70 \text{ m} + 0.12 \text{ m})/2 = \boxed{0.91 \text{ m}}$$

   b. The bottom edge of the mirror should be above the floor by

   $$h' = (1.70 \text{ m})/2 = \boxed{0.85 \text{ m}}$$

33. **REASONING AND SOLUTION** From the mirror equation, we have that

$$\frac{1}{d_i} = \frac{1}{f} - \frac{1}{d_o} = \frac{1}{17 \text{ cm}} - \frac{1}{38 \text{ cm}} \quad \text{so} \quad \boxed{d_i = +31 \text{ cm}}$$

---

34. **REASONING AND SOLUTION**
   a. The image distance is

$$\frac{1}{d_i} = \frac{1}{f} - \frac{1}{d_o} = \frac{1}{-2.0 \text{ m}} - \frac{1}{15 \text{ m}} \quad \text{so} \quad \boxed{d_i = -1.8 \text{ m}}$$

   b. Since the image distance is negative, the image is $\boxed{\text{virtual}}$.

   c. The height of the object is

$$h_i = -h_o \frac{d_i}{d_o} = -(1.6 \text{ m})\frac{-1.8 \text{ m}}{15 \text{ m}} = \boxed{0.19 \text{ cm}}$$

---

35. [SSM] **REASONING** This problem can be solved using the mirror equation, Equation 25.3.

   SOLUTION Using the mirror equation with $d_i = +26$ cm and $f = 12$ cm, we find

$$\frac{1}{d_o} = \frac{1}{f} - \frac{1}{d_i} = \frac{1}{12 \text{ cm}} - \frac{1}{26 \text{ cm}} \quad \text{or} \quad \boxed{d_o = +22 \text{ cm}}$$

---

36. **REASONING** The drawing shows two plane mirrors that intersect at an angle of 50°. An incident light ray reflects from one mirror and then the other. The various angles are labeled θ, α, β, γ, ε and ϕ.

   We wish to find the numerical value of the angle θ. We will do this using both the law of reflection, which states that when light is reflected from a surface, the angle of incidence is equal to the angle of reflection, and the fact that the sum of the interior angles of a triangle is 180°.

932 THE REFLECTION OF LIGHT: MIRRORS

SOLUTION Since the normal to the surface for the incident ray forms a 90° angle with the horizontal mirror, we know that $\gamma = 90° - \alpha$. Then, using the same reasoning, $\varepsilon = 90° - \beta$. Finally, from the diagram, we see that $\theta + \phi = 180°$, or, solving for θ, we find that $\theta = 180° - \phi$.

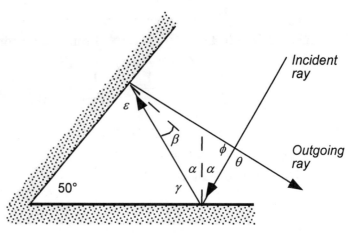

We now proceed to reduce this expression to angles and expressions of known values.
From the figure, $2\alpha + 2\beta + \phi = 180°$, so that $\phi = 180° - 2(\alpha + \beta)$. Therefore, we find that $\theta = 180° - \phi = 180° - 180° + 2(\alpha + \beta)$, or

$$\theta = 2(\alpha + \beta) \tag{1}$$

We now proceed to find the quantity $\alpha + \beta$. For the triangle formed by the intersection of the mirrors and the first reflected ray, $180° = 50° + \gamma + \varepsilon$. Since $\gamma = 90° - \alpha$ and $\varepsilon = 90° - \beta$ as shown above, this becomes

$$180° = 50° + (90° - \alpha) + (90° - \beta) = 50° + 180° - (\alpha + \beta)$$

or

$$\alpha + \beta = 50°$$

Substituting into Equation (1),

$$\theta = 2(\alpha + \beta) = 2 \times 50° = \boxed{100°}$$

---

37. **REASONING AND SOLUTION** To obtain the focal length, we will use the mirror equation. To do so, it is necessary to obtain values for the object and image distances. Since the image distance for virtual image 1 in Figure 25.21b is a negative quantity, $-d_i$ will be a positive quantity. Therefore, the distance between the readout device and image 1 is $d_o - d_i = 2.00$ m. Equation 25.4 gives the magnification m as

$$m = -d_i/d_o = 4.00 \quad \text{so that} \quad d_i = -4.00\, d_o$$

As a result

$$d_o - d_i = d_o - (-4.00\, d_o) = 2.00\text{ m} \quad \text{or} \quad d_o = 0.400\text{ m}$$

The image distance is

$$d_i = -4.00\, d_o = -4.00(0.400\text{ m}) = -1.60\text{ m}$$

We can now obtain the focal length from the mirror equation:

$$\frac{1}{d_o} + \frac{1}{d_i} = \frac{1}{0.400 \text{ m}} + \frac{1}{-1.60 \text{ m}} = \frac{1}{f} \quad \text{or} \quad \boxed{f = 0.533 \text{ m}}$$

---

38. **REASONING** The image in the plane mirror is located 15.0 cm behind the mirror. This is because the candle is located 15.0 cm in front of the mirror, and plane mirrors always produce images that are the same distance behind the mirror as the object is in front of it. We know that the image moves 7.0 cm farther away from the mirror, when the plane mirror replaces the convex mirror. Therefore, the image in the convex mirror must have been 15.0 cm − 7.0 cm = 8.0 cm behind the mirror. This means that the image distance for the convex mirror is −8.0 cm. Knowing both the object and image distances, we can use the mirror equation to calculate the focal length.

SOLUTION According to the mirror equation, we have

$$\frac{1}{f} = \frac{1}{d_o} + \frac{1}{d_i} = \frac{1}{15.0 \text{ cm}} + \frac{1}{(-8.0 \text{ cm})} = -0.058 \text{ cm}^{-1} \quad \text{or} \quad \boxed{f = -17 \text{ cm}}$$

---

39. **SSM** **REASONING** We need to know the focal length of the mirror and can obtain it from the mirror equation, Equation 25.3, as applied to the first object:

$$\frac{1}{d_{o1}} + \frac{1}{d_{i1}} = \frac{1}{14.0 \text{ cm}} + \frac{1}{-7.00 \text{ cm}} = \frac{1}{f} \quad \text{or} \quad f = -14.0 \text{ cm}$$

According to the magnification equation, Equation 25.4, the image height h_i is related to the object height h_o as follows: $h_i = m h_o = (-d_i / d_o) h_o$.

SOLUTION Applying this result to each object, we find that $h_{i2} = h_{i1}$, or

$$\left(\frac{-d_{i2}}{d_{o2}}\right) h_{o2} = \left(\frac{-d_{i1}}{d_{o1}}\right) h_{o1}$$

Therefore,

$$d_{i2} = d_{o2} \left(\frac{d_{i1}}{d_{o1}}\right) \left(\frac{h_{o1}}{h_{o2}}\right)$$

Using the fact that $h_{o2} = 2 h_{o1}$, we have

934 THE REFLECTION OF LIGHT: MIRRORS

$$d_{i2} = d_{o2}\left(\frac{d_{i1}}{d_{o1}}\right)\left(\frac{h_{o1}}{h_{o2}}\right) = d_{o2}\left(\frac{-7.00 \text{ cm}}{14.0 \text{ cm}}\right)\left(\frac{h_{o1}}{2h_{o1}}\right) = -0.250\, d_{o2}$$

Using this result in the mirror equation, as applied to the second object, we find that

$$\frac{1}{d_{o2}} + \frac{1}{d_{i2}} = \frac{1}{f}$$

or

$$\frac{1}{d_{o2}} + \frac{1}{-0.250\, d_{o2}} = \frac{1}{-14.0 \text{ cm}}$$

Therefore,

$$\boxed{d_{o2} = +42.0 \text{ cm}}$$

---

40. **REASONING AND SOLUTION** The magnification is $m = -d_i/d_o = +1/4$, so that $d_i = -(1/4)d_o$. Thus,

$$\frac{1}{f} = \frac{1}{d_o} + \frac{1}{d_i} = \frac{1}{d_o} + \frac{1}{\left(-\frac{1}{4}d_o\right)} = \frac{-3}{d_o}$$

Therefore, $f = -(1/3)d_o$, for the convex side. Since the radius of curvature is the same in each case, the focal length of the concave side is $f' = +\frac{1}{3}d_o'$. So,

$$\frac{1}{d_o'} + \frac{1}{d_i'} = \frac{1}{f'} = \frac{1}{\frac{1}{3}d_o'} \qquad \text{so that} \qquad \frac{1}{d_i'} = \frac{2}{d_o'}$$

The magnification of the concave side is, therefore, $m' = -d_i'/d_o' = \boxed{-1/2}$.

---

41. **REASONING AND SOLUTION** We can see from the diagram that $\tan\theta = x/(L/2)$. We also see that $\tan\theta = (L-x)/L$. Equating,

$$\frac{x}{\frac{1}{2}L} = \frac{L-x}{L} \qquad \text{or} \qquad x = \tfrac{1}{3}L$$

Therefore,

$$\theta = \tan^{-1}\left(\frac{\frac{1}{3}L}{\frac{1}{2}L}\right) = \boxed{33.7°}$$

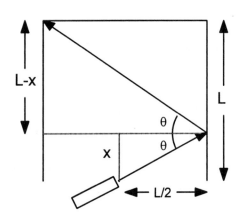

42. **CONCEPT QUESTIONS** a. According to the discussion about relative velocity in Section 3.4, it follows that $\mathbf{v}_{IY} = \mathbf{v}_{IM} + \mathbf{v}_{MY}$.

b. As you walk perpendicularly toward the stationary mirror, you perceive the mirror moving toward you in the opposite direction. Thus, $\mathbf{v}_{MY} = -\mathbf{v}_{YM}$.

c. The velocities $\mathbf{v}_{YM}$ and $\mathbf{v}_{IM}$ have the same magnitude. This is because the image in a plane mirror is always just as far behind the mirror as the object is in front of it. For instance, if you move 1 meter perpendicularly toward the mirror in 1 second, the magnitude of your velocity relative to the mirror is 1 m/s. But your image also moves 1 meter toward the mirror in the same time interval, so that the magnitude of its velocity relative to the mirror is also 1 m/s. The two velocities, however, have opposite directions.

SOLUTION According to the discussion in Section 3.4, we have

$$\mathbf{v}_{IY} = \mathbf{v}_{IM} + \mathbf{v}_{MY} = \mathbf{v}_{IM} - \mathbf{v}_{YM} \qquad (1)$$

Remembering that the magnitudes of both velocities $\mathbf{v}_{YM}$ and $\mathbf{v}_{IM}$ are the same and that the direction in which you walk is positive, we have

$$v_{YM} = +0.90 \text{ m/s} \quad \text{and} \quad v_{IM} = -0.90 \text{ m/s}$$

The velocity $\mathbf{v}_{IM}$ is negative, because its direction is opposite to the direction in which you walk. Substituting these values into Equation (1), we obtain

$$v_{IY} = (-0.90 \text{ m/s}) - (+0.90 \text{ m/s}) = \boxed{-1.80 \text{ m/s}}$$

---

43. **CONCEPT QUESTIONS** a. According to the discussion about relative velocity in Section 3.4, $\mathbf{v}_{IY} = \mathbf{v}_{IM} + \mathbf{v}_{MY}$.

b. As you walk toward the stationary mirror, you perceive the mirror moving toward you in the opposite direction. Thus, $\mathbf{v}_{MY} = -\mathbf{v}_{YM}$.

c. The velocity $\mathbf{v}_{YM}$ has the components v_{YMx} and v_{YMy}, while the velocity $\mathbf{v}_{IM}$ has the components v_{IMx} and v_{IMy}. The x direction is perpendicular to the mirror, and the two x components have the same magnitude. This is because the image in a plane mirror is always just as far behind the mirror as the object is in front of it. For instance, if an object moves 1 meter perpendicularly toward the mirror in 1 second, the magnitude of its velocity relative to the mirror is 1 m/s. But the image also moves 1 meter toward the mirror in the same time interval, so that the magnitude of its velocity relative to the mirror is also 1 m/s. The two x components, however, have opposite directions.

936 THE REFLECTION OF LIGHT: MIRRORS

The two y components have the same magnitude and the same direction. This is because an object moving parallel to a plane mirror has an image that remains at the same distance behind the mirror as the object is in front of it and moves in the same direction as the object.

SOLUTION According to the discussion in Section 3.4, we have

$$\mathbf{v}_{IY} = \mathbf{v}_{IM} + \mathbf{v}_{MY} = \mathbf{v}_{IM} - \mathbf{v}_{YM}$$

This vector equation is equivalent to two equations, one for the x components and one for the y components. For the x direction, we note that $v_{YMx} = -v_{IMx}$.

$$v_{IYx} = v_{IMx} - v_{YMx}$$
$$= -(0.90 \text{ m/s})\cos 50.0° - (0.90 \text{ m/s})\cos 50.0° = -1.2 \text{ m/s}$$

For the y direction, we note that $v_{YMy} = v_{IMy}$.

$$v_{IYy} = v_{IMy} - v_{YMy}$$
$$= (0.90 \text{ m/s})\sin 50.0° - (0.90 \text{ m/s})\sin 50.0° = 0 \text{ m/s}$$

Since the y component of the velocity $\mathbf{v}_{IY}$ is zero, the velocity of your image relative to you points in the $\boxed{-x \text{ direction and has a magnitude of 1.2 m/s}}$.

---

44. **CONCEPT QUESTIONS** a. For an image that is in front of a mirror, the image distance is positive.

    b. Given the image distance, the mirror equation can be used to determine the focal length, but to do so a value for the object distance is also needed.

    c. For an image that is inverted, the image height is negative.

    d. The object and image heights, together with a knowledge of whether the image is upright or inverted, allow you to calculate the magnification m. The magnification is given by Equation 25.4 as $m = -d_i/d_o$, where d_i and d_o are the image and object distances, respectively. Therefore, to determine the object distance, a value for the image distance is also needed.

    SOLUTION According to Equation 25.4, the magnification is

    $$m = \frac{h_i}{h_o} = -\frac{d_i}{d_o} \quad \text{or} \quad d_o = -\frac{d_i h_o}{h_i}$$

Substituting this result into the mirror equation, we obtain

$$\frac{1}{f} = \frac{1}{d_o} + \frac{1}{d_i} = -\frac{h_i}{d_i h_o} + \frac{1}{d_i} = \frac{1}{d_i}\left(-\frac{h_i}{h_o} + 1\right)$$

$$= \left(\frac{1}{13 \text{ cm}}\right)\left[-\frac{(-1.5 \text{ cm})}{(3.5 \text{ cm})} + 1\right] = 0.11 \text{ cm}^{-1} \quad \text{or} \quad \boxed{f = 9.1 \text{ cm}}$$

---

45. **CONCEPT QUESTIONS** a. As the object distance increases, ray 1 remains parallel to the principal axis. Therefore, after reflection it appears to come from the focal point of the mirror, no matter where the object is. Reflected ray 1 does not change.

b. Ray 3 is directed toward the center of curvature of the spherical mirror and, therefore, strikes the mirror perpendicularly. The reflected ray 3 travels back straight along the incident direction. This is so, no matter where the object is located. However, as the object distance increases, reflected ray 3 makes a smaller angle with respect to the principal axis.

c. Because of the behavior of the reflected rays 1 and 3, the dashed lines in Figure 25.22a intersect at a point that moves closer to the principal axis as the object distance increases. Thus, the height of the image decreases with increasing object distance.

SOLUTION The magnification is given by Equation 25.4 as $m = -d_i/d_o$, where d_i and d_o are the image and object distances, respectively. We can obtain the image distance from the mirror equation as follows:

$$\frac{1}{d_o} + \frac{1}{d_i} = \frac{1}{f} \quad \text{or} \quad \frac{1}{d_i} = \frac{1}{f} - \frac{1}{d_o} = \frac{d_o - f}{fd_o} \quad \text{or} \quad d_i = \frac{fd_o}{d_o - f}$$

Substituting this result into Equation 25.4 gives

$$m = -\frac{d_i}{d_o} = -\frac{fd_o/(d_o - f)}{d_o} = \frac{f}{f - d_o}$$

Using this result with the given values for the focal length and object distances, we find

Smaller object distance $\quad m = \dfrac{f}{f - d_o} = \dfrac{-27.0 \text{ cm}}{(-27.0 \text{ cm}) - (9.0 \text{ cm})} = \boxed{0.750}$

Greater object distance $\quad m = \dfrac{f}{f - d_o} = \dfrac{-27.0 \text{ cm}}{(-27.0 \text{ cm}) - (18.0 \text{ cm})} = \boxed{0.600}$

CHAPTER 26 | THE REFRACTION OF LIGHT: LENSES AND OPTICAL INSTRUMENTS

PROBLEMS

1. **SSM** **REASONING AND SOLUTION** The speed of light in benzene v is related to the speed of light in vacuum c by the index of refraction n. The index of refraction is defined by Equation 26.1 ($n = c/v$). According to Table 26.1, the index of refraction of benzene is 1.501. Therefore, solving for v, we have

$$v = \frac{c}{n} = \frac{3.00 \times 10^8 \text{ m/s}}{1.501} = \boxed{2.00 \times 10^8 \text{ m/s}}$$

2. **REASONING AND SOLUTION** Using Equation 26.1 for diamond and for ice, we have

$$v_d = c/n_d \quad \text{and} \quad v_i = c/n_i$$

Dividing the above equations reveals that

$$v_d/v_i = n_i/n_d = 1.309/2.419 = \boxed{0.5411}$$

3. **REASONING** The wavelength λ is related to the frequency f and speed v of the light in a material by Equation 16.1 ($\lambda = v/f$). The speed of the light in each material can be expressed using Equation 26.1 ($v = c/n$) and the refractive indices n given in Table 26.1. With these two equations, we can obtain the desired ratio.

SOLUTION Using Equations 16.1 and 26.1, we find

$$\lambda = \frac{v}{f} = \frac{c/n}{f} = \frac{c}{fn}$$

Using this result and recognizing that the frequency f and the speed c of light in a vacuum do not depend on the material, we obtain the ratio of the wavelengths as follows:

$$\frac{\lambda_{\text{alcohol}}}{\lambda_{\text{disulfide}}} = \frac{\left(\frac{c}{fn}\right)_{\text{alcohol}}}{\left(\frac{c}{fn}\right)_{\text{disulfide}}} = \frac{\frac{c}{f}\left(\frac{1}{n}\right)_{\text{alcohol}}}{\frac{c}{f}\left(\frac{1}{n}\right)_{\text{disulfide}}} = \frac{n_{\text{disulfide}}}{n_{\text{alcohol}}} = \frac{1.632}{1.362} = \boxed{1.198}$$

4. **REASONING** We can identify the substance in Table 26.1 if we can determine its index of refraction. The index of refraction n is equal to the speed of light c in a vacuum divided by the speed of light v in the substance, or $n = c/v$. According to Equation 16.1, however, the speed of light is related to its wavelength λ and frequency f via $v = f\lambda$. Combining these two equations by eliminating the speed v yields $n = c/(f\lambda)$.

SOLUTION The index of refraction of the substance is

$$n = \frac{c}{f\lambda} = \frac{2.998 \times 10^8 \text{ m/s}}{(5.403 \times 10^{14} \text{ Hz})(340.0 \times 10^{-9} \text{ m})} = 1.632$$

An examination of Table 26.1 shows that the substance is $\boxed{\text{carbon disulfide}}$.

5. **SSM WWW** **REASONING** Since the light will travel in glass at a constant speed v, the time it takes to pass perpendicularly through the glass is given by $t = d/v$, where d is the thickness of the glass. The speed v is related to the vacuum value c by Equation 26.1: $n = c/v$.

SOLUTION Substituting for v from Equation 26.1 and substituting values, we obtain

$$t = \frac{d}{v} = \frac{nd}{c} = \frac{(1.5)(4.0 \times 10^{-3} \text{ m})}{3.00 \times 10^8 \text{ m/s}} = \boxed{2.0 \times 10^{-11} \text{ s}}$$

6. **REASONING AND SOLUTION** If $n_B = c/v$, then $n_A = c/(1.25\, v)$, so

$$n_A/n_B = 1/1.25 = \boxed{0.800}$$

7. **REASONING AND SOLUTION** The speed of light in the vacuum is $c = (3.50 \text{ km})/t$. The speed of light in the liquid is $v = (2.50 \text{ km})/t$, so

$$n = c/v = (3.50)/(2.50) = \boxed{1.40}$$

8. **REASONING** Distance traveled is the speed times the travel time. Assuming that t is the time it takes for the light to travel through the two sheets, it would travel a distance of ct in a vacuum, where its speed is c. Thus, to find the desired distance, we need to determine the travel time t. This time is the sum of the travel times in each sheet. The travel time in each sheet is determined by the thickness of the sheet and the speed of the light in the material. The speed in the material is less than the speed in a vacuum and depends on the refractive index of the material.

SOLUTION In the ice of thickness d_i, the speed of light is v_i, and the travel time is $t_i = d_i/v_i$. Similarly, the travel time in the quartz sheet is $t_q = d_q/v_q$. Therefore, the desired distance ct is

$$ct = c(t_i + t_q) = c\left(\frac{d_i}{v_i} + \frac{d_q}{v_q}\right) = d_i \frac{c}{v_i} + d_q \frac{c}{v_q}$$

Since Equation 26.1 gives the refractive index as $n = c/v$ and since Table 26.1 gives the indices of refraction for ice and quartz as $n_i = 1.309$ and $n_q = 1.544$, the result just obtained can be written as follows:

$$ct = d_i \frac{c}{v_i} + d_q \frac{c}{v_q} = d_i n_i + d_q n_q = (2.0 \text{ cm})(1.309) + (1.1 \text{ cm})(1.544) = \boxed{4.3 \text{ cm}}$$

---

9. **SSM REASONING** We begin by using Snell's law (Equation 26.2: $n_1 \sin\theta_1 = n_2 \sin\theta_2$) to find the index of refraction of the material. Then we will use Equation 26.1, the definition of the index of refraction ($n = c/v$) to find the speed of light in the material.

SOLUTION From Snell's law, the index of refraction of the material is

$$n_2 = \frac{n_1 \sin\theta_1}{\sin\theta_2} = \frac{(1.000)\sin 63.0°}{\sin 47.0°} = 1.22$$

Then, from Equation 26.1, we find that the speed of light v in the material is

$$v = \frac{c}{n_2} = \frac{3.00 \times 10^8 \text{ m/s}}{1.22} = \boxed{2.46 \times 10^8 \text{ m/s}}$$

10. **REASONING AND SOLUTION** Applying Snell's law at the gas-solid interface gives the angle of refraction θ_2 to be

$$(1.00)\sin 35.0° = (1.55)\sin\theta_2 \quad \text{or} \quad \theta_2 = 21.7°$$

Since the refractive index of the liquid is the same as that of the solid, light is not refracted when it enters the liquid. Therefore, the light enters the liquid at an angle of $\boxed{21.7°}$.

---

11. **SSM REASONING** We will use the geometry of the situation to determine the angle of incidence. Once the angle of incidence is known, we can use Snell's law to find the index of refraction of the unknown liquid. The speed of light v in the liquid can then be determined.

SOLUTION From the drawing in the text, we see that the angle of incidence at the liquid-air interface is

$$\theta_1 = \tan^{-1}\left(\frac{5.00 \text{ cm}}{6.00 \text{ cm}}\right) = 39.8°$$

The drawing also shows that the angle of refraction is 90.0°. Thus, according to Snell's law (Equation 26.2: $n_1 \sin\theta_1 = n_2 \sin\theta_2$), the index of refraction of the unknown liquid is

$$n_1 = \frac{n_2 \sin\theta_2}{\sin\theta_1} = \frac{(1.000)(\sin 90.0°)}{\sin 39.8°} = 1.56$$

From Equation 26.1 ($n = c/v$), we find that the speed of light in the unknown liquid is

$$v = \frac{c}{n_1} = \frac{3.00 \times 10^8 \text{ m/s}}{1.56} = \boxed{1.92 \times 10^8 \text{ m/s}}$$

---

12. **REASONING AND SOLUTION** The actual height d of the diving board above the water can be obtained by using Equation 26.3. As usual, n_1 is the index of refraction of the medium (air) associated with the incident ray, and n_2 is that of the medium (water) associated with the refracted ray. Taking the refractive index of water from Table 26.1, we find

$$d = d'\left(\frac{n_1}{n_2}\right) = (4.0 \text{ m})\left(\frac{1.00}{1.33}\right) = \boxed{3.0 \text{ m}}$$

942 THE REFRACTION OF LIGHT: LENSES AND OPTICAL INSTRUMENTS

13. ***REASONING AND SOLUTION*** Snell's law gives

$$(1.45)\sin 65.0° = n_2 \sin 53.0°$$

Solving, we find that the index of refraction is $n_2 = \boxed{1.64}$.

---

14. ***REASONING*** When the incident light is in a vacuum, Snell's law, Equation 26.2, can be used to express the relation between the angle of incidence (35.0°), the (unknown) index of refraction n_2 of the glass and the (unknown) angle θ_2 of refraction for the light entering the slab: $(1.00)\sin 35.0° = n_2 \sin \theta_2$ When the incident light is in the liquid, we can again use Snell's law to express the relation between the index of refraction n_1 of the liquid, the angle of incidence (20.3°), the index of refraction n_2 of the glass, and the (unknown) angle of refraction θ_2: $n_1 \sin 20.3° = n_2 \sin \theta_2$. By equating these two equations, we can determine the index of refraction of the liquid.

SOLUTION Setting the two equations above equal to each other and solving for the index of refraction of the liquid gives

$$n_1 \sin 20.3° = (1.00)\sin 35.0° \quad \text{and} \quad n_1 = \frac{(1.00)\sin 35.0°}{\sin 20.3°} = \boxed{1.65}$$

---

15. **SSM** **WWW** The drawing at the right shows the geometry of the situation using the same notation as that in Figure 26.7. In addition to the text's notation, let t represent the thickness of the pane, let L represent the length of the ray in the pane, let x (shown twice in the figure) equal the displacement of the ray, and let the difference in angles $\theta_1 - \theta_2$ be given by ϕ.

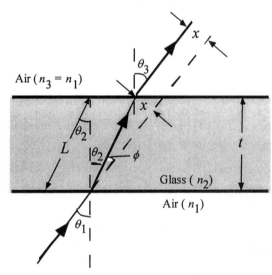

We wish to find the amount x by which the emergent ray is displaced relative to the incident ray. This can be done by applying Snell's law at each interface, and then making use of the geometric and trigonometric relations in the drawing.

SOLUTION If we apply Snell's law (see Equation 26.2) to the bottom interface we obtain $n_1 \sin \theta_1 = n_2 \sin \theta_2$. Similarly, if we apply Snell's law at the top interface where the ray emerges, we have $n_2 \sin \theta_2 = n_3 \sin \theta_3 = n_1 \sin \theta_3$. Comparing this with Snell's law at the bottom face, we see that $n_1 \sin \theta_1 = n_1 \sin \theta_3$, from which we can conclude that $\theta_3 = \theta_1$. Therefore, the emerging ray is parallel to the incident ray.

From the geometry of the ray and thickness of the pane, we see that $L \cos \theta_2 = t$, from which it follows that $L = t/\cos \theta_2$. Furthermore, we see that $x = L \sin \phi = L \sin(\theta_1 - \theta_2)$. Substituting for L, we find

$$x = L \sin(\theta_1 - \theta_2) = \frac{t \sin(\theta_1 - \theta_2)}{\cos \theta_2}$$

Before we can use this expression to determine a numerical value for x, we must find the value of θ_2. Solving the expression for Snell's law at the bottom interface for θ_2, we have

$$\sin \theta_2 = \frac{n_1 \sin \theta_1}{n_2} = \frac{(1.000)(\sin 30.0°)}{1.52} = 0.329 \quad \text{or} \quad \theta_2 = \sin^{-1} 0.329 = 19.2°$$

Therefore, the amount by which the emergent ray is displaced relative to the incident ray is

$$x = \frac{t \sin(\theta_1 - \theta_2)}{\cos \theta_2} = \frac{(6.00 \text{ mm}) \sin(30.0° - 19.2°)}{\cos 19.2°} = \boxed{1.19 \text{ mm}}$$

---

16. **REASONING AND SOLUTION**
a. Snell's law applied at the air-ice interface gives

$$\sin \theta_2 = (n_1/n_2) \sin \theta_1 = (1.000/1.309) \sin 45° = 0.54$$

$$\theta_2 = \sin^{-1}(0.54) = \boxed{33°}$$

b. This angle becomes the angle of incidence at the ice-water interface, so

$$\sin \theta_2 = (1.309/1.333) \sin 33° = 0.53 \quad \text{or} \quad \theta_2 = \sin^{-1}(0.53) = \boxed{32°}$$

---

17. **REASONING AND SOLUTION** The horizontal distance of the chest from the normal is found from Figure 26.5a to be $x = d \tan \theta_1$ and $x = d' \tan \theta_2$, where θ_1 is the angle from the dashed normal to the solid rays and θ_2 is the angle from the dashed normal to the dashed rays. Hence,

$$d' = (\tan \theta_1 / \tan \theta_2) d$$

Snell's law applied at the interface gives

$$n_1 \sin \theta_1 = n_2 \sin \theta_2$$

For small angles, $\sin \theta_1 \approx \tan \theta_1$ and $\sin \theta_2 \approx \tan \theta_2$, so

$$\tan \theta_1 / \tan \theta_2 \approx \sin \theta_1 / \sin \theta_2 = (n_2/n_1)$$

Now $d' = d (\tan \theta_1 / \tan \theta_2)$. Therefore,

$$\boxed{d' \approx d \left(\frac{n_2}{n_1} \right)}$$

---

18. **REASONING** Following the discussion in Conceptual Example 4, we have the drawing at the right to use as a guide. In this drawing the symbol d refers to depths in the water, while the symbol h refers to heights in the air above the water. Moreover, symbols with a prime denote apparent distances, and unprimed symbols denote actual distances. We will use Equation 26.3 to relate apparent distances to actual distances. In so doing, we will use the fact that the refractive index of air is essentially $n_{air} = 1$ and denote the refractive index of water by $n_w = 1.333$ (see Table 26.1).

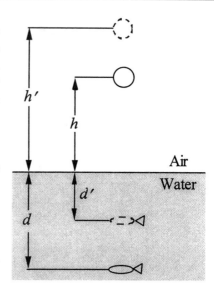

SOLUTION To the fish, the man appears to be a distance above the air-water interface that is given by Equation 26.3 as $h' = h(n_w/1)$. Thus, measured above the eyes of the fish, the man appears to be located at a distance of

$$h' + d = h \left(\frac{n_w}{1} \right) + d \qquad (1)$$

To the man, the fish appears to be a distance below the air-water interface that is given by Equation 26.3 as $d' = d(1/n_w)$. Thus, measured below the man's eyes, the fish appears to be located at a distance of

$$h + d' = h + d \left(\frac{1}{n_w} \right) \qquad (2)$$

Dividing Equation (1) by Equation (2) and using the fact that $h = d$, we find

$$\frac{h'+d}{h+d'} = \frac{h\left(\dfrac{n_w}{1}\right)+d}{h+d\left(\dfrac{1}{n_w}\right)} = \frac{n_w+1}{1+\dfrac{1}{n_w}} = n_w \qquad (3)$$

In Equation (3), $h'+d$ is the distance we seek, and $h+d'$ is given as 2.0 m. Thus, we find

$$h'+d = n_w(h+d') = (1.333)(2.0 \text{ m}) = \boxed{2.7 \text{ m}}$$

---

19. **REASONING AND SOLUTION** Using Equation 26.3, we have for the block in air

$$d_a' = (1/n_p)d$$

and for the block in water

$$d_w' = (n_w/n_p)d$$

Therefore, using the refractive index for water given in Table 26.1, we find

$$d_w' = n_w d_a' = (1.333)(1.6 \text{ cm}) = \boxed{2.1 \text{ cm}}$$

---

20. **REASONING AND SOLUTION** The light rays coming from the bottom of the beaker are refracted at two interfaces, the water-oil interface and the oil-air interface. When the rays enter the oil from the water, they appear to have originated from an apparent depth d' below the water-oil interface. This apparent depth is given by Equation 26.3 as

$$d' = d\left(\frac{n_{oil}}{n_{water}}\right) = (15.0 \text{ cm})\left(\frac{1.48}{1.33}\right) = 16.7 \text{ cm}$$

When the rays reach the top of the oil, a distance of 15.0 cm above the water, they can be regarded as having originated from a depth of 15.0 cm + 16.7 cm = 31.7 cm below the oil-air interface. When the rays enter the air, they are refracted again and appear to have come from an apparent depth d'' below the oil-air interface. This apparent depth is given by Equation 26.3 as

$$d'' = (31.7 \text{ cm})\left(\frac{n_{air}}{n_{oil}}\right) = (31.7 \text{ cm})\left(\frac{1.00}{1.48}\right) = \boxed{21.4 \text{ cm}}$$

21. **SSM** **REASONING** The drawing at the right shows the situation. As discussed in the text, when the observer is directly above, the apparent depth d' of the object is related to the actual depth by Equation 26.3:

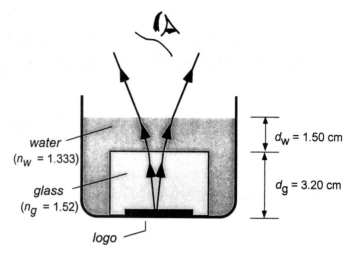

$$d' = d\left(\frac{n_2}{n_1}\right)$$

In this case, we must apply Equation 26.3 twice; once for the rays in the glass, and once again for the rays in the water.

SOLUTION We refer to the drawing for our notation and begin at the logo. To an observer *in the water* directly above the logo, the apparent depth of the logo is $d'_g = d_g(n_w/n_g)$. When viewed directly from above in air, the logo's apparent depth is, $d'_w = (d_w + d'_g)(n_{air}/n_w)$, where we have used the fact that when viewed from air, the logo's actual depth appears to be $d_w + d'_g$. Substituting the expression for d'_g into the expression for d'_w, we obtain

$$d'_w = (d_w + d'_g)\left(\frac{n_{air}}{n_w}\right) = d_w\left(\frac{n_{air}}{n_w}\right) + d_g\left(\frac{n_w}{n_g}\right)\left(\frac{n_{air}}{n_w}\right) = d_w\left(\frac{n_{air}}{n_w}\right) + d_g\left(\frac{n_{air}}{n_g}\right)$$

$$= (1.50 \text{ cm})\left(\frac{1.000}{1.333}\right) + (3.20 \text{ cm})\left(\frac{1.000}{1.52}\right) = \boxed{3.23 \text{ cm}}$$

---

22. **REASONING AND SOLUTION**
a. The apparent "depth" of the fish is given by

$$d' = (1.000/1.333)(15.0 \text{ cm}) = \boxed{11.3 \text{ cm}}$$

b. The image of the fish is located 15.0 cm behind the mirror. The apparent "depth" of the image is then

$$d' = (1.000/1.333)(45.0 \text{ cm}) = \boxed{33.8 \text{ cm}}$$

c. Since the apparent "depth" is inversely proportional to the index of refraction of the liquid, a $\boxed{\text{larger}}$ index of refraction may cause the image to appear in front of the mirror.

23. **SSM** *REASONING AND SOLUTION* According to Equation 26.4, the critical angle is related to the refractive indices n_1 and n_2 by $\sin\theta_c = n_2/n_1$, where $n_1 > n_2$. Solving for n_1, we find

$$n_1 = \frac{n_2}{\sin\theta_c} = \frac{1.000}{\sin 40.5°} = \boxed{1.54}$$

24. *REASONING AND SOLUTION* Using Equation 26.4 and taking the refractive index for carbon disulfide from Table 26.1, we obtain

$$\theta_c = \sin^{-1}(1.000/1.632) = \boxed{37.79°}$$

25. *REASONING AND SOLUTION*

a. The index of refraction n_2 of the liquid must match that of the glass, or $n_2 = \boxed{1.50}$.

b. When none of the light is transmitted into the liquid, the angle of incidence must be equal to or greater than the critical angle. According to Equation 26.4, the critical angle θ_c is given by $\sin\theta_c = n_2/n_1$, where n_2 is the index of refraction of the liquid and n_1 is that of the glass. Therefore,

$$n_2 = n_1 \sin\theta_c = (1.50)\sin 58.0° = \boxed{1.27}$$

If n_2 were larger than 1.27, the critical angle would also be larger, and light would be transmitted from the glass into the liquid. Thus, $n_2 = 1.27$ represents the largest index of refraction of the liquid such that none of the light is transmitted into the liquid.

26. *REASONING AND SOLUTION* Only the light which has an angle of incidence less than or equal θ_c can escape. This light leaves the source in a cone whose apex angle is $2\theta_c$. The radius of this cone at the surface of the water ($n = 1.333$, see Table 26.1) is $R = d\tan\theta_c$. Now

$$\theta_c = \sin^{-1}(1.000/1.333) = 48.6°$$

so

$$R = (2.2\text{ m})\tan 48.6° = \boxed{2.5\text{ m}}$$

948 THE REFRACTION OF LIGHT: LENSES AND OPTICAL INSTRUMENTS

27. **SSM** *REASONING AND SOLUTION* If a person's eyes are very close to the surface of the water, a light ray coming from the shark will be seen even when it is refracted through an angle of 90.0° as it enters the air. In this situation, the ray strikes the water-air interface at the critical angle. The critical angle θ_c is given by Equation 26.4 as

$$\theta_c = \sin^{-1}\left(\frac{1.00}{1.333}\right) = 48.6°$$

where we have used $n = 1.333$ for the refractive index of water (see Table 26.1). The horizontal distance x of the shark from the boat is related to the depth (4.5 m) of the shark and the critical angle by trigonometry:

$$x = (4.5 \text{ m}) \tan 48.6° = \boxed{5.1 \text{ m}}$$

If the shark is farther than 5.1 m from the boat, a light ray from the shark will strike the water-air interface at an angle that is greater than the critical angle. The ray will be totally reflected back into the water, and the person will not see the shark.

---

28. *REASONING AND SOLUTION*
a. Using Equation 26.4 and the refractive index for crown glass given in Table 26.1, we find that the critical angle for a crown glass-air interface is

$$\theta_c = \sin^{-1}(1.00/1.523) = 41.0°$$

The light will be totally reflected at point A since the incident angle of 60.0° is greater than θ_c. The incident angle at point B, however, is 30.0° and smaller than θ_c. The light will exit first at $\boxed{\text{point B}}$.

b. The critical angle for crown glass-water is

$$\theta_c = \sin^{-1}(1.333/1.523) = 61.1°$$

The incident angle at point A is less than this, so the light will first exit at $\boxed{\text{point A}}$.

29. **REASONING** Total internal reflection will occur at point P provided that the angle α in the drawing at the right exceeds the critical angle. This angle is determined by the angle θ_2 at which the light rays enter the quartz slab. We can determine θ_2 by using Snell's law of refraction and the incident angle, which is given as $\theta_1 = 34°$.

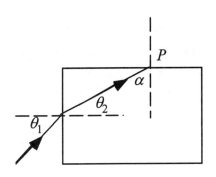

SOLUTION Using n for the refractive index of the fluid that surrounds the crystalline quartz slab and n_q for the refractive index of quartz and applying Snell's law give

$$n \sin\theta_1 = n_q \sin\theta_2 \quad \text{or} \quad \sin\theta_2 = \frac{n}{n_q} \sin\theta_1 \quad (1)$$

But when α equals the critical angle, we have from Equation 26.4 that

$$\sin\alpha = \sin\theta_c = \frac{n}{n_q} \quad (2)$$

According to the geometry in the drawing above, $\alpha = 90° - \theta_2$. As a result, Equation (2) becomes

$$\sin(90° - \theta_2) = \cos\theta_2 = \frac{n}{n_q} \quad (3)$$

Squaring Equation (3), using the fact that $\sin^2\theta_2 + \cos^2\theta_2 = 1$, and substituting from Equation (1), we obtain

$$\cos^2\theta_2 = 1 - \sin^2\theta_2 = 1 - \frac{n^2}{n_q^2} \sin^2\theta_1 = \frac{n^2}{n_q^2} \quad (4)$$

Solving Equation (4) for n and using the value given in Table 26.1 for the refractive index of crystalline quartz, we find

$$n = \frac{n_q}{\sqrt{1+\sin^2\theta_1}} = \frac{1.544}{\sqrt{1+\sin^2 34°}} = \boxed{1.35}$$

30. **REASONING** When the light enters the optical fiber, the angle of incidence θ_1 is related to the angle of refraction θ_2 by Snell's law, Equation 26.2, as $n_{air}\sin\theta_1 = n_{flint}\sin\theta_2$. As the drawing shows, the ray of light in the flint glass strikes the crown glass at an angle of incidence of $90° - \theta_2$. According to the problem statement, this angle is the critical angle for the interface between the two glasses.

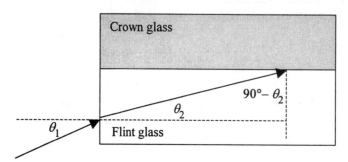

According to Equation 26.4, the critical angle is related to the indices of refraction by

$$\underbrace{\sin\theta_c}_{90°-\theta_2} = \frac{n_{crown}}{n_{flint}}$$

This relation and Snell's law can be used to find the angle of incidence.

SOLUTION The critical angle is

$$\theta_c = \sin^{-1}\left(\frac{n_{crown}}{n_{flint}}\right) = \sin^{-1}\left(\frac{1.523}{1.667}\right) = 66.01°$$

Since $\theta_c = 90° - \theta_2$, it follows that $\theta_2 = 90° - 66.01° = 23.99°$. Using Snell's law, $n_{air}\sin\theta_1 = n_{flint}\sin\theta_2$, we find that the angle of incidence is

$$\theta_1 = \sin^{-1}\left(\frac{n_{flint}\sin\theta_2}{n_{air}}\right) = \sin^{-1}\left(\frac{1.667\sin 23.99°}{1.000}\right) = \boxed{42.67°}$$

31. **SSM REASONING** Since the light reflected from the coffee table is completely polarized parallel to the surface of the glass, the angle of incidence must be the Brewster angle ($\theta_B = 56.7°$) for the air-glass interface. We can use Brewster's law (Equation 26.5: $\tan\theta_B = n_2/n_1$) to find the index of refraction n_2 of the glass.

SOLUTION Solving Brewster's law for n_2, we find that the refractive index of the glass is

$$n_2 = n_1 \tan\theta_B = (1.000)(\tan 56.7°) = \boxed{1.52}$$

---

32. **REASONING** Using the value given for the critical angle in Equation 26.4 ($\sin\theta_c = n_2/n_1$), we can obtain the ratio of the refractive indices. Then, using this ratio in Equation 26.5 (Brewster's law), we can obtain Brewster's angle θ_B.

SOLUTION From Equation 26.4, with $n_2 = n_{air} = 1$ and $n_2 = n_{liquid}$, we have

$$\sin\theta_c = \sin 39° = \frac{1}{n_{liquid}} \qquad (1)$$

According to Brewster's law,

$$\tan\theta_B = \frac{n_2}{n_1} = \frac{1}{n_{liquid}} \qquad (2)$$

Substituting Equation (2) into Equation (1), we find

$$\tan\theta_B = \frac{1}{n_{liquid}} = \sin 39° = 0.63 \qquad \text{or} \qquad \theta_B = \tan^{-1}(0.63) = \boxed{32°}$$

---

33. **REASONING AND SOLUTION**
a. Using Brewster's law (Equation 26.5) and taking values for the refractive indices from Table 26.1, we find

$$\theta_B = \tan^{-1}(1.333/1.000) = \boxed{53.12°}$$

b. Similarly, we find

$$\theta_B = \tan^{-1}(1.309/1.000) = \boxed{52.62°}$$

---

34. **REASONING AND SOLUTION** The incident angle must equal Brewster's angle for the light to be 100% polarized. Using the refractive index of carbon tetrachloride given in Table 26.1, we have

$$\theta_B = \tan^{-1}(1.461/1.000) = 55.61°$$

Snell's law applied at the interface gives

$$\sin\theta_2 = (1.000/1.461)\sin 55.61° \qquad \text{or} \qquad \theta_2 = \boxed{34.39°}$$

35. **SSM WWW REASONING** Brewster's law (Equation 26.5: $\tan \theta_B = n_2/n_1$) relates the angle of incidence θ_B at which the reflected ray is completely polarized parallel to the surface to the indices of refraction n_1 and n_2 of the two media forming the interface. We can use Brewster's law for light incident from above to find the ratio of the refractive indices n_2/n_1. This ratio can then be used to find the Brewster angle for light incident from below on the same interface.

SOLUTION The index of refraction for the medium in which the incident ray occurs is designated by n_1. For the light striking from above $n_2/n_1 = \tan \theta_B = \tan 65.0° = 2.14$. The same equation can be used when the light strikes from below if the indices of refraction are interchanged

$$\theta_B = \tan^{-1}\left(\frac{n_1}{n_2}\right) = \tan^{-1}\left(\frac{1}{n_2/n_1}\right) = \tan^{-1}\left(\frac{1}{2.14}\right) = \boxed{25.0°}$$

36. **REASONING** When light is incident at the Brewster angle, we know that the angle between the refracted ray and the reflected ray is 90°. This relation will allow us to determine the Brewster angle. By applying Snell's law to the incident and refracted rays, we can find the index of refraction of the glass.

SOLUTION The drawing shows the incident, reflected, and refracted rays.

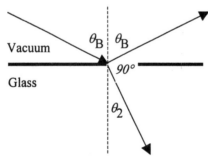

a. We see from the drawing that $\theta_B + 90° + \theta_2 = 180°$, so that $\theta_B = 90° - 29.9° = \boxed{60.1°}$.

b. Applying Snell's law at the vacuum/glass interface gives

$$n_{vacuum} \sin \theta_B = n_{glass} \sin \theta_2$$

$$n_{glass} = \left[\frac{(1.00)\sin 60.1°}{\sin 29.9°}\right] = \boxed{1.74}$$

37. ***REASONING AND SOLUTION*** From Snell's law we have

$$\sin\theta_B = \left(\frac{n_2 \sin\theta_2}{n_1}\right)$$

But from Brewster's law, Equation 26.5, $n_2/n_1 = \tan\theta_B$. Substituting this expression for n_2/n_1 into Snell's law, we see that

$$\sin\theta_B = \tan\theta_B \sin\theta_2 = \left(\frac{\sin\theta_B}{\cos\theta_B}\right)\sin\theta_2$$

This result shows that $\cos\theta_B = \sin\theta_2$. Since $\sin\theta_2 = \cos(90° - \theta_2)$, we have that $\cos\theta_B = \sin\theta_2 = \cos(90° - \theta_2)$. Thus, $\theta_B = 90° - \theta_2$, so $\theta_B + \theta_2 = 90°$, and

$$\boxed{\text{the reflected and refracted rays are perpendicular}}.$$

---

38. ***REASONING AND SOLUTION*** Snell's law for the yellow light is

$$(1.000) \sin 45.0° = n_y \sin\theta_2$$

and for the blue light is

$$(1.000) \sin\theta_1 = n_b \sin\theta_2$$

Dividing these equations gives

$$\sin\theta_1 = \frac{n_b}{n_y} \sin 45.0° = \left(\frac{1.684}{2.417}\right) \sin 45.0° \quad \text{or} \quad \theta_1 = \boxed{29.5°}$$

---

39. **SSM** ***REASONING*** Because the refractive index of the glass depends on the wavelength (i.e., the color) of the light, the rays corresponding to different colors are bent by different amounts in the glass. We can use Snell's law (Equation 26.2: $n_1 \sin\theta_1 = n_2 \sin\theta_2$) to find the angle of refraction for the violet ray and the red ray. The angle between these rays can be found by the subtraction of the two angles of refraction.

SOLUTION In Table 26.2 the index of refraction for violet light in crown glass is 1.538, while that for red light is 1.520. According to Snell's law, then, the sine of the angle of refraction for the violet ray in the glass is $\sin\theta_2 = (1.000/1.538) \sin 45.00° = 0.4598$, so that

$$\theta_2 = \sin^{-1}(0.4598) = 27.37°$$

Similarly, for the red ray, $\sin\theta_2 = (1.000/1.520)\sin 45.00° = 0.4652$, from which it follows that

$$\theta_2 = \sin^{-1}(0.4652) = 27.72°$$

Therefore, the angle between the violet ray and the red ray in the glass is

$$27.72° - 27.37° = \boxed{0.35°}$$

---

40. **REASONING** The angle of each refracted ray in the crown glass can be obtained from Snell's law (Equation 26.2) as $n_{diamond} \sin\theta_1 = n_{crown\ glass} \sin\theta_2$, where θ_1 is the angle of incidence and θ_2 is the angle of refraction.

SOLUTION The angles of refraction for the red and blue rays are:

Blue ray
$$\theta_2 = \sin^{-1}\left(\frac{n_{diamond}\sin\theta_1}{n_{crown\ glass}}\right) = \sin^{-1}\left[\frac{(2.444)\sin 35.00°}{1.531}\right] = 66.29°$$

Red ray
$$\theta_2 = \sin^{-1}\left(\frac{n_{diamond}\sin\theta_1}{n_{crown\ glass}}\right) = \sin^{-1}\left[\frac{2.410\sin 35.00°}{1.520}\right] = 65.43°$$

The angle between the blue and red rays is

$$\theta_{blue} - \theta_{red} = 66.29° - 65.43° = \boxed{0.86°}$$

---

41. **REASONING AND SOLUTION** Both the red and violet rays strike the prism at 0° and will have 0° angles of refraction. Both rays will then have 25.0° angle of incidence at the glass-air interface. Snell's law gives

Red ray $\sin\theta_R = (1.662/1.000)\sin 25.0°$ or $\theta_R = \boxed{44.6°}$

Violet ray $\sin\theta_V = (1.698/1.000)\sin 25.0°$ or $\theta_V = \boxed{45.9°}$

---

42. **REASONING AND SOLUTION** From geometry, the angle of incidence at the left face of the prism is $\theta_1 = 27.0°$. The angle of refraction θ_2 of the light entering the prism is given by Snell's law, $n_1 \sin\theta_1 = n_2 \sin\theta_2$. Thus,

$$\theta_2 = \sin^{-1}\left(\frac{1.48 \sin 27.0°}{1.31}\right) = 30.86°$$

Geometry can be used again to show that the angle of incidence θ_3 at the right face of the prism is $\theta_3 = 23.14°$. The angle of refraction θ_4 of the light entering the liquid is given by Snell's law

$$\theta_4 = \sin^{-1}\left(\frac{1.31 \sin 23.14°}{1.48}\right) = \boxed{20.4°}$$

---

43. **SSM** **REASONING** We can use Snell's law (Equation 26.2: $n_1 \sin \theta_1 = n_2 \sin \theta_2$) at each face of the prism. At the first interface where the ray enters the prism, $n_1 = 1.000$ for air and $n_2 = n_g$ for glass. Thus, Snell's law gives

$$(1)\sin 60.0° = n_g \sin \theta_2 \quad \text{or} \quad \sin \theta_2 = \frac{\sin 60.0°}{n_g} \quad (1)$$

We will represent the angles of incidence and refraction at the second interface as θ_1' and θ_2', respectively. Since the triangle is an equilateral triangle, the angle of incidence at the second interface, where the ray emerges back into air, is $\theta_1' = 60.0° - \theta_2$. Therefore, at the second interface, where $n_1 = n_g$ and $n_2 = 1.000$, Snell's law becomes

$$n_g \sin(60.0° - \theta_2) = (1)\sin \theta_2' \quad (2)$$

We can now use Equations (1) and (2) to determine the angles of refraction θ_2' at which the red and violet rays emerge into the air from the prism.

SOLUTION

Red Ray The index of refraction of flint glass at the wavelength of red light is $n_g = 1.662$. Therefore, using Equation (1), we can find the angle of refraction for the red ray as it enters the prism:

$$\sin \theta_2 = \frac{\sin 60.0°}{1.662} = 0.521 \quad \text{or} \quad \theta_2 = \sin^{-1} 0.521 = 31.4°$$

Substituting this value for θ_2 into Equation (2), we can find the angle of refraction at which the red ray emerges from the prism:

$$\sin \theta_2' = 1.662 \sin(60.0° - 31.4°) = 0.796 \quad \text{or} \quad \theta_2' = \sin^{-1} 0.796 = \boxed{52.7°}$$

Violet Ray For violet light, the index of refraction for glass is $n_g = 1.698$. Again using Equation (1), we find

$$\sin \theta_2 = \frac{\sin 60.0°}{1.698} = 0.510 \quad \text{or} \quad \theta_2 = \sin^{-1} 0.510 = 30.7°$$

Using Equation (2), we find

$$\sin \theta_2' = 1.698 \sin(60.0° - 30.7°) = 0.831 \quad \text{or} \quad \theta_2' = \sin^{-1} 0.831 = \boxed{56.2°}$$

---

44. ***REASONING AND SOLUTION*** Using the thin lens equation, we find

$$1/f = 1/d_o + 1/d_i = 1/(13 \text{ cm}) + 1/(-5.0 \text{ cm}) \quad \text{or} \quad f = \boxed{-8.1 \text{ cm}}$$

---

45. **SSM** ***REASONING AND SOLUTION*** Equation 26.6 gives the thin-lens equation which relates the object and image distances d_o and d_i, respectively, to the focal length f of the lens: $(1/d_o) + (1/d_i) = (1/f)$.

The optical arrangement is similar to that in Figure 26.26. The problem statement gives values for the focal length ($f = 50.0$ mm) and the maximum lens-to-film distance ($d_i = 275$ mm). Therefore, the maximum distance that the object can be located in front of the lens is

$$\frac{1}{d_o} = \frac{1}{f} - \frac{1}{d_i} = \frac{1}{50.0 \text{ mm}} - \frac{1}{275 \text{ mm}} \quad \text{or} \quad \boxed{d_o = 61.1 \text{ mm}}$$

---

46. ***REASONING AND SOLUTION*** The image distance for the first case is

$$1/d_i = 1/f - 1/d_o = 1/(200.0 \text{ mm}) - 1/(3.5 \times 10^3 \text{ mm}) \quad \text{or} \quad d_i = 212 \text{ mm}$$

and, similarly, for the second case it is $d_i = 201$ mm. Thus, the lens must be capable of moving through a distance of $212 \text{ mm} - 201 \text{ mm} = 11 \text{ mm}$ or $\boxed{0.011 \text{ m}}$.

47. **REASONING AND SOLUTION**
a. According to the thin-lens equation, we have

$$1/d_i = 1/f - 1/d_o = 1/(-25 \text{ cm}) - 1/(38 \text{ cm}) \quad \text{or} \quad d_i = \boxed{-15 \text{ cm}}$$

b. The image is $\boxed{\text{virtual}}$ since the image distance is negative.

---

48. **REASONING** The height h_i of the person's image on the film is related to his actual height h_o, the image distance d_i, and the object distance d_0 by the magnification equation, Equation 26.7. All these variables are known, except the image distance. Since the object distance and the focal length are also known, we can use the thin lens equation, Equation 26.6, to find the image distance.

SOLUTION The image height is

$$h_i = h_o\left(-\frac{d_i}{d_o}\right)$$

The image distance d_i can be obtained from the thin-lens equation

$$\frac{1}{d_i} = \frac{1}{f} - \frac{1}{d_o} = \frac{1}{85.0 \times 10^{-3} \text{ m}} - \frac{1}{16.0 \text{ m}} = 11.7 \text{ m}^{-1} \quad \text{or} \quad d_i = 0.0855 \text{ m}$$

The height of the image on the film is

$$h_i = h_o\left(-\frac{d_i}{d_o}\right) = (145 \text{ cm})\left(-\frac{0.0855 \text{ m}}{16.0 \text{ m}}\right) = \boxed{-0.775 \text{ cm}}$$

The minus sign tells us that the image is $\boxed{\text{inverted}}$ relative to the object.

---

49. **SSM** **REASONING** The ray diagram is constructed by drawing the paths of two rays from a point on the object. For convenience, we will choose the top of the object. The ray that is parallel to the principal axis will be refracted by the lens so that it passes through the focal point on the right of the lens. The ray that passes through the center of the lens passes through undeflected. The image is formed at the intersection of these two rays. In this case, the rays do not intersect on the right of the lens. However, if they are extended backwards they intersect on the left of the lens, locating a virtual, upright, and enlarged image.

SOLUTION
a. The ray-diagram, drawn to scale, is shown below.

958 THE REFRACTION OF LIGHT: LENSES AND OPTICAL INSTRUMENTS

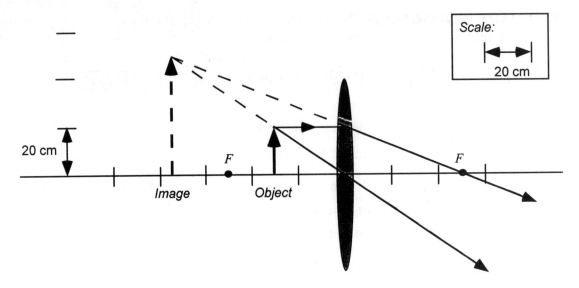

From the diagram, we see that the image distance is $\boxed{d_i = -75 \text{ cm}}$ and the magnification is $\boxed{+2.5}$. The negative image distance indicates that the image is virtual. The positive magnification indicates that the image is larger than the object.

b. From the thin-lens equation [Equation 26.6: $(1/d_o) + (1/d_i) = (1/f)$], we obtain

$$\frac{1}{d_i} = \frac{1}{f} - \frac{1}{d_o} = \frac{1}{50.0 \text{ m}} - \frac{1}{30.0 \text{ cm}} \quad \text{or} \quad \boxed{d_i = -75.0 \text{ cm}}$$

The magnification equation (Equation 26.7) gives the magnification to be

$$m = -\frac{d_i}{d_o} = -\frac{-75.0 \text{ cm}}{30.0 \text{ cm}} = \boxed{+2.50}$$

---

50. **REASONING** The distance from the lens to the screen, the image distance, can be obtained directly from the thin-lens equation, Equation 26.6, since the object distance and focal length are known. The width and height of the image on the screen can be determined by using Equation 26.7, the magnification equation.

SOLUTION
a. The distance d_i to the screen is

$$\frac{1}{d_i} = \frac{1}{f} - \frac{1}{d_o} = \frac{1}{105.00 \text{ mm}} - \frac{1}{108.00 \text{ mm}} = 2.646 \times 10^{-4} \text{ mm}^{-1}$$

so that $d_i = 3.78 \times 10^3 \text{ mm} = \boxed{3.78 \text{ m}}$.

b. According to the magnification equation, the width and height of the image on the screen are

Width $\quad h_i = h_o\left(-\dfrac{d_i}{d_o}\right) = (24.0 \text{ mm})\left(-\dfrac{3.78 \times 10^3 \text{ mm}}{108 \text{ mm}}\right) = -8.40 \times 10^2 \text{ mm}$

The width is $\boxed{8.40 \times 10^2 \text{ mm}}$.

Height $\quad h_i = h_o\left(-\dfrac{d_i}{d_o}\right) = (36.0 \text{ mm})\left(-\dfrac{3.78 \times 10^3 \text{ mm}}{108 \text{ mm}}\right) = -1.26 \times 10^3 \text{ mm}$

The height is $\boxed{1.26 \times 10^3 \text{ mm}}$.

---

51. **REASONING AND SOLUTION** Solving the thin-lens equation for the image distance yields

$$d_i = \dfrac{f d_o}{d_o - f}$$

The magnification is

$$m = -\dfrac{d_i}{d_o} = -\dfrac{f}{d_o - f}$$

a. For $f = 35.0$ mm and $d_o = 8.00$ m, the magnification is $m = -4.39 \times 10^{-3}$, and we find

$$h_i = (-4.39 \times 10^{-3})(1.80 \text{ m}) = \boxed{-7.90 \times 10^{-3} \text{ m}}$$

b. For $f = 150.0$ mm and $d_o = 8.00$ m, the magnification is $m = -1.91 \times 10^{-2}$ and we find

$$h_i = (-1.91 \times 10^{-2})(1.80 \text{ m}) = \boxed{-3.44 \times 10^{-2} \text{ m}}$$

---

52. **REASONING** A diverging lens always produces a virtual image, so that the image distance d_i is negative. Moreover, the object distance d_o is positive. Therefore, the distance between the object and the image is $d_o + d_i = 6.0$ cm, rather than $d_o - d_i = 6.0$ cm. The equation $d_o + d_i = 6.0$ cm and the thin-lens equation constitute two equations in two unknowns, and we will solve them simultaneously to obtain values for d_i and d_o.

960 THE REFRACTION OF LIGHT: LENSES AND OPTICAL INSTRUMENTS

SOLUTION
a. Solving the equation $d_o + d_i = 6.0$ cm for d_o, substituting into the thin-lens equation, and suppressing the units give

$$\frac{1}{d_o} + \frac{1}{d_i} = \frac{1}{f} \quad \text{or} \quad \frac{1}{6.0 - d_i} + \frac{1}{d_i} = \frac{1}{-3.0} \quad (1)$$

Grouping the terms on the left of Equation (1) over a common denominator, we have

$$\frac{d_i + 6.0 - d_i}{d_i(6.0 - d_i)} = \frac{6.0}{d_i(6.0 - d_i)} = \frac{1}{-3.0} \quad (2)$$

Cross-multiplying and rearranging in Equation (2) gives

$$d_i(6.0 - d_i) = -18 \quad \text{or} \quad d_i^2 - 6.0 d_i - 18 = 0 \quad (3)$$

Using the quadratic formula to solve Equation (3), we obtain

$$d_i = \frac{-(-6.0) \pm \sqrt{(-6.0)^2 - 4(1.00)(-18)}}{2(1.00)} = \boxed{-2.2 \text{ cm}}$$

We have discarded the positive root, because we know that d_i must be negative for the virtual image.

b. Using the fact that $d_o + d_i = 6.0$ cm, we find that the object distance is

$$d_o = 6.0 \text{ cm} - d_i = (6.0 \text{ cm}) - (-2.2 \text{ cm}) = \boxed{8.2 \text{ cm}}$$

---

53. **SSM** *REASONING* The magnification equation (Equation 26.7) relates the object and image distances d_o and d_i, respectively, to the relative size of the of the image and object: $m = -(d_i / d_o)$. We consider two cases: in case 1, the object is placed 18 cm in front of a diverging lens. The magnification for this case is given by m_1. In case 2, the object is moved so that the magnification m_2 is reduced by a factor of 2 compared to that in case 1. In other words, we have $m_2 = \frac{1}{2} m_1$. Using Equation 26.7, we can write this as

$$-\frac{d_{i2}}{d_{o2}} = -\frac{1}{2}\left(\frac{d_{i1}}{d_{o1}}\right) \quad (1)$$

This expression can be solved for d_{o2}. First, however, we must find a numerical value for d_{i1}, and we must eliminate the variable d_{i2}.

SOLUTION
The image distance for case 1 can be found from the thin-lens equation [Equation 26.6: $(1/d_o)+(1/d_i)=(1/f)$]. The problem statement gives the focal length as $f = -12$ cm. Since the object is 18 cm in front of the diverging lens, $d_{o1} = 18$ cm. Solving for d_{i1}, we find

$$\frac{1}{d_{i1}} = \frac{1}{f} - \frac{1}{d_{o1}} = \frac{1}{-12 \text{ cm}} - \frac{1}{18 \text{ cm}} \quad \text{or} \quad d_{i1} = -7.2 \text{ cm}$$

where the minus sign indicates that the image is virtual. Solving Equation (1) for d_{o2}, we have

$$d_{o2} = 2d_{i2}\left(\frac{d_{o1}}{d_{i1}}\right) \qquad (2)$$

To eliminate d_{i2} from this result, we note that the thin-lens equation applied to case 2 gives

$$\frac{1}{d_{i2}} = \frac{1}{f} - \frac{1}{d_{o2}} = \frac{d_{o2}-f}{fd_{o2}} \quad \text{or} \quad d_{i2} = \frac{fd_{o2}}{d_{o2}-f}$$

Substituting this expression for d_{i2} into Equation (2), we have

$$d_{o2} = \left(\frac{2fd_{o2}}{d_{o2}-f}\right)\left(\frac{d_{o1}}{d_{i1}}\right) \quad \text{or} \quad d_{o2} - f = 2f\left(\frac{d_{o1}}{d_{i1}}\right)$$

Solving for d_{o2}, we find

$$d_{o2} = f\left[2\left(\frac{d_{o1}}{d_{i1}}\right)+1\right] = (-12 \text{cm})\left[2\left(\frac{18 \text{ cm}}{-7.2 \text{ cm}}\right)+1\right] = \boxed{48 \text{ cm}}$$

---

54. **REASONING AND SOLUTION** The focal length of the lens can be obtained from the thin-lens equation as follows:

$$1/f = 1/(4.00 \text{ m}) + 1/(0.210 \text{ m}) \quad \text{or} \quad f = 0.200 \text{ m}$$

The same equation applied to the projector gives

$$1/d_o = 1/(0.200 \text{ m}) - 1/(0.500 \text{ m}) \quad \text{or} \quad d_o = \boxed{0.333 \text{ m}}$$

55. **SSM WWW REASONING** The optical arrangement is similar to that in Figure 26.26. We begin with the thin-lens equation, [Equation 26.6: $(1/d_o)+(1/d_i)=(1/f)$]. Since the distance between the moon and the camera is so large, the object distance d_o is essentially infinite, and $1/d_o = 1/\infty = 0$. Therefore the thin-lens equation becomes $1/d_i = 1/f$ or $d_i = f$. The diameter of the moon's imagine on the slide film is equal to the image height h_i, as given by the magnification equation (Equation 26.7: $h_i/h_o = -d_i/d_o$).

When the slide is projected onto a screen, the situation is similar to that in Figure 26.27. In this case, the thin-lens and magnification equations can be used in their usual forms.

SOLUTION
a. Solving the magnification equation for h_i gives

$$h_i = -h_o \frac{d_i}{d_o} = (-3.48 \times 10^6 \text{ m})\left(\frac{50.0 \times 10^{-3} \text{ m}}{3.85 \times 10^8 \text{ m}}\right) = -4.52 \times 10^{-4} \text{ m}$$

The diameter of the moon's image on the slide film is, therefore, $\boxed{4.52 \times 10^{-4} \text{ m}}$.

b. From the magnification equation, $h_i = -h_o(d_i/d_o)$. We need to find the ratio d_i/d_o. Beginning with the thin-lens equation, we have

$$\frac{1}{d_o}+\frac{1}{d_i}=\frac{1}{f} \quad \text{or} \quad \frac{1}{d_o}=\frac{1}{f}-\frac{1}{d_i} \quad \text{which leads to} \quad \frac{d_i}{d_o}=\frac{d_i}{f}-\frac{d_i}{d_i}=\frac{d_i}{f}-1$$

Therefore,

$$h_i = -h_o\left(\frac{d_i}{f}-1\right) = -(4.52 \times 10^{-4} \text{ m})\left(\frac{15.0 \text{ m}}{110.0 \times 10^{-3} \text{ m}}-1\right) = -6.12 \times 10^{-2} \text{ m}$$

The diameter of the image on the screen is $\boxed{6.12 \times 10^{-2} \text{ m}}$.

---

56. **REASONING AND SOLUTION** The thin-lens equation for the first situation is

$$1/d_o + 1/d_i = 1/f$$

The magnification equation can be solved for the image distance and used in the thin-lens equation, so that

$$1/d_o(1 - 1/m) = 1/f$$

In the second situation where f' = 50.0 mm

$$1/d_o'(1 - 1/m) = 1/f'$$

Dividing the equations, keeping in mind that the image should be the same size in both cases, that is, m is the same, we get

$$d_o'/d_o = f'/f = (50.0 \text{ mm})/(300.0 \text{ mm}) = 1/6.00$$

or

$$d_o' = (72 \text{ m})/(6.00) = \boxed{12 \text{ m}}$$

---

57. **REASONING AND SOLUTION** Let d represent the distance between the object and the screen. Then, $d_o + d_i = d$. Using this expression in the thin-lens equation gives

$$1/d_o + 1/(d - d_o) = 1/f \quad \text{or} \quad d_o^2 - dd_o + df = 0$$

With d = 125 cm and f = 25.0 cm, the quadratic formula yields solutions of

$$d_o = \boxed{+35 \text{ cm}} \quad \text{and} \quad d_o = \boxed{+90.5 \text{ cm}}$$

---

58. **REASONING AND SOLUTION** The thin-lens equation written for the first situation gives (suppressing the units)

$$\frac{1}{20.0} + \frac{1}{d_i} = \frac{1}{f} \quad (1)$$

The thin-lens equation written for the second situation gives

$$\frac{1}{16.0} + \frac{1}{d_i + 2.70} = \frac{1}{f} \quad (2)$$

Since the right hand sides of Equations (1) and (2) are the same, we have

$$\frac{1}{16.0} + \frac{1}{d_i + 2.70} = \frac{1}{20.0} + \frac{1}{d_i} \quad \text{or} \quad \frac{1}{16.0} - \frac{1}{20.0} = \frac{1}{d_i} - \frac{1}{d_i + 2.70}$$

Combining terms over common denominators gives

$$\frac{20.0 - 16.0}{(16.0)(20.0)} = \frac{d_i + 2.70 - d_i}{d_i(d_i + 2.70)} \quad \text{or} \quad \frac{1}{80.0} = \frac{2.70}{d_i(d_i + 2.70)}$$

Cross multiplying and rearranging terms gives

964 THE REFRACTION OF LIGHT: LENSES AND OPTICAL INSTRUMENTS

$$d_i^2 + 2.70 d_i - 216 = 0$$

Using the quadratic formula, we find

$$d_i = \frac{-2.70 \pm \sqrt{(2.70)^2 - 4(1.00)(-216)}}{2(1.00)} = +13.4 \text{ cm}$$

where we have chosen the positive root since the lens produces a real image. Substituting this value for the image distance into Equation (1), we find

$$\frac{1}{20.0 \text{ cm}} + \frac{1}{13.4 \text{ cm}} = \frac{1}{f} \quad \text{or} \quad f = \boxed{8.0 \text{ cm}}$$

---

59. **SSM** ***REASONING*** The problem can be solved using the thin-lens equation [Equation 26.6: $(1/d_o) + (1/d_i) = (1/f)$] twice in succession. We begin by using the thin lens-equation to find the location of the image produced by the converging lens; this image becomes the object for the diverging lens.

SOLUTION
a. The image distance for the converging lens is determined as follows:

$$\frac{1}{d_{i1}} = \frac{1}{f} - \frac{1}{d_{o1}} = \frac{1}{12.0 \text{ cm}} - \frac{1}{36.0 \text{ cm}} \quad \text{or} \quad d_{i1} = 18.0 \text{ cm}$$

This image acts as the object for the diverging lens. Therefore,

$$\frac{1}{d_{i2}} = \frac{1}{f} - \frac{1}{d_{o2}} = \frac{1}{-6.00 \text{ cm}} - \frac{1}{(30.0 \text{ cm} - 18.0 \text{ cm})} \quad \text{or} \quad d_{i2} = -4.00 \text{ cm}$$

Thus, the final image is located $\boxed{4.00 \text{ cm to the left of the diverging lens}}$.

b. The magnification equation (Equation 26.7: $h_i / h_o = -d_i / d_o$) gives

$$\underbrace{m_c = -\frac{d_{i1}}{d_{o1}} = -\frac{18.0 \text{ cm}}{36.0 \text{ cm}} = -0.500}_{\text{Converging lens}} \qquad \underbrace{m_d = -\frac{d_{i2}}{d_{o2}} = -\frac{-4.00 \text{ cm}}{12.0 \text{ cm}} = 0.333}_{\text{Diverging lens}}$$

Therefore, the overall magnification is given by the product $m_c m_d = \boxed{-0.167}$.

c. Since the final image distance is negative, we can conclude that the image is $\boxed{\text{virtual}}$.

d. Since the overall magnification of the image is negative, the image is $\boxed{\text{inverted}}$.

e. The magnitude of the overall magnification is less than one; therefore, the final image is $\boxed{\text{smaller}}$.

---

60. **REASONING** We will consider one lens at a time, using the thin-lens equation for each. The key to the solution is the fact that the image formed by the first lens serves as the object for the second lens.

    SOLUTION Using the thin-lens equation, we find the image distance for the first lens:

    $$\frac{1}{d_i} = \frac{1}{f} - \frac{1}{d_o} = \frac{1}{-8.0 \text{ cm}} - \frac{1}{4.0 \text{ cm}} \qquad \text{or} \qquad d_i = -2.7 \text{ cm}$$

    The negative value for d_i indicates that the image is virtual and located 2.7 cm to the left of the first lens. The lenses are 16 cm apart, so this image is located 2.7 cm + 16 cm = 18.7 cm from the second lens. Since this image serves as the object for the second lens, we can locate the image formed by the second lens with the aid of the thin-lens equation, with $d_o = 18.7$ cm:

    $$\frac{1}{d_i} = \frac{1}{f} - \frac{1}{d_o} = \frac{1}{-8.0 \text{ cm}} - \frac{1}{18.7 \text{ cm}} \qquad \text{or} \qquad d_i = \boxed{-5.6 \text{ cm}}$$

---

61. **REASONING AND SOLUTION** For the first image, the image distance is given by the thin-lens equation:

    $$\frac{1}{d_i} = \frac{1}{f} - \frac{1}{d_o} = \frac{1}{0.080 \text{ m}} - \frac{1}{0.040 \text{ m}} \qquad \text{or} \qquad d_i = -0.080 \text{ m}$$

    The object distance to the second lens is 0.080 m + 0.120 m = 0.200 m. For the final image, the image distance is

    $$\frac{1}{d_i} = \frac{1}{f} - \frac{1}{d_o} = \frac{1}{0.080 \text{ m}} - \frac{1}{(0.200 \text{ m})}$$

    $$d_i = \boxed{0.13 \text{ m to the right of the second lens}}$$

62. **REASONING** This is a two-lens problem, and so the image produced by the first lens acts as the object for the second lens. Since we know the final image distance and the focal point of the second (diverging) lens, we can determine the object distance for this lens by using the thin-lens equation. This object is the image produced by the first (converging) lens. By employing the thin-lens equation again, we can calculate how far the object is from the converging lens.

SOLUTION The focal length of the second lens is $f_2 = -28.0$ cm, and the image distance is $d_{i2} = -20.7$ cm. The minus sign arises because the image falls to the left of the lens, which, by convention, is a negative distance. According to the thin-lens equation, the object distance d_{o2} for the second lens is

$$\frac{1}{d_{o2}} = \frac{1}{f_2} - \frac{1}{d_{i2}} = \frac{1}{-28.0 \text{ cm}} - \frac{1}{-20.7 \text{ cm}} = 0.0126 \text{ cm}^{-1} \quad \text{or} \quad d_{o2} = 79.4 \text{ cm}$$

Since d_{o2} is positive, the object lies 79.4 cm to the left of the second lens. However, the first lens is 56.0 cm to the left of the second lens, so the separation between this object and the first lens is 79.4 cm − 56.0 cm = 23.4 cm. This object is the image produced by the first lens. However, the image distance is $d_{i1} = -23.4$ cm, since the image falls to the left of the first lens and, by convention, is a negative distance. Using the thin-lens equation, we find that the object distance d_{o1} for the first lens is

$$\frac{1}{d_{o1}} = \frac{1}{f_1} - \frac{1}{d_{i1}} = \frac{1}{24.0 \text{ cm}} - \frac{1}{-23.4 \text{ cm}} = 0.0844 \text{ cm}^{-1} \quad \text{or} \quad d_{o1} = \boxed{11.8 \text{ cm}}$$

---

63. **REASONING AND SOLUTION**
a. The image distance for the diverging lens is

$$1/d_i = 1/f - 1/d_o = 1/(-8.00 \text{ cm}) - 1/(12.0 \text{ cm}) \quad \text{or} \quad d_i = -4.80 \text{ cm}$$

This image serves as the object for the converging lens, so its focal length is

$$1/f' = 1/d_i' + 1/d_o' = 1/(-37.0 \text{ cm}) + 1/(12.8 \text{ cm}) \quad \text{or} \quad f' = \boxed{19.6 \text{ cm}}$$

b. The magnification of the diverging lens is

$$m = -(-4.80 \text{ cm}/12.0 \text{ cm}) = +0.400$$

and the magnification of the converging lens is

$$m' = -(-37.0 \text{ cm}/12.8 \text{ cm}) = +2.89$$

The overall magnification is, then,

$$mm' = (0.400)(2.89) = 1.16$$

and the final image height is

$$h_i' = mm' \, h_o = (1.16)(0.75 \text{ cm}) = \boxed{0.87 \text{ cm}}$$

---

64. ***REASONING AND SOLUTION*** Let d be the separation of the lenses. The first lens forms its image at

$$1/d_i = 1/f - 1/d_o = 1/(16.0 \text{ cm}) - 1/(20.0 \text{ cm}) \quad \text{or} \quad d_i = 80.0 \text{ cm}$$

This image serves as the object for the second lens, so the object distance for the second lens is $d - 80.0$ cm. According to the thin-lens equation, the reciprocal of the image distance for the second lens is

$$1/d_i' = 1/f - 1/(d - 80.0 \text{ cm}) \quad (1)$$

The magnification of the first lens is

$$m = -(80.0 \text{ cm})/(20.0 \text{ cm}) = -4.00$$

Since the overall magnification of the combination must be +1.000, the magnification of the second lens must be –0.250. Therefore, applying the magnification equation to the second lens, we have

$$d_i' = 0.250 \, (d - 80.0 \text{ cm}) \quad (2)$$

Substituting Equation (2) into Equation (1) and rearranging yields d = $\boxed{160.0 \text{ cm}}$.

---

65. **SSM** ***REASONING*** We begin by using the thin-lens equation [Equation 26.6: $(1/d_o) + (1/d_i) = (1/f)$] to locate the image produced by the lens. This image is then treated as the object for the mirror.

SOLUTION
a. The image distance from the diverging lens can be determined as follows:

$$\frac{1}{d_i} = \frac{1}{f} - \frac{1}{d_o} = \frac{1}{-8.00 \text{ cm}} - \frac{1}{20.0 \text{ cm}} \quad \text{or} \quad d_i = -5.71 \text{ cm}$$

The image produced by the lens is 5.71 cm to the left of the lens. The distance between this image and the concave mirror is 5.71 cm + 30.0 cm = 35.7 cm. The mirror equation [Equation 25.3: $(1/d_o)+(1/d_i)=(1/f)$] gives the image distance from the mirror:

$$\frac{1}{d_i} = \frac{1}{f} - \frac{1}{d_o} = \frac{1}{12.0 \text{ cm}} - \frac{1}{35.7 \text{ cm}} \quad \text{or} \quad \boxed{d_i = 18.1 \text{ cm}}$$

b. The image is $\boxed{\text{real}}$, because d_i is a positive number, indicating that the final image lies to the left of the concave mirror.

c. The image is $\boxed{\text{inverted}}$, because a diverging lens always produces an upright image, and the concave mirror produces an inverted image when the object distance is greater than the focal length of the mirror.

---

66. **REASONING AND SOLUTION**

a. The image distance for the first lens is

$$1/d_i = 1/f_1 - 1/d_o = 1/(9.00 \text{ cm}) - 1/(12.0 \text{ cm}) \quad \text{or} \quad d_i = 36 \text{ cm}$$

Since the lenses are separated by 18.0 cm, the value of $d_i = 36$ cm places the image 18 cm *to the right* of the second lens. This image serves as the object for the second lens. This is a case in which the object, being to the right of the lens, has a negative object distance, as indicated in the Reasoning Strategy given in Section 26.8. The image distance for the second lens is

$$1/d_i' = 1/f_2 - 1/d_o' = 1/(6.00 \text{ cm}) - 1/(-18 \text{ cm}) \quad \text{or} \quad d_i' = +4.50 \text{ cm}$$

The positive sign indicates that the final image lies $\boxed{4.50 \text{ cm to the right of the second lens}}$.

b. The magnification of the first lens is

$$m = -(36 \text{ cm})/(12 \text{ cm}) = -3.0$$

and the magnification of the second lens is

$$m' = -(4.50)/(-18.0) = +0.25$$

The overall magnification is then $mm' = \boxed{-0.75}$.

c. The final image is $\boxed{\text{real}}$ since its distance is positive.

d. The final image is ⬚inverted⬚ since the overall magnification is negative.

e. The final image is ⬚smaller⬚ than the object since the magnitude of the overall magnification is less than one.

---

67. **REASONING** We will apply the thin-lens equation to solve this problem. In doing so, we must be careful to take into account the fact that the lenses of the glasses are worn at a distance of 2.0 or 3.0 cm from her eyes.

SOLUTION
a. The object distance is 25.0 cm – 2.0 cm, since it is measured relative to the lenses, which are worn 2.0 cm from the eyes. As discussed in the text, the lenses form a virtual image located at the near point. The image distance must be negative for a virtual image, but the value is not –48.0 cm, because the glasses are worn 2.0 cm from the eyes. Instead, the image distance is –48.0 cm + 2.0 cm. Using the thin-lens equation, we can find the focal length as follows:

$$\frac{1}{d_o} + \frac{1}{d_i} = \frac{1}{25.0 \text{ cm} - 2.0 \text{ cm}} + \frac{1}{-48.0 \text{ cm} + 2.0 \text{ cm}} = \frac{1}{f} \quad \text{or} \quad f = \boxed{46.0 \text{ cm}}$$

b. Similarly, we find

$$\frac{1}{d_o} + \frac{1}{d_i} = \frac{1}{25.0 \text{ cm} - 3.0 \text{ cm}} + \frac{1}{-48.0 \text{ cm} + 3.0 \text{ cm}} = \frac{1}{f} \quad \text{or} \quad f = \boxed{43.0 \text{ cm}}$$

---

68. **REASONING AND SOLUTION**

a. The woman sees distant objects clearly, so she must be ⬚hyperopic⬚.

b. The thin-lens equation gives the refractive power to be

$$1/f = 1/d_o + 1/d_i = 1/(25 \text{ cm} - 2.0 \text{ cm}) + 1/(-65 \text{ cm} + 2.0 \text{ cm}) = 0.028 \text{ cm}^{-1} = \boxed{2.8 \text{ diopters}}.$$

---

69. **SSM REASONING** The far point is 5.0 m from the right eye, and 6.5 m from the left eye. For an object infinitely far away ($d_o = \infty$), the image distances for the corrective lenses are then –5.0 m for the right eye and –6.5 m for the left eye, the negative sign indicating that the images are virtual images formed to the left of the lenses. The thin-lens equation [Equation 26.6: $(1/d_o) + (1/d_i) = (1/f)$] can be used to find the focal length. Then, Equation 26.8 can be used to find the refractive power for the lens for each eye.

SOLUTION Since the object distance d_o is essentially infinite, $1/d_o = 1/\infty = 0$, and the thin-lens equation becomes $1/d_i = 1/f$, or $d_i = f$. Therefore, for the right eye, $f = -5.0$ m, and the refractive power is (see Equation 26.8)

[Right eye] $\quad$ Refractive power (in diopters) $= \dfrac{1}{f} = \dfrac{1}{(-5.0 \text{ m})} = \boxed{-0.20 \text{ diopters}}$

Similarly, for the left eye, $f = -6.5$ m, and the refractive power is

[Left eye] $\quad$ Refractive power (in diopters) $= \dfrac{1}{f} = \dfrac{1}{(-6.5 \text{ m})} = \boxed{-0.15 \text{ diopters}}$

---

70. **REASONING** The thin-lens equation, Equation 26.6, can be used to find the distance from the blackboard to her eyes (the object distance). The distance from her eye lens to the retina is the image distance, and the focal length of her lens is the reciprocal of the refractive power (see Equation 26.8). The magnification equation, Equation 26.7, can be used to find the height of the image on her retina.

SOLUTION
a. The thin-lens equation can be used to find the object distance d_o. However, we note from Equation 26.8 that $1/f = 57.50 \text{ m}^{-1}$ and $d_i = 0.01750$ m, so that

$$\frac{1}{d_o} = \frac{1}{f} - \frac{1}{d_i} = 57.50 \text{ m}^{-1} - \frac{1}{0.01750 \text{ m}} = 0.36 \text{ m}^{-1} \quad \text{or} \quad d_o = \boxed{2.8 \text{ m}}$$

b. The magnification equation can be used to find the height h_i of the image on the retina

$$h_i = h_o \left(-\frac{d_i}{d_o}\right) = (5.00 \text{ cm}) \left(-\frac{0.01750 \text{ m}}{2.8 \text{ m}}\right) = \boxed{-3.1 \times 10^{-2} \text{ cm}} \qquad (26.7)$$

---

71. **REASONING** Nearsightedness is corrected using diverging lenses to form a virtual image at the far point of the eye, as Section 26.10 discusses. The far point is given as 5.2 m, so we know that the image distance for the contact lenses is $d_i = -5.2$ m. The minus sign indicates that the image is virtual. The thin-lens equation can be used to determine the focal length.

SOLUTION According to the thin-lens equation, we have

$$\frac{1}{d_o} + \frac{1}{d_i} = \frac{1}{12.0 \text{ cm}} + \frac{1}{-5.2 \text{ cm}} = \frac{1}{f} \quad \text{or} \quad f = \boxed{-9.2 \text{ cm}}$$

72. **REASONING AND SOLUTION** The magnification is $h_i/h_o = -d_i/d_o$, so $h_o/d_o = -h_i/d_i$. Therefore, the size of the image on the retina, h_i, is

$$h_i = h_o(-d_i/d_o) = (2.0 \text{ mm})(-1.7 \text{ cm})/(25 \text{ cm}) = \boxed{-0.14 \text{ mm}} .$$

---

73. **REASONING AND SOLUTION** If the far point is 3.62 m from the eyes, the focal length of the lens is –3.62 m, or –362 cm. The near point of 25 cm then represents the virtual image distance, i.e., $d_i = -25$ cm. Therefore, using the thin-lens equation, we find

$$1/d_o = 1/f - 1/d_i = [1/(-362 \text{ cm})] - [1/(-25 \text{ cm})] \quad \text{or} \quad d_o = \boxed{26.9 \text{ cm}}$$

---

74. **REASONING AND SOLUTION** We need to determine the focal lengths for Bill's glasses and for Anne's glasses. Using the thin-lens equation we have

[Bill] $\quad 1/f_B = 1/d_o + 1/d_i = 1/(23.0 \text{ cm}) + 1/(-123 \text{ cm}) \quad$ or $\quad f_B = 28.3$ cm

[Anne] $\quad 1/f_A = 1/d_o + 1/d_i = 1/(23.0 \text{ cm}) + 1/(-73.0 \text{ cm}) \quad$ or $\quad f_A = 33.6$ cm

Now find d_o' for Bill and Anne when they switch glasses.

a. Anne:

$$1/d_o' = 1/f_B - 1/d_i = 1/(28.3 \text{ cm}) - 1/(-73 \text{ cm}) \quad \text{or} \quad d_o' = 20.4 \text{ cm}$$

Relative to the eyes, this becomes 20.4 cm + 2.00 cm = $\boxed{22.4 \text{ cm}}$.

b. Bill:

$$1/d_o' = 1/f_A - 1/d_i = 1/(33.6 \text{ cm}) - 1/(-123 \text{ cm}) \quad \text{or} \quad d_o' = 26.4 \text{ cm}$$

Relative to the eyes, this becomes 26.4 cm + 2.00 cm = $\boxed{28.4 \text{ cm}}$.

---

75. **REASONING AND SOLUTION** The far point of 6.0 m tells us that the focal length of the lens is $f = -6.0$ m.

a. The image distance can be found using

972 THE REFRACTION OF LIGHT: LENSES AND OPTICAL INSTRUMENTS

$$1/d_i = 1/f - 1/d_o = 1/(-6.0 \text{ m}) - 1/(18.0 \text{ m}) \quad \text{or} \quad d_i = \boxed{-4.5 \text{ m}}$$

b. The image size as obtained from the magnification is

$$h_i = h_o(-d_i/d_o) = (2.0 \text{ m})[-(-4.5 \text{ m})/(18.0 \text{ m})] = \boxed{0.50 \text{ m}}$$

---

76. REASONING AND SOLUTION

a. The angular size of an object is $\theta \approx h_o/d_o$, where h_o is the height of the object and d_o is the distance to the object. For the spectator watching the game live, we find

$$\theta \approx h_o/d_o = (1.9 \text{ m})/(75 \text{ m}) = \boxed{0.025 \text{ rad}}$$

b. Similarly for the TV viewer, we find

$$\theta \approx h_o/d_o = (0.12 \text{ m})/(3.0 \text{ m}) = \boxed{0.040 \text{ rad}}$$

c. Since the angular size of the player on the TV is greater than the angular size seen by the spectator, $\boxed{\text{the player looks larger on television}}$.

---

77. SSM REASONING
The angular size of a distant object in radians is approximately equal to the diameter of the object divided by the distance from the eye. We will use this definition to calculate the angular size of the quarter, and then, calculate the angular size of the sun; we can then form the ratio $\theta_{\text{quarter}} / \theta_{\text{sun}}$.

SOLUTION The angular sizes are

$$\theta_{\text{quarter}} \approx \frac{2.4 \text{ cm}}{70.0 \text{ cm}} = 0.034 \text{ rad} \quad \text{and} \quad \theta_{\text{sun}} \approx \frac{1.39 \times 10^9 \text{ m}}{1.50 \times 10^{11} \text{ m}} = 0.0093 \text{ rad}$$

Therefore, the ratio of the angular sizes is

$$\frac{\theta_{\text{quarter}}}{\theta_{\text{sun}}} = \frac{0.034 \text{ rad}}{0.0093 \text{ rad}} = \boxed{3.7}$$

---

78. REASONING
The angular magnification M of a magnifying glass is given by Equation 26.10 as

$$M = \frac{\theta'}{\theta} \approx \left(\frac{1}{f} - \frac{1}{d_i} \right) N$$

where $\theta' = 0.0380$ rad is the angular size of the final image produced by the magnifying glass, $\theta = 0.0150$ rad is the reference angular size of the object seen at the near point without the magnifying glass, and N is the near point of the eye. The largest possible angular magnification occurs when the image is at the near point of the eye, or $d_i = -N$, where the minus sign denotes that the image lies on the left side of the lens (the same side as the object). This equation can be solved to find the focal length of the magnifying glass.

SOLUTION Letting $d_i = -N$, and solving Equation 26.10 for the focal length f gives

$$f = \frac{N}{\frac{\theta'}{\theta} - 1} = \frac{21.0 \text{ cm}}{\frac{0.0380 \text{ rad}}{0.0150 \text{ rad}} - 1} = \boxed{13.7 \text{ cm}}$$

---

79. **REASONING AND SOLUTION** With the magnifying glass held so that the image is at the near point of the eye, we have from Equation 26.10 that $M = (N/f) + 1$. Solving for f,

$$f = N/(M-1) = (0.30 \text{ m})/(3.4 - 1) = \boxed{0.13 \text{ m}}$$

---

80. **REASONING AND SOLUTION** Using Equation 26.10, we have

$$M = \left(\frac{1}{f} - \frac{1}{d_i}\right) N$$

Solving for the image distance gives

$$d_i = \frac{fN}{N - fM} = \frac{(22 \text{ cm})(36 \text{ cm})}{(36 \text{ cm}) - (22 \text{ cm})(2.5)} = \boxed{-42 \text{ cm}}$$

---

81. **SSM** **REASONING** The angular magnification of a magnifying glass is given by Equation 26.10: $M \approx \left[(1/f) - (1/d_i)\right]N$, where N is the distance from the eye to the near-point. For maximum magnification, the closest to the eye that the image can be is at the near point, with $d_i = -N$ (where the minus sign indicates that the image lies to the left of the lens and is virtual). In this case, Equation 26.10 becomes $M_{max} \approx N/f + 1$. At minimum magnification, the image is as far from the eye as it can be ($d_i = -\infty$); this occurs when the object is placed at the focal point of the lens. In this case, Equation 26.10 simplifies to $M_{min} \approx N/f$.

Since the woman observes that for clear vision, the maximum angular magnification is 1.25 times larger than the minimum angular magnification, we have $M_{max} = 1.25 M_{min}$.

974 THE REFRACTION OF LIGHT: LENSES AND OPTICAL INSTRUMENTS

This equation can be written in terms of N and f using the above expressions, and then solved for f.

SOLUTION We have

$$\frac{N}{f} + 1 = 1.25 \frac{N}{f}$$

Solving for f, we find that

$$f = 0.25 N = (0.25)(25 \text{ cm}) = \boxed{6.3 \text{ cm}}$$

---

82. **REASONING AND SOLUTION** The information given allows us to determine the near point for this farsighted person. With $f = 45.4$ cm and $d_o = 25.0$ cm, we find from the thin-lens equation that

$$\frac{1}{d_i} = \frac{1}{f} - \frac{1}{d_o} = \frac{1}{45.4 \text{ cm}} - \frac{1}{25.0 \text{ cm}} \quad \text{or} \quad d_i = -55.6 \text{ cm}$$

Therefore, this person's near point, N, is 55.6 cm. We now need to find the focal length of the magnifying glass based on the near point for a normal eye, i.e., $M = N/f + 1$, where $N = 25.0$ cm. Thus,

$$f = N/(M-1) = (25.0 \text{ cm})/(7.50 - 1) = 3.85 \text{ cm}$$

We can now determine the maximum angular magnification for the farsighted person

$$M = N/f + 1 = (55.6 \text{ cm})/(3.85 \text{ cm}) + 1 = \boxed{15.4}$$

---

83. **REASONING AND SOLUTION** According to Equation 26.11, the angular magnification of the microscope is

$$M \approx -\frac{(L - f_e)N}{f_o f_e} = -\frac{(14.0 \text{ cm} - 2.5 \text{ cm})(25.0 \text{ cm})}{(0.50 \text{ cm})(2.5 \text{ cm})} = -2.3 \times 10^2$$

Now the new angle is

$$\theta' = M\theta = (-2.3 \times 10^2)(2.1 \times 10^{-5} \text{ rad}) = -4.8 \times 10^{-3} \text{ rad}$$

The magnitude of the angle is $\boxed{4.8 \times 10^{-3} \text{ rad}}$.

84. **REASONING AND SOLUTION** Using Equation 26.11, we find

$$f_e = \frac{LN}{N - Mf_o} = \frac{(18 \text{ cm})(25 \text{ cm})}{(25 \text{ cm}) - (-83)(1.5 \text{ cm})} = \boxed{3.0 \text{ cm}}$$

---

85. **SSM REASONING** The angular magnification of a compound microscope is given by Equation 26.11:

$$M \approx -\frac{(L - f_e)N}{f_o f_e}$$

where f_o is the focal length of the objective, f_e is the focal length of the eyepiece, and L is the separation between the two lenses. This expression can be solved for f_o, the focal length of the objective.

SOLUTION Solving for f_o, we find that the focal length of the objective is

$$f_o = -\frac{(L - f_e)N}{f_e M} = -\frac{(16.0 \text{ cm} - 1.4 \text{ cm})(25 \text{ cm})}{(1.4 \text{ cm})(-320)} = \boxed{0.81 \text{ cm}}$$

---

86. **REASONING AND SOLUTION** The angular magnification of the microscope when using the 100 diopter objective is

$$M_1 \approx -\frac{(L - f_e)N}{f_{o1} f_e}$$

and when using the 300 diopter objective it is

$$M_2 \approx -\frac{(L - f_e)N}{f_{o2} f_e}$$

Division of the equations results in

$$M_2/M_1 = f_{o1}/f_{o2} = (300 \text{ diopters})/(100 \text{ diopters}) = 3$$

Since the angular magnification of the 300 diopter objective is three times greater than that of the 100 diopter objective the angle will be $\boxed{9 \times 10^{-3} \text{ rad}}$.

---

87. **REASONING AND SOLUTION** First, we find the focal length of the magnifying glass. The maximum angular magnification of the magnifying glass is $M = N/f + 1$. Rearranging this expression gives

$$f = N/(M-1) = (25.0 \text{ cm})/(12.0-1) = 2.27 \text{ cm}$$

This lens is now used as the eyepiece for a microscope. Rearranging Equation 26.11, we can find the focal length of the objective of this microscope as follows:

$$f_o = -(L-f_e)N/(f_e M) = -(23.0 \text{ cm} - 2.27 \text{ cm})(25.0 \text{ cm})/[(2.27 \text{ cm})(-525)] = \boxed{0.435 \text{ cm}}$$

---

88. **REASONING** The angular magnification of a compound microscope is given by Equation 26.11. All the necessary data are given in the statement of the problem, so the angular magnification can be calculated directly. In order to find how far the object is from the objective, examine Figure 26.33a. The object distance d_{o1} for the first lens is related to its focal length f_1 and image distance d_{i1} by the thin-lens equation (Equation 26.6). From the drawing we see that the image distance is approximately equal to the distance L between the lenses minus the focal length f_e of the eyepiece, or $d_{i1} \approx L - f_e$. Thus, the object distance is given by

$$\frac{1}{d_{o1}} = \frac{1}{f_o} - \frac{1}{d_{i1}} \approx \frac{1}{f_o} - \frac{1}{L-f_e}$$

The magnification due to the objective is given by Equation 26.7 as $m_{\text{objective}} = -d_{i1}/d_{o1}$. Since both d_{i1} and d_{o1} are now known, the magnification can be evaluated.

SOLUTION

a. According to Equation 26.11, the angular magnification of the compound microscope is

$$M \approx -\frac{(L-f_e)N}{f_o f_e} = -\frac{(26.0 \text{ cm} - 6.50 \text{ cm})(35.0 \text{ cm})}{(3.50 \text{ cm})(6.50 \text{ cm})} = \boxed{30.0}$$

b. Using the thin-lens equation, we can determine the object distance from the objective as follows:

$$\frac{1}{d_{o1}} = \frac{1}{f_o} - \frac{1}{L-f_e} = \frac{1}{3.50 \text{ cm}} - \frac{1}{26.0 \text{ cm} - 6.50 \text{ cm}} = 0.234 \text{ cm}^{-1}$$

or $d_{o1} = \boxed{4.27 \text{ cm}}$

c. The magnification m of the objective is given by Equation 26.7 as

$$m_{\text{objective}} = -\frac{d_{i1}}{d_{o1}} = -\frac{26.0 \text{ cm} - 6.50 \text{ cm}}{4.27 \text{ cm}} = \boxed{-4.57}$$

89. **SSM** *REASONING AND SOLUTION* The angular magnification of an astronomical telescope, is given by Equation 26.12 as $M \approx -f_o/f_e$. Solving for the focal length of the eyepiece, we find

$$f_e \approx -\frac{f_o}{M} = -\frac{48.0 \text{ cm}}{(-184)} = \boxed{0.261 \text{ cm}}$$

---

90. *REASONING AND SOLUTION*
a. From Equation 26.12, we have $M = -f_o/f_e$, so that

$$f_o = -Mf_e = -(-155)(5.00 \times 10^{-3} \text{ m}) = \boxed{0.775 \text{ m}}$$

b. The length of the telescope is the sum of the focal lengths of the objective and eyepiece, so

$$L = f_o + f_e = 0.775 \text{ m} + 0.005 \text{ m} \quad \boxed{0.780 \text{ m}}$$

---

91. *REASONING* Knowing the angles subtended at the unaided eye and with the telescope will allow us to determine the angular magnification of the telescope. Then, since the angular magnification is related to the focal lengths of the eyepiece and the objective, we will use the known focal length of the eyepiece to determine the focal length of the objective.

SOLUTION From Equation 26.12, we have

$$M = \frac{\theta'}{\theta} = -\frac{f_o}{f_e}$$

where θ is the angle subtended by the unaided eye and θ' is the angle subtended when the telescope is used. We note that θ' is negative, since the telescope produces an inverted image. Thus, using Equation 26.12, we find

$$f_o = -\frac{f_e \theta'}{\theta} = -\frac{(0.032 \text{ m})(-2.8 \times 10^{-3} \text{ rad})}{8.0 \times 10^{-5} \text{ rad}} = \boxed{1.1 \text{ m}}$$

---

92. *REASONING AND SOLUTION* First we calculate the focal lengths of the eyepiece and objective lenses using Equation 26.8: Refractive power (RP) = 1/f.

[Objective] $\quad f_o = 1/(RP) = 1/(1.25 \text{ diopters}) = 0.800 \text{ m}$

[Eyepiece] $\quad f_e = 1/(RP) = 1/(250 \text{ diopters}) = 4.0 \times 10^{-3} \text{ m}$

978 THE REFRACTION OF LIGHT: LENSES AND OPTICAL INSTRUMENTS

Using Equation 26.12, we find that the angular magnification is

$$M = -f_o/f_e = -(0.800 \text{ m})/(4.0 \times 10^{-3} \text{ m}) = \boxed{-2.0 \times 10^2}$$

---

93. SSM REASONING AND SOLUTION

a. The lens with the largest focal length should be used for the objective of the telescope. Since the refractive power is the reciprocal of the focal length (in meters), the lens with the smallest refractive power is chosen as the objective, namely, the $\boxed{1.3\text{-diopter lens}}$.

b. According to Equation 26.8, the refractive power is related to the focal length f by Refractive power (in diopters) $= 1/[f(\text{in meters})]$. Since we know the refractive powers of the two lenses, we can solve Equation 26.8 for the focal lengths of the objective and the eyepiece. We find that $f_o = 1/(1.3 \text{ diopters}) = 0.77 \text{ m}$. Similarly, for the eyepiece, $f_e = 1/(11 \text{ diopters}) = 0.091 \text{ m}$. Therefore, the distance between the lenses should be

$$L \approx f_o + f_e = 0.77 \text{ m} + 0.091 \text{ m} = \boxed{0.86 \text{ m}}$$

c. The angular magnification of the telescope is given by Equation 26.12 as

$$M \approx -\frac{f_o}{f_e} = -\frac{0.77 \text{ m}}{0.091 \text{ m}} = \boxed{-8.5}$$

---

94. REASONING
The angular magnification M of an astronomical telescope is given by Equation 26.12 as $M \approx -f_o/f_e$, where f_o and f_e are the focal lengths of the objective and eyepiece, respectively. We are given a value for f_e, but not for f_o. However, an astronomical telescope is used to view objects that are very distant. Therefore, the image produced by the objective is located very near to its focal length (see Figure 26.42a). For the greatest possible magnification, the focal point of the eyepiece is also placed very close to the first image, so that $f_o \approx L - f_e$, where L is the distance between the lenses.

SOLUTION The angular magnification is

$$M \approx -\frac{f_o}{f_e} \approx -\frac{L - f_e}{f_e} = -\frac{1.25 \text{ m} - 0.050 \text{ m}}{0.050 \text{ m}} = \boxed{-24}$$

---

95. REASONING AND SOLUTION
a. The magnification is

$$M = -f_o/f_e = -(19.4 \text{ m})/(0.100 \text{ m}) = \boxed{-194}$$

b. The angular size of the crater is

$$\theta = h_o/d_o = (1500 \text{ m})/(3.77 \times 10^8 \text{ m}) = 4.0 \times 10^{-6} \text{ rad}$$

The angular magnification is, $M = \theta'/\theta$, so that

$$\theta' = M\theta = (-194)(4.0 \times 10^{-6} \text{ rad}) = -7.8 \times 10^{-4} \text{ rad}$$

Since $h_i' = \theta' f_e$, we have

$$h_i' = \theta' f_e = (-7.8 \times 10^{-4} \text{ rad})(0.100 \text{ m}) = \boxed{-7.8 \times 10^{-5} \text{ m}}$$

c. The apparent distance is shorter by a factor of 194, so

$$\text{Apparent Distance} = (3.77 \times 10^8 \text{ m})/(194) = \boxed{1.94 \times 10^6 \text{ m}}$$

---

96. ***REASONING AND SOLUTION*** Use the thin–lens equation to find the first image distance:

$$1/d_i = 1/f_o - 1/d_o = 1/(1.500 \text{ m}) - 1/(114.00 \text{ m}) \quad \text{or} \quad d_i = 1.520 \text{ m}$$

The magnification is

$$M = -d_i/d_o = -(1.520 \text{ m})/(114.00 \text{ m}) = -0.01333$$

Now use this "first image" as the object for the second lens,

$$1/d_i' = 1/f_e - 1/d_o' = 1/(0.070 \text{ m}) - 1/(0.050 \text{ m}) \quad \text{or} \quad d_i' = -0.18 \text{ m}$$

The magnification in the second case is

$$M' = -d_i'/d_o' = +3.6$$

The total linear magnification is therefore,

$$M_l = M \times M' = (-0.01333)(+3.6) = -0.048$$

However, we need the angular magnification, so

$$\theta' = -h_i'/d_i' \text{ and } \theta = h_o/(d_o + f_o + f_e)$$

where we'll use $h_o = 1$ m, $h_i' = -0.048$ m, $d_o + f_o + f_e = (114.00$ m $+ 1.500$ m $+ 0.070$ m$)$ $= 115.57$ m, and $d_i' = -0.18$ m. Therefore,

$$\theta' = -(-0.048 \text{ m})/(-0.18 \text{ m}) = -0.27 \text{ rad}$$

and

$$\theta = (1.0 \text{ m})/(115.57 \text{ m}) = 8.65 \times 10^{-3} \text{ rad}$$

So that,

$$M_{ang} = \theta'/\theta = (-0.27 \text{ rad})/(8.65 \times 10^{-3} \text{ rad}) = \boxed{-31}$$

---

97. **SSM REASONING** The ray diagram is constructed by drawing the paths of two rays from a point on the object. For convenience, we choose the top of the object. The ray that is parallel to the principal axis will be refracted by the lens and pass through the focal point on the right side. The ray that passes through the center of the lens passes through undeflected. The image is formed at the intersection of these two rays on the right side of the lens.

SOLUTION The following ray diagram (to scale) shows that $\boxed{d_i = 18 \text{ cm}}$ and reveals a real, inverted, and enlarged image.

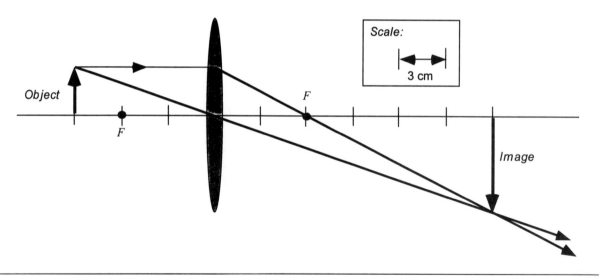

---

98. **REASONING AND SOLUTION** The angular magnification of a compound microscope is negative, and Equation 26.9 gives $M = \theta'/\theta$, so that

$$\theta' = M\theta = (-160)(4.0 \times 10^{-3} \text{ rad}) = -0.64 \text{ rad}$$

The magnitude of this angular size is $\boxed{0.64 \text{ rad}}$.

99. **REASONING AND SOLUTION** The critical angle for the glass-liquid interface must be larger than 65.0° if the light is to penetrate into the liquid. Using Equation 26.4, we have

$$n_L/n > \sin 65.0° \quad \text{or} \quad n_L > n \sin 65.0° = (1.60)\sin 65.0° = \boxed{1.45}$$

---

100. **REASONING AND SOLUTION** The angle of incidence is found from the drawing to be

$$\theta_1 = \tan^{-1}\left(\frac{8.0 \text{ m}}{2.5 \text{ m}}\right) = 73°$$

Snell's law gives the angle of refraction to be

$$\sin \theta_2 = (n_1/n_2)\sin \theta_1 = (1.000/1.333)\sin 73° = 0.72 \quad \text{or} \quad \theta_2 = 46°$$

The distance d is found from the drawing to be

$$d = 8.0 \text{ m} + (4.0 \text{ m})\tan \theta_2 = \boxed{12.1 \text{ m}}$$

---

101. **SSM REASONING AND SOLUTION** Since the far point is 220 cm, we know that the image distance is $d_i = -220$ cm, when the object is infinitely far from the lens ($d_o = \infty$). Thus, the thin-lens equation (Equation 26.6) becomes

$$\frac{1}{d_o} + \frac{1}{d_i} = \frac{1}{\infty} + \frac{1}{(-220 \text{ cm})} = \frac{1}{(-220 \text{ cm})} = \frac{1}{f} \quad \text{or} \quad \boxed{f = -220 \text{ cm}}$$

---

102. **REASONING AND SOLUTION** Using Equation 26.3, we find

$$d = (n_1/n_2)\, d' = (1.546/1.000)(2.5 \text{ cm}) = \boxed{3.9 \text{ cm}}$$

---

103. **REASONING** The height of the mountain's image is given by the magnification equation as $h_i = -h_o d_i/d_o$. To use this expression, however, we will need to know the image distance d_i, which can be determined using the thin-lens equation. Knowing the image distance, we can apply the expression for the image height directly to calculate the desired ratio.

SOLUTION According to the thin-lens equation, we have

$$\frac{1}{d_i} + \frac{1}{d_o} = \frac{1}{f} \tag{1}$$

For both pictures, the object distance d_o is very large compared to the focal length f. Therefore, $1/d_o$ is negligible compared to $1/f$, and the thin-lens equation indicates that $d_i = f$. As a result, the magnification equation indicates that the image height is given by

$$h_i = -\frac{h_o d_i}{d_o} = -\frac{h_o f}{d_o} \qquad (2)$$

Applying Equation (2) for the two pictures and noting that in each case the object height h_o and the focal length f are the same, we find

$$\frac{(h_i)_{5\,km}}{(h_i)_{14\,km}} = \frac{\left(-\frac{h_o f}{d_o}\right)_{5\,km}}{\left(-\frac{h_o f}{d_o}\right)_{14\,km}} = \frac{(d_o)_{14\,km}}{(d_o)_{5\,km}} = \frac{14\,km}{5.0\,km} = \boxed{2.8}$$

---

104. REASONING AND SOLUTION

a. Using the thin-lens equation, we obtain

$$1/d_i = 1/f - 1/d_o = 1/(-32\,\text{cm}) - 1/(+19\,\text{cm}) \qquad \text{or} \qquad d_i = \boxed{-12\,\text{cm}}$$

b. Using the magnification equation, we find

$$m = -(d_i/d_o) = -(-12\,\text{cm})/(19\,\text{cm}) = \boxed{+0.63}$$

c. The image is $\boxed{\text{virtual}}$ since d_i is negative.

d. The image is $\boxed{\text{upright}}$ since m is +.

e. The image is $\boxed{\text{reduced}}$ in size since $m < 1$.

---

105. SSM REASONING AND SOLUTION

a. We know from the law of reflection (Section 25.2), that the angle of reflection is equal to the angle of incidence, so the reflected ray is reflected at $\boxed{43°}$.

b. Snell's law of refraction (Equation 26.2: $n_1 \sin \theta_1 = n_2 \sin \theta_2$ can be used to find the angle of refraction. Table 26.1 indicates that the index of refraction of water is 1.333. Solving for θ_2 and substituting values, we find that

$$\sin \theta_2 = \frac{n_1 \sin \theta_1}{n_2} = \frac{(1.000)(\sin 43°)}{1.333} = 0.51 \quad \text{or} \quad \theta_2 = \sin^{-1} 0.51 = \boxed{31°}$$

---

106. **REASONING AND SOLUTION** Using Equation 26.10, with $(1/f) - (1/d_i) = 1/d_o$, we have

$$M = N/d_o = (72 \text{ cm})/(4.0 \text{ cm}) = \boxed{18}$$

---

107. **SSM** **REASONING AND SOLUTION** An optometrist prescribes contact lenses that have a focal length of 55.0 cm.

a. The focal length is positive (+55.0 cm); therefore, we can conclude that the lenses are $\boxed{\text{converging}}$.

b. As discussed in the text (see Section 26.10), farsightedness is corrected by converging lenses. Therefore, the person who wears these lens is $\boxed{\text{farsighted}}$.

c. If the lenses are designed so that objects no closer than 35.0 cm can be seen clearly, we have $d_o = 35.0$ cm. The thin-lens equation (Equation 26.6) gives the image distance:

$$\frac{1}{d_i} = \frac{1}{f} - \frac{1}{d_o} = \frac{1}{55.0 \text{ cm}} - \frac{1}{35.0 \text{ cm}} \quad \text{or} \quad d_i = -96.3 \text{ cm}$$

Thus, the near point is located $\boxed{96.3 \text{ cm}}$ from the eyes.

---

108. **REASONING AND SOLUTION** Using Snell's law, we have

$$n_A \sin 72° = n_B \sin 56° \quad \text{or} \quad \frac{n_A}{n_B} = \frac{\sin 56°}{\sin 72°} = \boxed{0.87}$$

---

109. **REASONING AND SOLUTION**
a. The sun is so far from the lens that the incident rays are nearly parallel to the principal axis, so the image distance d_i is nearly equal to the focal length of the lens. The magnification of the lens is

$$m = -\frac{d_i}{d_o} = -\frac{10.0 \times 10^{-2} \text{ m}}{1.50 \times 10^{11} \text{ m}} = -6.67 \times 10^{-13}$$

The image height h_i is

$$h_i = mh_o = (-6.67 \times 10^{-13})(1.39 \times 10^9 \text{ m}) = -9.27 \times 10^{-4} \text{ m}$$

The diameter of the sun's image on the paper is the magnitude of h_i, or 9.27×10^{-4} m. The area A of the image is

$$A = \tfrac{1}{4}\pi(9.27 \times 10^{-4} \text{ m})^2 = \boxed{6.74 \times 10^{-7} \text{ m}^2}$$

b. The intensity I of the light wave is the power P that strikes the paper perpendicularly divided by the illuminated area A (see Equation 16.8)

$$I = \frac{P}{A} = \frac{0.530 \text{ W}}{6.74 \times 10^{-7} \text{ m}^2} = \boxed{7.86 \times 10^5 \text{ W/m}^2}$$

---

110. **REASONING AND SOLUTION** We note that the object is placed 20.0 cm from the lens. Since the focal point of the lens is $f = -20.0$ cm, the object is situated at the focal point. In the scale drawing that follows, we locate the image using the two rays labeled a and b, which originate at the top of the object.

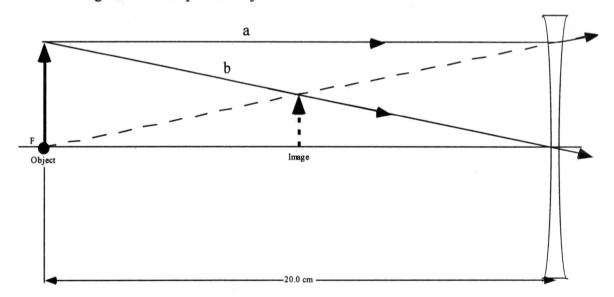

a. Measuring according to the scale used in the drawing, we find that the image is located 10.0 cm to the left of the lens. The lens is a diverging lens and forms a virtual image, so the image distance is $d_i = \boxed{-10.0 \text{ cm}}$.

b. Measuring the heights of the image and the object in the drawing, we find that the magnification is m = $\boxed{+0.500}$.

---

111. $\boxed{\text{SSM}}$ *REASONING AND SOLUTION*

a. A real image must be projected on the drum; therefore, the lens in the copier must be a $\boxed{\text{converging lens}}$.

b. If the document and its copy have the same size, but are inverted with respect to one another, the magnification equation (Equation 26.7) indicates that $m = -d_i/d_o = -1$. Therefore, $d_i/d_o = 1$ or $d_i = d_o$. Then, the thin-lens equation (Equation 26.6) gives

$$\frac{1}{d_i} + \frac{1}{d_o} = \frac{1}{f} = \frac{2}{d_o} \qquad \text{or} \qquad d_o = d_i = 2f$$

Therefore the document is located at a distance $\boxed{2f}$ from the lens.

c. Furthermore, the image is located at a distance of $\boxed{2f}$ from the lens.

---

112. *REASONING AND SOLUTION*

a. With the image at the near point of the eye, the angular magnification is M = N/f +1. Find the focal length for M = 6.0, using

$$f = N/(M - 1) = (25 \text{ cm})/(6.0 - 1) = 5.0 \text{ cm} = 0.050 \text{ m}$$

The refractive power of this lens is, therefore,

$$\text{Refractive power } = 1/f = 1/(0.050 \text{ m}) = \boxed{2.0 \times 10^1 \text{ diopters}}$$

b. When the image of the stamp is 45 cm from the eye,

$$M = (1/f - 1/d_i)N = [1/(0.050 \text{ m}) - 1/(-0.45 \text{ m})](0.25 \text{ m}) = \boxed{5.6}$$

---

113. $\boxed{\text{SSM}}$ *REASONING* A contact lens is placed directly on the eye. Therefore, the object distance, which is the distance from the book to the lens, is 25.0 cm. The near point can be determined from the thin-lens equation [Equation 26.6: $(1/d_o) + (1/d_i) = (1/f)$].

986 THE REFRACTION OF LIGHT: LENSES AND OPTICAL INSTRUMENTS

SOLUTION
a. Using the thin-lens equation, we have

$$\frac{1}{d_i} = \frac{1}{f} - \frac{1}{d_o} = \frac{1}{65.0 \text{ cm}} - \frac{1}{25.0 \text{ cm}} \quad \text{or} \quad d_i = -40.6 \text{ cm}$$

In other words, at age 40, the man's near point is 40.6 cm. Similarly, when the man is 45, we have

$$\frac{1}{d_i} = \frac{1}{f} - \frac{1}{d_o} = \frac{1}{65.0 \text{ cm}} - \frac{1}{29.0 \text{ cm}} \quad \text{or} \quad d_i = -52.4 \text{ cm}$$

and his near point is 52.4 cm. Thus, the man's near point has changed by $52.4 \text{ cm} - 40.6 \text{ cm} = \boxed{11.8 \text{ cm}}$.

b. With $d_o = 25.0$ cm and $d_i = -52.4$ cm, the focal length of the lens is found as follows:

$$\frac{1}{f} = \frac{1}{d_o} + \frac{1}{d_i} = \frac{1}{25.0 \text{ cm}} + \frac{1}{(-52.4 \text{ cm})} \quad \text{or} \quad \boxed{f = 47.8 \text{ cm}}$$

---

114. **REASONING** To find the distance through which the object must be moved, we must obtain the object distances for the two situations described in the problem. To do this, we combine the thin-lens equation and the magnification equation, since data for the magnification is given.

SOLUTION
a. Since the magnification is positive, the image is upright, and the object must be located within the focal point of the lens, as in Figure 26.28. When the magnification is negative and has a magnitude greater than one, the object must be located between the focal point and the point that is at a distance of twice the focal length from the lens, as in Figure 26.27. Therefore, the object should be moved $\boxed{\text{away from the lens}}$.

b. According to the thin-lens equation, we have

$$\frac{1}{d_i} + \frac{1}{d_o} = \frac{1}{f} \quad \text{or} \quad d_i = \frac{d_o f}{d_o - f} \quad (1)$$

According to the magnification equation, with d_i expressed as in Equation (2), we have

$$m = -\frac{d_i}{d_o} = -\frac{1}{d_o}\left(\frac{d_o f}{d_o - f}\right) = \frac{f}{f - d_o} \quad \text{or} \quad d_o = \frac{f(m-1)}{m} \quad (2)$$

Applying Equation (2) to the two cases described in the problem, we have

$$\left(d_o\right)_{+m} = \frac{f(m-1)}{m} = \frac{f(+4.0-1)}{+4.0} = \frac{3.0f}{4.0} \qquad (3)$$

$$\left(d_o\right)_{-m} = \frac{f(m-1)}{m} = \frac{f(-4.0-1)}{-4.0} = \frac{5.0f}{4.0} \qquad (4)$$

Subtracting Equation (3) from Equation (4), we find that the object must be moved away from the lens by an additional distance of

$$\left(d_o\right)_{-m} - \left(d_o\right)_{+m} = \frac{5.0f}{4.0} - \frac{3.0f}{4.0} = \frac{2.0f}{4.0} = \frac{0.30\text{ m}}{2.0} = \boxed{0.15\text{ m}}$$

---

115. **SSM** **REASONING** The angular magnification of a refracting telescope is 32 8000 times larger when you look through the correct end of the telescope than when you look through the wrong end. We wish to find the angular magnification, $M = -f_o/f_e$ (see Equation 26.12) of the telescope. Thus, we proceed by finding the ratio of the focal lengths of the objective and the eyepiece and using Equation 26.12 to find M.

SOLUTION When you look through the correct end of the telescope, the angular magnification of the telescope is $M_c = -f_o/f_e$. If you look through the wrong end, the roles of the objective and eyepiece lenses are interchanged, so that the angular magnification would be $M_w = -f_e/f_o$. Therefore,

$$\frac{M_c}{M_w} = \frac{-f_o/f_e}{-f_e/f_o} = \left(\frac{f_o}{f_e}\right)^2 = 32\,800 \qquad \text{or} \qquad \frac{f_o}{f_e} = \pm\sqrt{32\,800} = \pm 181$$

The angular magnification of the telescope is negative, so we choose the positive root and obtain $M = -f_o/f_e = -(+181) = \boxed{-181}$.

---

116. **REASONING AND SOLUTION** From the drawing we see that $d_o = x + f$ and $d_i = x' + f$. Substituting these two expressions into the thin-lens equation, we obtain

$$\frac{1}{d_o} + \frac{1}{d_i} = \frac{1}{x+f} + \frac{1}{x'+f} = \frac{1}{f}$$

Combining the terms on the left over a common denominator gives

988 THE REFRACTION OF LIGHT: LENSES AND OPTICAL INSTRUMENTS

$$\frac{x+f+x'+f}{(x+f)(x'+f)} = \frac{x+x'+2f}{(x+f)(x'+f)} = \frac{1}{f}$$

Cross-multiplying shows that

$$f(x+x'+2f) = (x+f)(x'+f)$$

Expanding and simplifying this result, we obtain

$$fx + fx' + 2f^2 = xx' + fx' + xf + f^2 \quad \text{or} \quad \boxed{xx' = f^2}$$

---

117. **REASONING AND SOLUTION** If the near point is 79.0 cm, then $d_i = -79.0$ cm, and $d_o = 25.0$ cm. Using the thin-lens equation, we find that the focal length of the correcting lens is

$$f = d_o d_i/(d_o + d_i) = (25.0 \text{ cm})(-79.0 \text{ cm})/(25.0 \text{ cm} - 79.0 \text{ cm}) = +36.6 \text{ cm}$$

a. The distance d_o' to the poster can be obtained as follows:

$$1/d_o' = 1/f - 1/d_i' = 1/(36.6 \text{ cm}) - 1/(-217 \text{ cm}) \quad \text{or} \quad d_o' = \boxed{31.3 \text{ cm}}$$

b. The image size is

$$h_i = h_o(-d_i'/d_o') = (0.350 \text{ m})[-(-217 \text{ cm})/(31.3 \text{ cm})] = \boxed{2.43 \text{ m}}$$

---

118. **CONCEPT QUESTIONS**
a. The refracted ray is physically possible. When light goes from a medium of lower index of refraction ($n = 1.4$) to one of higher index of refraction ($n = 1.6$), the refracted ray is bent toward the normal, as it does in part (a).

b. The refracted ray is physically not possible. When light goes from a medium of lower index of refraction ($n = 1.5$) to one of higher index of refraction ($n = 1.6$), the refracted ray must bend toward the normal, not away from it, as part (b) of the drawing shows.

c. The refracted ray is physically possible. When light goes from a medium of higher index of refraction ($n = 1.6$) to one of lower index of refraction ($n = 1.4$), the refracted ray bends away from the normal, as it does part (c) of the drawing.

d. The refracted ray is physically not possible. When the angle of incidence is 0°, the angle of refraction is also 0°, regardless of the indices of refraction.

SOLUTION

a. The angle of refraction θ_2 is given by Snell's law, Equation 26.2, as

$$\theta_2 = \sin^{-1}\left(\frac{n_1 \sin \theta_1}{n_2}\right) = \sin^{-1}\left[\frac{(1.4)\sin 55°}{1.6}\right] = \boxed{46°}$$

b. The actual angle of refraction is

$$\theta_2 = \sin^{-1}\left(\frac{n_1 \sin \theta_1}{n_2}\right) = \sin^{-1}\left[\frac{(1.5)\sin 55°}{1.6}\right] = \boxed{50°}$$

c. The angle of refraction is

$$\theta_2 = \sin^{-1}\left(\frac{n_1 \sin \theta_1}{n_2}\right) = \sin^{-1}\left[\frac{(1.6)\sin 55°}{1.4}\right] = \boxed{69°}$$

d. The actual angle of refraction is

$$\theta_2 = \sin^{-1}\left(\frac{n_1 \sin \theta_1}{n_2}\right) = \sin^{-1}\left[\frac{(1.6)\sin 0°}{1.4}\right] = \boxed{0°}$$

---

119. **CONCEPT QUESTION** When the light ray passes from *a* into *b*, it is bent toward the normal. According to the discussion in Section 26.2, this happens when the index of refraction of *b* is greater than that of *a*, or $n_b > n_a$. When the light passes from *b* into *c*, it is bent away from the normal. This means that the index of refraction of *c* is less than that of *b*, or $n_c < n_b$. The smaller the value of n_c, the greater is the angle of refraction. As can be seen from the drawing, the angle of refraction in material *c* is greater than the angle of incidence at the *a-b* interface. Applying Snell's law to the *a-b* and *b-c* interfaces gives $n_a \sin \theta_a = n_b \sin \theta_b = n_c \sin \theta_c$. Since θ_c is greater than θ_a, the equation $n_a \sin \theta_a = n_c \sin \theta_c$ shows that the index of refraction of *a* must be greater than that of *c*, $n_a > n_c$. Thus, the ordering of the indices of refraction, highest to lowest, is n_b, n_a, n_c.

SOLUTION The index of refraction for each medium can be evaluated from Snell's law, Equation 26.2:

a-b interface
$$n_b = \frac{n_a \sin \theta_a}{\sin \theta_b} = \frac{(1.20)\sin 50.0°}{\sin 45.0°} = \boxed{1.30}$$

b-c interface
$$n_c = \frac{n_b \sin \theta_b}{\sin \theta_c} = \frac{(1.30)\sin 45.0°}{\sin 56.7°} = \boxed{1.10}$$

As expected, the ranking of the indices of refraction, highest to lowest, is $n_b = 1.30$, $n_a = 1.20$, $n_c = 1.10$.

---

120. **CONCEPT QUESTION** Total internal reflection occurs only when light goes from a higher index material toward a lower index material (see Section 26.3). Since total internal reflection occurs at both the *a-b* and *a-c* interfaces, the index of refraction of material *a* is larger that that of either material *b* or *c*: $n_a > n_b$ and $n_a > n_c$. We now need to determine which index of refraction, n_b or n_c, is larger. The critical angle is given by Equation 26.4 as $\sin \theta_c = n_2/n_1$, where n_2 is the smaller index of refraction. Therefore, the larger the value of n_2, the larger the critical angle. It is evident from the drawing that the critical angle for the *a-c* interface is larger than the critical angle for the *a-b* interface. Therefore n_c must be larger than n_b. The ranking of the indices of refraction, largest to smallest, is: n_a, n_c, n_b.

    SOLUTION For the *a-b* interface, the critical angle is given by Equation 26.4 as $\sin \theta_c = n_b/n_a$. Therefore, the index of refraction for material *b* is

    $$n_b = n_a \sin \theta_c = (1.80) \sin 40.0° = \boxed{1.16}$$

    For the *a-c* interface, we note that the angle of incidence is $90.0° - 40.0° = 50.0°$. The index of refraction for material *c* is

    $$n_c = n_a \sin \theta_c = (1.80) \sin 50.0° = \boxed{1.38}$$

    As expected, the ranking of the indices of refraction, highest-to-lowest, is $n_a = 1.80$, $n_c = 1.38$, $n_b = 1.16$.

---

121. **CONCEPT QUESTION** Total internal reflection can occur only when light is traveling from a higher index material toward a lower index material. Thus, total internal reflection is possible when the material above or below a layer has a smaller index of refraction than the layer itself. With this criteria in mind, the table can be filled in as follows:

Layer	Is total internal reflection possible?	
	Top surface of layer	Bottom surface of layer
a	Yes	No
b	Yes	Yes
c	No	Yes

SOLUTION The critical angle for each interface at which total internal reflection is possible is obtained from Equation 26.4:

Layer a, top surface
$$\theta_c = \sin^{-1}\left(\frac{n_{air}}{n_a}\right) = \sin^{-1}\left(\frac{1.00}{1.30}\right) = 50.3°$$

$$\Delta\theta = 75.0° - 50.3° = \boxed{24.7°}$$

Layer b, top surface
$$\theta_c = \sin^{-1}\left(\frac{n_a}{n_b}\right) = \sin^{-1}\left(\frac{1.30}{1.50}\right) = 60.1°$$

$$\Delta\theta = 75.0° - 60.1° = \boxed{14.9°}$$

Layer b, bottom surface
$$\theta_c = \sin^{-1}\left(\frac{n_c}{n_b}\right) = \sin^{-1}\left(\frac{1.40}{1.50}\right) = 69.0°$$

$$\Delta\theta = 75.0° - 69.0° = \boxed{6.0°}$$

Layer c, bottom surface
$$\theta_c = \sin^{-1}\left(\frac{n_{air}}{n_c}\right) = \sin^{-1}\left(\frac{1.00}{1.40}\right) = 45.6°$$

$$\Delta\theta = 75.0° - 45.6° = \boxed{29.4°}$$

---

122. **CONCEPT QUESTIONS**
   a. A converging lens must be used, because a diverging lens cannot produce a real image.

   b. Since the image is one-half the size of the object and inverted relative to it, the image height h_i is related to the object height h_o by $h_i = -\frac{1}{2}h_o$, where the minus sign indicates that the image is inverted.

   c. According to the magnification equation, Equation 26.7, the image distance d_i is related to the object distance d_o by $d_i/d_o = -h_i/h_o$. But we know that $h_i/h_o = -\frac{1}{2}$, so $d_i/d_o = -\left(-\frac{1}{2}\right) = \frac{1}{2}$.

SOLUTION
a. Let d be the distance between the object and image, so that $d = d_o + d_i$. However, we know from the Concept Questions that $d_i = \frac{1}{2}d_o$, so $d = d_o + \frac{1}{2}d_o = \frac{3}{2}d_o$. The object distance is, therefore,

$$d_o = = \tfrac{2}{3}d = \tfrac{2}{3}(90.0 \text{ cm}) = \boxed{60.0 \text{ cm}}$$

b. The thin-lens equation, Equation 26.6, can be used to find the focal length f of the lens:

$$\frac{1}{f} = \frac{1}{d_o} + \frac{1}{d_i} = \frac{1}{d_o} + \frac{1}{\frac{1}{2}d_o} = \frac{3}{d_o}$$

$$f = \frac{d_o}{3} = \frac{60.0 \text{ cm}}{3} = \boxed{20.0 \text{ cm}}$$

---

123. **CONCEPT QUESTION** In part a of the drawing, the object lies inside the focal point of the converging (#1) lens. According to Figure 26.28, such an object produces a virtual image that lies to the left of the lens. This image act as the object for the diverging (#2) lens. Since a diverging lens always produces a virtual image that lies to the left of the lens, the final image lies to the left of the diverging lens. In part (b), the diverging (#1) lens produces a virtual image that lies to the left of the lens. This image act as the object for the converging (#2) lens. Since the object lies outside the focal point of the converging lens, the converging lens produces a real image that lies to the right of the lens (see Figure 26.26). Thus, the final image lies to the right of the converging lens.

SOLUTION

a. The focal length of lens #1 is $f_1 = 15.00$ cm, and the object distance is $d_{o1} = 10.0$ cm. The image distance d_{i1} produced by the first lens can be obtained from the thin-lens equation, Equation 26.6:

$$\frac{1}{d_{i1}} = \frac{1}{f_1} - \frac{1}{d_{o1}} = \frac{1}{15.00 \text{ cm}} - \frac{1}{10.00 \text{ cm}} = -3.33 \times 10^{-2} \text{ cm}^{-1} \text{ or } d_{i1} = -30.0 \text{ cm}$$

This image is located to the left of lens #1 and serves as the object for lens #2. Thus, the object distance for lens #2 is $d_{i2} = 30.0$ cm $+ 50.0$ cm $= 80.0$ cm. The image distance produced by lens #2 is

$$\frac{1}{d_{i2}} = \frac{1}{f_2} - \frac{1}{d_{o2}} = \frac{1}{-20.00 \text{ cm}} - \frac{1}{80.0 \text{ cm}} = -6.25 \times 10^{-2} \text{ cm}^{-1} \text{ or } d_{i2} = \boxed{-16.0 \text{ cm}}$$

The negative value for d_{i2} indicates that, as expected, the final image is to the left of lens #2.

b. The focal length of the lens #1 is $f_1 = -20.0$ cm, and the object distance is $d_{o1} = 10.00$ cm. The image distance d_{i1} produced by the first lens can be obtained from the thin-lens equation, Equation 26.6.

$$\frac{1}{d_{i1}} = \frac{1}{f_1} - \frac{1}{d_{o1}} = \frac{1}{-20.0 \text{ cm}} - \frac{1}{10.00 \text{ cm}} = -1.50 \times 10^{-1} \text{ cm}^{-1} \text{ or } d_{i1} = -6.67 \text{ cm}$$

This image is located to the left of lens #1 and serves as the object for lens #2. Thus the object distance for lens #2 is d_{i2} = 6.67 cm + 50.0 cm = 56.7 cm. The image distance produced by lens #2 is

$$\frac{1}{d_{i2}} = \frac{1}{f_2} - \frac{1}{d_{o2}} = \frac{1}{15.00 \text{ cm}} - \frac{1}{56.7 \text{ cm}} = 4.90 \times 10^{-2} \text{ cm}^{-1} \text{ or } d_{i2} = \boxed{20.4 \text{ cm}}$$

The positive value for d_{i2} indicates that, as expected, the final image is to the right of lens #2.

CHAPTER 27 INTERFERENCE AND THE WAVE NATURE OF LIGHT

PROBLEMS

1. **SSM** *REASONING AND SOLUTION* To decide whether constructive or destructive interference occurs, we need to determine the wavelength of the wave. For electromagnetic waves, Equation 16.1 can be written $f\lambda = c$, so that

$$\lambda = \frac{c}{f} = \frac{3.00 \times 10^8 \text{ m}}{536 \times 10^3 \text{ Hz}} = 5.60 \times 10^2 \text{ m}$$

Since the two wave sources are in phase, constructive interference occurs when the path difference is an integer number of wavelengths, and destructive interference occurs when the path difference is an odd number of half wavelengths. We find that the path difference is 8.12 km – 7.00 km = 1.12 km. The number of wavelengths in this path difference is $(1.12 \times 10^3 \text{ m})/(5.60 \times 10^2 \text{ m}) = 2.00$. Therefore, constructive interference occurs.

2. *REASONING AND SOLUTION*
   a. We know that $\theta = \sin^{-1}(m\lambda/d)$. Also, $m = 1$ for the first order bright fringe, so

   $$\theta = \sin^{-1}\left[\frac{(1)(630 \times 10^{-9} \text{ m})}{5.3 \times 10^{-5} \text{ m}}\right] = \boxed{0.68°}$$

   b. For the second-order bright fringe, $m = 2$ so $\theta = \boxed{1.4°}$.

   c. For the third-order bright fringe, $m = 3$ so $\theta = \boxed{2.0°}$.

3. *REASONING AND SOLUTION* For destructive interference, $(m + 1/2)\lambda = d \sin\theta$. For the second dark fringe, $m = 1$ so

$$\frac{d}{\lambda} = \frac{m + \tfrac{1}{2}}{\sin\theta} = \frac{1 + \tfrac{1}{2}}{\sin 5.4°} = \boxed{16}$$

4. **REASONING** Let ℓ_1 and ℓ_2 be the distances from source 1 and source 2, respectively. For constructive interference the condition is $\ell_2 - \ell_1 = m\lambda$, where λ is the wavelength and $m = 0, 1, 2, 3, \ldots$. For destructive interference the condition is $\ell_2 - \ell_1 = (m + \tfrac{1}{2})\lambda$, where $m = 0, 1, 2, 3, \ldots$. The fact that the wavelength is greater than the separation between the sources will allow us to decide which type of interference we have and to locate the two places in question.

SOLUTION a. Since the places we seek lie on the line between the two sources and since the separation between the sources is 4.00 m, we know that $\ell_2 - \ell_1$ cannot have a value greater than 4.00 m. Therefore, with a wavelength of 5.00 m, the only way constructive interference can occur is at a place where $m = 0$. But this would mean that $\ell_2 = \ell_1$, and there is only one place where this is true, namely, at the midpoint between the sources. But we know that there are two places. As a result, we conclude that destructive interference is occurring.

b. For destructive interference, the possible values for the integer m are $m = 0, 1, 2, 3, \ldots$, and we must decide which to use. The condition for destructive interference is $\ell_2 - \ell_1 = (m + \tfrac{1}{2})\lambda$. Since $\lambda = 5.00$ m, all values of m greater than or equal to one result in $\ell_2 - \ell_1$ being greater than 4.00 m, which we already know is not possible. Therefore, we conclude that $m = 0$, and the condition for destructive interference becomes $\ell_2 - \ell_1 = \tfrac{1}{2}\lambda$. We also know that $\ell_2 + \ell_1 = 4.00$ m, since the separation between the sources is 4.00 m. In other words, we have

$$\ell_2 - \ell_1 = \tfrac{1}{2}\lambda = \tfrac{1}{2}(5.00 \text{ m}) = 2.50 \text{ m} \quad \text{and} \quad \ell_2 + \ell_1 = 4.00 \text{ m}$$

Adding these two equations gives

$$2\ell_2 = 6.50 \text{ m} \quad \text{or} \quad \ell_2 = 3.25 \text{ m}$$

Since $\ell_2 + \ell_1 = 4.00$ m, it follows that

$$\ell_1 = 4.00 \text{ m} - \ell_2 = 4.00 \text{ m} - 3.25 \text{ m} = 0.75 \text{ m}$$

These results indicate that the two places are 3.25 m and 0.75 m from one of the speakers.

---

5. **SSM** **REASONING AND SOLUTION** The angular position θ of the bright fringes of a double slit are given by Equation 27.1 as $\sin\theta = m\lambda/d$, with the order of the fringe specified by $m = 0, 1, 2, 3, \ldots$. Solving for λ, we have

$$\lambda = \frac{d \sin\theta}{m} = \frac{(3.8 \times 10^{-5} \text{ m}) \sin 2.0°}{2} = 6.6 \times 10^{-7} \text{ m} = \boxed{660 \text{ nm}}$$

6. **REASONING AND SOLUTION** The angular position of a third order fringe is given by $\theta = \sin^{-1} 3\lambda/d$, and the position of the fringe on the screen is $y = L \tan \theta$.

For the third-order red fringe,

$$\theta = \sin^{-1}\left[\frac{3(665 \times 10^{-9} \text{ m})}{0.158 \times 10^{-3} \text{ m}}\right] = 0.724°$$

and

$$y = (2.24 \text{ m}) \tan 0.724° = 2.83 \times 10^{-2} \text{ m}$$

For the third-order green fringe, $\theta = 0.615°$ and $y = 2.40 \times 10^{-2}$ m.

The distance between the fringes is 2.83×10^{-2} m $- 2.40 \times 10^{-2}$ m $= \boxed{4.3 \times 10^{-3} \text{ m}}$.

---

7. **REASONING AND SOLUTION** From Example 1 we have that $y = L \tan \theta$. Since for small angles, $\tan \theta \approx \sin \theta = \lambda/d$, $y = (L/d)\lambda$. The first case gives

$$L/d = y/\lambda = (2.40 \times 10^{-2} \text{ m})/(475 \times 10^{-9} \text{ m}) = 5.05 \times 10^{4}$$

Then the fringe separation in the second case is

$$y = (5.05 \times 10^{4})(611 \times 10^{-9} \text{ m}) = \boxed{0.0309 \text{ m}}$$

---

8. **REASONING AND SOLUTION** The first-order orange fringes occur farther out from the center than do the first-order blue fringes. Therefore, the screen must be moved $\boxed{\text{toward the slits}}$ so that the orange fringes will appear on the screen. The distance between the screen and the slits is L, and the amount by which the screen must be moved toward the slits is $L_{\text{blue}} - L_{\text{orange}}$. We know that $L_{\text{blue}} = 0.500$ m, and must, therefore, determine L_{orange}. We begin with Equation 27.1, as it applies to first-order fringes, that is, $\sin \theta = \lambda/d$. Furthermore, as discussed in Example 1 in the text, $\tan \theta = y/L$, where y is the distance from the center of the screen to a fringe. Since the angle θ locating the fringes is small, $\sin \theta \approx \tan \theta$, and we have that

$$\frac{\lambda}{d} \approx \frac{y}{L} \quad \text{or} \quad L = \frac{yd}{\lambda}$$

Writing this result for L for both colors and dividing the two equations gives

$$\frac{L_{\text{blue}}}{L_{\text{orange}}} = \frac{yd/\lambda_{\text{blue}}}{yd/\lambda_{\text{orange}}} = \frac{\lambda_{\text{orange}}}{\lambda_{\text{blue}}}$$

where we have recognized that y is one-half the screen width (the same for both colors), and that the slit separation d is the same for both colors. Using the result above, we find that

$$L_{blue} - L_{orange} = L_{blue} - \left(\frac{\lambda_{blue}}{\lambda_{orange}}\right) L_{blue} = L_{blue}\left(1 - \frac{\lambda_{blue}}{\lambda_{orange}}\right)$$

$$= (0.500 \text{ m})\left[1 - \frac{471 \text{ nm}}{611 \text{ nm}}\right] = \boxed{0.115 \text{ m}}$$

---

9. **SSM** **WWW** *REASONING* The light that travels through the plastic has a different path length than the light that passes through the unobstructed slit. Since the center of the screen now appears dark, rather than bright, destructive interference, rather than constructive interference occurs there. This means that the difference between the number of wavelengths in the plastic sheet and that in a comparable thickness of air is $\frac{1}{2}$.

SOLUTION The wavelength of the light in the plastic sheet is given by Equation 27.3 as

$$\lambda_{plastic} = \frac{\lambda_{vacuum}}{n} = \frac{586 \times 10^{-9} \text{ m}}{1.60} = \boxed{366 \times 10^{-9} \text{ m}}$$

The number of wavelengths contained in a plastic sheet of thickness t is

$$N_{plastic} = \frac{t}{\lambda_{plastic}} = \frac{t}{366 \times 10^{-9} \text{ m}}$$

The number of wavelengths contained in an equal thickness of air is

$$N_{air} = \frac{t}{\lambda_{air}} = \frac{t}{586 \times 10^{-9} \text{ m}}$$

where we have used the fact that $\lambda_{air} \approx \lambda_{vacuum}$. Destructive interference occurs when the difference, $N_{plastic} - N_{air}$, in the number of wavelengths is $\frac{1}{2}$:

$$N_{plastic} - N_{air} = \frac{1}{2}$$

$$\frac{t}{366 \times 10^{-9} \text{ m}} - \frac{t}{586 \times 10^{-9} \text{ m}} = \frac{1}{2}$$

Solving this equation for t yields $t = 487 \times 10^{-9}$ m = $\boxed{487 \text{ nm}}$.

998 INTERFERENCE AND THE WAVE NATURE OF LIGHT

10. **REASONING AND SOLUTION** A phase shift equivalent to one-half a wavelength occurs at the top air-plastic interface, so the condition for destructive interference is $2t = m\lambda_{film}$. Since $\lambda_{film} = \lambda_{vacuum}/n$, the thickness of the film can be written as

$$t = \frac{m}{2}\left(\frac{\lambda_{vacuum}}{n}\right)$$

For $m = 1$:

$$t = \frac{1}{2}\left(\frac{589 \text{ nm}}{1.61}\right) = \boxed{183 \text{ nm}}$$

For $m = 2$:

$$t = \frac{2}{2}\left(\frac{589 \text{ nm}}{1.61}\right) = \boxed{366 \text{ nm}}$$

---

11. **SSM** **REASONING** To solve this problem, we must express the condition for destructive interference in terms of the film thickness t and the wavelength λ_{film} of the light as it passes through the magnesium fluoride coating. We must also take into account any phase changes that occur upon reflection.

SOLUTION Since the coating is intended to be nonreflective, its thickness must be chosen so that destructive interference occurs between waves 1 and 2 in the drawing. For destructive interference, the combined phase difference between the two waves must be an odd integer number of half wavelengths. The phase change for wave 1 is equivalent to one-half of a wavelength, since this light travels from a smaller refractive index ($n_{air} = 1.00$) toward a larger refractive index ($n_{film} = 1.38$).

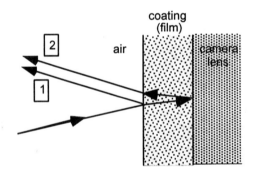

Similarly, there is a phase change when wave 2 reflects from the right surface of the film, since this light also travels from a smaller refractive index ($n_{film} = 1.38$) toward a larger one ($n_{lens} = 1.52$). Therefore, a phase change of one-half wavelength occurs at both boundaries, so the net phase change between waves 1 and 2 due to reflection is zero. Since wave 2 travels back and forth through the film and, and since the light is assumed to be at nearly normal incidence, the extra distance traveled by wave 2 compared to wave 1 is twice the film thickness, or $2t$. Thus, in this case, the minimum condition for destructive interference is

$$2t = \tfrac{1}{2}\lambda_{film}$$

The wavelength of light in the coating is

$$\lambda_{\text{film}} = \frac{\lambda_{\text{vacuum}}}{n} = \frac{565 \text{ nm}}{1.38} = 409 \text{ nm} \qquad (27.3)$$

Solving the above expression for t, we find that the minimum thickness that the coating can have is

$$t = \tfrac{1}{4}\lambda_{\text{film}} = \tfrac{1}{4}(409 \text{ nm}) = \boxed{102 \text{ nm}}$$

---

12. **REASONING AND SOLUTION** The condition for destructive interference is $2t = \lambda_{\text{film}}$, so $2t = \lambda_{\text{vacuum}}/n$.

    a. For removal of violet light, which makes the film look yellow,

    $$t = \frac{1}{2}\left(\frac{410 \text{ nm}}{1.40}\right) = \boxed{150 \text{ nm}}$$

    b. For removal of yellow light, which makes the film look violet,

    $$t = \frac{1}{2}\left(\frac{580 \text{ nm}}{1.40}\right) = \boxed{210 \text{ nm}}$$

---

13. **REASONING AND SOLUTION** The only change from Example 5 is $\lambda_{\text{film}} = \lambda_{\text{vacuum}}/n = (552 \text{ nm})/1.33 = 415 \text{ nm}$. Since $t = \tfrac{1}{2}\left(m + \tfrac{1}{2}\right)\lambda_{\text{film}}$,

$$m = \frac{2t}{\lambda_{\text{film}}} - \frac{1}{2} = \frac{2(4.10 \times 10^{-5} \text{ m})}{415 \times 10^{-9} \text{ m}} - \frac{1}{2} = 197$$

Since the first bright fringe occurs when $m = 0$, the number of bright fringes is $m + 1 = \boxed{198}$.

---

14. **REASONING** In air the index of refraction is nearly $n = 1$, while in the film it is greater than one. A phase change occurs whenever light travels through a material with a smaller refractive index toward a material with a larger refractive index and reflects from the boundary between the two. The phase change is equivalent to $\tfrac{1}{2}\lambda_{\text{film}}$, where λ_{film} is the wavelength in the film. This phase change occurs at the top surface of the film, where the light first strikes it. However, no phase change occurs when light that has penetrated the film reflects back upward from the bottom surface. This is because a phase change does not occur when light travels through a material with a larger refractive index toward a material with a smaller refractive index and reflects from the boundary between the two. Thus, to

evaluate destructive interference correctly, we must consider a net phase change of $\frac{1}{2}\lambda_{film}$ due to reflection as well as the extra distance traveled by the light within the film.

SOLUTION Figure 27.42 shows the soap film and the two rays of light that represent the interfering light waves. At nearly perpendicular incidence, ray 2 travels a distance of $2t$ further than ray 1, where t is the thickness of the film. In addition, the net phase change for the two rays is $\frac{1}{2}\lambda_{film}$, as discussed in the reasoning section. We must combine this amount with the extra travel distance to determine the condition for destructive interference. For destructive interference, the combined total must be an odd-integer number of half-wavelengths in the film:

$$\underbrace{2t}_{\substack{\text{Extra distance}\\\text{traveled by}\\\text{ray 2}}} + \underbrace{\tfrac{1}{2}\lambda_{film}}_{\substack{\text{Half wavelength}\\\text{net phase change}\\\text{due to reflection}}} = \underbrace{\tfrac{1}{2}\lambda_{film}, \tfrac{3}{2}\lambda_{film}, \tfrac{5}{2}\lambda_{film}, \dots}_{\substack{\text{Condition for}\\\text{destructive}\\\text{interference}}}$$

Subtracting the term $\frac{1}{2}\lambda_{film}$ from the left side of this equation and from each term on the right side, we obtain

$$2t = 0, \lambda_{film}, 2\lambda_{film}, \dots$$

The minimum nonzero thickness is $t = \lambda_{film}/2$. But the wavelength in the film is related to the vacuum-wavelength according to Equation 27.3: $\lambda_{film} = \lambda_{vacuum}/n$. Thus, the minimum nonzero thickness is $t = \lambda_{vacuum}/(2n)$. Applying this result to both regions of the film allows us to obtain the desired ratio:

$$\frac{t_{magenta}}{t_{yellow}} = \frac{\lambda_{vacuum, green}/(2n)}{\lambda_{vacuum, blue}/(2n)} = \frac{\lambda_{vacuum, green}}{\lambda_{vacuum, blue}} = \frac{555 \text{ nm}}{469 \text{ nm}} = \boxed{1.18}$$

---

15. **SSM WWW REASONING** To solve this problem, we must express the condition for constructive interference in terms of the film thickness t and the wavelength λ_{film} of the light in the soap film. We must also take into account any phase changes that occur upon reflection.

SOLUTION For the reflection at the top film surface, the light travels from air, where the refractive index is smaller ($n = 1.00$), toward the film, where the refractive index is larger ($n = 1.33$). Associated with this reflection there is a phase change that is equivalent to one-half of a wavelength. For the reflection at the bottom film surface, the light travels from the film, where the refractive index is larger ($n = 1.33$), toward air, where the refractive index is smaller ($n = 1.00$). Associated with this reflection, there is no phase change. As a result of these two reflections, there is a net phase change that is equivalent to one-half of a wavelength. To obtain the condition for constructive interference, this net phase change must be added to the phase change that arises because of the film thickness t, which is traversed twice by the light that penetrates it. For constructive interference we find that

$$2t + \tfrac{1}{2}\lambda_{film} = \lambda_{film}, 2\lambda_{film}, 3\lambda_{film}, \ldots$$

or

$$2t = \left(m + \tfrac{1}{2}\right)\lambda_{film}, \quad \text{where } m = 0, 1, 2, \ldots$$

Equation 27.3 indicates that $\lambda_{film} = \lambda_{vacuum}/n$. Using this expression and the fact that $m = 0$ for the minimum thickness t, we find that the condition for constructive interference becomes

$$2t = \left(m + \tfrac{1}{2}\right)\lambda_{film} = \left(0 + \tfrac{1}{2}\right)\left(\frac{\lambda_{vacuum}}{n}\right)$$

or

$$t = \frac{\lambda_{vacuum}}{4n} = \frac{611 \text{ nm}}{4(1.33)} = \boxed{115 \text{ nm}}$$

---

16. **REASONING** The number m of bright fringes in an air wedge is discussed in Example 5, where it is shown that m is related to the thickness t of the film and the wavelength λ_{film} of the light within the film by (see Figure 27.13)

$$\underbrace{2t}_{\substack{\text{Extra distance} \\ \text{traveled by} \\ \text{wave 2}}} + \underbrace{\tfrac{1}{2}\lambda_{film}}_{\substack{\text{Half-wavelength} \\ \text{shift due to reflection} \\ \text{of wave 1}}} = m\lambda_{film} \quad \text{where} \quad m = 1, 2, 3, \ldots$$

For a given thickness, we can solve this equation for the number of bright fringes.

SOLUTION Since the light is traveling in a film of air, $\lambda_{film} = \lambda_{vacuum} = 550$ nm. The number m of bright fringes is

$$m = \frac{2t}{\lambda_{vacuum}} + \frac{1}{2} = \frac{2(1.37 \times 10^{-5} \text{ m})}{550 \times 10^{-9} \text{ m}} + \frac{1}{2} = 50.3$$

Thus, there are $\boxed{50 \text{ bright fringes}}$.

---

17. **REASONING AND SOLUTION** The condition for destructive interference is $2t = \lambda_{film} = \lambda_{vacuum}/n$, so that

$$nt = \lambda_{vacuum}/2 = (660 \text{ nm})/2 = 330 \text{ nm}$$

The condition for constructive interference is $2t = (m + 1/2)\lambda_{film} = (m + 1/2)(\lambda_{vacuum}/n)$, so

1002 INTERFERENCE AND THE WAVE NATURE OF LIGHT

$$\lambda_{vacuum} = \frac{2nt}{m+\frac{1}{2}} = \frac{2(330 \text{ nm})}{m+\frac{1}{2}}$$

For $m = 0$, $\lambda_{vacuum} = 1320$ nm, for $m = 1$, $\lambda_{vacuum} = 440$ nm, and for $m = 2$, $\lambda_{vacuum} = 264$ nm.

The only visible wavelength that will give constructive interference is $\boxed{440 \text{ nm}}$.

---

18. **REASONING AND SOLUTION**
    a. The condition for constructive interference is

    $$2t = m\lambda_{film} = m(\lambda_{vacuum}/n_{water})$$

    For $\lambda_{vacuum} = 432$ nm,

    $$t = \frac{m}{2}\left(\frac{432 \text{ nm}}{1.33}\right)$$

    For $\lambda_{vacuum} = 648$ nm,

    $$t = \frac{m'}{2}\left(\frac{648 \text{ nm}}{1.33}\right)$$

    Equating the two expressions for t, we find that $m/m' = 1.50$. For minimum thickness, this means that $m = 3$ and $m' = 2$. Then

    $$t = \frac{m}{2}\left(\frac{432 \text{ mn}}{1.33}\right) = \frac{3}{2}\left(\frac{432 \text{ nm}}{1.33}\right) = \boxed{487 \text{ nm}}$$

    b. The condition for destructive interference is $2t = (m + 1/2)(\lambda_{vacuum}/n_{water})$. Thus,

    $$\lambda_{vacuum} = \frac{2n_{water}t}{m+\frac{1}{2}} = \frac{2(1.33)(487 \times 10^{-9} \text{ m})}{m+\frac{1}{2}}$$

    The wavelength lying within the specified range is for $m = 2$, so that

    $$\lambda_{vacuum} = \boxed{518 \text{ nm}}$$

---

19. **SSM** *REASONING* This problem can be solved by using Equation 27.4 for the value of the angle θ when $m = 1$ (first dark fringe).

SOLUTION

a. When the slit width is $W = 1.8 \times 10^{-4}$ m and $\lambda = 675$ nm $= 675 \times 10^{-9}$ m, we find, according to Equation 27.4,

$$\theta = \sin^{-1}\left(m\frac{\lambda}{W}\right) = \sin^{-1}\left[(1)\frac{675 \times 10^{-9} \text{ m}}{1.8 \times 10^{-4} \text{ m}}\right] = \boxed{0.21°}$$

b. Similarly, when the slit width is $W = 1.8 \times 10^{-6}$ m and $\lambda = 675 \times 10^{-9}$ m, we find

$$\theta = \sin^{-1}\left[(1)\frac{675 \times 10^{-9} \text{ m}}{1.8 \times 10^{-6} \text{ m}}\right] = \boxed{22°}$$

---

20. **REASONING AND SOLUTION**

a. We know that $\theta = \sin^{-1}(\lambda/W)$ for the first ($m = 1$) dark fringe. Therefore,

$$\theta = \sin^{-1}\left(\frac{660 \times 10^{-9} \text{ m}}{0.91 \text{ m}}\right) = \boxed{4.2 \times 10^{-5} \text{ degrees}}$$

b. The wavelength of the sound wave is $\lambda = v/f = (343 \text{ m/s})/(440 \text{ Hz}) = 0.78$ m. For the sound wave,

$$\theta = \sin^{-1}\left(\frac{0.78 \text{ m}}{0.91 \text{ m}}\right) = \boxed{59°}$$

---

21. **REASONING AND SOLUTION** We have $\theta = \sin^{-1}(2\lambda/W)$ for the second dark fringe.

a. So

$$\theta = \sin^{-1}\left[\frac{2(430 \times 10^{-9} \text{ m})}{2.1 \times 10^{-6} \text{ m}}\right] = \boxed{24°}$$

b. And

$$\theta = \sin^{-1}\left[\frac{2(660 \times 10^{-9} \text{ m})}{2.1 \times 10^{-6} \text{ m}}\right] = \boxed{39°}$$

---

22. **REASONING AND SOLUTION** For the first-order dark fringe in a single-slit diffraction pattern, Equation 27.4 with $m = 1$ indicates that $\sin\theta = \lambda/W$, or $W = \lambda/\sin\theta$. Using this result, we find that

$$\frac{W_A}{W_B} = \frac{\lambda/\sin\theta_A}{\lambda/\sin\theta_B} = \frac{\sin\theta_B}{\sin\theta_A} = \frac{\sin 56°}{\sin 34°} = \boxed{1.5}$$

1004 INTERFERENCE AND THE WAVE NATURE OF LIGHT

23. **SSM WWW** ***REASONING*** According to Equation 27.4, the angles at which the dark fringes occur are given by $\sin\theta = m\lambda/W$, where W is the slit width. In Figure 27.24, we see from the trigonometry of the situation that $\tan\theta = y/L$. Therefore, the latter expression can be used to determine the angle θ, and Equation 27.4 can be used to find the wavelength λ.

SOLUTION The angle θ is given by

$$\theta = \tan^{-1}\left(\frac{y}{L}\right) = \tan^{-1}\left(\frac{3.5 \times 10^{-3}\text{ m}}{4.0\text{ m}}\right) = 0.050°$$

The wavelength of the light is

$$\lambda = W\sin\theta = (5.6 \times 10^{-4}\text{ m})\sin 0.050° = 4.9 \times 10^{-7}\text{ m} = \boxed{490\text{ nm}}$$

---

24. ***REASONING AND SOLUTION*** The width of the bright fringe that is next to the central bright fringe is $y_2 - y_1$, where y_1 is the distance from the center of the central bright fringe to the first-order dark fringe, and y_2 is the analogous distance for the second-order dark fringe. To find the fringe width, we follow the approach outlined in Example 6 in the text. According to Equation 27.4, the dark fringes for single slit diffraction are located by $\sin\theta = m\lambda/W$. For first-order fringes, $m = 1$, and $\sin\theta_1 = \lambda/W$. For second-order fringes, $m = 2$, and $\sin\theta_2 = 2\lambda/W$. Therefore, we find that

$$\theta_1 = \sin^{-1}\left(\frac{\lambda}{W}\right) = \sin^{-1}\left(\frac{480 \times 10^{-9}\text{ m}}{2.0 \times 10^{-5}\text{ m}}\right) = 1.4°$$

$$\theta_2 = \sin^{-1}\left(\frac{2\lambda}{W}\right) = \sin^{-1}\left[\frac{2(480 \times 10^{-9}\text{ m})}{2.0 \times 10^{-5}\text{ m}}\right] = 2.8°$$

From Example 6 we know that the distance y is given by $y = L\tan\theta$, where L is the distance between the slit and the screen. Therefore, it follows that

$$y_2 - y_1 = L(\tan\theta_2 - \tan\theta_1) = (0.50\text{ m})(\tan 2.8° - \tan 1.4°) = \boxed{0.012\text{ m}}$$

---

25. **SSM** ***REASONING AND SOLUTION*** Using Equation 27.4 for the first-order dark fringes ($m = 1$) and referring to Figure 27.24 in the text, we see that

$$\sin\theta = (1)\frac{\lambda}{W} = \frac{y}{\sqrt{L^2 + y^2}}$$

Since the distance L between the slit and the screen equals the width $2y$ of the central bright fringe, this equation becomes

$$\frac{\lambda}{W} = \frac{y}{\sqrt{(2y)^2 + y^2}} = \frac{1}{\sqrt{5}} = \boxed{0.447}$$

---

26. **REASONING** The width of the central bright fringe is defined by the location of the first dark fringe on either side of it. According to Equation 27.4, the angle θ locating the first dark fringe can be obtained from $\sin\theta = \lambda/W$, where λ is the wavelength of the light and W is the width of the slit. According to the drawing, $\tan\theta = y/L$, where y is half the width of the central bright fringe and L is the distance between the slit and the screen.

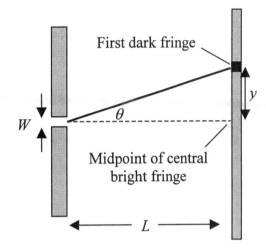

SOLUTION Since the angle θ is small, we can use the fact that $\sin\theta \approx \tan\theta$. Since $\sin\theta = \lambda/W$ and $\tan\theta = y/L$, we have

$$\frac{\lambda}{W} = \frac{y}{L} \quad \text{or} \quad W = \frac{\lambda L}{y}$$

Applying this result to both slits gives

$$\frac{W_2}{W_1} = \frac{\lambda L / y_2}{\lambda L / y_1} = \frac{y_1}{y_2}$$

$$W_2 = W_1 \frac{y_1}{y_2} = (3.2 \times 10^{-5} \text{ m}) \frac{\frac{1}{2}(1.2 \text{ cm})}{\frac{1}{2}(1.9 \text{ cm})} = \boxed{2.0 \times 10^{-5} \text{ m}}$$

---

27. **REASONING AND SOLUTION** It is given that $2y = 450\,W$ and $L = 18\,000\,W$. We know $\lambda/W = \sin\theta$. Now $\sin\theta \approx \tan\theta = y/L$, so

$$\frac{\lambda}{W} = \frac{y}{L} = \frac{225\,W}{18\,000\,W} = \boxed{0.013}$$

28. **REASONING AND SOLUTION** The minimum angular separation of the cars must be $\theta_{min} = 1.22\, \lambda/D$, and the separation of the cars is $y = L\theta_{min} = 1.22\, L\lambda/D$.

a. For red light, $\lambda = 665$ nm, and

$$y = (1.22)(8690 \text{ m})(665 \times 10^{-9} \text{ m})/(2.00 \times 10^{-3} \text{ m}) = \boxed{3.53 \text{ m}}$$

b. For violet light, $\lambda = 405$ nm, and

$$y = (1.22)(8690 \text{ m})(405 \times 10^{-9} \text{ m})/(2.00 \times 10^{-3} \text{ m}) = \boxed{2.15 \text{ m}}$$

29. **REASONING AND SOLUTION**
a. The minimum angular separation of the points must be $\theta_{min} = 1.22\, \lambda/D$, where $\lambda = 550$ nm. Thus,

$$\theta_{min} = \frac{1.22\lambda}{D} = \frac{1.22(550 \times 10^{-9} \text{ m})}{2.0 \times 10^{-3} \text{ m}} = 3.4 \times 10^{-4} \text{ rad}$$

The distance at which the points can be resolved is

$$L = \frac{y}{\theta_{min}} = \frac{0.0738 \text{ m}}{3.4 \times 10^{-4} \text{ rad}} = \boxed{220 \text{ m}}$$

b. $\boxed{\text{No}}$

30. **REASONING** The drawing shows the two asteroids that are a distance r from the Hubble Space Telescope and separated by a distance s. If s is much smaller than r, the angle θ (measured in radians) is given by $\theta = s/r$. The smallest angle that can just be resolved by the telescope is given by Equation 27.6 as

$$\theta_{min} = 1.22 \frac{\lambda}{D}$$

where λ is the wavelength of the light and D is the diameter of the telescope's aperture. These two relations can be used to find the separation distance s.

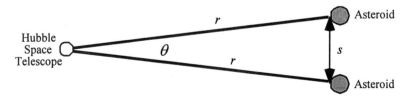

SOLUTION The distance between the asteroids is

$$s = r\theta = r\left(\frac{1.22\lambda}{D}\right) = (2.0 \times 10^{10} \text{ m})\left[\frac{1.22(550 \times 10^{-9} \text{ m})}{2.4 \text{ m}}\right] = \boxed{5.6 \times 10^{3} \text{ m}}$$

---

31. **SSM** **REASONING AND SOLUTION** According to Rayleigh's criterion (Equation 27.6), the minimum angular separation of the two objects is

$$\theta_{min} \text{ (in rad)} \approx 1.22\frac{\lambda}{D} = 1.22\left(\frac{565 \times 10^{-9} \text{ m}}{1.02 \text{ m}}\right) = 6.76 \times 10^{-7} \text{ rad}$$

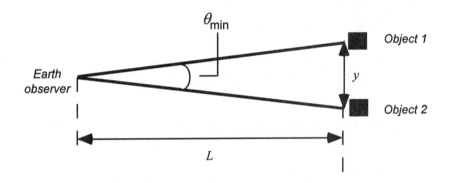

Therefore, from the figure above, the separation y of the two objects is

$$y = L\theta_{min} = (3.75 \times 10^{4} \text{ m})(6.76 \times 10^{-7} \text{ rad}) = \boxed{0.0254 \text{ m}}$$

---

32. **REASONING** At the aperture the star and its planet must subtend an angle at least as large as that given by the Rayleigh criterion, which is $\theta_{min} \approx 1.22\ \lambda/D$, where λ is the wavelength of the light and D is the diameter of the aperture. The angle θ_{min} is given by this criterion in radians. We can obtain the angle subtended at the telescope aperture by using the separation s between the planet and the star and the distance L of the star from the earth. According to Equation 8.1, the angle in radians is $\theta_{min} = s/L$.

SOLUTION Using the Rayleigh criterion and Equation 8.1, we have

$$\frac{s}{L} \approx 1.22\frac{\lambda}{D} \quad \text{or} \quad D \approx \frac{1.22\ \lambda L}{s} = \frac{1.22(550 \times 10^{-9} \text{ m})(4.2 \times 10^{17} \text{ m})}{1.2 \times 10^{11} \text{ m}} = \boxed{2.3 \text{ m}}$$

1008 INTERFERENCE AND THE WAVE NATURE OF LIGHT

33. **REASONING AND SOLUTION**
  a. Equation 27.6 ($\theta_{min} = 1.22 \lambda / D$) gives the minimum angle θ_{min} that two point objects can subtend at an aperture of diameter D and still be resolved. The angle must be measured in radians. For a dot separation s and a distance L between the painting and the eye, Equation 8.1 gives the angle in radians as $\theta_{min} = s/L$. Therefore, we find that

$$\frac{1.22 \lambda}{D} = \frac{s}{L} \quad \text{or} \quad L = \frac{sD}{1.22 \lambda}$$

With $\lambda = 550$ nm and a pupil diameter of $D = 2.5$ mm, the distance L is

$$L = \frac{sD}{1.22 \lambda} = \frac{(1.5 \times 10^{-3} \text{ m})(2.5 \times 10^{-3} \text{ m})}{(1.22)(550 \times 10^{-9} \text{ m})} = \boxed{5.6 \text{ m}}$$

  b. The calculation here is similar to that in part a, except that $n = 1.00$ and $D = 25$ mm for the camera. Therefore, the distance L for the camera is

$$L = \frac{sD}{1.22 \lambda} = \frac{(1.5 \times 10^{-3} \text{ m})(25 \times 10^{-3} \text{ m})}{(1.22)(550 \times 10^{-9} \text{ m})} = \boxed{56 \text{ m}}$$

---

34. **REASONING AND SOLUTION** The minimum angular separation of the mice is $\theta_{min} = 1.22 \lambda/D$, where $\lambda = 550$ nm. Thus,

$$\theta_{min} = 1.22 \lambda/D = 1.22(550 \times 10^{-9} \text{ m})/(6.0 \times 10^{-3} \text{ m}) = 1.12 \times 10^{-4} \text{ rad}$$

The distance at which the mice can just be resolved as separate objects is

$$y = \frac{0.010 \text{ m}}{\tan \theta_{min}} = \frac{0.010 \text{ m}}{1.12 \times 10^{-4}} = 89 \text{ m}$$

So the eagle moves a distance of $d = 176$ m $- 89$ m $= 87$ m at a speed of 17 m/s. Therefore, the amount of time which passes is

$$t = \frac{d}{v} = \frac{87 \text{ m}}{17 \text{ m/s}} = \boxed{5.1 \text{ s}}$$

---

35. [SSM] [WWW] **REASONING** When light of wavelength λ passes through a circular opening of diameter D, a circular diffraction pattern results, and Equation 27.5 [$\sin \theta = 1.22(\lambda/D)$] locates the first circular dark fringe relative to the central bright spot. Therefore, Equation 27.5 can be used to find the angular separation θ of the central bright spot and the first dark fringe. The geometry of the situation can be used to find the diameter of the central bright spot on the moon.

SOLUTION From Equation 27.5,
$$\theta = \sin^{-1}\left(\frac{1.22\lambda}{D}\right) = \sin^{-1}\left[\frac{1.22(694.3 \times 10^{-9}\ \text{m})}{0.20\ \text{m}}\right] = 4.2 \times 10^{-6}\ \text{rad}$$

From the figure below, we see that $2\theta = \dfrac{d}{R}$.

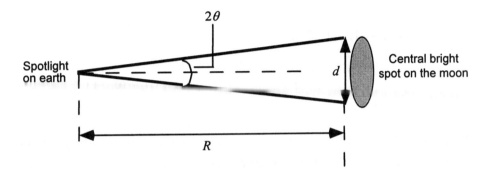

Therefore, the diameter of the spot on the moon is

$$d = 2\theta R = 2(4.2 \times 10^{-6}\ \text{rad})(3.77 \times 10^{8}\ \text{m}) = \boxed{3.2 \times 10^{3}\ \text{m}}$$

---

36. *REASONING AND SOLUTION*

a. Equation 27.6 $(\theta_{\min} = 1.22\,\lambda/D)$ gives the minimum angle $\theta_{\min}$ that two point objects can subtend at an aperture of diameter D and still be resolved. The angle must be measured in radians. For a separation s between the two circles and a distance L between the concentric arrangement and the camera, Equation 8.1 gives the angle in radians as $\theta_{\min} = s/L$. Therefore, we find that

$$\frac{1.22\,\lambda}{D} = \frac{s}{L} \quad \text{or} \quad L = \frac{sD}{1.22\,\lambda}$$

Since $s = 0.040\ \text{m} - 0.010\ \text{m} = 0.030\ \text{m}$, we calculate that

$$L = \frac{sD}{1.22\,\lambda} = \frac{(0.030\ \text{m})(12.5 \times 10^{-3}\ \text{m})}{(1.22)(555 \times 10^{-9}\ \text{m})} = \boxed{550\ \text{m}}$$

b. The calculation here is similar to that in part a, except that the separation s is between one side of a diameter of the small circle and the other side, or $s = 0.020\ \text{m}$:

$$L = \frac{sD}{1.22\,\lambda} = \frac{(0.020\ \text{m})(12.5 \times 10^{-3}\ \text{m})}{(1.22)(555 \times 10^{-9}\ \text{m})} = \boxed{370\ \text{m}}$$

37. **SSM** *REASONING AND SOLUTION* According to Equation 27.7, the angles that correspond to the first-order ($m = 1$) maximum for the two wavelengths in question are:

a. for $\lambda = 660$ nm $= 660 \times 10^{-9}$ m,

$$\theta = \sin^{-1}\left(m\frac{\lambda}{d}\right) = \sin^{-1}\left[(1)\left(\frac{660 \times 10^{-9} \text{ m}}{1.1 \times 10^{-6} \text{ m}}\right)\right] = \boxed{37°}$$

b. for $\lambda = 410$ nm $= 410 \times 10^{-9}$ m,

$$\theta = \sin^{-1}\left(m\frac{\lambda}{d}\right) = \sin^{-1}\left[(1)\left(\frac{410 \times 10^{-9} \text{ m}}{1.1 \times 10^{-6} \text{ m}}\right)\right] = \boxed{22°}$$

---

38. *REASONING* For a diffraction grating the angle θ that locates a bright fringe can be found using Equation 27.7, $\sin\theta = m\lambda/d$, where λ is the wavelength, d is the separation between the grating slits, and the order m is $m = 0, 1, 2, 3, \ldots$. By applying this relation to both cases, we will be able to determine the unknown wavelength, because the order and the slit separation are the same for both.

SOLUTION We will use λ_1 and λ_2 to denote the known and unknown wavelengths, respectively. The corresponding angles are θ_1 and θ_2. Applying Equation 27.7 to the two cases, we have

$$\sin\theta_1 = \frac{m\lambda_1}{d} \quad \text{and} \quad \sin\theta_2 = \frac{m\lambda_2}{d}$$

Dividing the expression for case 2 by the expression for case 1 gives

$$\frac{\sin\theta_2}{\sin\theta_1} = \frac{m\lambda_2/d}{m\lambda_1/d} = \frac{\lambda_2}{\lambda_1}$$

Solving for λ_2, we find

$$\lambda_2 = \lambda_1 \frac{\sin\theta_2}{\sin\theta_1} = (420 \text{ nm})\frac{\sin 41°}{\sin 26°} = \boxed{630 \text{ nm}}$$

---

39. *REASONING AND SOLUTION* We know that

$$\tan\theta = y/L = (8.94 \times 10^{-2} \text{ m})/(0.625 \text{ m}) \quad \text{so} \quad \theta = 8.14°$$

Therefore,

$$\lambda = d\sin\theta = (4.17 \times 10^{-6} \text{ m})\sin 8.14° = \boxed{5.90 \times 10^{-7} \text{ m}}$$

40. **REASONING AND SOLUTION** We have

$$d = \frac{3\lambda}{\sin\theta} = \frac{3(621 \times 10^{-9}\text{ m})}{\sin 18.0°} = 6.03 \times 10^{-6}\text{ m}$$

The number of lines per centimeter is

$$1/d = 1/(6.03 \times 10^{-4}\text{ cm}) = \boxed{1660 \text{ lines/cm}}$$

41. [SSM] **REASONING AND SOLUTION** The geometry of the situation is shown below.

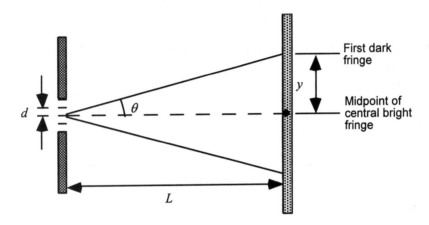

From the geometry, we have

$$\tan\theta = \frac{y}{L} = \frac{0.60\text{ mm}}{3.0\text{ mm}} = 0.20 \quad \text{or} \quad \theta = 11.3°$$

Then, solving Equation 27.7 with $m = 1$ for the separation d between the slits, we have

$$d = \frac{m\lambda}{\sin\theta} = \frac{(1)(780 \times 10^{-9}\text{ m})}{\sin 11.3°} = \boxed{4.0 \times 10^{-6}\text{ m}}$$

42. **REASONING AND SOLUTION** For the first order, $\lambda/d = \sin 16.0° = 0.2756$. For the third order

$$\theta = \sin^{-1}\left(3\frac{\lambda}{d}\right) = \sin^{-1}[3(0.2756)] = \boxed{55.8°}$$

43. **SSM REASONING** The angle θ that locates the first-order maximum produced by a grating with 3300 lines/cm is given by Equation 27.7, $\sin\theta = m\lambda/d$, with the order of the fringes given by $m = 0, 1, 2, 3, \ldots$ Any two of the diffraction patterns will overlap when their angular positions are the same.

SOLUTION Since the grating has 3300 lines/cm, we have

$$d = \frac{1}{3300 \text{ lines/cm}} = 3.0\times10^{-4} \text{ cm} = 3.0\times10^{-6} \text{ m}$$

a. In first order, $m = 1$; therefore, for violet light,

$$\theta = \sin^{-1}\left(m\frac{\lambda}{d}\right) = \sin^{-1}\left[(1)\left(\frac{410\times10^{-9} \text{ m}}{3.0\times10^{-6} \text{ m}}\right)\right] = \boxed{7.9°}$$

Similarly for red light,

$$\theta = \sin^{-1}\left(m\frac{\lambda}{d}\right) = \sin^{-1}\left[(1)\left(\frac{660\times10^{-9} \text{ m}}{3.0\times10^{-6} \text{ m}}\right)\right] = \boxed{13°}$$

b. Repeating the calculation for the second order maximum ($m = 2$), we find that

($m = 2$)	
for violet	$\theta = 16°$
for red	$\theta = 26°$

c. Repeating the calculation for the third order maximum ($m = 3$), we find that

($m = 3$)	
for violet	$\theta = 24°$
for red	$\theta = 41°$

d. Comparisons of the values for θ calculated in parts (a), (b) and (c) show that the $\boxed{\text{second and third orders overlap}}$.

---

44. **REASONING** The maximum number of bright fringes that can be seen on either side of the central maximum occurs when the angle is $\theta = 90.0°$. Equation 27.7 gives the relationship between the angle and the wavelength as

$$\underbrace{\sin 90.0°}_{=1} = m\frac{\lambda}{d}$$

Because there are only three bright fringes, $m = 3$. We can use the equation above to find the separation d between the slits, and then the number of lines per centimeter for the grating.

SOLUTION The separation between the slits is

$$d = \frac{m\lambda}{\sin\theta} = \frac{3(510 \times 10^{-9} \text{ m})}{\sin 90.0°} = 1.53 \times 10^{-6} \text{ m}$$

The number of lines per centimeter is

$$\frac{1}{d} = \frac{1}{1.53 \times 10^{-4} \text{ cm}} = \boxed{6540 \text{ lines/cm}}$$

---

45. **REASONING AND SOLUTION**
a. The angular positions of the specified orders are equal, so $\lambda/d_A = 2\lambda/d_B$, or

$$\frac{d_B}{d_A} = \boxed{2}$$

b. Similarly, we have for the m_A order of grating A and the m_B order of grating B that $m_A\lambda/d_A = m_B\lambda/d_B$, so $m_A = m_B/2$.

The next highest orders which overlap are

$$m_B = \boxed{4}, \; m_A = \boxed{2} \quad \text{and} \quad m_B = \boxed{6}, \; m_A = \boxed{3}$$

---

46. **REASONING AND SOLUTION** The interference minima are given by Equation 27.2 with $m = 0$, since we are asked for the smallest angle:

$$\theta = \sin^{-1}[\lambda/(2d)], \quad \text{where} \quad \lambda = v/f = (343 \text{ m/s})/(80.0 \text{ Hz}) = 4.29 \text{ m}$$

Now

$$\theta = \sin^{-1}[(4.29 \text{ m})/(14.0 \text{ m})] = \boxed{17.8°}$$

---

47. **SSM REASONING** The slit separation d is given by Equation 27.1 with $m = 1$; namely $d = \lambda/\sin\theta$. As shown in Example 1 in the text, the angle θ is given by $\theta = \tan^{-1}(y/L)$.

SOLUTION The angle θ is

$$\theta = \tan^{-1}\left(\frac{0.037 \text{ m}}{4.5 \text{ m}}\right) = 0.47°$$

Therefore, the slit width d is

$$d = \frac{\lambda}{\sin\theta} = \frac{490 \times 10^{-9} \text{ m}}{\sin 0.47°} = \boxed{6.0 \times 10^{-5} \text{ m}}$$

1014 INTERFERENCE AND THE WAVE NATURE OF LIGHT

48. **REASONING AND SOLUTION** The position of the first minimum is $y = L \tan \theta$ (see Example 6), so the width of the central maximum is $2y = 2L \tan \theta$. We know that $\theta = \sin^{-1}(\lambda/W)$. For $\lambda = 635$ nm and $W = 4.30 \times 10^{-5}$ m, $\theta = 0.846°$ and

$$2y = 2(1.32 \text{ m})(\tan 0.846°) = \boxed{0.0390 \text{ m}}$$

49. **REASONING AND SOLUTION** This problem is analogous to Example 3 in the text, where the condition for destructive interference is obtained as $2t = m\lambda_{film}$. For the minimum non-zero thickness, $m = 1$. In addition, the wavelength in the film is given by Equation 27.3: $\lambda_{film} = \lambda_{vacuum}/n$. Therefore, green light is removed when the minimum non-zero thickness is

$$t = \frac{m}{2}\lambda_{film} = \frac{1}{2}\left(\frac{\lambda_{vacuum}}{n}\right) = \frac{1}{2}\left(\frac{551 \text{ nm}}{1.33}\right) = \boxed{207 \text{ nm}}$$

50. **REASONING AND SOLUTION** The position of the third dark fringe ($m = 2$) is given by $\sin \theta = \left(m + \tfrac{1}{2}\right)\frac{\lambda}{d} = \left(2 + \tfrac{1}{2}\right)\frac{\lambda}{d}$. The position of the fourth bright fringe ($m = 4$) is given by $\sin \theta = m\frac{\lambda'}{d} = 4\frac{\lambda'}{d}$. Equating these two expressions and solving for λ' yields

$$\lambda' = \frac{\left(2 + \tfrac{1}{2}\right)\lambda}{4} = \frac{\left(2 + \tfrac{1}{2}\right)(645 \text{ nm})}{4} = \boxed{403 \text{ nm}}$$

51. **SSM** **REASONING** According to Rayleigh's criterion, the two taillights must be separated by a distance s sufficient to subtend an angle $\theta_{min} \approx 1.22\lambda/D$ at the pupil of the observer's eye. Recalling that this angle must be expressed in radians, we relate θ_{min} to the distances s and L.

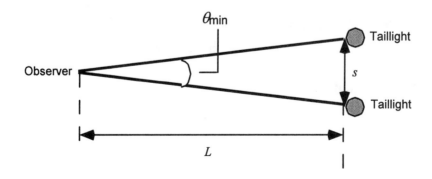

SOLUTION The wavelength λ is 660 nm. Therefore, we have from Equation 27.6

$$\theta_{min} \approx 1.22 \frac{\lambda}{D} = 1.22\left(\frac{660 \times 10^{-9} \text{ m}}{7.0 \times 10^{-3} \text{ m}}\right) = 1.2 \times 10^{-4} \text{ rad}$$

According to Equation 8.1, the distance L between the observer and the taillights is

$$L = \frac{s}{\theta_{min}} = \frac{1.2 \text{ m}}{1.2 \times 10^{-4} \text{ rad}} = \boxed{1.0 \times 10^4 \text{ m}}$$

---

52. **REASONING AND SOLUTION** Equation 27.7 gives the angles that define the positions of the principal maxima of a diffraction grating in air as $\sin\theta = m\lambda_{air}/d$. But, the wavelength in air is related to that in a vacuum by $\lambda_{air} = \lambda_{vacuum}/n_{air} = \lambda_{vacuum}/1$. Thus, we have

$$\sin 33° = \frac{m\lambda_{vacuum}}{d} \tag{1}$$

When the diffraction grating is immersed in water, the analogous relation is

$$\sin\theta_{water} = \frac{m\lambda_{water}}{d} = \frac{m\left(\frac{\lambda_{vacuum}}{n_{water}}\right)}{d} = \frac{m\left(\frac{\lambda_{vacuum}}{1.33}\right)}{d} \tag{2}$$

Dividing Equation (2) by Equation (1) and algebraically simplifying, we get

$$\frac{\sin\theta_{water}}{\sin 33°} = \frac{1}{1.33}$$

Thus,

$$\sin\theta_{water} = \frac{\sin 33°}{1.33} \quad \text{or} \quad \boxed{\theta_{water} = 24°}$$

---

53. **REASONING AND SOLUTION**
a. The minimum angular separation of the cells must be $\theta_{min} = 1.22\ \lambda/D$. The separation of the cells is

$$y = L\theta_{min} = 1.22\ L\lambda/D$$

Now $D = f$, the focal length of the lens. Also, $L = f$, so

$$\boxed{y = 1.22\lambda}$$

b. The wavelength must be $\boxed{\text{shorter}}$ to resolve cells that are closer together.

54. REASONING AND SOLUTION

The position of a particular wavelength on the screen is

$$y = L \tan \theta \quad \text{where} \quad \theta = \sin^{-1}(m\lambda/d)$$

a. In the first order ($m = 1$), $\theta = \sin^{-1}(\lambda/d)$. For $\lambda = 410$ nm,

$$\theta = \sin^{-1}(410 \times 10^{-9} \text{ m})/(2.2 \times 10^{-6} \text{ m}) = 11°$$

and

$$y = (3.2 \text{ m}) \tan 11° = 0.62 \text{ m}$$

For $\lambda = 660$ nm, we find similarly that $\theta = 17°$ and $y = 0.98$ m.

The width of the spectrum is 0.98 m – 0.62 m = $\boxed{0.36 \text{ m}}$

b. In the second order ($m = 2$), $\theta = \sin^{-1}(2\lambda/d)$. For $\lambda = 410$ nm, we find that $\theta = 22°$ and $y = 1.3$ m.

For $\lambda = 660$ nm, we find that $\theta = 37°$ and $y = 2.4$ m.

The width of the spectrum is 2.4 m – 1.3 m = $\boxed{1.1 \text{ m}}$

55. [SSM] REASONING

For a diffraction grating, the angular position θ of a principal maximum on the screen is given by Equation 27.7 as $\sin\theta = m\lambda/d$ with $m = 0, 1, 2, 3, \ldots$

SOLUTION When the fourth-order principal maximum of light A exactly overlaps the third-order principal maximum of light B, we have

$$\sin\theta_A = \sin\theta_B$$

$$\frac{4\lambda_A}{d} = \frac{3\lambda_B}{d} \quad \text{or} \quad \frac{\lambda_A}{\lambda_B} = \boxed{\frac{3}{4}}$$

56. REASONING AND SOLUTION

a. The condition for constructive interference is $2t = \frac{1}{2}\lambda_{film} = \frac{1}{2}\left(\frac{\lambda_{vacuum}}{n_{gasoline}}\right)$. Therefore,

$$t = \frac{\lambda_{vacuum}}{4n_{gasoline}} = \frac{580 \times 10^{-9} \text{ m}}{4(1.40)} = \boxed{1.0 \times 10^{-7} \text{ m}}$$

b. Now, the condition for constructive interference is $2t = \lambda_{film} = \lambda_{vacuum}/n_{gasoline}$, so

$$t = \frac{\lambda_{vacuum}}{2n_{gasoline}} = \frac{580 \times 10^{-9} \text{ m}}{2(1.40)} = \boxed{2.1 \times 10^{-7} \text{ m}}$$

---

57. **REASONING AND SOLUTION** The last maximum formed by the grating corresponds to $\theta = 90.0°$, so that

$$m = \frac{d}{\lambda} \sin 90.0° = \frac{d}{\lambda} = \frac{1.78 \times 10^{-6} \text{ m}}{471 \times 10^{-9} \text{ m}} = 3.78$$

where we have used $d = \frac{1}{5620 \text{ lines/cm}} = 1.78 \times 10^{-4} \text{ cm} = 1.78 \times 10^{-6} \text{ m}$.

Thus, the last maximum formed by the grating is $m = 3$. This maximum lies at

$$\theta = \sin^{-1}\left(m\frac{\lambda}{d}\right) = \sin^{-1}\left[3\left(\frac{1}{3.78}\right)\right] = 52.5°$$

The distance from the center of the screen to the $m = 3$ maximum is

$$y = (0.750 \text{ m}) \tan 52.5° = 0.977 \text{ m}$$

The screen must have a width of $2y = \boxed{1.95 \text{ m}}$

---

58. **REASONING AND SOLUTION** The condition for destructive interference is $2t = m\lambda$. The thickness t of the "air wedge" at the 100th dark fringe is related to the radius R of the fringe and the radius of curvature r of the lens in the following way.

If s is the length of the circular arc along the lens from the center to the fringe, then the angle subtended by this arc is $\theta = s/r$. Since r is large, the arc is almost straight and is the hypotenuse of a right triangle in the air wedge of sides t and R, and angle $(1/2)\theta$ opposite t.

In the right triangle
$$(1/2)\theta \approx \tan(1/2)\theta = t/R \quad \text{so} \quad R = 2tr/s$$

Now if $(1/2)\theta$ is very small, then $s \approx R$, so

$$R^2 = 2tr = (100)\lambda r = (100)(654 \times 10^{-9} \text{ m})(10.0 \text{ m})$$

so
$$\boxed{R = 0.0256 \text{ m}}$$

59. **CONCEPT QUESTIONS** a. The loudspeakers are in-phase sources of identical sound waves. The waves from one speaker travel a distance ℓ_1 in reaching point A and the waves from the second speaker travel a distance ℓ_2. The condition that leads to constructive interference is $\ell_2 - \ell_1 = m\lambda$, where λ is the wavelength of the waves and $m = 0, 1, 2, 3 \ldots$ In other words, the two distances are the same or differ by an integer number of wavelengths. Point A is the midpoint of a side of the square, so that the distances ℓ_1 and ℓ_2 are the same, and constructive interference occurs.

b. As you walk toward the corner, the waves from one of the sources travels a greater distance in reaching you than does the other wave. This difference in the distances increases until it reaches one half of a wavelength, at which spot destructive interference occurs and you hear no sound. As you walk on, the difference in distances continues to increase, and you gradually hear a louder and louder sound. Ultimately, at the corner, the difference in distances becomes one wavelength, constructive interference occurs, and you hear a maximally loud sound.

c. The general condition that leads to constructive interference is $\ell_2 - \ell_1 = m\lambda$, where $m = 0, 1, 2, 3 \ldots$ There are many possible values for m, and we are being asked what the specific value is. At point A the value is $m = 0$. As you walk from point A toward either empty corner, the maximal loudness that indicates constructive interference does not occur again until you arrive at the corner. Thus, the next possibility for m applies at the corner; in other words, $m = 1$.

SOLUTION Consider the constructive interference that occurs at either empty corner. Using L to denote the length of a side of the square and taking advantage of the Pythagorean theorem, we have

$$\ell_1 = L \quad \text{and} \quad \ell_2 = \sqrt{L^2 + L^2} = \sqrt{2}\, L$$

The specific condition for the constructive interference at the empty corner is

$$\ell_2 - \ell_1 = \sqrt{2}\, L - L = \lambda$$

Solving for the wavelength of the waves gives

$$\lambda = L(\sqrt{2} - 1) = (4.6 \text{ m})(\sqrt{2} - 1) = \boxed{1.9 \text{ m}}$$

60. **CONCEPT QUESTIONS** a. Each bright fringe is located by an angle θ that is determined by Equation 27.1, $\sin \theta = m\lambda/d$, where λ is the wavelength, d is the separation between the slits, and $m = 0, 1, 2, 3, \ldots$. Thus, the angular position of each fringe does not depend on where the screen is located. In Figure 27.8, for instance, the $m = 3$ bright fringe has a fixed angular position as the screen is moved. To ensure that this fringe falls off the screen, the

screen must be moved so that the line corresponding to its angular position will not intersect the screen. This means moving the screen to the right.

b. Considering the discussion in Question (a), fewer bright fringes will lie on the screen when the distance L in Figure 27.8 is larger.

SOLUTION The drawing shows the double slit and the screen, along with the angle θ that locates the upper edge of the screen relative to its midpoint. y denotes one half of the screen width. For version A of the setup, the drawing indicates that

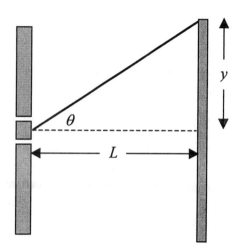

$$\tan\theta = \frac{y}{L} \quad \text{or} \quad \theta = \tan^{-1}\left(\frac{y}{L}\right) = \tan^{-1}\left(\frac{0.10 \text{ m}}{0.35 \text{ m}}\right) = 16°$$

But Equation 27.1 indicates that $\sin\theta = m\lambda/d$. Solving this expression for m and using $\theta = 16°$, we find

$$m = \frac{d\sin\theta}{\lambda} = \frac{(1.4\times10^{-5}\text{ m})\sin 16°}{625\times10^{-9}\text{ m}} = 6.2$$

Thus, for version A, there are $\boxed{6}$ bright fringes on the screen, not counting the central bright fringe. For version B, we find

$$\theta = \tan^{-1}\left(\frac{y}{L}\right) = \tan^{-1}\left(\frac{0.10 \text{ m}}{0.50 \text{ m}}\right) = 11°$$

$$m = \frac{d\sin\theta}{\lambda} = \frac{(1.4\times10^{-5}\text{ m})\sin 11°}{625\times10^{-9}\text{ m}} = 4.3$$

For this version, there are $\boxed{4}$ bright fringes on the screen, again not counting the central bright fringe.

---

61. *CONCEPT QUESTIONS* a. In air the index of refraction is nearly $n = 1$, while in the film it is $n = 1.33$. A phase change occurs whenever light travels through a material with a smaller refractive index toward a material with a larger refractive index and reflects from the boundary between the two. The phase change is equivalent to $\frac{1}{2}\lambda_{\text{film}}$, where λ_{film} is the wavelength in the film.

b. In the film the index of refraction is $n = 1.33$, while in the glass it is $n = 1.52$. This situation is like that in Question (a) and, once again, the phase change is equivalent to $\frac{1}{2}\lambda_{film}$, where λ_{film} is the wavelength in the film.

c. The wavelength of the light in the film (refractive index = n) is given by Equation 27.3 as $\lambda_{film} = \lambda_{vacuum}/n$. Since the refractive index of the film is $n = 1.33$, the wavelength in the film is less than the wavelength in a vacuum.

SOLUTION Both the light reflected from the air-film interface and from the film-glass interface experience phase changes, each of which is equivalent to $\frac{1}{2}\lambda_{film}$. In other words, there is no net phase change between the waves reflected from the two interfaces, and only the extra travel distance of the light within the film leads to the destructive interference. The extra distance is $2t$, where t is the film thickness. The condition for destructive interference in this case is

$$\underbrace{2t}_{\text{Extra travel distance in film}} = \underbrace{\tfrac{1}{2}\lambda_{film}, \tfrac{3}{2}\lambda_{film}, \tfrac{5}{2}\lambda_{film}, \ldots = \left(m+\tfrac{1}{2}\right)\lambda_{film}}_{\text{Condition for destructive interference}} \quad \text{where } m = 0, 1, 2, 3, \ldots$$

Solving for the wavelength gives

$$\lambda_{film} = \frac{2t}{m+\tfrac{1}{2}} \qquad m = 0, 1, 2, 3, \ldots$$

According to Equation 27.3, we have $\lambda_{film} = \lambda_{vacuum}/n$. Using this substitution in our result for λ_{film}, we obtain

$$\lambda_{vacuum} = \frac{2nt}{m+\tfrac{1}{2}} \qquad m = 0, 1, 2, 3, \ldots$$

For the first four values of m and the given values for n and t, we find

$m = 0$ $\lambda_{vacuum} = \dfrac{2nt}{m+\tfrac{1}{2}} = \dfrac{2(1.33)(465 \text{ nm})}{0+\tfrac{1}{2}} = 2470$ nm

$m = 1$ $\lambda_{vacuum} = 825$ nm

$m = 2$ $\lambda_{vacuum} = 495$ nm

$m = 3$ $\lambda_{vacuum} = 353$ nm

The range of visible wavelengths (in vacuum) extends from 380 to 750 nm. Therefore, the only visible wavelength in which the film appears dark due to destructive interference is $\boxed{495 \text{ nm}}$.

62. **CONCEPT QUESTIONS** a. Since the width of the central bright fringe on the screen is defined by the locations of the dark fringes on either side, we consider Equation 27.4, which specifies the angle θ that determines the location of a dark fringe. This equation is $\sin\theta = m\lambda/W$, where λ is the wavelength, W is the slit width, and $m = 1, 2, 3, \ldots$. For a given slit width, the angle θ increases as the wavelength increases. This means that the dark fringe on either side of the central bright fringe moves further out from the center. Thus, the width of the central bright fringe increases with increasing wavelength.

b. For a given wavelength, the angle θ decreases as the slit width increases. This means that the dark fringe on either side of the central bright fringe moves more toward the center. Thus, the width of the central bright fringe decreases with increasing slit width.

c. The fact that the width of the central bright fringe does not change means that the positions of the dark fringes to either side also do not change. In other words, the angle θ remains constant. According to Equation 27.4, the condition that must be satisfied for this to happen is that the ratio λ/W of the wavelength to the slit width remains constant.

SOLUTION Since the width of the central bright fringe on the screen remains constant, the angular position θ of the dark fringes must also remain constant. Thus, according to Equation 27.4, we have

$$\underbrace{\sin\theta = \frac{m\lambda_1}{W_1}}_{\text{Case 1}} \quad \text{and} \quad \underbrace{\sin\theta = \frac{m\lambda_2}{W_2}}_{\text{Case 2}}$$

Since θ is the same for each case, it follows that

$$\frac{m\lambda_1}{W_1} = \frac{m\lambda_2}{W_2}$$

The term m can be eliminated algebraically from this result, so that solving for W_2 gives

$$W_2 = \frac{W_1 \lambda_2}{\lambda_1} = \frac{(2.3\times 10^{-6}\text{ m})(740\text{ nm})}{510\text{ nm}} = \boxed{3.3\times 10^{-6}\text{ m}}$$

63. **CONCEPT QUESTIONS** a. The Rayleigh criterion is $\theta_{\min} \approx 1.22\,\lambda/D$, where λ is the wavelength of the light and D is the diameter of the pupil in this case. The largest value for $\theta_{\min}$ corresponds to the largest wavelength, which is the wavelength of the red light.

b. The smallest value for $\theta_{\min}$ corresponds to the smallest wavelength, which is the wavelength of the blue light.

c. In the Rayleigh criterion the angle θ_{min} is expressed in radians. According to Equation 8.1, the angle in radians is $\theta_{min} = s/L$.

d. The maximum allowable dot separation should be chosen so that neither the red, the green, nor the blue dots can be resolved separately. This means that the distance should be determined by the blue dots. A separation that is slightly smaller than s_{blue} will automatically be smaller than s_{red} and s_{green}. This is because the θ_{min} values for red and green light are larger than for blue light. In other words, if the blue dots can't be resolved separately, then neither can the red dots nor the green dots.

SOLUTION Using the Rayleigh criterion and Equation 8.1, we have

$$\frac{s}{L} \approx 1.22 \frac{\lambda}{D} \quad \text{or} \quad s \approx \frac{1.22 \lambda L}{D}$$

The maximum allowable dot separation, then, is

$$s_{blue} \approx \frac{1.22 \lambda L}{D} = \frac{1.22(470 \times 10^{-9} \text{ m})(0.40 \text{ m})}{2.0 \times 10^{-3} \text{ m}} = \boxed{1.1 \times 10^{-4} \text{ m}}$$

64. **CONCEPT QUESTIONS** a. Since the principal maxima are closer together for grating A, the light spreads out less as it passes through grating A. Thus, grating B diffracts the light to a greater extent than grating A.

b. For a diffraction grating, the angle θ that locates a principal maximum can be found using Equation 27.7, $\sin \theta = m\lambda/d$, where λ is the wavelength, d is the separation between the grating slits, and the order m is $m = 0, 1, 2, 3, \ldots$. Greater diffraction means that the angle θ is greater. In turn, greater values for θ correspond to smaller values for d. Therefore, since greater diffraction occurs for grating B, it also has the smaller slit separation.

c. The number of lines per meter is the reciprocal of the slit separation d. Therefore, a smaller value for d means that the number of lines per meter is greater. As a result, grating B has the greater number of lines per meter.

SOLUTION The drawing shows the angle θ that locates a principal maximum on the screen, along with the separation L between the grating and the screen and the distance y from the midpoint of the screen. It follows from the drawing that $y = L \tan \theta$, which becomes $y = L \sin \theta$, since we are dealing with small angles. But $\sin \theta = m\lambda/d$, according to Equation 27.7, so that our expression for y becomes

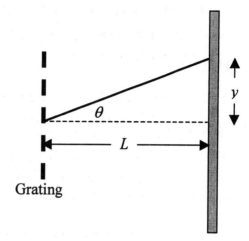

$$y = \frac{Lm\lambda}{d}$$

The separation between adjacent principal maxima is

$$\Delta y = \frac{L(m+1)\lambda}{d} - \frac{Lm\lambda}{d} = \frac{L\lambda}{d}$$

But the number of lines per meter N is $1/d$, so that this result becomes $\Delta y = L\lambda N$. Applying this expression to gratings A and B gives

$$\Delta y_A = L\lambda N_A \quad \text{and} \quad \Delta y_B = L\lambda N_B$$

Forming the ratio of these two expressions, we have

$$\frac{\Delta y_B}{\Delta y_A} = \frac{L\lambda N_B}{L\lambda N_A} = \frac{N_B}{N_A}$$

$$N_B = \frac{N_A \Delta y_B}{\Delta y_A} = \frac{(2000 \text{ m}^{-1})(3.2 \text{ cm})}{2.7 \text{ cm}} = \boxed{2400 \text{ m}^{-1}}$$

CHAPTER 28 | SPECIAL RELATIVITY

PROBLEMS

1. **SSM REASONING** Since the "police car" is moving relative to the earth observer, the earth observer measures a greater time interval Δt between flashes. Since both the proper time Δt_0 (as observed by the officer) and the dilated time Δt (as observed by the person on earth) are known, the speed of the "police car" relative to the observer can be determined from the time dilation relation, Equation 28.1.

   SOLUTION According to Equation 28.1, the dilated time interval between flashes is $\Delta t = \Delta t_0 / \sqrt{1-(v^2/c^2)}$, where Δt_0 is the proper time. Solving for the speed v, we find

   $$v = c\sqrt{1-\left(\frac{\Delta t_0}{\Delta t}\right)^2} = (3.0 \times 10^8 \text{ m/s})\sqrt{1-\left(\frac{1.5 \text{ s}}{2.5 \text{ s}}\right)^2} = \boxed{2.4 \times 10^8 \text{ m/s}}$$

2. **REASONING** The total time for the trip is one year. This time is the proper time interval Δt_0, because it is measured by an observer (the astronaut) who is at rest relative to the beginning and ending events (the times when the trip started and ended) and who sees them at the same location in spacecraft. On the other hand, the astronaut measures the clocks on earth to run at the dilated time interval Δt, which is the time interval of one hundred years. The relation between the two time intervals is given by Equation 28.1, which can be used to find the speed of the spacecraft.

   SOLUTION The dilated time interval Δt is related to the proper time interval Δt_0 by $\Delta t = \Delta t_0 / \sqrt{1-(v^2/c^2)}$. Solving this equation for the speed v of the spacecraft yields

   $$v = c\sqrt{1-\left(\frac{\Delta t_0}{\Delta t}\right)^2} = c\sqrt{1-\left(\frac{1 \text{ yr}}{100 \text{ yr}}\right)^2} = \boxed{0.999\,95c} \qquad (28.1)$$

3. **SSM REASONING AND SOLUTION** The period Δt_0 of the rotating radar antenna is given by Equation 10.4 as

   $$\Delta t_0 = \frac{2\pi}{\omega_0} = \frac{2\pi}{0.25 \text{ rad/s}} = 25 \text{ s}$$

This period, or time interval, is a proper time interval, because an earth-based observer measures the beginning and ending of each revolution at the same place. An observer moving relative to the earth measures a dilated time interval given by Equation 28.1 as

$$\Delta t = \frac{\Delta t_0}{\sqrt{1 - \frac{v^2}{c^2}}} = \frac{25 \text{ s}}{\sqrt{1 - \frac{(0.80 \, c)^2}{c^2}}} = 42 \text{ s}$$

The angular speed ω of the rotating antenna as measured by the moving observer is, therefore,

$$\omega = \frac{2\pi}{\Delta t} = \frac{2\pi}{42 \text{ s}} = \boxed{0.15 \text{ rad/s}}$$

---

4. **REASONING AND SOLUTION** The time interval for 8.0 breaths in your rest frame is $\Delta t_0 = 1.0$ min. The same time interval as measured by someone on the earth is given by Equation 28.1:

$$\Delta t = \frac{\Delta t_0}{\sqrt{1 - v^2/c^2}} = \frac{1.0 \text{ min}}{\sqrt{1 - (0.975c/c)^2}} = \frac{1.0 \text{ min}}{\sqrt{1 - (0.975)^2}} = 4.5 \text{ min}$$

Therefore, your breathing rate as measured by an earth-based observer is

$$\frac{8.0 \text{ breaths}}{4.5 \text{ min}} = \boxed{1.8 \text{ breaths per min}}$$

---

5. **REASONING AND SOLUTION** According to Equations 10.4 and 10.11, the period Δt_0 of oscillation on earth is

$$\Delta t_0 = 2\pi \sqrt{\frac{m}{k}} = 2\pi \sqrt{\frac{5.00 \text{ kg}}{49.3 \text{ N/m}}} = 2.00 \text{ s}$$

Equation 28.1 can be used to find the dilated period Δt measured by the moving observer

$$\Delta t = \frac{\Delta t_0}{\sqrt{1 - \left(\frac{v}{c}\right)^2}} = \frac{2.00 \text{ s}}{\sqrt{1 - \left(\frac{2.80 \times 10^8 \text{ m/s}}{3.00 \times 10^8 \text{ m/s}}\right)^2}} = \boxed{5.57 \text{ s}}$$

1026 SPECIAL RELATIVITY

6. **REASONING AND SOLUTION** Using Equation 28.1 with

$$\sqrt{1-\left(\frac{v}{c}\right)^2} \approx 1-\frac{1}{2}\left(\frac{v}{c}\right)^2$$

we have

$$\Delta t_0 = \Delta t \left[1-\frac{1}{2}\left(\frac{v}{c}\right)^2\right]$$

The difference between Δt and Δt_0 is

$$\text{Time difference} = \Delta t - \Delta t_0 = \Delta t \left(\frac{1}{2}\right)\left(\frac{v}{c}\right)^2$$

$$\text{Time difference} = (15 \text{ days})\left(\frac{8.64 \times 10^4 \text{ s}}{1 \text{ day}}\right)\left(\frac{1}{2}\right)\left(\frac{7800 \text{ m/s}}{3.00 \times 10^8 \text{ m/s}}\right)^2 = \boxed{4.4 \times 10^{-4} \text{ s}}$$

---

7. **REASONING AND SOLUTION** The proper time is the time it takes for the bacteria to double its number, i.e., $\Delta t_0 = 24.0$ hours. For the earth based sample to grow to 256 bacteria, it would take 8 days ($2^n = 256$ or $n = 8$). The "doubling time" for the space culture would be

$$\Delta t = \frac{\Delta t_0}{\sqrt{1-\left(\frac{v}{c}\right)^2}} = \frac{24.0 \text{ h}}{\sqrt{1-\left(\frac{0.866\,c}{c}\right)^2}} = 48.0 \text{ h} \quad \text{or} \quad 2 \text{ days}$$

In eight earth days, the space bacteria would undergo $n' = \left(\frac{1}{2}\right)8 = 4$ "doublings". The number of space bacteria is

$$\text{Number of Space Bacteria} = 2^{n'} = 2^4 = \boxed{16}$$

---

8. **REASONING AND SOLUTION** The proper length is $L_0 = 4.2 \times 10^6$ m, so the length measured by the UFO voyagers traveling at $v = 0.70c$ is, using Equation 28.2,

$$L = L_0\sqrt{1-\left(\frac{v}{c}\right)^2} = (4.2 \times 10^6 \text{ m})\sqrt{1-\left(\frac{0.70\,c}{c}\right)^2} = \boxed{3.0 \times 10^6 \text{ m}}$$

9. **SSM** ***REASONING AND SOLUTION*** The length L_0 that the person measures for the UFO when it lands is the proper length, since the UFO is at rest with respect to the person. Therefore, from Equation 28.2 we have

$$L_0 = \frac{L}{\sqrt{1-\frac{v^2}{c^2}}} = \frac{230 \text{ m}}{\sqrt{1-\frac{(0.90\,c)^2}{c^2}}} = \boxed{530 \text{ m}}$$

10. ***REASONING*** The distance between earth and the center of the galaxy is the proper length L_0, because it is the distance measured by an observer who is at rest relative to the earth and the center of the galaxy. A person on board the spaceship is moving with respect to them and measures a contracted length L that is related to the proper length by Equation 28.2 as $L = L_0\sqrt{1-(v^2/c^2)}$. The contracted distance is also equal to the product of the spaceship's speed v the time interval measured by a person on board the spaceship. This time interval is the proper time interval Δt_0 because the person on board the spaceship measures the beginning and ending events (the times when the trip starts and ends) at the same location relative to a coordinate system fixed to the spaceship. Thus, the contracted distance is also $L = v\Delta t_0$. By setting the two expressions for L equal to each other, we can find the how long the trip will take according to a clock on board the spaceship.

SOLUTION Setting $L = L_0\sqrt{1-(v^2/c^2)}$ equal to $L = v\Delta t_0$ and solving for the proper time interval Δt_0 gives

$$\Delta t_0 = \frac{L_0}{v}\sqrt{1-(v^2/c^2)}$$

$$= \frac{(23\,000 \text{ ly})\left(\frac{9.47 \times 10^{15} \text{ m}}{1 \text{ ly}}\right)}{0.9990\,(3.00 \times 10^8 \text{ m/s})}\sqrt{1-\left[\frac{(0.9990c)^2}{c^2}\right]}\left(\frac{1 \text{ yr}}{3.16 \times 10^7 \text{ s}}\right) = \boxed{1.0 \times 10^3 \text{ yr}}$$

11. **SSM** **WWW** ***REASONING AND SOLUTION*** The diameter D of the planet, as measured by a moving spacecraft, is given in terms of the proper diameter D_0 by Equation 28.2. Taking the ratio of the diameter D_A of the planet measured by spaceship A to the diameter D_B measured by spaceship B, we find

$$\frac{D_A}{D_B} = \frac{D_0\sqrt{1-\frac{v_A^2}{c^2}}}{D_0\sqrt{1-\frac{v_B^2}{c^2}}} = \frac{\sqrt{1-\frac{(0.60\,c)^2}{c^2}}}{\sqrt{1-\frac{(0.80\,c)^2}{c^2}}} = \boxed{1.3}$$

---

12. **REASONING AND SOLUTION** Since the spacecraft is moving parallel to the length of the space pad, only the length, but not the width, of the pad is contracted. Taking $L = 1800$ m in Equation 28.2 we can solve for the proper length, i.e.,

$$L_0 = \frac{L}{\sqrt{1-\left(\frac{v}{c}\right)^2}} = \frac{1800\text{ m}}{\sqrt{1-\left(\frac{0.85\,c}{c}\right)^2}} = 3400\text{ m}$$

The dimensions of the landing pad, according to the engineer who built it, are

$$\text{Dimensions} = \boxed{3400\text{ m} \times 1500\text{ m}}$$

---

13. **REASONING AND SOLUTION** Since the proper length is the same in either case, we have $(L_0)_A = (L_0)_B$, or using Equation 28.2,

$$L_A \Big/ \sqrt{1-v_A^2/c^2} = L_B \Big/ \sqrt{1-v_B^2/c^2}$$

Therefore,

$$L_B = L_A \sqrt{\frac{1-v_B^2/c^2}{1-v_A^2/c^2}} = L_A\sqrt{\frac{1-(0.80\,c/c)^2}{1-(0.60\,c/c)^2}}$$

or

$$L_B = (8.0\text{ light years})\sqrt{\frac{1-(0.80)^2}{1-(0.60)^2}} = \boxed{6.0\text{ light years}}$$

---

14. **REASONING AND SOLUTION** The side x will have its length contracted as viewed from the rocket. Taking the proper length to be $L_0 = x$, we can find the dilated length $L = x'$ from Equation 28.2:

$$x' = x\sqrt{1-\left(\frac{v}{c}\right)^2} = x\sqrt{1-\left(\frac{0.730\,c}{c}\right)^2} = 0.683\,x$$

The new angle, can be found using $\tan \theta' = y/x' = y/(0.683\ x) = \tan 30.0°/(0.683)$,

$$\boxed{\theta' = 40.2°}$$

---

15. **SSM WWW** *REASONING* Only the sides of the rectangle that lie in the direction of motion will experience length contraction. In order to make the rectangle look like a square, each side must have a length of $L = 2.0$ m. Thus, we move along the long side, taking the proper length to be $L_0 = 3.0$ m. We can solve for the speed using Equation 28.2. Then, with this speed, we can use the relation for length contraction to find L for the short side as we move along it.

SOLUTION From Equation 28.2, $L = L_0 \sqrt{1 - (v^2/c^2)}$, we find that

$$v = c\sqrt{1 - \left(\frac{L}{L_0}\right)^2} = c\sqrt{1 - \left(\frac{2.0\ \text{m}}{3.0\ \text{m}}\right)^2} = 0.75\ c$$

Moving at this speed along the short side, we take $L_0 = 2.0$ m and find L:

$$L = L_0\sqrt{1 - \left(\frac{v}{c}\right)^2} = (2.0\ \text{m})\sqrt{1 - \left(\frac{0.75\ c}{c}\right)^2} = 1.3\ \text{m}$$

The observed dimensions of the rectangle are, therefore, $\boxed{3.0\ \text{m} \times 1.3\ \text{m}}$, since the long side is not contracted due to motion along the short side.

---

16. *REASONING AND SOLUTION*
a. The proper distance is $L_0 = 12.0$ light years (12.0 ly). First we determine the distance traveled as viewed by each twin:

[Twin A] $\quad L_A = L_0\sqrt{1 - \left(\frac{v_A}{c}\right)^2} = (12.0\ \text{ly})\sqrt{1 - \left(\frac{0.900c}{c}\right)^2} = 5.23\ \text{ly}$

[Twin B] $\quad L_B = L_0\sqrt{1 - \left(\frac{v_B}{c}\right)^2} = (12.0\ \text{ly})\sqrt{1 - \left(\frac{0.500c}{c}\right)^2} = 10.4\ \text{ly}$

Then, we determine the time to reach the planet for each twin:

[Twin A] $t_A = L_A/v_A = (5.23 \text{ ly})/(0.900c) = 5.81$ years

[Twin B] $t_B = L_B/v_B = (10.4 \text{ ly})/(0.500c) = 20.8$ years

Next, we determine the age of each twin when each arrives at the planet:

[Twin A] 19.0 years + 5.81 years = 24.8 years

[Twin B] 19.0 years + 20.8 years = 39.8 years

However, twin A has to wait for twin B to arrive. As seen from earth,

$$\Delta t_A = (12.0 \text{ ly})/(0.900c) = 13.3 \text{ y}$$
$$\Delta t_B = (12.0 \text{ ly})/(0.500c) = 24.0 \text{ y}$$

Therefore, twin A must wait another 24.0 y − 13.3 y = 10.7 y for twin B to arrive. When twin B arrives, twin A has an age of 24.8 y + 10.7 y = 35.5 y. The difference between their ages when they meet is, then,

$$\text{Age difference} = 39.8 \text{ y} - 35.5 \text{ y} = \boxed{4.3 \text{ y}}$$

b. Thus, $\boxed{\text{twin B, traveling at } 0.500\,c,}$ is older.

---

17. **REASONING AND SOLUTION** The magnitude of the nonrelativistic momentum of a particle is given by Equation 7.2 as $p_0 = mv$. The magnitude of relativistic momentum of a particle is given by Equation 28.3:

$$p = \frac{mv}{\sqrt{1-v^2/c^2}} = \frac{p_0}{\sqrt{1-v^2/c^2}}$$

When the magnitude of the relativistic momentum of a particle is three times the magnitude of its nonrelativistic momentum, we have

$$3p_0 = \frac{p_0}{\sqrt{1-v^2/c^2}} \quad \text{or} \quad \sqrt{1-v^2/c^2} = \frac{1}{3}$$

Squaring and rearranging the last expression gives

$$1 - \frac{v^2}{c^2} = \frac{1}{9} \quad \text{or} \quad \frac{v^2}{c^2} = 1 - \frac{1}{9} = \frac{8}{9}$$

Taking the square root, we find

$$v = \sqrt{\tfrac{8}{9}}\, c = 0.943c = \boxed{2.83 \times 10^8 \text{ m/s}}$$

---

18. **REASONING AND SOLUTION** According to Equation 28.3, the magnitude of the momentum is

$$p = \frac{mv}{\sqrt{1-\left(\dfrac{v}{c}\right)^2}} = \frac{(1550 \text{ kg})(15.0 \text{ m/s})}{\sqrt{1-\left(\dfrac{15.0 \text{ m/s}}{25.0 \text{ m/s}}\right)^2}} = \boxed{2.91 \times 10^4 \text{ kg} \cdot \text{m/s}}$$

---

19. **SSM REASONING AND SOLUTION** The magnitude of the relativistic momentum p of the rocket is given by Equation 28.3, $p = mv/\sqrt{1-(v^2/c^2)}$. Solving this expression for the speed v of the rocket gives

$$v = \frac{p}{\sqrt{m^2 + \dfrac{p^2}{c^2}}} = \frac{3.15 \times 10^{13} \text{ kg} \cdot \text{m/s}}{\sqrt{(1.40 \times 10^5 \text{ kg})^2 + \dfrac{(3.15 \times 10^{13} \text{ kg} \cdot \text{m/s})^2}{(3.00 \times 10^8 \text{ m/s})^2}}} = \boxed{1.80 \times 10^8 \text{ m/s}}$$

---

20. **REASONING** The height of the woman as measured by the observer is given by Equation 28.2 as $h = h_0 \sqrt{1-(v/c)^2}$, where h_0 is her proper height. In order to use this equation, we must determine the speed v of the woman relative to the observer. We are given the magnitude of her relativistic momentum, so we can determine v from p.

SOLUTION According to Equation 28.3 $p = mv/\sqrt{1-v^2/c^2}$, so $mv = p\sqrt{1-v^2/c^2}$
Squaring both sides, we have

$$m^2 v^2 = p^2(1 - v^2/c^2) \qquad \text{or} \qquad m^2 v^2 + p^2 \frac{v^2}{c^2} = p^2$$

$$v^2 \left(m^2 + \frac{p^2}{c^2} \right) = p^2 \qquad \text{or} \qquad v^2 = \frac{p^2}{m^2 + \dfrac{p^2}{c^2}}$$

Solving for v and substituting values, we have

$$v = \frac{p}{\sqrt{m^2 + \frac{p^2}{c^2}}} = \frac{2.0 \times 10^{10} \text{ kg} \cdot \text{m/s}}{\sqrt{(55 \text{ kg})^2 + \left(\frac{2.0 \times 10^{10} \text{ kg} \cdot \text{m/s}}{3.00 \times 10^8 \text{ m/s}}\right)^2}} = 2.3 \times 10^8 \text{ m/s}$$

Then, the height that the observer measures for the woman is

$$h = h_0 \sqrt{1 - \left(\frac{v}{c}\right)^2} = (1.6 \text{ m}) \sqrt{1 - \left(\frac{2.3 \times 10^8 \text{ m/s}}{3.0 \times 10^8 \text{ m/s}}\right)^2} = \boxed{1.0 \text{ m}}$$

---

21. **REASONING AND SOLUTION** The total momentum of the man/woman system is conserved, since no net external force acts on the system. Therefore, the final total momentum $p_m + p_w$ must equal the initial total momentum, which is zero. As a result, $p_m = -p_w$ where Equation 28.3 must be used for the momenta p_m and p_w. Thus, we find

$$\frac{m_m v_m}{\sqrt{1 - (v_m/c)^2}} = -\frac{m_w v_w}{\sqrt{1 - (v_w/c)^2}}$$

We know that $m_m = 88$ kg, $m_w = 54$ kg, and $v_w = +2.5$ m/s. Remember that c has the hypothetical value of 3.0 m/s. Solving for v_m, we find $\boxed{v_m = -2.0 \text{ m/s}}$.

---

22. **REASONING AND SOLUTION** The mass equivalent is given by $E_0 = KE = mc^2$ or

$$m = \frac{KE}{c^2} = \frac{7.8 \times 10^{-13} \text{ J}}{(3.00 \times 10^8 \text{ m/s})^2} = \boxed{8.7 \times 10^{-30} \text{ kg}} \quad (28.5)$$

---

23. **SSM REASONING** According to the work-energy theorem, Equation 6.3, the work that must be done on the electron to accelerate it from rest to a speed of $0.990c$ is equal to the kinetic energy of the electron when it is moving at $0.990c$.

SOLUTION Using Equation 28.6, we find that

$$KE = mc^2\left(\frac{1}{\sqrt{1-(v^2/c^2)}} - 1\right)$$

$$= (9.11\times 10^{-31}\text{ kg})(3.00\times 10^8\text{ m/s})^2\left(\frac{1}{\sqrt{1-(0.990\,c)^2/c^2}} - 1\right) = \boxed{5.0\times 10^{-13}\text{ J}}$$

---

24. **REASONING** The mass m of the aspirin is related to its rest energy E_0 by Equation 28.5, $E_0 = mc^2$. Since it requires 1.1×10^8 J to operate the car for twenty miles, we can calculate the number of miles that the car can go on the energy that is equivalent to the mass of one tablet.

SOLUTION The number N of miles the car can go on one aspirin tablet is

$$N = \frac{E_0}{(1.1\times 10^8\text{ J})/(20.0\text{ mi})} = \frac{mc^2}{(1.1\times 10^8\text{ J})/(20.0\text{ mi})}$$

$$= \frac{(325\times 10^{-3}\text{ kg})(3.0\times 10^8\text{ m/s})^2}{(1.1\times 10^8\text{ J})/(20.0\text{ mi})} = \boxed{5.3\times 10^9\text{ mi}}$$

---

25. **REASONING AND SOLUTION** The energy E_0 produced in one year is the product of the power P generated and the time t, $E_0 = Pt$. This energy is equivalent to an amount of mass m given by Equation 28.5 as $E_0 = mc^2$. Thus, we have that

$$mc^2 = Pt \quad\text{or}\quad m = \frac{Pt}{c^2}$$

The mass of nuclear fuel consumed in one year (3.15×10^7 s) is

$$m = \frac{Pt}{c^2} = \frac{(3.0\times 10^9\text{ W})(3.15\times 10^7\text{ s})}{(3.00\times 10^8\text{ m/s})^2} = \boxed{1.1\text{ kg}}$$

1034 SPECIAL RELATIVITY

26. **REASONING AND SOLUTION**

a. The energy Q needed to heat the water is

$$Q = cm\,\Delta T = [4186\ \text{J/(kg·C°)}](4.0\ \text{kg})(60.0\ °\text{C} - 20.0\ °\text{C}) = \boxed{6.7 \times 10^5\ \text{J}} \qquad (12.4)$$

b. Using $\Delta E_0 = (\Delta m)c^2$, we have

$$\Delta m = \Delta E_0/c^2 = (6.7 \times 10^5\ \text{J})/(3.00 \times 10^8\ \text{m/s})^2 = \boxed{7.4 \times 10^{-12}\ \text{kg}} \qquad (28.5)$$

---

27. **SSM REASONING AND SOLUTION**

a. In Section 28.6 it is shown that when the speed of a particle is $0.01c$ (or less), the relativistic kinetic energy becomes nearly equal to the nonrelativistic kinetic energy. Since the speed of the particle here is $0.001c$, the ratio of the relativistic kinetic energy to the nonrelativistic kinetic energy is $\boxed{1.0}$.

b. Taking the ratio of the relativistic kinetic energy, Equation 28.6, to the nonrelativistic kinetic energy, $\frac{1}{2}mv^2$, we find that

$$\frac{mc^2\left(\dfrac{1}{\sqrt{1-(v^2/c^2)}} - 1\right)}{\frac{1}{2}mv^2} = 2\left(\frac{c}{v}\right)^2\left(\frac{1}{\sqrt{1-(v^2/c^2)}} - 1\right)$$

$$= 2\left(\frac{c}{0.970\,c}\right)^2\left(\frac{1}{\sqrt{1-(0.970\,c)^2/c^2}} - 1\right) = \boxed{6.6}$$

---

28. **REASONING AND SOLUTION**

a. The kinetic energy is provided by the work done to accelerate the electron. We have

$$KE = W = qV = (1.6 \times 10^{-19}\ \text{C})(2.40 \times 10^7\ \text{V}) = \boxed{3.8 \times 10^{-12}\ \text{J}} \qquad (19.4)$$

b. We need to calculate the speed of the electron. From Equation 28.6 we find that

$$v = c\sqrt{1 - \frac{1}{\left(\dfrac{KE}{mc^2}+1\right)^2}} = \boxed{0.9998c}$$

29. **REASONING** The relativistic momentum is given by Equation 28.3 as $p = mv/\sqrt{1-v^2/c^2}$, while the total energy of an object is given by Equation 28.4 as $E = mc^2/\sqrt{1-v^2/c^2}$. We wish to use these expressions to show that $E^2 = p^2c^2 + (mc^2)^2$.

SOLUTION We begin by squaring both sides of Equation 28.4 to obtain

$$E^2 = \frac{(mc^2)^2}{1-v^2/c^2} = (mc^2)^2 \frac{1}{1-v^2/c^2}$$

We then add zero to the numerator of the right hand side in the form $0 = -v^2/c^2 + v^2/c^2$. We next regroup terms and simplify algebraically:

$$E^2 = (mc^2)^2 \left(\frac{1-v^2/c^2 + v^2/c^2}{1-v^2/c^2}\right) = (mc^2)^2 \left(\frac{1-v^2/c^2}{1-v^2/c^2}\right) + \left[\frac{(mc^2)^2(v^2/c^2)}{1-v^2/c^2}\right]$$

$$E^2 = (mc^2)^2 + \left(\frac{m^2c^2v^2}{1-v^2/c^2}\right) = (mc^2)^2 + \frac{(mv)^2}{1-v^2/c^2}c^2$$

Rearranging and substituting $p = mv/\sqrt{1-v^2/c^2}$ from Equation 28.3 show that this is equal to

$$E^2 = \left(\frac{mv}{\sqrt{1-v^2/c^2}}\right)^2 c^2 + (mc^2)^2 = p^2c^2 + (mc^2)^2$$

Therefore, we have shown that $\boxed{E^2 = p^2c^2 + (mc^2)^2}$.

---

30. **REASONING** Let's define the following relative velocities:

v_{AB} = velocity of galaxy **A** relative to galaxy **B**
v_{AE} = velocity of galaxy **A** relative to Earth
v_{EB} = velocity of Earth relative to galaxy **B**

These three velocities are related by the velocity-addition formula, Equation 28.7:

$$v_{AB} = \frac{v_{AE} + v_{EB}}{1 + \frac{v_{AE}v_{EB}}{c^2}}$$

Let's assume that galaxy A is moving to the right, along the +x direction, so its velocity relative to earth is $v_{AE} = +0.75c$. Galaxy B is moving along the –x axis with a speed of 0.55c relative to earth, so its velocity is $v_{BE} = -0.55c$. According to the velocity-addition formula, we need to know v_{EB}, not v_{BE}. However, they are related by $v_{EB} = -v_{BE}$, so that $v_{EB} = -(-0.55c) = +0.55c$.

SOLUTION The velocity of galaxy A relative to galaxy B is

$$v_{AB} = \frac{v_{AE} + v_{EB}}{1 + \frac{v_{AE}v_{EB}}{c^2}} = \frac{+0.75c + 0.55c}{1 + \frac{(+0.75c)(+0.55c)}{c^2}} = +0.920c$$

The speed of galaxy A relative to galaxy B is $\boxed{0.920c}$.

---

31. **SSM** **REASONING** Let's define the following relative velocities, assuming that the spaceship and exploration vehicle are moving in the positive direction.

v_{ES} = velocity of **E**xploration vehicle relative to the **S**paceship.
v_{EO} = velocity of **E**xploration vehicle relative to an **O**bserver on earth = +0.70c
v_{SO} = velocity of **S**paceship relative to an **O**bserver on earth = +0.50c

The velocity v_{ES} can be determined from the velocity-addition formula, Equation 28.7:

$$v_{ES} = \frac{v_{EO} + v_{OS}}{1 + \frac{v_{EO}v_{OS}}{c^2}}$$

The velocity v_{OS} of the observer on earth relative to the spaceship is not given. However, we know that v_{OS} is the negative of v_{SO}, so $v_{OS} = -v_{SO} = -(+0.50c) = -0.50c$.

SOLUTION The velocity of the exploration vehicle relative to the spaceship is

$$v_{ES} = \frac{v_{EO} + v_{OS}}{1 + \frac{v_{EO}v_{OS}}{c^2}} = \frac{+0.70c + (-0.50c)}{1 + \frac{(+0.70c)(-0.50c)}{c^2}} = \boxed{+0.31c}$$

32. **REASONING AND SOLUTION** Let's define the following relative velocities, assuming that the direction away from the earth is the positive direction:

v_{IE} = velocity of **I**ons relative to **E**arth

v_{IS} = velocity of **I**ons relative to the **S**paceship = $-0.80c$. This velocity is negative because the spaceship is moving away from the earth (in the positive direction), and the ions are emitted from the engine in an opposite direction (the negative direction) relative to the spaceship.

v_{SE} = velocity of **S**paceship relative to **E**arth = $+0.70c$

These velocities are related by the velocity-addition formula, Equation 28.7:

$$v_{IE} = \frac{v_{IS} + v_{SE}}{1 + \frac{v_{IS} v_{SE}}{c^2}} = \frac{-0.80c + 0.70c}{1 + \frac{(-0.80c)(+0.70c)}{c^2}} = \boxed{-0.23c}$$

---

33. **SSM WWW** **REASONING** Since the crew is initially at rest relative to the escape pod, the length of 45 m is the proper length L_0 of the pod. The length of the escape pod as determined by an observer on earth can be obtained from the relation for length contraction given by Equation 28.2, $L = L_0 \sqrt{1 - (v_{PE}^2/c^2)}$. The quantity v_{PE} is the speed of the escape pod relative to the earth, which can be found from the velocity-addition formula, Equation 28.7. The following are the relative velocities, assuming that the direction away from the earth is the positive direction:

v_{PE} = velocity of the escape **P**od relative to **E**arth.

v_{PR} = velocity of escape **P**od relative to the **R**ocket = $-0.55c$. This velocity is negative because the rocket is moving away from the earth (in the positive direction), and the escape pod is moving in an opposite direction (the negative direction) relative to the rocket.

v_{RE} = velocity of **R**ocket relative to **E**arth = $+0.75c$

These velocities are related by the velocity-addition formula, Equation 28.7.

SOLUTION The relative velocity of the escape pod relative to the earth is

$$v_{PE} = \frac{v_{PR} + v_{RE}}{1 + \frac{v_{PR} v_{RE}}{c^2}} = \frac{-0.55c + 0.75c}{1 + \frac{(-0.55c)(+0.75c)}{c^2}} = +0.34c$$

The speed of the pod relative to the earth is $0.34c$. The length of the pod as determined by an observer on earth is

$$L = L_0\sqrt{1 - \frac{v_{PE}^2}{c^2}} = (45 \text{ m})\sqrt{1 - \frac{(0.34c)^2}{c^2}} = \boxed{42 \text{ m}}$$

---

34. **REASONING** We define the following relative velocities, assuming that the rocket approaching the earth from the right is traveling in the positive direction:

v_{RL} = velocity of the **Right** rocket relative to the **L**eft rocket
v_{RE} = velocity of the **R**ight rocket relative to the person on **E**arth = $+0.75c$
v_{LE} = velocity of the **L**eft rocket relative to the person on **E**arth = $-0.65c$

The velocity v_{RL} can be found from the velocity-addition formula, Equation 28.7:

$$v_{RL} = \frac{v_{RE} + v_{EL}}{1 + \frac{v_{RE}v_{EL}}{c^2}}$$

The velocity v_{RE} is given, but v_{EL}, the velocity of the earth relative to the left rocket, is not. However, we know that v_{EL} is the negative of v_{LE}, so $v_{EL} = -v_{LE} = -(-0.65c) = +0.65c$.

SOLUTION The velocity of the right rocket relative to the left rocket is

$$v_{RL} = \frac{v_{RE} + v_{EL}}{1 + \frac{v_{RE}v_{EL}}{c^2}} = \frac{+0.75c + 0.65c}{1 + \frac{(+0.75c)(+0.65c)}{c^2}} = +0.94c$$

The relative speed is $\boxed{0.94c}$.

35. **REASONING AND SOLUTION**

a. The following relative velocities are pertinent to this problem. It is assumed that particle 1 is moving in the positive direction, and, therefore, particle 2 is moving in the negative direction.

$v_{P_1P_2}$ = velocity of particle 1 (P_1) relative to particle 2 (P_2)

v_{P_1L} = velocity of particle 1 (P_1) relative to an observer in the Laboratory
$= +2.10 \times 10^8$ m/s

v_{P_2L} = velocity of particle 2 (P_2) relative to an observer in the Laboratory
$= -2.10 \times 10^8$ m/s

The velocity $v_{P_1P_2}$ can be obtained from the velocity-addition formula, Equation 28.7:

$$v_{P_1P_2} = \frac{v_{P_1L} + v_{LP_2}}{1 + \frac{v_{P_1L} v_{LP_2}}{c^2}}$$

The velocity v_{P_1L} is given, but v_{LP_2}, the velocity of the laboratory observer relative to particle 2, is not. However, we know that v_{LP_2} is the negative of v_{P_2L}, so $v_{LP_2} = -v_{P_2L} = -(-2.10 \times 10^8 \text{ m/s}) = +2.10 \times 10^8$ m/s.

SOLUTION The velocity of the right rocket relative to the left rocket is

$$v_{P_1P_2} = \frac{v_{P_1L} + v_{LP_2}}{1 + \frac{v_{P_1L} v_{LP_2}}{c^2}} = \frac{+2.10 \times 10^8 \text{ m/s} + 2.10 \times 10^8 \text{ m/s}}{1 + \frac{(+2.10 \times 10^8 \text{ m/s})(+2.10 \times 10^8 \text{ m/s})}{c^2}} = \boxed{+2.82 \times 10^8 \text{ m/s}}$$

b. The relativistic momentum is given by Equation 28.3, where the speed is that determined in part a, that is, 0.940c. Therefore,

$$p = \frac{mv_{P_1P_2}}{\sqrt{1 - \left(\frac{v_{P_1P_2}}{c}\right)^2}} = \frac{(2.16 \times 10^{-25} \text{ kg})(+2.82 \times 10^8 \text{ m/s})}{\sqrt{1 - \left(\frac{2.82 \times 10^8 \text{ m/s}}{3.00 \times 10^8 \text{ m/s}}\right)^2}} = \boxed{1.8 \times 10^{-16} \text{ kg} \cdot \text{m/s}}$$

1040 SPECIAL RELATIVITY

36. **REASONING AND SOLUTION** The total energy for each particle is given by

$$E = \frac{mc^2}{\sqrt{1-\left(\frac{v}{c}\right)^2}} = \frac{(9.11\times 10^{-31}\text{ kg})(3.00\times 10^8\text{ m/s})^2}{\sqrt{1-\left(\frac{0.20c}{c}\right)^2}} = 8.4\times 10^{-14}\text{ J} \quad (28.4)$$

The annihilation energy is twice this value, so $E' = 2E = \boxed{1.7\times 10^{-13}\text{ J}}$.

---

37. **SSM WWW REASONING** All standard meter sticks at rest have a length of 1.00 m for observers who are at rest with respect to them. Thus, 1.00 m is the proper length L_0 of the meter stick. When the meter stick moves with speed v relative to an earth-observer, its length $L = 0.500$ m will be a contracted length. Since both L_0 and L are known, v can be found directly from Equation 28.2, $L = L_0\sqrt{1-(v^2/c^2)}$.

SOLUTION Solving Equation 28.2 for v, we find that

$$v = c\sqrt{1-\left(\frac{L}{L_0}\right)^2} = (3.00\times 10^8\text{ m/s})\sqrt{1-\left(\frac{0.500\text{ m}}{1.00\text{ m}}\right)^2} = \boxed{2.60\times 10^8\text{ m/s}}$$

---

38. **REASONING** The expression for time dilation is, according to Equation 28.1,

$$\Delta t = \frac{\Delta t_0}{\sqrt{1-v^2/c^2}}$$

For a given event, it relates the proper time interval Δt_0 to the time interval Δt that would be measured by an observer moving at a speed v relative to the frame of reference in which the event takes place.

We must consider two situations; in the first situation, the Klingon spacecraft has a speed of $0.75c$ with respect to the earth. In the second situation, the craft has a speed of $0.94c$ relative to the earth. We will refer to these two situations as A and B, respectively.

Since the proper time interval always has the same value, $(\Delta t_0)_A = (\Delta t_0)_B$. We can express both sides of this expression using Equation 28.1. The result can be solved for Δt_B.

SOLUTION Use of Equation 28.1 gives

$$\Delta t_A \sqrt{1 - v_A^2/c^2} = \Delta t_B \sqrt{1 - v_B^2/c^2}$$

$$\Delta t_B = \Delta t_A \frac{\sqrt{1 - v_A^2/c^2}}{\sqrt{1 - v_B^2/c^2}} = \Delta t_A \sqrt{\frac{1 - (v_A/c)^2}{1 - (v_B/c)^2}} = (37.0 \text{ h}) \sqrt{\frac{1 - (0.75c/c)^2}{1 - (0.94c/c)^2}} = \boxed{72 \text{ h}}$$

---

39. **REASONING AND SOLUTION**
a. According to Equation 7.2, the magnitude of the classical momentum of the spaceship is

$$p_0 = mv = (2.0 \times 10^7 \text{ kg})(0.85)(3.00 \times 10^8 \text{ m/s}) = \boxed{5.1 \times 10^{15} \text{ kg} \cdot \text{m/s}}$$

b. The magnitude of the relativistic momentum of the ship is given by Equation 28.3:

$$p = \frac{mv}{\sqrt{1 - v^2/c^2}} = \frac{p_0}{\sqrt{(1 - (0.85c/c)^2}} = \frac{5.1 \times 10^{15} \text{ kg} \cdot \text{m/s}}{\sqrt{(1 - (0.85)^2}} = \boxed{9.7 \times 10^{15} \text{ kg} \cdot \text{m/s}}$$

---

40. **REASONING AND SOLUTION**
a. According to the second postulate of special relativity, all observers measure the speed of light to be c, regardless of their velocities relative to each other. Therefore, the aliens aboard the hostile spacecraft see the photons of the laser approach $\boxed{\text{at the speed of light, } c}$.

b. To find the velocity of the ions relative to the aliens, we define the relative velocities as follows. (The direction of the ions and the intergalactic cruiser is taken to be the positive direction.)

v_{IS} = velocity of the **I**ons relative to the alien **S**pacecraft
v_{IC} = velocity of the **I**ons relative to the intergalactic **C**ruiser = $+0.950c$
v_{CS} = velocity of the intergalactic **C**ruiser relative to the alien **S**pacecraft = $+0.800c$

These velocities are related by the velocity-addition formula, Equation 28.7. The velocity of the ions relative to the alien spacecraft is:

$$v_{IS} = \frac{v_{IC} + v_{CS}}{1 + \frac{v_{IC} v_{CS}}{c^2}} = \frac{+0.950c + 0.800c}{1 + \frac{(+0.950c)(+0.800c)}{c^2}} = \boxed{+0.994c}$$

c. The aliens see the laser light (photons) moving with respect to the cruiser at a speed

$$U = 1.000c - 0.800c = \boxed{0.200c}$$

d. The aliens see the ions moving away from the cruiser at speed

$$U' = 0.994c - 0.800c = \boxed{0.194c}$$

---

41. **SSM REASONING** The total linear momentum of the system is conserved, since no net external force acts on the system. Therefore, the final total momentum $p_1 + p_2$ of the two fragments must equal the initial total momentum, which is zero since the particle is initially at rest. As a result, $p_1 = -p_2$, where Equation 28.3 must be used for the magnitudes of the momenta p_1 and p_2. Thus, we find

$$\frac{m_1 v_1}{\sqrt{1-(v_1^2/c^2)}} = \frac{-m_2 v_2}{\sqrt{1-(v_2^2/c^2)}}$$

SOLUTION Letting fragment 2 be the more-massive fragment, we have that $v_1 = 0.800c$, $m_1 = 1.67 \times 10^{-27}$ kg, and $m_2 = 5.01 \times 10^{-27}$ kg. Squaring both sides of the above equation, rearranging terms, substituting the known values for v_1, m_1, and m_2, we find that

$$\frac{v_2^2}{1-(v_2^2/c^2)} = \frac{m_1^2 v_1^2}{m_2^2 \left[1-(v_1^2/c^2)\right]} = \frac{(1.67 \times 10^{-27} \text{ kg})^2 (0.800\,c)^2}{(5.01 \times 10^{-27} \text{ kg})^2 \left[1-(0.800\,c/c)^2\right]} = 0.1975\,c^2$$

Therefore,

$$v_2^2 = 0.1975\,c^2 \left[1-(v_2^2/c^2)\right] = 0.1975\,c^2 - 0.1975\,v_2^2$$

Solving for v_2 gives

$$v_2 = \pm\sqrt{\frac{0.1975\,c^2}{1.1975}} = \pm 0.406\,c$$

We reject the positive root, since then both fragments would be moving in the same direction after the break-up and the system would have a non-zero momentum. According to the principle of conservation of momentum, the total momentum after the break-up must be zero, just as it was before the break-up. The momentum of the system will be zero only if the velocity v_2 is opposite to the velocity v_1. Hence, we chose the negative root and $\boxed{v_2 = -0.406\,c}$.

42. **REASONING AND SOLUTION** Any change in the energy of a system is equivalent to a change in the mass of the system, according to $\Delta E_0 = (\Delta m)c^2$ (see Section 28.6). Therefore, the change in mass of a system when its energy changes by an amount ΔE_0 is given by

$$\Delta m = \frac{\Delta E_0}{c^2} \qquad (1)$$

We want to know how close two stationary electrons have to be positioned so that their total mass is twice what it is when the electrons are very far apart. Let $m_{\text{total}} = 2m_e$ represent the total mass of the system when the electrons are very far apart. Then, when the electrons have been brought together so that their total mass becomes $2m_{\text{total}}$, we have

$$m_{\text{total}} + \Delta m = 2m_{\text{total}} \qquad (2)$$

When the two electrons are very far apart, their electric potential energy is zero. When they are brought together so that their separation is r, their electric potential energy is

$$\text{EPE} = \frac{ke^2}{r}$$

where e is the charge on the electron (see Equations 19.3 and 19.6). Therefore, the change in energy of the system is

$$\Delta E_0 = \text{EPE} - 0 = \frac{ke^2}{r} \qquad (3)$$

Combining Equations (1) and (2), we have

$$m_{\text{total}} + \frac{\Delta E_0}{c^2} = 2m_{\text{total}}$$

Substituting the right hand side of Equation (3) for ΔE_0, we have

$$m_{\text{total}} + \frac{1}{c^2}\left(\frac{ke^2}{r}\right) = 2m_{\text{total}}$$

Further simplification and solving for r leads to

$$\frac{1}{c^2}\left(\frac{ke^2}{r}\right) = m_{\text{total}}$$

1044 SPECIAL RELATIVITY

$$r = \frac{ke^2}{m_{total}c^2} = \frac{ke^2}{2m_ec^2} = \frac{(8.99\times 10^9 \text{ N}\cdot\text{m}^2/\text{C}^2)(1.60\times 10^{-19}\text{ C})^2}{2(9.11\times 10^{-31}\text{ kg})(3.00\times 10^8\text{ m/s})^2} = \boxed{1.40\times 10^{-15}\text{ m}}$$

43. **REASONING AND SOLUTION** The length L_M of the man's rocket as measured by the woman is related to the proper length L_{0M} of the man's rocket by the length-contraction formula

$$L_M = L_{0M}\sqrt{1-\frac{v^2}{c^2}}$$

However, the problem states that L_M is also equal to L_{0W}, the proper length of the woman's rocket. Setting the expression above for L_M equal to L_{0W} and solving for L_{0M}, we obtain

$$L_{0M} = \frac{L_{0W}}{\sqrt{1-\frac{v^2}{c^2}}}$$

The length L_W of the woman's rocket as measured by the man is related to the proper length L_{0W} of the woman's rocket by the length-contraction formula

$$L_W = L_{0W}\sqrt{1-\frac{v^2}{c^2}}$$

Dividing this expression for L_W into the one directly above it for L_{0M} gives

$$\frac{L_{0M}}{L_W} = \frac{\frac{L_{0W}}{\sqrt{1-(v/c)^2}}}{L_{0W}\sqrt{1-(v/c)^2}} = \frac{1}{1-(v/c)^2} = \frac{1}{1-(0.940c)^2/c^2} = \boxed{8.59}$$

44. **CONCEPT QUESTIONS**
a. The Martian measures the proper time interval Δt_0, because the Martian measures the beginning and ending events (the times when the trip starts and ends) at the same location relative to a coordinate system fixed to the spaceship.

b. The given distance between Mars and Venus is the distance as measured by a person on earth. That person is at rest relative to the two planets and, hence, measures the proper length. The Martian, who is moving relative to the planets, does not measure the proper length, but measures a contracted length.

c. According to the Martian, the time of the trip Δt_o is equal to the contracted length that he measures divided by the speed v of the spaceship.

SOLUTION
a. The contracted length L measured by the Martian is related to the proper length L_o by Equation 28.2 as

$$L = L_o\sqrt{1-\frac{v^2}{c^2}} = (1.20\times 10^{11}\text{ m})\sqrt{1-\frac{(0.80c)^2}{c^2}} = \boxed{0.72\times 10^{11}\text{ m}}$$

b. The time of the trip as measured by the Martian is

$$\Delta t_o = \frac{L}{v} = \frac{0.72\times 10^{11}\text{ m}}{0.80(3.00\times 10^8\text{ m/s})} = \boxed{3.0\times 10^2\text{ s}}$$

---

45. **CONCEPT QUESTION** In special relativity the momentum of a particle is given by Equation 28.3 as $p = mv/\sqrt{1-(v^2/c^2)}$. Because of the $\sqrt{1-(v^2/c^2)}$ term in the denominator, doubling the particle's speed more than doubles its momentum. On the other hand, halving the particle's mass halves its momentum. Therefore, increasing the speed by a certain factor has a greater effect on the momentum than decreasing the mass by the same factor. Therefore, particle c, having the greatest speed, has the greatest momentum, even though it has the smallest mass. The ranking, largest first, of the momenta is: c, b, and a.

SOLUTION The momenta of the three particles are:

Particle a $\quad p = \dfrac{mv}{\sqrt{1-\dfrac{v^2}{c^2}}} = \dfrac{(1.20\times 10^{-8}\text{ kg})(0.200)(3.00\times 10^8\text{ m/s})}{\sqrt{1-\dfrac{(0.200c)^2}{c^2}}} = \boxed{0.735\text{ kg}\cdot\text{m/s}}$

Particle b $\quad p = \dfrac{mv}{\sqrt{1-\dfrac{v^2}{c^2}}} = \dfrac{(\tfrac{1}{2}\times 1.20\times 10^{-8}\text{ kg})(2\times 0.200)(3.00\times 10^8\text{ m/s})}{\sqrt{1-\dfrac{(2\times 0.200c)^2}{c^2}}} = \boxed{0.786\text{ kg}\cdot\text{m/s}}$

Particle c $\quad p = \dfrac{mv}{\sqrt{1-\dfrac{v^2}{c^2}}} = \dfrac{(\tfrac{1}{4}\times 1.20\times 10^{-8}\text{ kg})(4\times 0.200)(3.00\times 10^8\text{ m/s})}{\sqrt{1-\dfrac{(4\times 0.200c)^2}{c^2}}} = \boxed{1.20\text{ kg}\cdot\text{m/s}}$

1046 SPECIAL RELATIVITY

As expected, the ranking (largest first) is c, b, a.

---

46. CONCEPT QUESTIONS

a. In order to change a certain mass of ice at 0 °C into liquid water at 0 °C heat must be added, and heat is a form of energy. Therefore, the energy of the liquid water is greater than that of the ice. According to special relativity, energy and mass are equivalent. Since the liquid water has the greater energy, it also has the greater mass.

b. Heat must also be added to boil water into steam. Following the same type of reasoning as in the answer to question (a), we conclude that the steam has the greater mass.

c. The amount of heat Q that must be supplied to change the phase of m kilograms of a substance is given by Equation 12.5 as $Q = mL$, where L is the latent heat of the substance. Since the latent heat of vaporization L_v for water is greater than the latent heat of fusion L_f (see Table 12.3), the change in mass is greater when liquid water turns into steam at 100 °C than when ice turns into liquid water at 0 °C.

SOLUTION The change in mass Δm associated with a change in rest energy ΔE_o is given by Equation 28.5 as $\Delta m = \Delta E_o / c^2$. The change in rest energy is the heat Q that must be added to change the phase of the water, so that $Q = mL$. Thus, the change in mass is $\Delta m = Q/c^2 = mL/c^2$.

a. According to Table 12.3, the latent heat of fusion for water is $L_f = 3.35 \times 10^5$ J/kg. The change in mass associated with the ice to liquid-water phase change at 0 °C is, then,

$$\Delta m = \frac{mL_f}{c^2} = \frac{(2.00 \text{ kg})(3.35 \times 10^5 \text{ J/kg})}{(3.00 \times 10^8 \text{ m/s})^2} = \boxed{7.44 \times 10^{-12} \text{ kg}}$$

b. According to Table 12.3, the latent heat of vaporization for water is $L_f = 2.26 \times 10^6$ J/kg. The change in mass associated with the liquid-water to steam phase change at 100 °C is

$$\Delta m = \frac{mL_v}{c^2} = \frac{(2.00 \text{ kg})(2.26 \times 10^6 \text{ J/kg})}{(3.00 \times 10^8 \text{ m/s})^2} = \boxed{5.02 \times 10^{-11} \text{ kg}}$$

As expected, the change in mass for the liquid-to-steam phase change is greater than that for the ice-to-liquid phase change.

47. CONCEPT QUESTIONS

a. Yes. If the speeds of your car and the truck are much less than the speed of light, the relative speed at which the truck approaches you is the same in both situations. Let's suppose that you are traveling due east, which is taken to be the positive direction, and the truck is traveling due west, which is the negative direction. The relative velocities are:

v_{TC} = velocity of the **Truck** relative to the **Car**

v_{TG} = velocity of the **Truck** relative to the **Ground** = –35 m/s

v_{CG} = velocity of the **Car** relative to the **Ground** = +25 m/s. (Note that the velocity v_{GC} of the **Ground** relative to the **Car** is $v_{GC} = -v_{CG} = -25$ m/s.

The velocity v_{TC} of the truck with respect to the car is equal to the velocity v_{TG} of the truck with respect to the ground plus the velocity v_{GC} of the ground with respect to the car: $v_{TC} = v_{TG} + v_{GC}$. When $v_{TG} = -35$ m/s and $v_{GC} = -25$ m/s, the velocity of the truck relative to the car is $v_{TC} = -60$ m/s, where the minus sign indicates that the relative velocity is westward.

When $v_{TG} = -55$ m/s and $v_{CG} = +5.0$ m/s, the relative velocity of the truck with respect to the car is still $v_{TC} = v_{TG} + v_{GC} = -55$ m/s – 5 m/s = –60 m/s. In either case, the speed is the magnitude of v_{TC}, or 60 m/s.

b. The relative velocities and, hence, the relative speeds would not be the same for the two situations if the speeds were comparable to the speed of light. According to special relativity, the correct relation is the velocity-addition formula, Equation 28.7:

$$v_{TC} = \frac{v_{TG} + v_{GC}}{1 + \frac{v_{TG}v_{GC}}{c^2}}$$

Because of the presence of the term $v_{TG}v_{GC}/c^2$ in the denominator, different results are obtained when $v_{TG} = -35$ m/s and $v_{GC} = -25$ m/s than when $v_{TG} = -55$ m/s and $v_{GC} = -5.0$ m/s.

SOLUTION

a. When $v_{TG} = -35$ m/s and $v_{GC} = -25$ m/s, the velocity of the truck relative to the car is

$$v_{TC} = \frac{v_{TG} + v_{GC}}{1 + \frac{v_{TG}v_{GC}}{c^2}} = \frac{-35 \text{ m/s} - 25 \text{ m/s}}{1 + \frac{(-35 \text{ m/s})(-25 \text{ m/s})}{(65 \text{ m/s})^2}} = -49.7 \text{ m/s}$$

The speed of the truck relative to the car is the magnitude of this result, or $\boxed{49.7 \text{ m/s}}$.

b. When $v_{TG} = -55$ m/s and $v_{GC} = -5.0$ m/s, the velocity of the truck relative to the car is

$$v_{TC} = \frac{v_{TG} + v_{GC}}{1 + \frac{v_{TG} v_{GC}}{c^2}} = \frac{-55 \text{ m/s} - 5.0 \text{ m/s}}{1 + \frac{(-55 \text{ m/s})(-5.0 \text{ m/s})}{(65 \text{ m/s})^2}} = -56.3 \text{ m/s}$$

The speed of the truck relative to the car is the magnitude of this result, or $\boxed{56.3 \text{ m/s}}$.

CHAPTER 29 | PARTICLES AND WAVES

PROBLEMS

1. **SSM REASONING** The energy of the photon is related to its frequency by Equation 29.2, $E = hf$. Equation 16.1, $v = f\lambda$, relates the frequency and the wavelength for any wave.

   SOLUTION Combining Equations 29.2 and 16.1, and noting that the speed of a photon is c, the speed of light in a vacuum, we have

   $$\lambda = \frac{c}{f} = \frac{c}{(E/h)} = \frac{hc}{E} = \frac{(6.63 \times 10^{-34} \text{ J} \cdot \text{s})(3.0 \times 10^8 \text{ m/s})}{6.4 \times 10^{-19} \text{ J}} = 3.1 \times 10^{-7} \text{ m} = \boxed{310 \text{ nm}}$$

2. **REASONING AND SOLUTION** The frequency of a photon is $f = E/h$. For $E = 3.3 \times 10^{-16}$ J, $f = 5.0 \times 10^{17}$ Hz, which is in the $\boxed{\text{X-ray}}$ region of the spectrum. For $E = 1.3 \times 10^{-20}$ J, $f = 2.0 \times 10^{13}$ Hz, which is in the $\boxed{\text{infrared}}$ region.

3. **REASONING AND SOLUTION** The energy of a single photon is

   $$E = hf = (6.63 \times 10^{-34} \text{ J} \cdot \text{s})(98.1 \times 10^6 \text{ Hz}) = 6.50 \times 10^{-26} \text{ J}$$

   The number of photons emitted per second is

   $$\frac{\text{Power radiated}}{\text{Energy per photon}} = \frac{5.0 \times 10^4 \text{ W}}{6.50 \times 10^{-26} \text{ J}} = \boxed{7.7 \times 10^{29} \text{ photons/s}}$$

4. **REASONING** The photons of this wave must carry at least enough energy to equal the work function. Then the electrons are ejected with zero kinetic energy. Since the energy of a photon is $E = hf$ according to Equation 29.2, where f is the frequency of the wave, we have that $W_0 = hf$. Equation 16.1 relates the frequency to the wavelength λ according to $f = c/\lambda$, where c is the speed of light. Thus, it follows that $W_0 = hc/\lambda$.

   SOLUTION Using Equations 29.2 and 16.1, we find that

   $$W_0 = \frac{hc}{\lambda} = \frac{(6.63 \times 10^{-34} \text{ J} \cdot \text{s})(3.00 \times 10^8 \text{ m/s})}{485 \times 10^{-9} \text{ m}} = 4.10 \times 10^{-19} \text{ J}$$

1050 PARTICLES AND WAVES

Since $1 \text{ eV} = 1.60 \times 10^{-19}$ J, it follows that

$$W_0 = (4.10 \times 10^{-19} \text{ J})\left(\frac{1 \text{ eV}}{1.60 \times 10^{-19} \text{ J}}\right) = \boxed{2.56 \text{ eV}}$$

---

5. **[SSM] *REASONING*** According to Equation 29.3, the work function W_0 is related to the photon energy hf and the maximum kinetic energy KE_{max} by $W_0 = hf - \text{KE}_{\text{max}}$. This expression can be used to find the work function of the metal.

SOLUTION KE_{max} is 6.1 eV. The photon energy (in eV) is, according to Equation 29.2,

$$hf = (6.63 \times 10^{-34} \text{ J} \cdot \text{s})(3.00 \times 10^{15} \text{ Hz})\left(\frac{1 \text{ eV}}{1.60 \times 10^{-19} \text{ J}}\right) = 12.4 \text{ eV}$$

The work function is, therefore,

$$W_0 = hf - \text{KE}_{\text{max}} = 12.4 \text{ eV} - 6.1 \text{ eV} = \boxed{6.3 \text{ eV}}$$

---

6. ***REASONING AND SOLUTION*** In the first case, the energy of the incident photon is given by Equation 29.3 as

$$hf = \text{KE}_{\text{max}} + W_0 = 0.68 \text{ eV} + 2.75 \text{ eV} = 3.43 \text{ eV}$$

In the second case, a rearrangement of Equation 29.3 yields

$$\text{KE}_{\text{max}} = hf - W_0 = 3.43 \text{ eV} - 2.17 \text{ eV} = \boxed{1.26 \text{ eV}}$$

---

7. ***REASONING AND SOLUTION*** The energy of an FM photon is given by

$$E_{\text{FM}} = hf = (6.63 \times 10^{-34} \text{ J} \cdot \text{s})(91.9 \times 10^6 \text{ Hz}) = 6.09 \times 10^{-26} \text{ J}$$

The energy of an AM photon is

$$E_{\text{AM}} = (6.63 \times 10^{-34} \text{ J} \cdot \text{s})(665 \times 10^3 \text{ Hz}) = 4.41 \times 10^{-28} \text{ J}$$

The number of AM photons needed to equal the energy of one FM photon is

$$E_{\text{FM}}/E_{\text{AM}} = \boxed{138}$$

8. ***REASONING AND SOLUTION*** The work function of the material is

$$W_0 = \frac{hc}{\lambda} - \text{KE}_{\text{max}} = \frac{hc}{\lambda} - \tfrac{1}{2}mv_{\text{max}}^2$$

$$= \frac{(6.63 \times 10^{-34}\ \text{J} \cdot \text{s})(3.00 \times 10^8\ \text{m/s})}{238 \times 10^{-9}\ \text{m}} - \tfrac{1}{2}(9.11 \times 10^{-31}\ \text{kg})(3.75 \times 10^5\ \text{m/s})^2$$

$$= (7.72 \times 10^{-19}\ \text{J})\left(\frac{1\ \text{eV}}{1.60 \times 10^{-19}\ \text{J}}\right) = 4.82\ \text{eV}$$

The metal with this work function is $\boxed{\text{gold}}$.

---

9. **SSM** ***REASONING AND SOLUTION*** The number of photons per second, N, entering the owl's eye is $N = SA/E$, where S is the intensity of the beam, A is the area of the owl's pupil, and E is the energy of a single photon. Assuming that the owl's pupil is circular, $A = \pi r^2 = \pi\left(\tfrac{1}{2}d\right)^2$, where d is the diameter of the owl's pupil. Combining Equations 29.2 and 16.1, we have $E = hf = hc/\lambda$. Therefore,

$$N = \frac{SA\lambda}{hc} = \frac{(5.0 \times 10^{-13}\ \text{W/m}^2)\pi\left[\tfrac{1}{2}(8.5 \times 10^{-3}\ \text{m})\right]^2 (510 \times 10^{-9}\ \text{m})}{(6.63 \times 10^{-34}\ \text{J} \cdot \text{s})(3.0 \times 10^8\ \text{m/s})} = \boxed{73\ \text{photons/s}}$$

---

10. ***REASONING AND SOLUTION*** The work function of the material (using $\lambda = 196$ nm) is found from

$$W_0 = hf = hc/\lambda = 1.01 \times 10^{-18}\ \text{J}$$

The maximum kinetic energy of the ejected electron is (using $\lambda = 141$ nm)

$$\text{KE}_{\text{max}} = hf - W_0 = hc/\lambda - W_0 = 3.96 \times 10^{-19}\ \text{J}$$

The speed of the electron is then

$$v = \sqrt{\frac{2(\text{KE}_{\text{max}})}{m}} = \sqrt{\frac{2(3.96 \times 10^{-19}\ \text{J})}{9.11 \times 10^{-31}\ \text{kg}}} = \boxed{9.32 \times 10^5\ \text{m/s}}$$

---

11. **SSM** **WWW** ***REASONING*** The heat required to melt the ice is given by $Q = mL_f$, where m is the mass of the ice and L_f is the latent heat of fusion for water (see Section 12.8). Since, according to Equation 29.2, each photon carries an energy of $E = hf$, the energy content of N photons is $E_{\text{Total}} = Nhf$. According to Equation 16.1, $f = c/\lambda$, so we have

$$E_{Total} = \frac{Nhc}{\lambda}$$

If we assume that all of the photon energy is used to melt the ice, then, $E_{Total} = Q$, so that

$$\underbrace{\frac{Nhc}{\lambda}}_{E_{Total}} = \underbrace{mL_f}_{Q}$$

This expression may be solved for N to determine the required number of photons.

SOLUTION
a. We find that

$$N = \frac{mL_f \lambda}{hc} = \frac{(2.0 \text{ kg})(33.5 \times 10^4 \text{ J/kg})(620 \times 10^{-9} \text{ m})}{(6.63 \times 10^{-34} \text{ J} \cdot \text{s})(3.00 \times 10^8 \text{ m/s})} = \boxed{2.1 \times 10^{24} \text{ photons}}$$

b. The number N' of molecules in 2.0-kg of water is

$$N' = (2.0 \text{ kg})\left(\frac{1 \text{ mol}}{18 \times 10^{-3} \text{ kg}}\right)\left(\frac{6.022 \times 10^{23} \text{ molecules}}{1 \text{ mol}}\right) = 6.7 \times 10^{25} \text{ molecules}$$

Therefore, on average, the number of water molecules that one photon converts from the ice phase to the liquid phase is

$$\frac{N'}{N} = \frac{6.7 \times 10^{25} \text{ molecules}}{2.1 \times 10^{24} \text{ photons}} = \boxed{32 \text{ molecules/photon}}$$

---

12. **REASONING AND SOLUTION**
a. According to Equation 24.5b, the electric field can be found from $E = \sqrt{S/(\varepsilon_0 c)}$. The intensity S of the beam is

$$S = \frac{\text{Energy per unit time}}{A} = \frac{Nhf}{A} = \frac{Nh}{A}\left(\frac{c}{\lambda}\right)$$

$$= \frac{(1.30 \times 10^{18} \text{ photons/s})(6.63 \times 10^{-34} \text{ J} \cdot \text{s})}{\pi(1.00 \times 10^{-3} \text{ m})^2}\left(\frac{3.00 \times 10^8 \text{ m/s}}{514.5 \times 10^{-9} \text{ m}}\right)$$

$$= \boxed{1.60 \times 10^5 \text{ W/m}^2}$$

where N is the number of photons per second emitted. Then,

$$E = \sqrt{S/(\varepsilon_0 c)} = \boxed{7760 \text{ N/C}}$$

b. According to Equation 24.3, the average magnetic field is

$$B = E/c = \boxed{2.59 \times 10^{-5} \text{ T}}$$

---

13. **REASONING AND SOLUTION** We know that

$$p = \frac{h}{\lambda} = \frac{6.63 \times 10^{-34} \text{ J} \cdot \text{s}}{0.13 \text{ m}} = \boxed{5.1 \times 10^{-33} \text{ kg} \cdot \text{m/s}}$$

---

14. **REASONING AND SOLUTION** The momentum of the photon is $p = h/\lambda$ and that of the electron is $p = mv$. Equating and solving for the wavelength of the photon,

$$\lambda = \frac{h}{mv} = \frac{6.63 \times 10^{-34} \text{ J} \cdot \text{s}}{(9.11 \times 10^{-31} \text{ kg})(2.0 \times 10^5 \text{ m/s})} = \boxed{3.6 \times 10^{-9} \text{ m}}$$

---

15. [SSM] **REASONING AND SOLUTION**

a. According to Equation 29.6, the magnitude of the momentum of the incident photon is

$$p = \frac{h}{\lambda} = \frac{6.626 \times 10^{-34} \text{ J} \cdot \text{s}}{0.3120 \times 10^{-9} \text{ m}} = \boxed{2.124 \times 10^{-24} \text{ kg} \cdot \text{m/s}}$$

b. The wavelength of the scattered photon is, from Equation 29.7,

$$\lambda' = \lambda + \frac{h}{mc}(1 - \cos\theta)$$

where θ is the scattering angle. Combining this expression with Equation 29.6, we find that the magnitude of the momentum of the scattered photon is

$$p' = \frac{h}{\lambda'} = \frac{h}{\lambda + \left(\dfrac{h}{mc}\right)(1 - \cos\theta)}$$

$$= \frac{6.626 \times 10^{-34} \text{ J} \cdot \text{s}}{0.3120 \times 10^{-9} \text{ m} + \left[\dfrac{6.626 \times 10^{-34} \text{ J} \cdot \text{s}}{(9.109 \times 10^{-31} \text{ kg})(2.998 \times 10^8 \text{ m/s})}\right](1 - \cos 135.0°)}$$

$$= \boxed{2.096 \times 10^{-24} \text{ kg} \cdot \text{m/s}}$$

1054   PARTICLES AND WAVES

16. **REASONING** The change in the wavelength in Compton's experiment is given by Equation 29.7 as $\lambda' - \lambda = \frac{h}{mc}(1 - \cos\theta)$. In this problem, we do not know the value of the incident wavelength λ. However, we do know that it is the same for the two given values of the wavelength λ' of the scattered photons. Thus, we can apply Equation 29.7 for each of the given values of λ' and eliminate the incident wavelength from our calculation.

SOLUTION Applying Equation 29.7 for each of the given wavelength values, we have

$$\lambda_1' - \lambda = \frac{h}{mc}(1 - \cos\theta_1) \quad \text{and} \quad \lambda_2' - \lambda = \frac{h}{mc}(1 - \cos\theta_2)$$

Subtracting the first equation from the second gives

$$\lambda_2' - \lambda - (\lambda_1' - \lambda) = \frac{h}{mc}(1 - \cos\theta_2) - \frac{h}{mc}(1 - \cos\theta_1)$$

$$\lambda_2' - \lambda_1' = \frac{h}{mc}(-\cos\theta_2 + \cos\theta_1)$$

$$= \frac{(6.63 \times 10^{-34} \text{ J} \cdot \text{s})}{(9.11 \times 10^{-31} \text{ kg})(3.00 \times 10^8 \text{ m/s})}(-\cos 70.0° + \cos 30.0°)$$

$$= \boxed{1.27 \times 10^{-12} \text{ m}}$$

---

17. **REASONING AND SOLUTION** Since there are no external forces that act on the system, conservation of linear momentum applies. Since the photon is scattered at $\theta = 180°$, the collision is "head-on" and all motion occurs along the horizontal direction, which we will take as the x axis. For an initially stationary electron, the conservation of linear momentum states that:

$$\underbrace{p_0}_{\substack{\text{Momentum} \\ \text{of incident} \\ \text{photon}}} = \underbrace{p_f}_{\substack{\text{Momentum} \\ \text{of scattered} \\ \text{photon}}} + \underbrace{p_{\text{electron}}}_{\substack{\text{Momentum} \\ \text{of recoil} \\ \text{electron}}}$$

If we take the right to be the positive direction, the associated scalar equation is

$$p_0 = -p_f + p_{\text{electron}} \quad \text{or} \quad p_{\text{electron}} = p_0 + p_f$$

Using Equation 29.6, we can write the expression for the momentum gained by the electron as

$$p_{\text{electron}} = \frac{h}{\lambda_0} + \frac{h}{\lambda_f} = h\left(\frac{1}{\lambda_0} + \frac{1}{\lambda_f}\right)$$

Substituting numerical values, we have

$$p_{electron} = (6.626 \times 10^{-34} \text{ J} \cdot \text{s})\left(\frac{1}{0.2800 \times 10^{-9} \text{ m}} + \frac{1}{0.2849 \times 10^{-9} \text{ m}}\right)$$

$$= \boxed{4.692 \times 10^{-24} \text{ kg} \cdot \text{m/s}}$$

18. **REASONING AND SOLUTION**
   a. We have

$$\lambda = \lambda' - (h/mc)(1 - \cos 163°) = \boxed{0.1819 \text{ nm}}$$

   b. For the incident photon

$$E = hf = hc/\lambda = \boxed{1.092 \times 10^{-15} \text{ J}}$$

   c. For the scattered photon

$$E' = hf' = hc/\lambda' = \boxed{1.064 \times 10^{-15} \text{ J}}$$

   d. The kinetic energy of the recoil electron is, therefore,

$$KE = E - E' = \boxed{2.8 \times 10^{-17} \text{ J}}$$

19. [SSM] [WWW] **REASONING** The change in wavelength that occurs during Compton scattering is given by Equation 29.7:

$$\lambda' - \lambda = \frac{h}{mc}(1 - \cos\theta) \quad \text{or} \quad (\lambda' - \lambda)_{max} = \frac{h}{mc}(1 - \cos 180°) = \frac{2h}{mc}$$

$(\lambda' - \lambda)_{max}$ is the maximum change in the wavelength, and to calculate it we need a value for the mass m of a nitrogen molecule. This value can be obtained from the mass per mole M of nitrogen (N_2) and Avogadro's number N_A, according to $m = M/N_A$ (see Section 14.1).

SOLUTION Using a value of $M = 0.028$ kg/mol, we obtain the following result for the maximum change in the wavelength:

$$(\lambda' - \lambda)_{max} = \frac{2h}{mc} = \frac{2h}{\left(\frac{M}{N_A}\right)c} = \frac{2(6.63 \times 10^{-34} \text{ J} \cdot \text{s})}{\left(\frac{0.028 \text{ kg/mol}}{6.02 \times 10^{23} \text{ mol}^{-1}}\right)(3.00 \times 10^8 \text{ m/s})}$$

$$= \boxed{9.50 \times 10^{-17} \text{ m}}$$

20. **REASONING** Since the net external force acting on the system (the photon and the electron) is zero, the conservation of linear momentum applies. In addition, there are no nonconservative forces, so the conservation of total energy applies as well. Since the photon scatters at an angle of $\theta = 180.0°$ in Figure 29.10, the collision is "head-on." Thus, the motion takes place entirely along the horizontal direction, which we will take as the x axis, with the right as being the positive direction.

The conservation of linear momentum gives rise to Equation 29.7, which relates the difference $\lambda' - \lambda$ between the scattered and incident X-ray photon wavelengths to the scattering angle θ of the electron as

$$\lambda' - \lambda = \frac{h}{mc}(1 - \cos\theta) = \frac{h}{mc}(1 - \cos 180.0°) = \frac{2h}{mc} \qquad (1)$$

The conservation of total energy is written as

$$\underbrace{\frac{hc}{\lambda}}_{\substack{\text{Energy} \\ \text{of incident} \\ \text{photon}}} + \underbrace{0}_{\substack{\text{Initial} \\ \text{kinetic} \\ \text{energy} \\ \text{of electron}}} = \underbrace{\frac{hc}{\lambda'}}_{\substack{\text{Energy} \\ \text{of scattered} \\ \text{photon}}} + \underbrace{\tfrac{1}{2}mv^2}_{\substack{\text{Final} \\ \text{kinetic} \\ \text{energy} \\ \text{of electron}}} \qquad (2)$$

Equations (1) and (2) will permit us to find the wavelength λ of the incident X-ray photon.

SOLUTION Solving Equation (1) for λ' and substituting the result into Equation (2) gives

$$\frac{hc}{\lambda} = \frac{hc}{\frac{2h}{mc} + \lambda} + \tfrac{1}{2}mv^2$$

Algebraically rearranging this result, we obtain a quadratic equation for λ:

$$\lambda^2 + \underbrace{\left(\frac{2h}{mc}\right)}_{4.85 \times 10^{-12} \text{ m}} \lambda - \underbrace{\frac{2h^2}{m\left(\tfrac{1}{2}mv^2\right)}}_{9.70 \times 10^{-20} \text{ m}^2} = 0$$

where we have used $h = 6.63 \times 10^{-34}$ J·s, $m = 9.11 \times 10^{-31}$ kg, $c = 3.00 \times 10^8$ m/s, and $v = 4.67 \times 10^6$ m/s. Solving this quadratic equation for λ, we obtain

$$\boxed{\lambda = 3.09 \times 10^{-10} \text{ m}}$$

21. ***REASONING AND SOLUTION*** We know that $\lambda = h/mv$. Solving for the mass yields

$$m = \frac{h}{\lambda v} = \frac{6.63 \times 10^{-34} \text{ J} \cdot \text{s}}{(8.4 \times 10^{-14} \text{ m})(1.2 \times 10^6 \text{ m/s})} = \boxed{6.6 \times 10^{-27} \text{ kg}}$$

22. ***REASONING AND SOLUTION*** The de Broglie wavelength λ is given by Equation 29.8 as $\lambda = h/p$, where p is the magnitude of the momentum of the particle. The magnitude of the momentum is $p = mv$, where m is the mass and v is the speed of the particle. Using this expression in Equation 29.8, we find that

$$\lambda = \frac{h}{mv} \quad \text{or} \quad v = \frac{h}{m\lambda} = \frac{6.63 \times 10^{-34} \text{ J} \cdot \text{s}}{(1.67 \times 10^{-27} \text{ kg})(0.282 \times 10^{-9} \text{ m})} = \boxed{1.41 \times 10^3 \text{ m/s}}$$

23. **SSM** ***REASONING*** In order for the person to diffract to the same extent as the sound wave, the de Broglie wavelength of the person must be equal to the wavelength of the sound wave.

SOLUTION
a. Therefore,

$$\lambda_{\text{sound}} = \lambda_{\text{person}}$$

$$\lambda_{\text{sound}} = \frac{h}{m_{\text{person}} v_{\text{person}}}$$

Solving for v_{person}, and using the relation $\lambda_{\text{sound}} = v_{\text{sound}} / f_{\text{sound}}$ (Equation 16.1), we have

$$v_{\text{person}} = \frac{h}{m_{\text{person}} (v_{\text{sound}} / f_{\text{sound}})} = \frac{h f_{\text{sound}}}{m_{\text{person}} v_{\text{sound}}}$$

$$= \frac{(6.63 \times 10^{-34} \text{ J} \cdot \text{s})(128 \text{ Hz})}{(55.0 \text{ kg})(343 \text{ m/s})} = \boxed{4.50 \times 10^{-36} \text{ m/s}}$$

b. At the speed calculated in part (a), the time required for the person to move a distance of one meter is

$$t = \frac{x}{v} = \frac{1.0 \text{ m}}{4.50 \times 10^{-36} \text{ m/s}} \underbrace{\left(\frac{1.0 \text{ h}}{3600 \text{ s}}\right) \left(\frac{1 \text{ day}}{24.0 \text{ h}}\right) \left(\frac{1 \text{ year}}{365.25 \text{ days}}\right)}_{\text{Factors to convert seconds to years}} = \boxed{7.05 \times 10^{27} \text{ years}}$$

1058   PARTICLES AND WAVES

24. **REASONING AND SOLUTION**
   a. We know $E = hc/\lambda$ for a photon. The energy of the photon is

$$E = 5.0 \text{ eV}\left(\frac{1.60 \times 10^{-19} \text{ J}}{1 \text{ eV}}\right) = 8.0 \times 10^{-19} \text{ J}$$

The wavelength is

$$\lambda = \frac{hc}{E} = \frac{(6.63 \times 10^{-34} \text{ J·s})(3.00 \times 10^8 \text{ m/s})}{8.0 \times 10^{-19} \text{ J}} = \boxed{2.5 \times 10^{-7} \text{ m}}$$

   b. The speed of the 5.0-eV electron is

$$v = \sqrt{\frac{2E}{m}} = \sqrt{\frac{2(8.0 \times 10^{-19} \text{ J})}{9.11 \times 10^{-31} \text{ kg}}} = 1.3 \times 10^6 \text{ m/s}$$

The de Broglie wavelength is

$$\lambda = \frac{h}{mv} = \frac{6.63 \times 10^{-34} \text{ J·s}}{(9.11 \times 10^{-31} \text{ kg})(1.3 \times 10^6 \text{ m/s})} = \boxed{5.6 \times 10^{-10} \text{ m}}$$

___

25. **REASONING AND SOLUTION**   The average kinetic energy of a helium atom is

$$\text{KE} = (3/2)kT = (3/2)(1.38 \times 10^{-23} \text{ J/K})(293 \text{ K}) = 6.07 \times 10^{-21} \text{ J}$$

The speed of the atom corresponding to the average kinetic energy is

$$v = \sqrt{\frac{2(\text{KE})}{m}} = \sqrt{\frac{2(6.07 \times 10^{-21} \text{ J})}{6.65 \times 10^{-27} \text{ kg}}} = 1.35 \times 10^3 \text{ m/s}$$

The deBroglie wavelength is

$$\lambda = \frac{h}{mv} = \frac{6.63 \times 10^{-34} \text{ J·s}}{(6.65 \times 10^{-27} \text{ kg})(1.35 \times 10^3 \text{ m/s})} = \boxed{7.38 \times 10^{-11} \text{ m}}$$

___

26. **REASONING AND SOLUTION**   The de Broglie wavelength of the neutrons is (see Equation 29.8)

$$\lambda = \frac{h}{p} = \frac{h}{mv} = \frac{6.63 \times 10^{-34} \text{ J·s}}{(1.67 \times 10^{-27} \text{ kg})(2.80 \times 10^3 \text{ m/s})} = 1.42 \times 10^{-10} \text{ m}$$

According to Rayleigh's criterion, the minimum angle θ_{min} between the two objects, such that they are just resolved using these neutrons, is

$$\theta_{min} = 1.22 \frac{\lambda}{D} = 1.22 \left(\frac{1.42 \times 10^{-10} \text{ m}}{0.100 \times 10^{-3} \text{ m}} \right) = \boxed{1.73 \times 10^{-6} \text{ rad}}$$

---

27. **SSM** *REASONING AND SOLUTION* The de Broglie wavelength λ of the woman is given by Equation 29.8 as $\lambda = h/p$, where p is the magnitude of her momentum. The magnitude of the momentum is $p = mv$, where m is the woman's mass and v is her speed. According to Equation 3.6b of the equations of kinematics, the speed v is given by $v = \sqrt{2a_y y}$, since the woman jumps from rest. In this expression, $a_y = -9.80 \text{ m/s}^2$ and $y = -9.5 \text{ m}$. With these considerations we find that

$$\lambda = \frac{h}{mv} = \frac{h}{m\sqrt{2a_y y}} = \frac{6.63 \times 10^{-34} \text{ J} \cdot \text{s}}{(41 \text{ kg})\sqrt{2(-9.80 \text{ m/s}^2)(-9.5 \text{ m})}} = \boxed{1.2 \times 10^{-36} \text{ m}}$$

---

28. *REASONING* The de Broglie wavelength is given by $\lambda = h/p$ according to Equation 29.8, where h is Planck's constant and p is the magnitude of the momentum. Thus, to evaluate the wavelength, we need information about the momentum. While we do not have values for the momentum of particle B before the collision or for the object that moves away after the collision, we do know that the total momentum is conserved. We know this, because it is given that no external forces act on the particles. As Section 7.2 discusses, the total momentum is conserved in such circumstances. This fact will allow us to solve the problem.

SOLUTION The de Broglie wavelength of the object that moves off after the collision is given by Equation 29.8 as $\lambda_{after} = h/p_{after}$. Since momentum is mass times velocity, the magnitude of the momentum that the object has after the collision is $p_{after} = (m_A + m_B)v$, where v is the common speed of the two particles. We can evaluate this momentum be using the law of conservation of momentum, which indicates that the total momentum after the collision is the same as it is before the collision. Before the collision only particle B is moving, so that the total momentum at that time has a value of just $m_B v_{0B}$, where v_{0B} is the initial velocity of particle B. Momentum conservation, then, dictates that

$$\underbrace{(m_A + m_B)v}_{\text{Total momentum after collision}} = \underbrace{m_B v_{0B}}_{\text{Total momentum before collision}}$$

Using this result, we find that the desired de Broglie wavelength is

$$\lambda_{\text{after}} = \frac{h}{p_{\text{after}}} = \frac{h}{(m_A + m_B)v} = \frac{h}{m_B v_{0B}}$$

But the term on the far right is just the given de Broglie wavelength of the incident particle B. Therefore, we conclude that $\boxed{\lambda_{\text{after}} = 2.0 \times 10^{-34} \text{ m}}$.

---

29. **REASONING** The only force acting on the moving charge is the conservative electric force. Therefore, the total energy of the electron at all points in its trajectory as it is accelerated through the potential difference V remains the same. In other words, if A and B are the end points of the trajectory, then

$$\underbrace{KE_A + EPE_A}_{\text{Total energy at A}} = \underbrace{KE_B + EPE_B}_{\text{Total energy at B}}$$

Since the electron starts from rest $KE_A = 0$, and we have

$$EPE_A - EPE_B = \tfrac{1}{2} m v_B^2$$

According to Equation 19.4,

$$EPE_B - EPE_A = q(V_B - V_A) = qV = -eV$$

Combining these results, we find

$$\tfrac{1}{2} m v_B^2 = eV$$

Solving for V, we have

$$V = \frac{m v_B^2}{2e}$$

We can express the speed v_B of the electron in terms of its de Broglie wavelength:

$$\lambda = \frac{h}{p} = \frac{h}{m v_B} \quad \text{or} \quad v_B = \frac{h}{m\lambda}$$

The expression for V becomes

$$V = \frac{m}{2e}\left(\frac{h}{m\lambda}\right)^2 = \frac{1}{2me}\left(\frac{h}{\lambda}\right)^2$$

SOLUTION Given the data in the problem statement and the fact that the mass and the magnitude of the charge of the electron are, respectively, $m = 9.11 \times 10^{-31}$ kg and $e = 1.6 \times 10^{-19}$ C, we find that the potential difference V is

$$V = \frac{1}{2(9.11 \times 10^{-31} \text{ kg})(1.6 \times 10^{-19} \text{ C})}\left(\frac{6.63 \times 10^{-34} \text{ J} \cdot \text{s}}{1.0 \times 10^{-11} \text{ m}}\right)^2 = \boxed{1.5 \times 10^4 \text{ V}}$$

30. **REASONING AND SOLUTION** The energy of the photon is $E = hf = hc/\lambda_{photon}$, while the kinetic energy of the particle is KE $= (1/2)mv^2 = h^2/(2m\lambda^2)$. Equating the two energies and rearranging the result gives $\lambda_{photon}/\lambda = (2mc/h)\lambda$. Now the speed of the particle is $v = 0.050c$, so $\lambda = h/(0.050\ mc)$, and

$$\lambda_{photon}/\lambda = 2/0.050 = \boxed{4.0 \times 10^1}$$

---

31. **SSM WWW** **REASONING AND SOLUTION** According to the uncertainty principle, the minimum uncertainty in the momentum can be determined from $\Delta p_y \Delta y = h/(2\pi)$. Since $p_y = mv_y$, it follows that $\Delta p_y = m\Delta v_y$. Thus, the minimum uncertainty in the velocity of the oxygen molecule is given by

$$\Delta v_y = \frac{h}{2\pi m \Delta y} = \frac{6.63 \times 10^{-34}\ \text{J} \cdot \text{s}}{2\pi(5.3 \times 10^{-26}\ \text{kg})(0.25 \times 10^{-3}\ \text{m})} = \boxed{8.0 \times 10^{-6}\ \text{m/s}}$$

---

32. **REASONING** We know that the object is somewhere on the line. Therefore, the uncertainty in the object's position is $\Delta y = 2.5$ m. The minimum uncertainty in the object's momentum is Δp_y and is specified by the Heisenberg uncertainty principle (Equation 29.10) in the form $(\Delta p_y)(\Delta y) = h/(2\pi)$. Since momentum is mass m times velocity v, the uncertainty in the velocity Δv is related to the uncertainty in the momentum by $\Delta v = (\Delta p_y)/m$.

SOLUTION
a. Using the uncertainty principle, we find the minimum uncertainty in the momentum as follows:

$$(\Delta p_y)(\Delta y) = \frac{h}{2\pi}$$

$$\Delta p_y = \frac{h}{2\pi \Delta y} = \frac{6.63 \times 10^{-34}\ \text{J} \cdot \text{s}}{2\pi(2.5\ \text{m})} = \boxed{4.2 \times 10^{-35}\ \text{kg} \cdot \text{m/s}}$$

b. For a golf ball this uncertainty in momentum corresponds to an uncertainty in velocity that is given by

$$\Delta v_y = \frac{\Delta p_y}{m} = \frac{4.2 \times 10^{-35}\ \text{kg} \cdot \text{m/s}}{0.045\ \text{kg}} = \boxed{9.3 \times 10^{-34}\ \text{m/s}}$$

c. For an electron this uncertainty in momentum corresponds to an uncertainty in velocity that is given by

$$\Delta v_y = \frac{\Delta p_y}{m} = \frac{4.2 \times 10^{-35}\ \text{kg} \cdot \text{m/s}}{9.11 \times 10^{-31}\ \text{kg}} = \boxed{4.6 \times 10^{-5}\ \text{m/s}}$$

1062 PARTICLES AND WAVES

33. **REASONING AND SOLUTION** The minimum uncertainty Δp in the momentum of the proton is given by the Heisenberg uncertainty principle (Equation 29.10) as $\Delta p = h/(2\pi \Delta d)$ where Δd is the uncertainty in position; Therefore,

$$\Delta p = \frac{h}{2\pi \Delta d} = \frac{6.63 \times 10^{-34} \text{ J} \cdot \text{s}}{2\pi(5.5 \times 10^{-15} \text{ m})} = \boxed{1.9 \times 10^{-20} \text{ kg} \cdot \text{m/s}}$$

34. **REASONING AND SOLUTION** We know from Equation 29.8 that

$$h = \lambda p = \lambda m v = (0.90 \text{ m})(82 \text{ kg})(0.50 \text{ m/s}) = \boxed{37 \text{ J} \cdot \text{s}}$$

35. [SSM] [WWW] **REASONING AND SOLUTION** If we assume that the number of particles that strikes the screen outside the bright fringe is negligible, the particles that pass through the slit will hit the screen somewhere in the central bright fringe. The angular width of the central bright fringe is equal to 2θ, where θ is shown in Figure 29.15. From this figure we see that

$$\theta = \tan^{-1}\left(\frac{\Delta p_y}{p_x}\right)$$

According to Equation 29.10, the minimum uncertainty in Δp_y occurs when

$$(\Delta p_y)(\Delta y) = \frac{h}{2\pi}$$

Since the electron can pass anywhere through the width W of the slit, the uncertainty in the y position of the electron is $\Delta y = W$. Thus, the minimum uncertainty can be written as

$$(\Delta p_y)W = \frac{h}{2\pi} \quad \text{or} \quad \Delta p_y = \frac{h}{2\pi W}$$

Using the fact that $p_x = h/\lambda$, we find

$$\theta = \tan^{-1}\left(\frac{h}{p_x 2\pi W}\right) = \tan^{-1}\left[\frac{h}{(h/\lambda)2\pi W}\right] = \tan^{-1}\left(\frac{\lambda}{2\pi W}\right)$$

$$= \tan^{-1}\left[\frac{633 \times 10^{-9} \text{ m}}{2\pi(0.200 \times 10^{-3} \text{ m})}\right] = 2.89 \times 10^{-2} \text{ degrees}$$

Therefore, the minimum range of angles over which the particles spread is $\boxed{-0.0289° \leq \theta \leq +0.0289°}$.

36. **REASONING AND SOLUTION** The uncertainty in the position is $\Delta y = \lambda = h/mv$. Now, since $\Delta p \Delta y = m\Delta v \Delta y = h/2\pi$, we have $m\Delta v \Delta y = m\Delta v(h/mv) = h\Delta v/v = h/2\pi$. Solving for Δv yields

$$\Delta v = v/2\pi = (4.5 \times 10^5 \text{ m/s})/2\pi = \boxed{7.2 \times 10^4 \text{ m/s}}$$

37. **REASONING AND SOLUTION** The de Broglie wavelength is

$$\lambda = \frac{h}{mv} = \frac{6.63 \times 10^{-34} \text{ J} \cdot \text{s}}{(1.3 \times 10^{-4} \text{ kg})(0.020 \text{ m/s})} = \boxed{2.6 \times 10^{-28} \text{ m}}$$

38. **REASONING** The energy of a photon of frequency f is, according to Equation 29.2, $E = hf$, where h is Planck's constant. Since the frequency and wavelength are related by $f = c/\lambda$ (see Equation 16.1), the energy of a photon can be written in terms of the wavelength as $E = hc/\lambda$. These expressions can be solved for both the wavelength and the frequency.

SOLUTION
a. The wavelength of the photon is

$$\lambda = \frac{hc}{E} = \frac{(6.63 \times 10^{-34} \text{ J} \cdot \text{s})(3.00 \times 10^8 \text{ m/s})}{1.22 \times 10^{-18} \text{ J}} = \boxed{1.63 \times 10^{-7} \text{ m}}$$

b. Using the answer from part (a), we find that the frequency of the photon is

$$f = \frac{c}{\lambda} = \frac{3.00 \times 10^8 \text{ m/s}}{1.63 \times 10^{-7} \text{ m}} = \boxed{1.84 \times 10^{15} \text{ Hz}}$$

Alternatively, we could use Equation 29.2 directly to obtain the frequency:

$$f = \frac{E}{h} = \frac{1.22 \times 10^{-18} \text{ J}}{6.63 \times 10^{-34} \text{ J} \cdot \text{s}} = \boxed{1.84 \times 10^{15} \text{ Hz}}$$

c. The wavelength and frequency values shown in Figure 24.9 indicate that this photon corresponds to electromagnetic radiation in the $\boxed{\text{ultraviolet region}}$ of the electromagnetic spectrum.

1064 PARTICLES AND WAVES

39. **SSM** *REASONING AND SOLUTION* The de Broglie wavelength λ is given by Equation 29.8 as $\lambda = h/p$, where p is the magnitude of the momentum of the particle. The magnitude of the momentum is $p = mv$, where m is the mass and v is the speed of the particle. Using this expression in Equation 29.8, we find that $\lambda = h/(mv)$, or

$$v = \frac{h}{m\lambda} = \frac{6.63 \times 10^{-34} \text{ J} \cdot \text{s}}{(1.67 \times 10^{-27} \text{ kg})(1.30 \times 10^{-14} \text{ m})} = 3.05 \times 10^7 \text{ m/s}$$

The kinetic energy of the proton is

$$\text{KE} = \tfrac{1}{2}mv^2 = \tfrac{1}{2}(1.67 \times 10^{-27} \text{ kg})(3.05 \times 10^7 \text{ m/s})^2 = \boxed{7.77 \times 10^{-13} \text{ J}}$$

---

40. *REASONING AND SOLUTION* The de Broglie wavelength is given by Equation 29.8 as $\lambda = h/p$. Since momentum is mass times velocity, or $p = mv$, it follows that

$$\frac{\lambda_{\text{electron}}}{\lambda_{\text{proton}}} = \frac{\dfrac{h}{m_{\text{electron}}v}}{\dfrac{h}{m_{\text{proton}}v}} = \frac{m_{\text{proton}}}{m_{\text{electron}}} = \frac{1.67 \times 10^{-27} \text{ kg}}{9.11 \times 10^{-31} \text{ kg}} = \boxed{1830}$$

---

41. *REASONING AND SOLUTION* Equation 29.6 gives the magnitude p of the photon momentum as $p = h/\lambda$, where λ is the wavelength. Using this expression for both the red and the violet light, we find that

$$\frac{p_{\text{violet}}}{p_{\text{red}}} = \frac{h/\lambda_{\text{violet}}}{h/\lambda_{\text{red}}} = \frac{\lambda_{\text{red}}}{\lambda_{\text{violet}}} = \frac{730 \text{ nm}}{380 \text{ nm}} = \boxed{1.9}$$

---

42. *REASONING AND SOLUTION* Equation 29.3 gives

$$\text{KE}_{\text{max}} = hf - W_o = \frac{hc}{\lambda} - W_o$$

$$= \frac{(6.63 \times 10^{-34} \text{ J} \cdot \text{s})(3.00 \times 10^8 \text{ m/s})}{215 \times 10^{-9} \text{ m}} - (3.68 \text{ eV})\left(\frac{1.60 \times 10^{-19} \text{ J}}{1 \text{ eV}}\right) = 3.36 \times 10^{-19} \text{ J}$$

Converting to electron volts

$$\text{KE}_{\text{max}} = (3.36 \times 10^{-19} \text{ J})\left(\frac{1 \text{ eV}}{1.60 \times 10^{-19} \text{ J}}\right) = \boxed{2.10 \text{ eV}}$$

43. **SSM** *REASONING* The width of the central bright fringe in the diffraction patterns will be identical when the electrons have the same de Broglie wavelength as the wavelength of the photons in the red light. The de Broglie wavelength of one electron in the beam is given by Equation 29.8, $\lambda_{electron} = h/p$, where $p = mv$.

SOLUTION Following the reasoning described above, we find

$$\lambda_{red\ light} = \lambda_{electron}$$

$$\lambda_{red\ light} = \frac{h}{m_{electron} v_{electron}}$$

Solving for the speed of the electron, we have

$$v_{electron} = \frac{h}{m_{electron} \lambda_{red\ light}} = \frac{6.63 \times 10^{-34}\ \text{J} \cdot \text{s}}{(9.11 \times 10^{-31}\ \text{kg})(661 \times 10^{-9}\ \text{m})} = \boxed{1.10 \times 10^3\ \text{m/s}}$$

---

44. *REASONING* We will first calculate the potential energy of the system at each of the two separations, and then find the energy difference for the two configurations. Since the electric potential energy lost by the system is carried off by a photon that is emitted during the process, the energy difference must be equal to the energy of the photon. The wavelength of the photon can then by found using Equation 29.2 with Equation 16.1: $E = hc/\lambda$.

SOLUTION The initial potential energy of the system is (see Equations 19.3 and 19.6)

$$\text{EPE}_1 = eV_1 = e\left(\frac{kq}{r_1}\right)$$

$$= (1.6 \times 10^{-19}\ \text{C}) \left[\frac{(8.99 \times 10^9\ \text{N} \cdot \text{m}^2/\text{C}^2)(8.30 \times 10^{-6}\ \text{C})}{0.420\ \text{m}} \right] = 2.84 \times 10^{-14}\ \text{J}$$

The final potential energy is

$$\text{EPE}_2 = eV_2 = (1.6 \times 10^{-19}\ \text{C}) \left[\frac{(8.99 \times 10^9\ \text{N} \cdot \text{m}^2/\text{C}^2)(8.30 \times 10^{-6}\ \text{C})}{1.58\ \text{m}} \right] = 7.56 \times 10^{-15}\ \text{J}$$

The energy difference, and therefore the energy of the emitted photon, is

$$\Delta E = \text{EPE}_1 - \text{EPE}_2 = 2.84 \times 10^{-14}\ \text{J} - 7.56 \times 10^{-15}\ \text{J} = 2.08 \times 10^{-14}\ \text{J}$$

The wavelength of this photon is

$$\lambda = \frac{hc}{\Delta E} = \frac{(6.63 \times 10^{-34} \text{ J·s})(3.00 \times 10^8 \text{ m/s})}{2.08 \times 10^{-14} \text{ J}} = \boxed{9.56 \times 10^{-12} \text{ m}}$$

---

45. **REASONING AND SOLUTION** The momentum of an incident photon is given by $p = h/\lambda$. The total momentum carried by the photons to the surface in a time Δt is

$$P = Nh\Delta t/\lambda, \quad \text{where} \quad N = 3.0 \times 10^{18} \text{ photons/s}$$

The impulse momentum principle gives the average force exerted on the wall to be $F = \Delta P/\Delta t$.

a. $\Delta P = 2P$, since the photons reflect from the mirror, so

$$F = 2Nh/\lambda = 2(3.0 \times 10^{18} \text{ photons /s})(6.63 \times 10^{-34} \text{ J·s})/(395 \times 10^{-9} \text{ m}) = \boxed{1.0 \times 10^{-8} \text{ N}}$$

b. $\Delta P = P$, since the photons are totally absorbed by the wall, so $F = Nh/\lambda$ and

$$F = \boxed{5.0 \times 10^{-9} \text{ N}}$$

---

46. **CONCEPT QUESTIONS** a. Kinetic energy is $\frac{1}{2}mv^2$, so that KE_{max} is greater when the speed is greater, that is, for case A.

b. The kinetic energy of the electrons is the energy that the photons carry in excess of that required to do the work of removing the electrons from the surface. To eject the electrons with a greater value of KE_{max}, then, requires that the energy of the photons must be greater to begin with. Thus, the photon energy in case A is greater than in case B. Equation 29.2 indicates that the photon energy is $E = hf$, where h is Planck's constant and f is the frequency. Therefore, since the energy E is greater in case A, the frequency must also be greater.

c. Equation 16.1 relates the frequency to the wavelength λ according to $f = c/\lambda$, where c is the speed of light. Frequency is inversely proportional to wavelength, so the wavelength in case A is smaller, because the frequency is greater.

SOLUTION According to Equation 29.3 and Equation 16.1 ($f = c/\lambda$), we have

$$hf = KE_{max} + W_0 \quad \text{or} \quad \frac{hc}{\lambda} = \frac{1}{2}mv_{max}^2 + W_0$$

Solving this expression for the wavelength gives

$$\lambda = \frac{hc}{\frac{1}{2}mv_{max}^2 + W_0}$$

$$\lambda_A = \frac{(6.63 \times 10^{-34} \text{ J} \cdot \text{s})(3.00 \times 10^8 \text{ m/s})}{\frac{1}{2}(9.11 \times 10^{-31} \text{ kg})(7.30 \times 10^5 \text{ m/s})^2 + 4.80 \times 10^{-19} \text{ J}} = \boxed{2.75 \times 10^{-7} \text{ m}}$$

$$\lambda_B = \frac{(6.63 \times 10^{-34} \text{ J} \cdot \text{s})(3.00 \times 10^8 \text{ m/s})}{\frac{1}{2}(9.11 \times 10^{-31} \text{ kg})(5.00 \times 10^5 \text{ m/s})^2 + 4.80 \times 10^{-19} \text{ J}} = \boxed{3.35 \times 10^{-7} \text{ m}}$$

---

47. **CONCEPT QUESTIONS** a. According to Equation 12.4, the necessary heat is $Q = c_{\text{specific heat}} m \Delta T$.

b. Infrared photons have a longer wavelength λ than do visible photons, as can be seen in Figure 24.9. Equation 29.2 indicates that the photon energy is $E = hf$, where h is Planck's constant and f is the frequency. Equation 16.1 relates the frequency to the wavelength λ according to $f = c/\lambda$, where c is the speed of light. Thus, the photon energy is $E = hc/\lambda$. This result indicates that an infrared photon carries a smaller amount of energy than a visible photon, because the infrared photon has a greater wavelength.

c. To deliver a given amount of energy to the glass, a greater number of infrared photons will be needed, because each infrared photon carries less energy than a visible photon.

SOLUTION The total energy Q delivered by N photons is NE, where E is the energy carried by one photon. Since $E = hc/\lambda$, according to Equations 29.2 and 16.1, we have $Q = NE = Nhc/\lambda$. But Equation 12.4 indicates that $Q = c_{\text{specific heat}} m \Delta T$. As a result, we find that

$$c_{\text{specific heat}} m \Delta T = \frac{Nhc}{\lambda} \quad \text{or} \quad N = \frac{\lambda c_{\text{specific heat}} m \Delta T}{hc}$$

Applying this result to each type of photon, we obtain

$$N_{\text{infrared}} = \frac{(6.0 \times 10^{-5} \text{ m})[840 \text{ J}/(\text{kg} \cdot \text{C}°)](0.50 \text{ kg})(2.0 \text{ C}°)}{(6.63 \times 10^{-34} \text{ J} \cdot \text{s})(3.00 \times 10^8 \text{ m/s})} = \boxed{2.5 \times 10^{23}}$$

$$N_{\text{visible}} = \frac{(4.7 \times 10^{-7} \text{ m})[840 \text{ J}/(\text{kg} \cdot \text{C}°)](0.50 \text{ kg})(2.0 \text{ C}°)}{(6.63 \times 10^{-34} \text{ J} \cdot \text{s})(3.00 \times 10^8 \text{ m/s})} = \boxed{2.0 \times 10^{21}}$$

48. **CONCEPT QUESTIONS** a. Before the scattering, the electron is at rest and has no momentum. Thus, the total initial momentum consists only of the photon momentum, which points along the +x axis. The total initial momentum has no y component.

b. Since the total momentum is conserved, the total momentum after the scattering must be the same as it was before and, therefore, has no y component.

c. The total momentum after the scattering is the sum of the momentum of the scattered photon and that of the scattered electron, and it has an x component. But the scattered photon is moving along the −y axis, so its momentum has no x component. Therefore, the momentum of the electron must have an x component.

d. The total momentum after the scattering is the sum of the momentum of the scattered photon and that of the scattered electron, and it has no y component. But the scattered photon is moving along the −y axis, so its momentum points along the −y axis. Therefore, this contribution to the total final momentum must be cancelled by part of the momentum of the scattered electron, which must have a component along the +y axis.

SOLUTION Since the total momentum is conserved and since the scattered photon has no momentum in the x direction, the momentum of the scattered electron must have an x component that equals the momentum of the incident photon. According to Equation 29.6, the momentum of the incident photon is $p = h/\lambda$, where h is Planck's constant and λ is the wavelength. Therefore, the momentum of the scattered electron has a component in the +x direction that is

$$p_x = \frac{h}{\lambda} = \frac{6.63 \times 10^{-34} \text{ J} \cdot \text{s}}{9.00 \times 10^{-12} \text{ m}} = \boxed{7.37 \times 10^{-23} \text{ kg} \cdot \text{m/s}}$$

The momentum of the scattered electron has a component in the +y direction that cancels the momentum of the scattered photon that points along the −y direction. To find the momentum of the scattered photon, we first need to determine its wavelength, which we can do using Equation 29.7:

$$\lambda' = \lambda + \frac{h}{mc}(1 - \cos\theta)$$
$$= 9.00 \times 10^{-12} \text{ m} + \frac{6.63 \times 10^{-34} \text{ J} \cdot \text{s}}{(9.11 \times 10^{-31} \text{ kg})(3.00 \times 10^8 \text{ m/s})} = 1.14 \times 10^{-11} \text{ m}$$

Again using Equation 29.6, we find that the momentum of the scattered electron has a component in the +y direction that is

$$p_y = \frac{h}{\lambda} = \frac{6.63 \times 10^{-34} \text{ J} \cdot \text{s}}{1.14 \times 10^{-11} \text{ m}} = \boxed{5.82 \times 10^{-23} \text{ kg} \cdot \text{m/s}}$$

49. **CONCEPT QUESTIONS** a. According to Equation 27.1, the angle θ that locates the first-order bright fringes is specified by $\sin\theta = \lambda/d$, where λ is the wavelength and d is the separation between the slits. For a given slit separation, greater values of the wavelength lead to greater values of the angle θ. Since θ_B is greater than θ_A, it follows, then, that λ_B is greater than λ_A.

b. The wavelength of the electron is the de Broglie wavelength, which is given by Equation 29.8 as $\lambda = h/p$, where h is Planck's constant and p is the magnitude of the momentum of the electron. Greater values of the wavelength correspond to smaller values of the momentum. Since λ_B is greater than λ_A, it follows that p_B is smaller than p_A.

SOLUTION Using Equation 27.1 and 29.8, we find that the angle locating the first-order bright fringes is specified by

$$\sin\theta = \frac{\lambda}{d} = \frac{h}{pd}$$

Applying this result to cases A and B, we find

$$\frac{\sin\theta_A}{\sin\theta_B} = \frac{h/(p_A d)}{h/(p_B d)} = \frac{p_B}{p_A} \quad \text{or} \quad p_B = \frac{p_A \sin\theta_A}{\sin\theta_B}$$

$$p_B = \frac{(1.2\times 10^{-22}\ \text{kg}\cdot\text{m/s})\sin(1.6\times 10^{-4}\ \text{degrees})}{\sin(4.0\times 10^{-4}\ \text{degrees})} = \boxed{4.8\times 10^{-23}\ \text{kg}\cdot\text{m/s}}$$

CHAPTER 30 | *THE NATURE OF THE ATOM*

PROBLEMS

1. **SSM** **REASONING** Assuming that the hydrogen atom is a sphere of radius r_{atom}, its volume V_{atom} is given by $\frac{4}{3}\pi r_{atom}^3$. Similarly, if the radius of the nucleus is $r_{nucleus}$, the volume $V_{nucleus}$ is given by $\frac{4}{3}\pi r_{nucleus}^3$.

    SOLUTION
    a. According to the given data, the nuclear dimensions are much smaller than the orbital radius of the electron; therefore, we can treat the nucleus as a point about which the electron orbits. The electron is normally at a distance of about 5.3×10^{-11} m from the nucleus, so we can treat the atom as a sphere of radius $r_{atom} = 5.3 \times 10^{-11}$ m. The volume of the atom is

    $$V_{atom} = \frac{4}{3}\pi r_{atom}^3 = \frac{4}{3}\pi (5.3 \times 10^{-11} \text{ m})^3 = \boxed{6.2 \times 10^{-31} \text{ m}^3}$$

    b. Similarly, since the nucleus has a radius of approximately $r_{nucleus} = 1 \times 10^{-15}$ m, its volume is

    $$V_{nucleus} = \frac{4}{3}\pi r_{nucleus}^3 = \frac{4}{3}\pi (1 \times 10^{-15} \text{ m})^3 = \boxed{4 \times 10^{-45} \text{ m}^3}$$

    c. The percentage of the atomic volume occupied by the nucleus is

    $$\frac{V_{nucleus}}{V_{atom}} \times 100\% = \frac{r_{nucleus}^3}{r_{atom}^3} \times 100\% = \left(\frac{1 \times 10^{-15} \text{ m}}{5.3 \times 10^{-11} \text{ m}}\right)^3 \times 100\% = \boxed{7 \times 10^{-13} \%}$$

2. **REASONING** According to the discussion in Conceptual Example 1, the radius of the electron orbit is about 1×10^5 times as great as the radius of the nucleus. In the scale model, the radius of the electron's orbit must also be 1×10^5 times as great as the radius of ball.

    SOLUTION Since the ball that represents the nucleus has a radius of 3.2 cm, the distance between the nucleus and the nearest electron in the model must be

    $$(3.2 \text{ cm})(1 \times 10^5) = (3.2 \times 10^5 \text{ cm})\left(\frac{1 \text{ mi}}{1.61 \times 10^5 \text{ cm}}\right) = \boxed{2 \text{ mi}}$$

3. **REASONING AND SOLUTION** The density ρ is the mass m per unit volume V or $\rho = m/V$. Since the volume of a sphere is $V = 4\pi r^3/3$, the density can be written as $\rho = 3m/(4\pi r^3)$. We use m_p and m_e to denote the mass of a proton and the mass of an electron, respectively. Considering that the hydrogen nucleus is a single proton with a radius of r_p, it follows that the desired density ratio is

$$\frac{\rho_{nucleus}}{\rho_{atom}} = \frac{3m_p/(4\pi r_p^3)}{3(m_p + m_e)/(4\pi r_{atom}^3)} = \frac{m_p r_{atom}^3}{(m_p + m_e) r_p^3}$$

$$= \frac{(1.67 \times 10^{-27} \text{ kg})(5.3 \times 10^{-11} \text{ m})^3}{(1.67 \times 10^{-27} \text{ kg} + 9.11 \times 10^{-31} \text{ kg})(1.0 \times 10^{-15} \text{ m})^3} = \boxed{1.5 \times 10^{14}}$$

---

4. **REASONING AND SOLUTION** The de Broglie wavelength is given by $\lambda = h/(mv)$. In order to find the speed v, we use the fact that the kinetic energy is given by $KE = mv^2/2$, Equation 6.2. Thus,

$$v = \sqrt{\frac{2KE}{m}} = \sqrt{\frac{2(7.00 \times 10^{-13} \text{ J})}{6.64 \times 10^{-27} \text{ kg}}} = 1.45 \times 10^7 \text{ m/s}$$

Now

$$\lambda = \frac{h}{mv} = \frac{6.626 \times 10^{-34} \text{ J} \cdot \text{s}}{(6.64 \times 10^{-27} \text{ kg})(1.45 \times 10^7 \text{ m/s})} = \boxed{6.88 \times 10^{-15} \text{ m}} \quad (29.8)$$

---

5. **SSM REASONING** The copper nucleus has a total charge $Ze = 29 \times 1.6 \times 10^{-19}$ C, and radius $r_{nucleus} = 4.8 \times 10^{-15}$ m. The required work can be found from Equation 19.1: $W_{AB} = EPE_A - EPE_B$. According to Equation 19.3, $EPE = Vq$, where $q = e$ for a proton. Therefore, $W_{AB} = V_A e - V_B e$. The proton is initially at infinity, so $V_A = 0$ V and $V_B = kZe/r_{nucleus}$, according to Equation 19.6. Thus, $W_{AB} = -kZe^2/r_{nucleus}$.

SOLUTION The work done by the electric force is, therefore,

$$W_{AB} = \frac{-kZe^2}{r_{nucleus}} = \frac{-(8.99 \times 10^9 \text{ N} \cdot \text{m}^2/\text{C}^2)(29)(1.6 \times 10^{-19} \text{ C})^2}{4.8 \times 10^{-15} \text{ m}} = -1.39 \times 10^{-12} \text{ J}$$

$$W_{AB} = (-1.39 \times 10^{-12} \text{ J}) \underbrace{\left(\frac{1.0 \text{ eV}}{1.6 \times 10^{-19} \text{ J}}\right)}_{\text{Converts from joules to electron volts}} = \boxed{-8.7 \times 10^6 \text{ eV}}$$

---

6. **REASONING AND SOLUTION** The distance of closest approach can be obtained by setting KE = EPE or KE = $(kZe/r)(2e)$, and solving for r to get $r = kZ(2e^2)/(\text{KE})$.

$$r = (8.99 \times 10^9 \text{ N} \cdot \text{m}^2/\text{C}^2)(79)2(1.602 \times 10^{-19} \text{ C})^2/(5.0 \times 10^{-13} \text{ J}) = \boxed{7.3 \times 10^{-14} \text{ m}}$$

---

7. **SSM** **WWW** **REASONING AND SOLUTION** The radii for Bohr orbits are given by Equation 30.10:

$$r_n = (5.29 \times 10^{-11} \text{ m}) \frac{n^2}{Z}$$

The Bohr radii for a doubly ionized lithium atom Li^{2+} are given by Equation 30.10 with $Z = 3$; therefore, the radius of the $n = 5$ Bohr orbit in a doubly ionized lithium atom Li^{2+} is

$$r_5 = (5.29 \times 10^{-11} \text{ m}) \frac{5^2}{3} = \boxed{4.41 \times 10^{-10} \text{ m}}$$

---

8. **REASONING AND SOLUTION** The first excited state occurs for $n = 2$ and has an energy given by Equation 30.13:

$$E_2 = -(13.6 \text{ eV}) \frac{Z^2}{n^2} = -(13.6 \text{ eV}) \frac{1^2}{2^2} = -3.40 \text{ eV}$$

After absorbing 2.86 eV of energy, the electron has an energy

$$E_n = -3.40 \text{ eV} + 2.86 \text{ eV} = -0.54 \text{ eV}$$

Again using Equation 30.13, we find $E_n = -0.54 \text{ eV} = -(13.6 \text{ eV})(1^2/n^2)$. Solving for n, we obtain

$$n = \sqrt{\frac{13.6 \text{ eV}}{0.54 \text{ eV}}} = \boxed{5}$$

9. **REASONING AND SOLUTION**
   a. The longest wavelength in the Pfund series occurs for the transition $n = 6$ to $n = 5$, so that

$$\frac{1}{\lambda} = R\left(\frac{1}{5^2} - \frac{1}{n^2}\right) = (1.097 \times 10^7 \text{ m}^{-1})\left(\frac{1}{5^2} - \frac{1}{6^2}\right) \quad \text{or} \quad \boxed{\lambda = 7458 \text{ nm}}$$

   b. The shortest wavelength occurs when $1/n^2 = 0$, so that

$$\frac{1}{\lambda} = R\left(\frac{1}{5^2} - \frac{1}{n^2}\right) = (1.097 \times 10^7 \text{ m}^{-1})\left(\frac{1}{5^2}\right) \quad \text{or} \quad \boxed{\lambda = 2279 \text{ nm}}$$

   c. The lines in the Pfund series occur in the $\boxed{\text{infrared region}}$.

10. **REASONING** The atomic number for helium is $Z = 2$. The ground state is the $n = 1$ state, the first excited state is the $n = 2$ state, and the second excited state is the $n = 3$ state. With $Z = 2$ and $n = 3$, we can use Equation 30.10 to find the radius of the ion.

    SOLUTION The radius of the second excited state is

$$r_3 = (5.29 \times 10^{-11} \text{ m})\frac{n^2}{Z} = (5.29 \times 10^{-11} \text{ m})\frac{3^2}{2} = \boxed{2.38 \times 10^{-10} \text{ m}} \quad (30.10)$$

11. **SSM REASONING** According to Equation 30.14, the wavelength λ emitted by the hydrogen atom when it makes a transition from the level with n_i to the level with n_f is given by

$$\frac{1}{\lambda} = \frac{2\pi^2 m k^2 e^4}{h^3 c}(Z^2)\left(\frac{1}{n_f^2} - \frac{1}{n_i^2}\right) \quad \text{with} \quad n_i, n_f = 1, 2, 3, \ldots \quad \text{and} \quad n_i > n_f$$

where $2\pi^2 m k^2 e^4/(h^3 c) = 1.097 \times 10^7 \text{ m}^{-1}$ and $Z = 1$ for hydrogen. Once the wavelength for the particular transition in question is determined, Equation 29.2 ($E = hf = hc/\lambda$) can be used to find the energy of the emitted photon.

1074   THE NATURE OF THE ATOM

SOLUTION In the Paschen series, $n_f = 3$. Using the above expression with $Z = 1$, $n_i = 7$ and $n_f = 3$, we find that

$$\frac{1}{\lambda} = (1.097 \times 10^7 \text{ m}^{-1})(1^2)\left(\frac{1}{3^2} - \frac{1}{7^2}\right) \quad \text{or} \quad \lambda = 1.005 \times 10^{-6} \text{ m}$$

The photon energy is

$$E = \frac{hc}{\lambda} = \frac{(6.63 \times 10^{-34} \text{ J} \cdot \text{s})(3.00 \times 10^8 \text{ m/s})}{1.005 \times 10^{-6} \text{ m}} = \boxed{1.98 \times 10^{-19} \text{ J}}$$

---

12. **REASONING** Since the atom emits two photons as it returns to the ground state, one is emitted when the electron falls from $n = 3$ to $n = 2$, and the other is emitted when it subsequently drops from $n = 2$ to $n = 1$. The wavelengths of the photons emitted during these transitions are given by Equation 30.14 with the appropriate values for the initial and final numbers, n_i and n_f.

SOLUTION The wavelengths of the photons are

$n = 3$ to $n = 2$
$$\frac{1}{\lambda} = (1.097 \times 10^7 \text{ m}^{-1})(1)^2 \left(\frac{1}{2^2} - \frac{1}{3^2}\right) = 1.524 \times 10^6 \text{ m}^{-1} \quad (30.14)$$

$$\lambda = \boxed{6.56 \times 10^{-7} \text{ m}}$$

$n = 2$ to $n = 1$
$$\frac{1}{\lambda} = (1.097 \times 10^7 \text{ m}^{-1})(1)^2 \left(\frac{1}{1^2} - \frac{1}{2^2}\right) = 8.228 \times 10^6 \text{ m}^{-1} \quad (30.14)$$

$$\lambda = \boxed{1.22 \times 10^{-7} \text{ m}}$$

---

13. **REASONING** The Bohr expression as it applies to any one-electron species of atomic number Z, is given by Equation 30.13: $E_n = -(13.6 \text{ eV})(Z^2/n^2)$. For certain values of the quantum number n, this expression predicts equal electron energies for singly ionized helium He^+ ($Z = 2$) and doubly ionized lithium Li^+ ($Z = 3$). As stated in the problem, the quantum number n is different for the equal energy states for each species.

SOLUTION For equal energies, we can write

$$(E_n)_{He} = (E_n)_{Li} \quad \text{or} \quad -(13.6 \text{ eV})\frac{Z_{He}^2}{n_{He}^2} = -(13.6 \text{ eV})\frac{Z_{Li}^2}{n_{Li}^2}$$

Simplifying, this becomes

$$\frac{Z_{He}^2}{n_{He}^2} = \frac{Z_{Li}^2}{n_{Li}^2} \quad \text{or} \quad \frac{4}{n_{He}^2} = \frac{9}{n_{Li}^2}$$

Thus,

$$n_{He} = n_{Li}\sqrt{\frac{4}{9}} = n_{Li}\left(\frac{2}{3}\right)$$

Therefore, the value of the helium energy level is equal to the lithium energy level for any value of n_{He} that is two-thirds of n_{Li}. For quantum numbers less than or equal to 9, an equality in energy levels will occur for $n_{He} = 2, 4, 6$ corresponding to $n_{Li} = 3, 6, 9$. The results are summarized in the following table.

n_{He}	n_{Li}	Energy
2	3	−13.6 eV
4	6	−3.40 eV
6	9	−1.51 eV

14. **REASONING AND SOLUTION** The energy of the hydrogen atom, $Z = 1$, in electron volts is given by Equation 30.13, $E_n = -13.6/n^2$, which requires a value for the quantum number n. A value for n can be obtained from Equation 30.10, $r_n = (5.29 \times 10^{-11} \text{ m})n^2$, which gives the radius of the hydrogen atom in meters.

$$n^2 = \frac{r_n}{5.29 \times 10^{-11} \text{ m}} = \frac{4.761 \times 10^{-10} \text{ m}}{5.29 \times 10^{-11} \text{ m}} = 9 \quad \text{or} \quad n = 3$$

$$E_n = -\frac{13.6 \text{ eV}}{n^2} = -\frac{13.6}{3^2} = \boxed{-1.51 \text{ eV}}$$

15. **SSM REASONING AND SOLUTION** For the Paschen series, $n_f = 3$. The range of wavelengths occurs for values of $n_i = 4$ to $n_i = \infty$. Using Equation 30.14, we find that the shortest wavelength occurs for $n_i = \infty$ and is given by

$$\frac{1}{\lambda} = (1.097 \times 10^7 \text{ m}^{-1})(1)^2 \left(\frac{1}{n_f^2} - \frac{1}{n_i^2}\right) = (1.097 \times 10^7 \text{ m}^{-1})\left(\frac{1}{3^2}\right) \quad \text{or} \quad \underbrace{\lambda = 8.204 \times 10^{-7} \text{ m}}_{\text{Shortest wavelength in Paschen series}}$$

The longest wavelength in the Paschen series occurs for $n_i = 4$ and is given by

$$\frac{1}{\lambda} = \left(1.097 \times 10^7 \text{ m}^{-1}\right)\left(\frac{1}{3^2} - \frac{1}{4^2}\right) \quad \text{or} \quad \underbrace{\lambda = 1.875 \times 10^{-6} \text{ m}}_{\text{Longest wavelength in Paschen series}}$$

For the Brackett series, $n_f = 4$. The range of wavelengths occurs for values of $n_i = 5$ to $n_i = \infty$. Using Equation 30.14, we find that the shortest wavelength occurs for $n_i = \infty$ and is given by

$$\frac{1}{\lambda} = \left(1.097 \times 10^7 \text{ m}^{-1}\right)(1)^2\left(\frac{1}{n_f^2} - \frac{1}{n_i^2}\right) = \left(1.097 \times 10^7 \text{ m}^{-1}\right)\left(\frac{1}{4^2}\right) \quad \text{or} \quad \underbrace{\lambda = 1.459 \times 10^{-6} \text{ m}}_{\text{Shortest wavelength in Brackett series}}$$

The longest wavelength in the Brackett series occurs for $n_i = 5$ and is given by

$$\frac{1}{\lambda} = \left(1.097 \times 10^7 \text{ m}^{-1}\right)\left(\frac{1}{4^2} - \frac{1}{5^2}\right) \quad \text{or} \quad \underbrace{\lambda = 4.051 \times 10^{-6} \text{ m}}_{\text{Longest wavelength in Brackett series}}$$

Since the longest wavelength in the Paschen series falls within the Brackett series, the wavelengths of the two series overlap.

---

16. **REASONING AND SOLUTION**
According to Newton's second law, as expressed in Equation 30.6, $mv^2 = kZe^2/r$, which says that KE = $-(1/2)$ (EPE). Since KE + EPE = E (total energy), we have $-(1/2)$ (EPE) + EPE = E, or EPE = $2E$ = $2(-4.90 \text{ eV})$ = $\boxed{-9.80 \text{ eV}}$.

Since KE = $-(1/2)$ (EPE), we have KE = $-(1/2)(-9.80 \text{ eV})$ = $\boxed{+4.90 \text{ eV}}$.

---

17. **SSM REASONING** Singly ionized helium, He^+, is a hydrogen-like species with $Z = 2$. The wavelengths of the series of lines produced when the electron makes a transition from higher energy levels into the $n_f = 4$ level are given by Equation 30.14 with $Z = 2$ and $n_f = 4$:

$$\frac{1}{\lambda} = \left(1.097 \times 10^7 \text{ m}^{-1}\right)(2^2)\left(\frac{1}{4^2} - \frac{1}{n_i^2}\right)$$

SOLUTION Solving this expression for n_i gives

$$n_i = \left[\frac{1}{4^2} - \frac{1}{4\lambda(1.097 \times 10^7 \text{ m}^{-1})}\right]^{-1/2}$$

Evaluating this expression at the limits of the range for λ, we find that $n_i = 19.88$ for $\lambda = 380$ nm, and $n_i = 5.58$ for $\lambda = 750$ nm. Therefore, the values of n_i for energy levels from which the electron makes the transitions that yield wavelengths in the range between 380 nm and 750 nm are $\boxed{6 \leq n_i \leq 19}$.

---

18. **REASONING** For each species, the wavelengths appearing in the line spectra can be calculated using Equation 30.14:

$$\frac{1}{\lambda} = RZ^2\left(\frac{1}{n_f^2} - \frac{1}{n_i^2}\right)$$

The shortest wavelength for a given series of lines occurs when an electron is in an initial energy level with a principal quantum number of $n_i = \infty$. Therefore, $1/n_i^2 = 1/\infty = 0$, and the shortest wavelength is given by

$$\frac{1}{\lambda} = \frac{RZ^2}{n_f^2}$$

In this expression, the value of n_f is the same for the series for Li^{2+} and for Be^{3+}. The term R is the Rydberg constant and is also the same for both ionic species. Thus, rearranging this equation gives

$$\lambda Z^2 = \frac{n_f^2}{R} = \text{constant}$$

which is the basis for our solution.

SOLUTION Since λZ^2 is the same for each species, we have

$$\left(\lambda Z^2\right)_{\text{Li}} = \left(\lambda Z^2\right)_{\text{Be}} \quad \text{or} \quad \lambda_{\text{Be}} = \frac{\left(\lambda Z^2\right)_{\text{Li}}}{Z_{\text{Be}}^2} = \frac{(40.5 \text{ nm})(3)^2}{(4)^2} = \boxed{22.8 \text{ nm}}$$

1078 THE NATURE OF THE ATOM

19. **REASONING AND SOLUTION** We need to use Equation 30.14 to find the spacing between the longest and next-to-the longest wavelengths in the Balmer series. In order to do this, we need to first find these two wavelengths.

Longest:

$$\frac{1}{\lambda_1} = R\left(\frac{1}{n_f^2} - \frac{1}{n_i^2}\right) = \left(1.097 \times 10^7 \text{ m}^{-1}\right)\left(\frac{1}{2^2} - \frac{1}{3^2}\right) \quad \text{or} \quad \lambda_1 = 656.3 \text{ nm}$$

Next-to-longest:

$$\frac{1}{\lambda_2} = R\left(\frac{1}{n_f^2} - \frac{1}{n_i^2}\right) = \left(1.097 \times 10^7 \text{ m}^{-1}\right)\left(\frac{1}{2^2} - \frac{1}{4^2}\right) \quad \text{or} \quad \lambda_2 = 486.2 \text{ nm}$$

Equation 27.7 states that $\sin\theta = m\lambda/d$. Using the small angle approximation, $\sin\theta \approx \tan\theta \approx \theta = y/L$, so that $y/L = m\lambda/d$. The position of the fringe due to the longest wavelength is $y_1 = m\lambda_1 L/d$. For the next-to-longest, $y_2 = m\lambda_2 L/d$. The difference in the positions on the screen is therefore, $y_1 - y_2 = (mL/d)(\lambda_1 - \lambda_2)$ which gives

$$d = (mL)(\lambda_1 - \lambda_2)/(y_1 - y_2)$$

$$d = (1)(0.81 \text{ m})(656.3 \times 10^{-9} \text{ m} - 486.2 \times 10^{-9} \text{ m})/(0.0300 \text{ m})$$

$$d = 4.59 \times 10^{-6} \text{ m} = 4.59 \times 10^{-4} \text{ cm}$$

This spacing between adjacent lines corresponds to a number of lines per centimeter that is

Number of lines per cm = $1/(4.59 \times 10^{-4} \text{ cm})$ = $\boxed{2180 \text{ lines/cm}}$

---

20. **REASONING AND SOLUTION** The shortest wavelength, λ_s, occurs when n_i = infinity, so that $1/n_i = 0$. In that case Equation 30.14 becomes

$$\frac{1}{\lambda_s} = \frac{RZ^2}{n_f^2} \quad \text{or} \quad RZ^2 = \frac{n_f^2}{\lambda_s}$$

The longest wavelength in the series occurs when $n_i = n_f + 1$.

$$\frac{1}{\lambda_1} = RZ^2\left(\frac{1}{n_f^2} - \frac{1}{n_i^2}\right) = \left(\frac{n_f^2}{\lambda_s}\right)\left(\frac{1}{n_f^2} - \frac{1}{n_i^2}\right)$$

A little algebra gives n_f as follows:

$$\lambda_l/\lambda_s = (41.02 \times 10^{-9} \text{ m})/(22.79 \times 10^{-9} \text{ m}) = (n_f + 1)^2/(2n_f + 1)$$

or

$$n_f^2 - 1.600\, n_f - 0.800 = 0$$

Solving the quadratic equation yields only one positive root, which is $n_f = 2$. Therefore, $1/\lambda_s = (1/22.79 \times 10^{-9} \text{ m}) = RZ^2/n_f^2 = RZ^2/4$, which gives $Z = 4$. As a result, the next-to-the-longest wavelength is:

$$1/\lambda = R(4)^2(1/2^2 - 1/4^2) \quad \text{or} \quad \boxed{\lambda = 30.39 \text{ nm}}$$

---

21. **REASONING** The orbital quantum number ℓ has values of $0, 1, 2, \ldots, (n-1)$, according to the discussion in Section 30.5. Since $\ell = 5$, we can conclude, therefore, that $n \geq 6$. This knowledge about the principal quantum number n can be used with Equation 30.13, $E_n = -(13.6 \text{ eV})Z^2/n^2$, to determine the smallest value for the total energy E_n.

SOLUTION The smallest value of E_n (i.e., the most negative) occurs when $n = 6$. Thus, using $Z = 1$ for hydrogen, we find

$$E_n = -(13.6 \text{ eV})\frac{Z^2}{n^2} = -(13.6 \text{ eV})\frac{1^2}{6^2} = \boxed{-0.378 \text{ eV}}$$

---

22. **REASONING AND SOLUTION** The largest value of ℓ for $n = 4$ is $\ell = n - 1 = 3$. The largest value of m_ℓ for $\ell = 3$ is $m_\ell = 3$. Since m_ℓ runs from minus its largest value to plus that value in integer steps, we have

$$m_\ell = \boxed{-3, -2, -1, 0, +1, +2, +3}$$

---

23. **SSM WWW REASONING** The values that ℓ can have depend on the value of n, and only the following integers are allowed: $\ell = 0, 1, 2, \ldots (n-1)$. The values that m_ℓ can have depend on the value of ℓ, with only the following positive and negative integers being permitted: $m_\ell = -\ell, \ldots -2, -1, 0, +1, +2, \ldots +\ell$.

SOLUTION Thus, when $n = 6$, the possible values of ℓ are 0, 1, 2, 3, 4, 5. Now when $m_\ell = 2$, the possible values of ℓ are 2, 3, 4, 5, ... These two series of integers overlap for

1080 THE NATURE OF THE ATOM

the integers 2, 3, 4, and 5. Therefore, the possible values for the orbital quantum number ℓ that this electron could have are $\boxed{\ell = 2, 3, 4, 5}$.

---

24. **REASONING**
a. The ground state is the $n = 1$ state, the first excited state is the $n = 2$ state, and the second excited state is the $n = 3$ state. The total energy (in eV) of a hydrogen atom in the $n = 3$ state is given by Equation 30.13.

b. According to quantum mechanics, the magnitude L of the angular momentum is given by Equation 30.15 as $L = \sqrt{\ell(\ell+1)}(h/2\pi)$, where ℓ is the orbital quantum number. The discussion in Section 30.5 indicates that the maximum value that ℓ can have is one less than the principal quantum number, so that $\ell_{max} = n - 1$.

c. Equation 30.16 gives the z-component L_z of the angular momentum as $L_z = m_\ell(h/2\pi)$, where m_ℓ is the magnetic quantum number. According to the discussion in Section 30.5, the maximum value that m_ℓ can attain is when it is equal to the orbital quantum number, which is ℓ_{max}.

SOLUTION
a. The total energy of the hydrogen atom is given by Equation 30.13. Using $n = 3$, we have

$$E_3 = -\frac{(13.6\,\text{eV})(1)^2}{3^2} = \boxed{-1.51\,\text{eV}}$$

b. The maximum orbital quantum number is $\ell_{max} = n - 1 = 3 - 1 = 2$. The maximum angular momentum L_{max} has a magnitude given by Equation 30.15:

$$L_{max} = \sqrt{\ell_{max}(\ell_{max}+1)}\,\frac{h}{2\pi} = \sqrt{2(2+1)}\,\frac{6.63 \times 10^{-34}\,\text{J}\cdot\text{s}}{2\pi} = \boxed{2.58 \times 10^{-34}\,\text{J}\cdot\text{s}}$$

c. The maximum value for the z-component L_z of the angular momentum is (with $m_\ell = \ell_{max} = 2$)

$$L_z = m_\ell\,\frac{h}{2\pi} = (2)\frac{6.63 \times 10^{-34}\,\text{J}\cdot\text{s}}{2\pi} = \boxed{2.11 \times 10^{-34}\,\text{J}\cdot\text{s}}$$

---

25. **REASONING AND SOLUTION** The magnitude of the angular momentum in a hydrogen atom is given by Equation 30.15:

$$L = \sqrt{\ell(\ell+1)}\left(\frac{h}{2\pi}\right)$$

Solving for ℓ and using the given value for L to show that $4\pi^2 L^2 / h^2 = 12$, we find that

$$\ell^2 + \ell - 4\pi^2 L^2/h^2 = \ell^2 + \ell - 12 = 0 \quad \text{or} \quad \ell = -4 \text{ or } +3$$

Since ℓ must be positive, we take $\ell = 3$ for this case. The z-component of the angular momentum is given by $L_z = m_\ell h/(2\pi)$. We need to determine the possible values of m_ℓ for the value of ℓ found above. We know that m_ℓ can have values $m_\ell = 0, \pm 1, \pm 2, ..., \pm \ell$. Therefore, the possible values of m_ℓ are $m_\ell = 3, 2, 1, 0, -1, -2, -3$. With $h = 6.626 \times 10^{-34}$ J·s, the seven possible values for L_z are, therefore,

$$L_z = 3\left(\frac{h}{2\pi}\right) = \boxed{3.16 \times 10^{-34} \text{ J·s}} \quad \text{or} \quad L_z = -3\left(\frac{h}{2\pi}\right) = \boxed{-3.16 \times 10^{-34} \text{ J·s}}$$

$$L_z = 2\left(\frac{h}{2\pi}\right) = \boxed{2.11 \times 10^{-34} \text{ J·s}} \quad \text{or} \quad L_z = -2\left(\frac{h}{2\pi}\right) = \boxed{-2.11 \times 10^{-34} \text{ J·s}}$$

$$L_z = 1\left(\frac{h}{2\pi}\right) = \boxed{1.05 \times 10^{-34} \text{ J·s}} \quad \text{or} \quad L_z = -1\left(\frac{h}{2\pi}\right) = \boxed{-1.05 \times 10^{-34} \text{ J·s}}$$

$$L_z = 0\left(\frac{h}{2\pi}\right) = \boxed{0 \text{ J·s}}$$

---

26. **REASONING AND SOLUTION**
a. For the angular momentum, Bohr's value is given by Equation 30.8, with $n = 1$,

$$L_n = \frac{nh}{2\pi} = \boxed{\frac{h}{2\pi}}$$

According to quantum theory, the angular momentum is given by Equation 30.15. For $n = 1, \ell = 0$

$$L = \sqrt{\ell(\ell+1)}\left(\frac{h}{2\pi}\right) = \sqrt{0(0+1)}\left(\frac{h}{2\pi}\right) = \boxed{0 \text{ J·s}}$$

b. For $n = 3$; Bohr theory gives

$$L_n = \frac{nh}{2\pi} = \boxed{\frac{3h}{2\pi}}$$

while quantum mechanics gives

1082   THE NATURE OF THE ATOM

[n = 3, ℓ = 0] $L = \sqrt{\ell(\ell+1)}\left(\dfrac{h}{2\pi}\right) = \sqrt{0(0+1)}\left(\dfrac{h}{2\pi}\right) = \boxed{0\text{ J}\cdot\text{s}}$

[n = 3, ℓ = 1] $L = \sqrt{\ell(\ell+1)}\left(\dfrac{h}{2\pi}\right) = \sqrt{1(1+1)}\left(\dfrac{h}{2\pi}\right) = \boxed{\dfrac{\sqrt{2}h}{2\pi}}$

[n = 3, ℓ = 2] $L = \sqrt{\ell(\ell+1)}\left(\dfrac{h}{2\pi}\right) = \sqrt{2(2+1)}\left(\dfrac{h}{2\pi}\right) = \boxed{\dfrac{\sqrt{6}h}{2\pi}}$

---

27. **SSM WWW REASONING** Let θ denote the angle between the angular momentum L and its z-component L_z. We can see from the figure at the right that $L_z = L\cos\theta$. Using Equation 30.16 for L_z and Equation 30.15 for L, we have

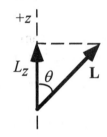

$$\cos\theta = \dfrac{L_z}{L} = \dfrac{m_\ell h/(2\pi)}{\sqrt{\ell(\ell+1)}[h/(2\pi)]} = \dfrac{m_\ell}{\sqrt{\ell(\ell+1)}}$$

The smallest value for θ corresponds to the largest value of $\cos\theta$. For a given value of ℓ, the largest value for $\cos\theta$ corresponds to the largest value for m_ℓ. But the largest possible value for m_ℓ is $m_\ell = \ell$. Therefore, we find that

$$\cos\theta = \dfrac{\ell}{\sqrt{\ell(\ell+1)}} = \sqrt{\dfrac{\ell}{\ell+1}}$$

SOLUTION The smallest value for θ corresponds to the largest value for ℓ. When the electron is in the $n = 5$ state, the largest allowed value of ℓ is $\ell = 4$; therefore, we see that

$$\cos\theta = \sqrt{\dfrac{\ell}{\ell+1}} = \sqrt{\dfrac{4}{4+1}} \quad \text{or} \quad \theta = \cos^{-1}\left(\sqrt{4/5}\right) = \boxed{26.6°}$$

---

28. **REASONING AND SOLUTION**

$$\boxed{1s^2\ 2s^2\ 2p^6\ 3s^2\ 3p^6\ 4s^2\ 3d^{10}\ 4p^6\ 5s^2\ 4d^1}$$

---

29. **REASONING AND SOLUTION** According to Figure 30.17, the energy sublevel with $n = 4$, $\ell = 0$, which corresponds to the notation 4s, is lower in energy than the $n = 3$, $\ell = 2$, (3d) energy sublevel. Thus, the 4s energy sublevel will be filled before the 3d energy

sublevel. Therefore, using Figure 30.17 as a guide, we find that the ground state electronic configuration of arsenic ($Z = 33$) is

$$\boxed{1s^2\ 2s^2\ 2p^6\ 3s^2\ 3p^6\ 4s^2\ 3d^{10}\ 4p^3}$$

---

30. **REASONING**
   a. The maximum number of electrons that the $n = 1$, $\ell = 0$ subshell can contain is 2 (see Figure 30.17). If the atom is in its ground state, the third electron must go into the $n = 2$, $\ell = 0$ subshell. For this subshell, the magnetic quantum number can only be zero ($m_\ell = 0$), and the electron can have a spin quantum number of either $m_s = +\tfrac{1}{2}$ or $-\tfrac{1}{2}$.

   b. If the atom is in its first excited state, the third electron must go into the $n = 2$, $\ell = 1$ subshell. For $\ell = 1$, there are three possibilities for m_ℓ: $m_\ell = +1, 0,$ and -1. For each value of m_ℓ, the electron can have a spin quantum number of either $m_s = +\tfrac{1}{2}$ or $-\tfrac{1}{2}$.

   SOLUTION
   a. According to the discussion in the Reasoning section, the third electron can have one of two possibilities for the four quantum numbers: $\boxed{n = 2, \ell = 0, m_\ell = 0, m_s = +\tfrac{1}{2}}$ and $\boxed{n = 2, \ell = 0, m_\ell = 0, m_s = -\tfrac{1}{2}}$.

   b. When the third electron is in the first excited state, there are six possibilities for the four quantum numbers:

   $\boxed{n = 2, \ell = 1, m_\ell = +1, m_s = +\tfrac{1}{2}}$, $\boxed{n = 2, \ell = 1, m_\ell = +1, m_s = -\tfrac{1}{2}}$,

   $\boxed{n = 2, \ell = 1, m_\ell = 0, m_s = +\tfrac{1}{2}}$, $\boxed{n = 2, \ell = 1, m_\ell = 0, m_s = -\tfrac{1}{2}}$,

   $\boxed{n = 2, \ell = 1, m_\ell = -1, m_s = +\tfrac{1}{2}}$, $\boxed{n = 2, \ell = 1, m_\ell = -1, m_s = -\tfrac{1}{2}}$.

---

31. **SSM REASONING** In the theory of quantum mechanics, there is a selection rule that restricts the initial and final values of the orbital quantum number ℓ. The selection rule states that when an electron makes a transition between energy levels, the value of ℓ may not remain the same or increase or decrease by more than one. In other words, the rule requires that $\Delta \ell = \pm 1$.

   SOLUTION
   a. For the transition $2s \rightarrow 1s$, the electron makes a transition from the 2s state ($n = 2$, $\ell = 0$) to the 1s state ($n = 1$, $\ell = 0$). Since the value of ℓ is the same in both states, $\Delta \ell = 0$, and we can conclude that this energy level transition is $\boxed{\text{not permitted}}$.

b. For the transition 2p → 1s, the electron makes a transition from the 2p state ($n = 2$, $\ell = 1$) to the 1s state ($n = 1$, $\ell = 0$). The value of ℓ changes so that $\Delta\ell = 0 - 1 = -1$, and we can conclude that this energy level transition is $\boxed{\text{permitted}}$.

c. For the transition 4p → 2p, the electron makes a transition from the 4p state ($n = 4$, $\ell = 1$) to the 2p state ($n = 2$, $\ell = 1$). Since the value of ℓ is the same in both states, $\Delta\ell = 0$, and we can conclude that this energy level transition is $\boxed{\text{not permitted}}$.

d. For the transition 4s → 2p, the electron makes a transition from the 4s state ($n = 4$, $\ell = 0$) to the 2p state ($n = 2$, $\ell = 1$). The value of ℓ changes so that $\Delta\ell = 1 - 0 = +1$, and we can conclude that this energy level transition is $\boxed{\text{permitted}}$.

e. For the transition 3d → 3s, the electron makes a transition from the 3d state ($n = 3$, $\ell = 2$) to the 3s state ($n = 3$, $\ell = 0$). The value of ℓ changes so that $\Delta\ell = 0 - 2 = -2$, and we can conclude that this energy level transition is $\boxed{\text{not permitted}}$.

---

32. **REASONING AND SOLUTION** Following the order of filling of subshells given in Figure 30.17, the required condition is found for the following electron configuration:

$$1s^2\, 2s^2\, 2p^6\, 3s^2\, 3p^6\, 4s^2\, 3d^{10}\, 4p^6\, 5s^2$$

In this configuration there are 10 electrons in the 3d subshell. In each of the 1s, 2s, 3s, 4s, and 5s subshells there are 2 electrons, for a total of 10. To identify the atom, we note that the total number of electrons in all of the subshells is 38. According to the periodic table, the atom with $Z = 38$ is $\boxed{\text{strontium, Sr}}$.

---

33. **SSM REASONING** This problem is similar to Example 10 in the text. We use Equation 30.14 with the initial value of n being $n_i = 2$, and the final value being $n_f = 1$. As in Example 10, we use a value of Z that is one less than the atomic number of the atom in question (in this case, a value of $Z = 41$ rather than 42); this accounts approximately for the shielding effect of the single K-shell electron in canceling out the attraction of one nuclear proton.

SOLUTION Using Equation 30.14, we obtain

$$\frac{1}{\lambda} = (1.097 \times 10^7 \text{ m}^{-1})(41)^2 \left(\frac{1}{1^2} - \frac{1}{2^2}\right) \quad \text{or} \quad \boxed{\lambda = 7.230 \times 10^{-11} \text{ m}}$$

---

34. **REASONING AND SOLUTION** From Equation 30.14 we have

$$\frac{1}{\lambda} = R(Z-1)^2 \left(\frac{1}{n_f^2} - \frac{1}{n_i^2} \right)$$

Solving for Z we obtain

$$Z = \sqrt{\frac{1/(\lambda R)}{\left(1/n_f^2\right)-\left(1/n_i^2\right)}} + 1 = \sqrt{\frac{\left[1/\left(4.5 \times 10^{-9}\text{ m}\right)\left(1.097 \times 10^7\text{ m}^{-1}\right)\right]}{\left(1/1^2\right)-\left(1/2^2\right)}} + 1 = 6.2$$

Therefore, the element is likely to be $\boxed{\text{carbon}}$ (Z = 6).

---

35. **REASONING AND SOLUTION** 25.0% of the incident kinetic energy means that the energy available for producing X-rays is

$$E = (0.250)(1.60 \times 10^{-19}\text{ C})(45.0 \times 10^3\text{ V}) = 1.80 \times 10^{-15}\text{ J}$$

The wavelength is then

$$\lambda = hc/E = (6.63 \times 10^{-34}\text{ J·s})(3.00 \times 10^8\text{ m/s})/(1.80 \times 10^{-15}\text{ J}) = \boxed{1.11 \times 10^{-10}\text{ m}}$$

---

36. **REASONING AND SOLUTION** To knock a K-shell electron out of a copper atom, we take $n_f = 1$ and $n_i =$ infinity. Because of shielding, take Z = 28 (rather than Z = 29), and use the Bohr model to find $E = hc/\lambda = hcRZ^2/n_f^2$.

$$E = (6.63 \times 10^{-34}\text{ J s})(3.00 \times 10^8\text{ m/s})(1.097 \times 10^7\text{ m}^{-1})(28)^2 = 1.71 \times 10^{-15}\text{ J}$$

The minimum potential difference needed is

$$V = E/e = (1.71 \times 10^{-15}\text{ J})/(1.60 \times 10^{-19}\text{ C}) = \boxed{10\ 700\text{V}}$$

---

37. **SSM REASONING** In the spectrum of X-rays produced by the tube, the cutoff wavelength λ_0 and the voltage V of the tube are related according to Equation 30.17, $V = hc/(e\lambda_0)$. Since the voltage is increased from zero until the K_α X-ray just appears in the spectrum, it follows that $\lambda_0 = \lambda_\alpha$ and $V = hc/(e\lambda_\alpha)$. Using Equation 30.14 for $1/\lambda_\alpha$, we find that

$$V = \frac{hc}{e\lambda_\alpha} = \frac{hcR(Z-1)^2}{e}\left(\frac{1}{1^2}-\frac{1}{2^2}\right)$$

1086   THE NATURE OF THE ATOM

In this expression we have replaced Z with $Z-1$, in order to account for shielding, as explained in Example 10 in the text.

SOLUTION The desired voltage is, then,

$$V = \frac{(6.63 \times 10^{-34} \text{ J} \cdot \text{s})(3.00 \times 10^8 \text{ m/s})(1.097 \times 10^7 \text{ m}^{-1})(47-1)^2}{(1.60 \times 10^{-19} \text{ C})}\left(\frac{1}{1^2} - \frac{1}{2^2}\right) = \boxed{21\,600 \text{ V}}$$

---

38. **REASONING** The kinetic energy of each electron is given by Equation 6.2 as $\frac{1}{2}mv^2$. If all of this energy goes into producing an X-ray photon, the photon has the highest possible energy. The energy E of a photon is related to its frequency f by Equation 29.2 as $E = hf$. We can express the frequency in terms of the wavelength λ by using Equation 16.1, $f = c/\lambda$. Combining these relations gives

$$\underbrace{\tfrac{1}{2}mv^2}_{\text{Kinetic energy}} = hf = \frac{hc}{\lambda}$$
$$\text{of a photon}$$

SOLUTION Solving the expression above for the speed of an electron just before it strikes the target, we find that

$$v = \sqrt{\frac{2hc}{m\lambda}} = \sqrt{\frac{2(6.63 \times 10^{-34} \text{ J} \cdot \text{s})(3.00 \times 10^8 \text{ m/s})}{(9.11 \times 10^{-31} \text{ kg})(1.20 \times 10^{-10} \text{ m})}} = \boxed{6.03 \times 10^7 \text{ m/s}}$$

---

39. **REASONING AND SOLUTION**

$$E = hf = \frac{hc}{\lambda} = \frac{(6.63 \times 10^{-34} \text{ J} \cdot \text{s})(3.00 \times 10^8 \text{ m/s})}{193 \times 10^{-9} \text{ m}} = \boxed{1.03 \times 10^{-18} \text{ J}}$$

---

40. **REASONING**
a. Average power is energy per unit time (see Equation 6.10b). Thus, the energy of a pulse is the product of the average power and its time of duration.

b. The number of photons in each pulse is the energy of a pulse divided by the energy of a single photon. According to Equations 29.2 and 16.1, the energy of a photon is equal to hc/λ, where λ is its wavelength.

SOLUTION

a. The energy E of a pulse is the product of the average power $\overline{P}$ and its duration time t:

$$E = \overline{P}t = (5.00 \times 10^{-3} \text{ W})(25.0 \times 10^{-3} \text{ s}) = \boxed{1.25 \times 10^{-4} \text{ J}}$$

b. The number N of photons in each pulse is the energy of a pulse divided by the energy of a single photon:

$$N = \frac{E}{\left(\dfrac{hc}{\lambda}\right)} = \frac{1.25 \times 10^{-4} \text{ J}}{\dfrac{(6.63 \times 10^{-34} \text{ J} \cdot \text{s})(3.00 \times 10^8 \text{ m/s})}{633 \times 10^{-9} \text{ m}}} = \boxed{3.98 \times 10^{14}}$$

---

41. **SSM** **WWW** *REASONING* The number of photons emitted by the laser will be equal to the total energy carried in the beam divided by the energy per photon.

SOLUTION The total energy carried in the beam is, from the definition of power,

$$E_{\text{total}} = Pt = (1.5 \text{ W})(0.050 \text{ s}) = 0.075 \text{ J}$$

The energy of a single photon is given by Equations 29.2 and 16.1 as

$$E_{\text{photon}} = hf = \frac{hc}{\lambda} = \frac{(6.63 \times 10^{-34} \text{ J} \cdot \text{s})(3.00 \times 10^8 \text{ m/s})}{514 \times 10^{-9} \text{ m}} = 3.87 \times 10^{-19} \text{ J}$$

where we have used the fact that 514 nm = 514×10^{-9} m. Therefore, the number of photons emitted by the laser is

$$\frac{E_{\text{total}}}{E_{\text{photon}}} = \frac{0.075 \text{ J}}{3.87 \times 10^{-19} \text{ J/photon}} = \boxed{1.9 \times 10^{17} \text{ photons}}$$

---

42. *REASONING AND SOLUTION* According to Equations 29.2 and 16.1, $E = hc/\lambda$, the energy of a single photon of the dye laser is

$$E = \frac{hc}{\lambda} = \frac{(6.63 \times 10^{-34} \text{ J} \cdot \text{s})(3.00 \times 10^8 \text{ m/s})}{585 \times 10^{-9} \text{ m}} = 3.40 \times 10^{-19} \text{ J}$$

while the energy of a single photon from the CO_2 laser is

$$E = \frac{hc}{\lambda} = \frac{(6.63 \times 10^{-34} \text{ J} \cdot \text{s})(3.00 \times 10^8 \text{ m/s})}{1.06 \times 10^{-5} \text{ m}} = 1.88 \times 10^{-20} \text{ J}$$

The ratio of these two photon energies is

$$\frac{3.40 \times 10^{-19} \text{ J}}{1.88 \times 10^{-20} \text{ J}} = 18.1$$

We can only have a whole number of photons, therefore, the minimum number of photons that the carbon dioxide laser must produce to deliver at least as much energy to a target as does a single photon from the dye laser is $\boxed{19}$.

---

43. **REASONING AND SOLUTION** For either laser, the number of photons is given by the energy E produced divided by the energy per photon. Since power P is energy per unit time, the energy produced in a time t is $E = Pt$. For a photon frequency f the energy per photon is hf. But the photon frequency f and wavelength λ are related to the speed of light c by $f = c/\lambda$. Therefore, the energy per photon is hc/λ. The number of photons is, then,

$$\text{Number of photons} = \frac{\text{Energy produced}}{\text{Energy per photon}} = \frac{Pt}{hc/\lambda} = \frac{Pt\lambda}{hc}$$

Since each laser produces the same number of photons, it follows that

$$\left(\frac{Pt\lambda}{hc}\right)_{\text{He/Ne}} = \left(\frac{Pt\lambda}{hc}\right)_{\text{SS}} \quad \text{or} \quad t_{\text{He/Ne}} = \frac{(Pt\lambda)_{\text{SS}}}{(P\lambda)_{\text{He/Ne}}}$$

where the "He/Ne" and "SS" refer to the helium/neon and solid state lasers, respectively. The time for the helium/neon laser is, then,

$$t_{\text{He/Ne}} = \frac{(Pt\lambda)_{\text{SS}}}{(P\lambda)_{\text{He/Ne}}} = \frac{(1.0 \times 10^{14} \text{ W})(1.1 \times 10^{-11} \text{ s})(1060 \text{ nm})}{(1.0 \times 10^{-3} \text{ W})(633 \text{ nm})} = 1.8 \times 10^6 \text{ s}$$

This is a time of $(1.8 \times 10^6 \text{ s})\left[(1 \text{ day})/(8.64 \times 10^4 \text{ s})\right] = \boxed{21 \text{ days}}$.

---

44. **REASONING AND SOLUTION** According to Figure 30.17, the energy sublevel with $n = 4$, $\ell = 0$, which corresponds to the notation 4s, is lower in energy than the $n = 3$, $\ell = 2$, (3d) energy sublevel. Thus, the 4s energy sublevel will be filled before the 3d energy sublevel. Therefore, using Figure 30.17 as a guide, we find that the ground state electronic configuration of manganese ($Z = 25$) is

$$\boxed{1s^2\ 2s^2\ 2p^6\ 3s^2\ 3p^6\ 4s^2\ 3d^5}$$

---

45. **SSM REASONING AND SOLUTION** In the ground state, the electron in a hydrogen atom is in the $n = 1$ energy level. To ionize such a hydrogen atom, the electron must be taken from the $n = 1$ level to the level for which n is infinity. The electromagnetic radiation needed to accomplish this corresponds to a Lyman line in which the final state n is taken to be infinity. Equation 30.1 gives the reciprocal wavelength for the Lyman series. With $n = \infty$, Equation 30.1 reveals that for the longest wavelength

$$\frac{1}{\lambda} = R\left(\frac{1}{1^2} - \frac{1}{n^2}\right) = (1.097 \times 10^7\ \text{m}^{-1})\left(\frac{1}{1^2} - \frac{1}{\infty^2}\right) \quad \text{or} \quad \boxed{\lambda = 91.2\ \text{nm}}$$

---

46. **REASONING AND SOLUTION** According to Equation 30.4, the energy for a K_α X-ray photon is $E_2 - E_1 = hf$. But since the frequency f is related to the wavelength λ and the speed of light c by $f = c/\lambda$, it follows that $E_2 - E_1 = hc/\lambda$. Using the Bohr expression for $1/\lambda$ given in Equation 30.14, we find that the photon energy is given by

$$E_2 - E_1 = hcR(Z-1)^2\left(\frac{1}{1^2} - \frac{1}{2^2}\right)$$

In this expression we have replaced Z with $Z - 1$, in order to account for shielding, as explained in Example 10 in the text. The photon energy is, then,

$$E_2 - E_1 = (6.63 \times 10^{-34}\ \text{J} \cdot \text{s})(3.00 \times 10^8\ \text{m/s})(1.097 \times 10^7\ \text{m}^{-1})(82-1)^2\left(\frac{1}{1^2} - \frac{1}{2^2}\right)$$

$$= \boxed{1.07 \times 10^{-14}\ \text{J}}$$

---

47. **REASONING AND SOLUTION** The table below lists the possible sets of the four quantum numbers that correspond to the electrons in a completely filled 4f subshell:

n	ℓ	m_ℓ	m_s
4	3	3	1/2
4	3	3	−1/2
4	3	2	1/2
4	3	2	−1/2
4	3	1	1/2
4	3	1	−1/2

1090 THE NATURE OF THE ATOM

4	3	0	1/2
4	3	0	−1/2
4	3	−1	1/2
4	3	−1	−1/2
4	3	−2	1/2
4	3	−2	−1/2
4	3	−3	1/2
4	3	−3	−1/2

---

48. **REASONING AND SOLUTION**

a. The energy difference between the two K_β energy levels is $E = hf = hc/\lambda$.

$$E = (6.63 \times 10^{-34} \text{ J·s})(3.00 \times 10^8 \text{ m/s})/(1.84 \times 10^{-11} \text{ m}) = \boxed{1.08 \times 10^{-14} \text{ J}}$$

b. Converting to electron volts, we have

$$(1.08 \times 10^{-14} \text{ J})(1 \text{ eV})/(1.60 \times 10^{-19} \text{ J}) = \boxed{6.75 \times 10^4 \text{ eV}}$$

---

49. **SSM REASONING** According to the Bohr model, the energy E_n (in eV) of the electron in an orbit is given by Equation 30.13: $E_n = -13.6(Z^2/n^2)$. In order to find the principal quantum number of the state in which the electron in a doubly ionized lithium atom Li^{2+} has the same total energy as a ground state electron in a hydrogen atom, we equate the right hand sides of Equation 30.13 for the hydrogen atom and the lithium ion. This gives

$$-(13.6)\left(\frac{Z^2}{n^2}\right)_H = -(13.6)\left(\frac{Z^2}{n^2}\right)_{Li} \quad \text{or} \quad n_{Li}^2 = \left(\frac{n^2}{Z^2}\right)_H Z_{Li}^2$$

This expression can be evaluated to find the desired principal quantum number.

SOLUTION For hydrogen, $Z = 1$, and $n = 1$ for the ground state. For lithium Li^{2+}, $Z = 3$. Therefore,

$$n_{Li}^2 = \left(\frac{n^2}{Z^2}\right)_H Z_{Li}^2 = \left(\frac{1^2}{1^2}\right)(3^2) \quad \text{or} \quad \boxed{n_{Li} = 3}$$

---

50. **REASONING** The energy levels and radii of a hydrogenic species of atomic number Z are given by Equations 30.13 and 30.10, respectively: $E_n = -(13.6 \text{ eV})(Z^2/n^2)$ and

$r_n = (5.29 \times 10^{-11} \text{ m})(n^2/Z)$. We can use Equation 30.13 to find the value of Z for the unidentified ionized atom and then calculate the radius of the $n = 5$ orbit using Equation 30.10.

SOLUTION Solving Equation 30.13 for atomic number Z of the unknown species, we have

$$Z = \sqrt{\frac{E_n n^2}{-13.6 \text{ eV}}} = \sqrt{\frac{(-30.6 \text{ eV})(2)^2}{-13.6 \text{ eV}}} = 3$$

Therefore, the radius of the $n = 5$ orbit is

$$r_5 = (5.29 \times 10^{-11} \text{ m})\left(\frac{5^2}{3}\right) = \boxed{4.41 \times 10^{-10} \text{ m}}$$

---

51. **REASONING AND SOLUTION** We can first determine the maximum magnetic quantum number using Equation 30.16, i.e., $L_z = m_\ell h/(2\pi)$, so that

$$m_\ell = 2\pi L_z/h = 2\pi (2.11 \times 10^{-34} \text{ J·s})/(6.626 \times 10^{-34} \text{ J·s}) = 2$$

Therefore, we can have $\ell \geq 2$, and $n \geq 3$. From Equation 30.13, we have

[$n = 3$] $E = (-13.6 \text{ eV})/3^2 = \boxed{-1.51 \text{ eV}}$

[$n = 4$] $E = (-13.6 \text{ eV})/4^2 = \boxed{-0.850 \text{ eV}}$

[$n = 5$] $E = (-13.6 \text{ eV})/5^2 = \boxed{-0.544 \text{ eV}}$

---

52. **REASONING AND SOLUTION** From the diagram we see that an integral number of half wavelengths fit into the "box". That is, $n\lambda/2 = L$, where $n = 1, 2, 3, \ldots$ Using the de Broglie equation yields $\lambda = h/(mv)$. Combining these two expressions yields, $nh/(2mv) = L$, and rearranging gives the velocity as $v = nh/(2mL)$. Finally, the kinetic energy can be written

$$KE = \tfrac{1}{2}mv^2 = \tfrac{1}{2}m\left(\frac{nh}{2mL}\right)^2$$

$$\boxed{KE = \frac{n^2 h^2}{8mL^2} \quad \text{where} \quad n = 1, 2, 3, \ldots}$$

53. **SSM REASONING** A wavelength of 410.2 nm is emitted by the hydrogen atoms in a high-voltage discharge tube. This transition lies in the visible region (380–750 nm) of the hydrogen spectrum. Thus, we can conclude that the transition is in the Balmer series and, therefore, that $n_f = 2$. The value of n_i can be found using Equation 30.14, according to which the Balmer series transitions are given by

$$\frac{1}{\lambda} = R\,(1^2)\left(\frac{1}{2^2} - \frac{1}{n_i^2}\right) \qquad n = 3, 4, 5, \ldots$$

This expression may be solved for n_i for the energy transition that produces the given wavelength.

SOLUTION Solving for n_i, we find that

$$n_i = \frac{1}{\sqrt{\dfrac{1}{2^2} - \dfrac{1}{R\lambda}}} = \frac{1}{\sqrt{\dfrac{1}{2^2} - \dfrac{1}{(1.097 \times 10^7 \text{ m}^{-1})(410.2 \times 10^{-9} \text{ m})}}} = 6$$

Therefore, the initial and final states are identified by $\boxed{n_i = 6 \text{ and } n_f = 2}$.

---

54. **REASONING AND SOLUTION**
a. To find an expression for v_n, we use Equation 30.8, $L_n = mv_n r_n = nh/(2\pi)$, and substitute for r_n from Equation 30.9:

$$mv_n\left(\frac{h^2 n^2}{4\pi^2 mke^2 Z}\right) = \frac{nh}{2\pi} \qquad \text{or} \qquad \boxed{v_n = \frac{2\pi ke^2 Z}{nh}}$$

b. For the hydrogen atom ($Z = 1$) in the $n = 1$ orbit,

$$v_n = 2\pi (8.99 \times 10^9 \text{ N·m}^2/\text{C}^2)(1.602 \times 10^{-19} \text{ C})^2/(6.626 \times 10^{-34} \text{ J·s}) = \boxed{2.19 \times 10^6 \text{ m/s}}$$

c. For the $n = 2$ orbit of hydrogen,

$$v_n = 2\pi ke^2 Z/(2h) = \boxed{1.09 \times 10^6 \text{ m/s}}$$

d. The speeds found in parts (b) and (c) are well below the speed of light, 3.0×10^8 m/s, and $\boxed{\text{are consistent with ignoring relativistic effects}}$.

55. CONCEPT QUESTIONS

a. The ionization energy for a given state is the energy needed to remove the electron completely from the atom. The removed electron has no kinetic energy and no electric potential energy, so its total energy is zero.

b. The ionization energy for a given excited state is less than the ionization energy for the ground state. In the excited state the electron already has part of the energy necessary to achieve ionization, so less energy is required to ionize the atom from the excited state than from the ground state.

SOLUTION

a. The energy of the nth state in the hydrogen atom is given by Equation 30.13 as $E_n = -(13.6 \text{ eV})Z^2/n^2$. When $n = \infty$, $E_\infty = 0 \text{ J}$, and when $n = 4$, $E_4 = -(13.6 \text{ eV})(1)^2/4^2 = -0.850 \text{ eV}$. The difference in energies between these two states is the ionization energy:

$$\text{Ionization energy} = E_\infty - E_4 = \boxed{0.850 \text{ eV}}$$

b. In the same manner, it can be shown that the ionization energy for the $n = 1$ state is 13.6 eV. The ratio of the ionization energies is

$$\frac{0.850 \text{ eV}}{13.6 \text{ eV}} = \boxed{0.0625}$$

---

56. CONCEPT QUESTIONS

a. No. The orbital quantum number ℓ can have any integer value from 0 up to $n - 1$. If for example, $n = 4$, ℓ can have the values 0, 1, 2, and 3.

b. No. The orbital quantum number ℓ can be zero or have only positive values [see Question (a)].

c. No. The magnetic quantum number m_ℓ can only have integer values, including 0, from $-\ell$ to $+\ell$. For instance, if $\ell = 2$, m_ℓ can have the values $-2, -1, 0, +1$, and $+2$.

d. Yes. The magnetic quantum number can have negative values [see Question (c)].

SOLUTION Of the five states listed in the table, three are not possible. The ones that are not possible, and the reasons they are not possible, are:

State	Reason

1094   THE NATURE OF THE ATOM

(a)   The quantum number ℓ must be less than n

(c)   The quantum number m_ℓ must be less than or equal to ℓ.

(d)   The quantum number ℓ cannot be negative.

---

57. **CONCEPT QUESTIONS**
   a. The orbital quantum number ℓ can have any integer value from 0 up to $n - 1$. If, for example, $n = 4$, ℓ can have the values 0, 1, 2, and 3.

   b. Angular momentum L is determined by the orbital quantum number ℓ according to $L = \sqrt{\ell(\ell+1)}\, h/(2\pi)$. Therefore, a given value of L implies a given value of ℓ. The magnetic quantum number m_ℓ can have any integer value, including 0, from $-\ell$ to $+\ell$. For instance, if $\ell = 2$, m_ℓ can have the values $-2, -1, 0, +1$, and $+2$.

   SOLUTION Of the five subshell configurations, three are not allowed. The ones that are not allowed, and the reasons they are not allowed, are:

   (b) The principal quantum number is $n = 2$ and the orbital quantum number is $\ell = 2$ (the d subshell). Since ℓ must be less than n, this subshell configuration is not permitted.

   (c) This subshell has 4 electrons. According to the Pauli exclusion principle, only two electrons can be in the s ($\ell = 0$) subshell, however. Therefore, this subshell configuration is not allowed.

   (d) This subshell has 8 electrons. According to the Pauli exclusion principle, only six electrons can be in the p ($\ell = 1$) subshell. Therefore, this subshell configuration is not allowed.

---

58. **CONCEPT QUESTIONS**
   a. The cutoff wavelength λ_0 depends only on the voltage V across the X-ray tube, according to Equation 30.17; $\lambda_0 = hc/(eV)$. Since the voltage does not change, the cutoff wavelength remains the same, independent of the target material.

   b. The wavelength of the K_α photon is given by Equation 30.14, with $n_i = 2$ and $n_f = 1$. The wavelength λ is proportional to $1/(Z-1)^2$, so the wavelength decreases for larger values of Z. Since silver has a larger value of Z than molybdenum, the wavelength of the K_α photon decreases when silver replaces molybdenum.

SOLUTION
a. The cutoff wavelength λ_0 is

$$\lambda_0 = \frac{hc}{eV} = \frac{(6.63 \times 10^{-34} \text{ J} \cdot \text{s})(3.00 \times 10^8 \text{ m/s})}{(1.60 \times 10^{-19} \text{ C})(35.0 \times 10^3 \text{ V})} = \boxed{3.55 \times 10^{-11} \text{ m}} \quad (30.17)$$

b. The K_α photon is emitted when an electron drops from the $n_i = 2$ to the $n_f = 1$ level. The wavelengths emitted by molybdenum and silver are:

Molybdenum $\quad \frac{1}{\lambda} = (1.097 \times 10^7 \text{ m}^{-1})(42-1)^2 \left(\frac{1}{1^2} - \frac{1}{2^2}\right) = 1.38 \times 10^{10} \text{ m}^{-1} \quad (30.14)$

$$\lambda = \boxed{7.23 \times 10^{-11} \text{ m}}$$

Silver $\quad \frac{1}{\lambda} = (1.097 \times 10^7 \text{ m}^{-1})(47-1)^2 \left(\frac{1}{1^2} - \frac{1}{2^2}\right) = 1.74 \times 10^{10} \text{ m}^{-1} \quad (30.14)$

$$\lambda = \boxed{5.74 \times 10^{-11} \text{ m}}$$

---

59. **CONCEPT QUESTIONS**
   a. The external source of energy must "pump" the electrons from the ground state E_0 to the metastable state E_2.

   b. The population inversion occurs between the metastable state E_2 and the one below it, E_1.

   c. The lasing action occurs between the two states that have the population inversion, the E_2 and E_1 states.

SOLUTION
a. From the drawing, we see that the energy required to raise an electron from the E_0 state to the E_2 state is $\boxed{0.289 \text{ eV}}$.

b. The lasing action occurs between the E_2 and E_1 states, and so the energy E of the emitted photon is the difference between them; $E = E_2 - E_1$. According to Equations 29.2 and 16.1, the wavelength λ of the photon is related to its energy via $\lambda = hc/E$, so that

THE NATURE OF THE ATOM

$$\lambda = \frac{hc}{E_2 - E_1} = \frac{(6.63 \times 10^{-34} \text{ J} \cdot \text{s})(3.00 \times 10^8 \text{ m/s})}{(0.289 \text{ eV} - 0.165 \text{ eV})\left(\frac{1.60 \times 10^{-19} \text{ J}}{1 \text{ eV}}\right)} = \boxed{1.00 \times 10^{-5} \text{ m}}$$

c. An examination of Figure 24.9 shows that the photon lies in the $\boxed{\text{infrared}}$ region of the electromagnetic spectrum.

CHAPTER 31 | NUCLEAR PHYSICS AND RADIOACTIVITY

PROBLEMS

1. **SSM** **REASONING** For an element whose chemical symbol is X, the symbol for the nucleus is $^A_Z X$ where A represents the number of protons and neutrons (the nucleon number) and Z represents the number of protons (the atomic number) in the nucleus.

   SOLUTION
   a. The symbol $^{195}_{78} X$ indicates that the nucleus in question contains $Z = 78$ protons, and $N = A - Z = 195 - 78 = \boxed{117 \text{ neutrons}}$. From the periodic table, we see that $Z = 78$ corresponds to $\boxed{\text{platinum, Pt}}$.

   b. Similar reasoning indicates that the nucleus in question is $\boxed{\text{sulfur, S}}$, and the nucleus contains $N = A - Z = 32 - 16 = \boxed{16 \text{ neutrons}}$.

   c. Similar reasoning indicates that the nucleus in question is $\boxed{\text{copper, Cu}}$, and the nucleus contains $N = A - Z = 63 - 29 = \boxed{34 \text{ neutrons}}$.

   d. Similar reasoning indicates that the nucleus in question is $\boxed{\text{boron, B}}$, and the nucleus contains $N = A - Z = 11 - 5 = \boxed{6 \text{ neutrons}}$.

   e. Similar reasoning indicates that the nucleus in question is $\boxed{\text{plutonium, Pu}}$, and the nucleus contains $N = A - Z = 239 - 94 = \boxed{145 \text{ neutrons}}$.

2. **REASONING AND SOLUTION**
   a. An element X can be represented by $^A_Z X$, where A is the nucleon number, and Z is the atomic number (number of protons). In addition, $A = Z + N$, where N is the number of neutrons in the nucleus. Therefore, $^{18}_{8} O$ has $\boxed{8 \text{ protons and } 10 \text{ neutrons}}$.

   b. Similarly, $^{120}_{50} Sn$ has $\boxed{50 \text{ protons and } 70 \text{ neutrons}}$.

3. **REASONING AND SOLUTION** Solving Equation 31.2 for A gives

$$A = r^3/(1.2 \times 10^{-15} \text{ m})^3$$

If r is doubled, then A will increase by a factor of $2^3 = \boxed{8}$

---

4. **REASONING** For an element whose chemical symbol is X, the symbol for the nucleus is ^A_ZX, where A represents the total number of protons and neutrons (the nucleon number) and Z represents the number of protons in the nucleus (the atomic number). The number of neutrons N is related to A and Z by Equation 31.1: $A = Z + N$.

SOLUTION For the nucleus $^{208}_{82}\text{Pb}$, we have $Z = 82$ and $A = 208$.

a. The net electrical charge of the nucleus is equal to the total number of protons multiplied by the charge on a single proton. Since the $^{208}_{82}\text{Pb}$ nucleus contains 82 protons, the net electrical charge of the $^{208}_{82}\text{Pb}$ nucleus is

$$q_{net} = (82)(+1.60 \times 10^{-19} \text{ C}) = \boxed{+1.31 \times 10^{-17} \text{ C}}$$

b. The number of neutrons is $N = A - Z = 208 - 82 = \boxed{126}$.

c. By inspection, the number of nucleons is $A = \boxed{208}$.

d. The approximate radius of the nucleus can be found from Equation 31.2, namely

$$r = (1.2 \times 10^{-15} \text{ m}) A^{1/3} = (1.2 \times 10^{-15} \text{ m})(208)^{1/3} = \boxed{7.1 \times 10^{-15} \text{ m}}$$

e. The nuclear density is the mass per unit volume of the nucleus. The total mass of the nucleus can be found by multiplying the mass $m_{nucleon}$ of a single nucleon by the total number A of nucleons in the nucleus. Treating the nucleus as a sphere of radius r, the nuclear density is

$$\rho = \frac{m_{total}}{V} = \frac{m_{nucleon} A}{\frac{4}{3}\pi r^3} = \frac{m_{nucleon} A}{\frac{4}{3}\pi\left[(1.2 \times 10^{-15} \text{ m}) A^{1/3}\right]^3} = \frac{m_{nucleon}}{\frac{4}{3}\pi (1.2 \times 10^{-15} \text{ m})^3}$$

Therefore,

$$\rho = \frac{1.67 \times 10^{-27} \text{ kg}}{\frac{4}{3}\pi (1.2 \times 10^{-15} \text{ m})^3} = \boxed{2.3 \times 10^{17} \text{ kg/m}^3}$$

5. **SSM WWW** **REASONING AND SOLUTION** The surface area of a sphere is Area $= 4\pi r^2$. But according to Equation 31.2, the radius of a nucleus in meters is $r = (1.2 \times 10^{-15} \text{ m}) A^{1/3}$, where A is the nucleon number. With this expression for r, the surface area becomes Area $= 4\pi (1.2 \times 10^{-15} \text{ m})^2 A^{2/3}$. The ratio of the largest to the smallest surface area is, then,

$$\frac{\text{Largest area}}{\text{Smallest area}} = \frac{4\pi (1.2 \times 10^{-15} \text{ m})^2 A_{\text{largest}}^{2/3}}{4\pi (1.2 \times 10^{-15} \text{ m})^2 A_{\text{smallest}}^{2/3}} = \frac{209^{2/3}}{1^{2/3}} = \boxed{35.2}$$

---

6. **REASONING** The nucleus is roughly spherical, so its volume is $V = \frac{4}{3}\pi r^3$. The radius r is given by Equation 31.2 as $r \approx (1.2 \times 10^{-15} \text{ m}) A^{1/3}$, where A is the atomic mass number or nucleon number. Therefore, the volume is given by $V = \frac{4}{3}\pi (1.2 \times 10^{-15} \text{ m})^3 A$ and is proportional to A. We can apply this expression to the unknown nucleus and to the nickel nucleus, knowing that the ratio of the two volumes is 2:1. This ratio provides the solution we seek.

SOLUTION Applying the expression for the volume to each nucleus gives

$$\frac{V}{V_{\text{Ni}}} = \frac{\frac{4}{3}\pi (1.2 \times 10^{-15} \text{ m})^3 A}{\frac{4}{3}\pi (1.2 \times 10^{-15} \text{ m})^3 (60)} = \frac{A}{60} = 2 \quad \text{or} \quad A = 120$$

The nucleon number is equal to the number of neutrons N plus the number of protons or the atomic number Z, so $A = N + Z$. Therefore, the atomic number for the unknown nucleus is

$$Z = A - N = 120 - 70 = 50$$

The unknown nucleus, then, is $^{120}_{50}\text{X}$. Reference to the periodic table reveals that the element that has an atomic number of $Z = 50$ is tin (Sn). Thus, $\boxed{^A_Z\text{X} = {}^{120}_{50}\text{Sn}}$.

---

7. **REASONING AND SOLUTION** For isotope X, $A_X = Z + N = Z + Z = 2Z$ (equal number of protons and neutrons, $Z = N$). For isotope Y, $A_Y = Z + N' = Z + 2N = Z + 2Z = 3Z$. The radii are related by

$$\frac{r_Y}{r_X} = \left(\frac{A_Y}{A_X}\right)^{1/3} = \left(\frac{3Z}{2Z}\right)^{1/3} = \left(\frac{3}{2}\right)^{1/3} = \boxed{1.14}$$

1100 NUCLEAR PHYSICS AND RADIOACTIVITY

8. **REASONING AND SOLUTION**
   a. The volume of the nucleus is found using

   $$V = (4/3)\pi r^3 = (4/3)\pi(1.2 \times 10^{-15} \text{ m})^3 A = (7.2 \times 10^{-45} \text{ m}^3) A$$

   The mass of the nucleus is $m = (1.67 \times 10^{-27} \text{ kg}) A$, so the density of the nucleus is

   $$\rho = m/V = \boxed{2.3 \times 10^{17} \text{ kg/m}^3}$$

   b. The mass is
   $$m = \rho V = (2.3 \times 10^{17} \text{ kg/m}^3)(4/3)\pi(2.3 \times 10^{-3} \text{ m})^3 = \boxed{1.2 \times 10^{10} \text{ kg}}$$

   c. The number of supertankers would be $N = (1.2 \times 10^{10} \text{ kg})/(1.5 \times 10^{8} \text{ kg}) = \boxed{80}$

---

9. **SSM** **WWW** **REASONING** According to Equation 31.2, the radius of a nucleus in meters is $r = (1.2 \times 10^{-15} \text{ m}) A^{1/3}$, where A is the nucleon number. If we treat the neutron star as a uniform sphere, its density (Equation 11.1) can be written as

   $$\rho = \frac{M}{V} = \frac{M}{\frac{4}{3}\pi r^3}$$

   Solving for the radius r, we obtain,

   $$r = \sqrt[3]{\frac{M}{\frac{4}{3}\pi \rho}}$$

   This expression can be used to find the radius of a neutron star of mass M and density ρ.

   SOLUTION As discussed in Conceptual Example 1, nuclear densities have the same approximate value in all atoms. If we consider a uniform spherical nucleus, then the density of nuclear matter is approximately given by

   $$\rho = \frac{M}{V} \approx \frac{A \times (\text{mass of a nucleon})}{\frac{4}{3}\pi r^3} = \frac{A \times (\text{mass of a nucleon})}{\frac{4}{3}\pi \left[(1.2 \times 10^{-15} \text{ m}) A^{1/3}\right]^3}$$

   $$= \frac{1.67 \times 10^{-27} \text{ kg}}{\frac{4}{3}\pi(1.2 \times 10^{-15} \text{ m})^3} = 2.3 \times 10^{17} \text{ kg/m}^3$$

   Substituting values into the expression for r determined above, we have

   $$r = \sqrt[3]{\frac{(0.40)(1.99 \times 10^{30} \text{ kg})}{\frac{4}{3}\pi(2.3 \times 10^{17} \text{ kg/m}^3)}} = \boxed{9.4 \times 10^{3} \text{ m}}$$

10. ***REASONING AND SOLUTION***
    a. The total mass of the separated nucleons is

    $$m = 27(1.007\ 825\ \text{u}) + 32(1.008\ 665\ \text{u}) = 59.488\ 555\ \text{u}$$

    The mass defect is then

    $$\Delta m = 59.488\ 555\ \text{u} - 58.933\ 198\ \text{u} = \boxed{0.555\ 357\ \text{u}}$$

    b. Expressed in kilograms we have

    $$\Delta m = (0.555\ 357\ \text{u})\left(\frac{1.6605 \times 10^{-27}\ \text{kg}}{1\ \text{u}}\right) = \boxed{9.2217 \times 10^{-28}\ \text{kg}}$$

---

11. **SSM** ***REASONING AND SOLUTION*** The symbol $^{16}_{8}\text{O}$ indicates that the oxygen nucleus contains $A = 16$ nucleons. According to the binding energy per nucleon curve in Figure 31.6, the corresponding binding energy per nucleon (for $A = 16$) is 8.00 MeV/nucleon. Therefore, the total binding energy of $^{16}_{8}\text{O}$ is

    $$(16\ \text{nucleons})\left(\frac{8.00\ \text{MeV}}{\text{nucleon}}\right) = \boxed{128\ \text{MeV}}$$

---

12. ***REASONING AND SOLUTION*** $^{27}_{13}\text{Al}$ contains

    $$Z = 13\ \text{protons and}\ N = A - Z = 27 - 13 = 14\ \text{neutrons}$$

    The mass of a hydrogen atom is 1.007 825 u. The mass of an aluminum atom is given as 26.981 539 u. The mass of a neutron is given in Table 31.1 as 1.008 665 u. Thus, the mass defect Δm is

    $$\Delta m = 13\ (1.007\ 825\ \text{u}) + 14\ (1.008\ 665\ \text{u}) - 26.981\ 539\ \text{u} = 0.241\ 496\ \text{u}$$

    Since 1 u = 931.5 MeV, this mass defect corresponds to a binding energy of

    $$(0.241\ 496\ \text{u})\left(\frac{931.5\ \text{MeV}}{1\ \text{u}}\right) = \boxed{225.0\ \text{MeV}}$$

---

13. ***REASONING AND SOLUTION*** The earth revolves around the sun, and the two represent a bound system that has a binding energy of 2.6×10^{33} J. Therefore, to completely separate the earth and the sun so that they are infinitely far apart requires an amount of energy equal to the binding energy, namely $\Delta E_0 = 2.6 \times 10^{33}$ J.

Any change in the energy of a system causes a change in the mass of the system according to $\Delta E_0 = (\Delta m)c^2$ (see Section 28.6). Therefore, the change in mass of the earth-sun system when its energy changes by an amount $\Delta E_0 = 2.6 \times 10^{33}$ J is given by

$$\Delta m = \frac{\Delta E_0}{c^2} = \frac{2.6 \times 10^{33} \text{ J}}{(3.0 \times 10^8 \text{ m/s})^2} = 2.9 \times 10^{16} \text{ kg}$$

Therefore, the difference between the mass of the separated system and that of the bound system is $\boxed{2.9 \times 10^{16} \text{ kg}}$.

---

14. **REASONING** The atomic mass given for $^{206}_{82}$Pb includes the 82 electrons in the neutral atom. Therefore, when computing the mass defect, we must account for these electrons. We do so by using the atomic mass of 1.007 825 u for the hydrogen atom $^1_1$H, which also includes the single electron, instead of the atomic mass of a proton. To obtain the binding energy in MeV, we will use the fact that 1 u is equivalent to 931.5 MeV.

SOLUTION
a. Noting that the number of neutrons is 206 – 82 = 124, we can obtain the mass defect Δm as follows:

$$\Delta m = \underbrace{82(1.007\,825 \text{ u})}_{\substack{\text{82 free hydrogen atoms} \\ \text{(protons plus electrons)}}} + \underbrace{124(1.008\,665 \text{ u})}_{\text{124 free neutrons}} - \underbrace{205.974\,440 \text{ u}}_{\substack{\text{Intact lead atom} \\ \text{(including 82 electrons)}}} = \boxed{1.741\,670 \text{ u}}$$

b. Since 1 u is equivalent to 931.5 MeV, the binding energy is

$$\text{Binding energy} = (1.741\,670 \text{ u})\left(\frac{931.5 \text{ MeV}}{1 \text{ u}}\right) = \boxed{1622 \text{ MeV}}$$

c. The binding energy per nucleon is

$$\text{Binding energy per nucleon} = \frac{\text{Binding energy}}{\text{Number of nucleons}}$$

$$= \frac{1622 \text{ MeV}}{206} = \boxed{7.87 \text{ MeV}}$$

---

15. **[SSM] REASONING** Since we know the difference in binding energies for the two isotopes, we can determine the corresponding mass defect. Also knowing that the isotope with the larger binding energy contains one more neutron than the other isotope gives us enough information to calculate the atomic mass difference between the two isotopes.

SOLUTION The mass defect corresponding to a binding energy difference of 5.03 MeV is

$$(5.03 \text{ MeV})\left(\frac{1 \text{ u}}{931.5 \text{ MeV}}\right) = 0.005\,40 \text{ u}$$

Since the isotope with the larger binding energy has one more neutron ($m = 1.008\,665$ u) than the other isotope, the difference in atomic mass between the two isotopes is

$$1.008\,665 \text{ u} - 0.005\,40 \text{ u} = \boxed{1.003\,27 \text{ u}}$$

---

16. **REASONING AND SOLUTION**
    a. The mass defect is

    $$\Delta m = 13.005\,738 \text{ u} + 1.008\,665 \text{ u} - 14.003\,074 \text{ u} = 0.011\,329 \text{ u}$$

    The binding energy of the neutron is then

    $$(0.011\,329 \text{ u})\left(\frac{931.5 \text{ MeV}}{1 \text{ u}}\right) = \boxed{10.55 \text{ MeV}}$$

    b. The mass defect is

    $$\Delta m = 13.003\,355 \text{ u} + 1.007\,825 \text{ u} - 14.003\,074 \text{ u} = 0.008\,106 \text{ u}$$

    The binding energy of the proton is then

    $$(0.008\,106 \text{ u})\left(\frac{931.5 \text{ MeV}}{1 \text{ u}}\right) = \boxed{7.55 \text{ MeV}}$$

    c. The $\boxed{\text{neutron}}$ is more tightly bound, since it has the larger binding energy.

---

17. **REASONING AND SOLUTION**
    a. The decay process looks like $\boxed{{}^{212}_{84}\text{Po} \rightarrow {}^{208}_{82}\text{Pb} + {}^{4}_{2}\text{He}}$.

    b. In this case $\boxed{{}^{232}_{92}\text{U} \rightarrow {}^{228}_{90}\text{Th} + {}^{4}_{2}\text{He}}$

---

18. **REASONING AND SOLUTION**
    a. The decay reaction is: ${}^{242}_{94}\text{Pu} \rightarrow {}^{A}_{Z}\text{X} + {}^{4}_{2}\text{He}$. Therefore, $242 = A + 4$, so that $A = 238$. In addition, $94 = Z + 2$, so that $Z = 92$. Thus, the daughter nucleus is $\boxed{{}^{238}_{92}\text{U}}$.

b. The decay reaction is $^{24}_{11}\text{Na} \rightarrow ^{A}_{Z}\text{X} + ^{0}_{-1}\text{e}$. Therefore, $24 = A$. In addition, $11 = Z - 1$, so that $Z = 12$. Thus the daughter nucleus is $\boxed{^{24}_{12}\text{Mg}}$.

c. The decay reaction is $^{13}_{7}\text{N} \rightarrow ^{A}_{Z}\text{X} + ^{0}_{1}\text{e}$. Therefore, $13 = A$. In addition, $7 = Z + 1$, so that $Z = 6$. Thus, the daughter nucleus is $\boxed{^{13}_{6}\text{C}}$.

---

19. **REASONING AND SOLUTION**

    a. The decay process looks like $\boxed{^{14}_{6}\text{C} \rightarrow ^{14}_{7}\text{N} + ^{0}_{-1}\text{e}}$

    b. For the next reaction $\boxed{^{212}_{82}\text{Pb} \rightarrow ^{212}_{83}\text{Bi} + ^{0}_{-1}\text{e}}$

---

20. **REASONING AND SOLUTION**

    a. The reaction looks like $\boxed{^{14}_{6}\text{C} \rightarrow ^{14}_{7}\text{N} + ^{0}_{-1}\text{e}}$

    b. The mass change during the reaction is

    $$\Delta m = 14.003\,241\text{ u} - 14.003\,074\text{ u} = 0.000\,167\text{ u}$$

    Since 1 u = 931.5 MeV, the energy released during the reaction is

    $$E = (0.000\,167\text{ u}) \left(\frac{931.5\text{ MeV}}{1\text{ u}} \right) = \boxed{0.156\text{ MeV}}$$

---

21. [SSM] **REASONING AND SOLUTION** The general form for β^- decay is

    $$\underbrace{^{A}_{Z}\text{P}}_{\substack{\text{Parent}\\\text{nucleus}}} \rightarrow \underbrace{^{A}_{Z+1}\text{D}}_{\substack{\text{Daughter}\\\text{nucleus}}} + \underbrace{^{0}_{-1}\text{e}}_{\substack{\beta^-\text{ particle}\\\text{(electron)}}}$$

    Therefore, the β^- decay process for $^{35}_{16}\text{S}$ is $\boxed{^{35}_{16}\text{S} \rightarrow ^{35}_{17}\text{Cl} + ^{0}_{-1}\text{e}}$.

---

22. **REASONING AND SOLUTION** The mass of the products is

    $$m = 222.017\,57\text{ u} + 4.002\,60\text{ u} = 226.020\,17\text{ u}$$

The mass defect for the decay is $\Delta m = 226.025\,40\text{ u} - 226.020\,17\text{ u} = 0.005\,23\text{ u}$, which corresponds to an energy of

$$(0.005\,23\text{ u})\left(\frac{931.5\text{ MeV}}{1\text{ u}}\right) = \boxed{4.87\text{ MeV}}$$

---

23. **[SSM] [WWW] REASONING AND SOLUTION** The general form for β^+ decay is

$$\underbrace{{}^{A}_{Z}\text{P}}_{\text{Parent nucleus}} \rightarrow \underbrace{{}^{A}_{Z-1}\text{D}}_{\text{Daughter nucleus}} + \underbrace{{}^{0}_{+1}\text{e}}_{\substack{\beta^+ \text{ particle}\\ \text{(positron)}}}$$

a. Therefore, the β^+ decay process for ${}^{18}_{9}\text{F}$ is $\boxed{{}^{18}_{9}\text{F} \rightarrow {}^{18}_{8}\text{O} + {}^{0}_{+1}\text{e}}$.

b. Similarly, the β^+ decay process for ${}^{15}_{8}\text{O}$ is $\boxed{{}^{15}_{8}\text{O} \rightarrow {}^{15}_{7}\text{N} + {}^{0}_{+1}\text{e}}$.

---

24. **REASONING** Electric charge must be conserved, and the number of nucleons must be conserved. Applying these two conservation principles, we will be able to obtain two equations containing the two unknown quantities N_α and N_β.

SOLUTION The overall decay process can be written as follows:

$${}^{220}_{86}\text{Rn} \rightarrow {}^{208}_{82}\text{Pb} + N_\alpha\left({}^{4}_{2}\text{He}\right) + N_\beta\left({}^{0}_{-1}\text{e}\right)$$

Since electric charge must be conserved, the 86 protons on the left must equal the total number of protons on the right, the result being

$$86 = 82 + N_\alpha(2) + N_\beta(-1)$$

Since the nucleon number must be conserved also, the 220 nucleons on the left must equal the total number of nucleons on the right, so that

$$220 = 208 + N_\alpha(4) + N_\beta(0)$$

Solving this result for the number of α particles gives $\boxed{N_\alpha = 3}$. Substituting $N_\alpha = 3$ into the conservation-of-charge equation, we find that the number of β^- particles is $\boxed{N_\beta = 2}$.

25. **REASONING AND SOLUTION** The decay reaction is $^{239}_{94}\text{Pu} \rightarrow {}^{235}_{92}\text{U} + {}^{4}_{2}\text{He}$. The mass defect of the reaction is

$$\Delta m = 239.052\ 16\ \text{u} - 4.002\ 603\ \text{u} - 235.043\ 924\ \text{u} = 0.005\ 63\ \text{u}$$

This corresponds to an energy release of

$$(0.005\ 63\ \text{u})\left(\frac{931.5\ \text{MeV}}{1\ \text{u}}\right) = 5.24\ \text{MeV}$$

This energy is assumed to be kinetic energy of the α particle (mass $m = 6.44 \times 10^{-27}$ kg, see Example 2), so the speed of the α particle is

$$v = \sqrt{\frac{2(\text{KE})}{m}} = \sqrt{\frac{2(5.24 \times 10^6\ \text{eV})\left(\frac{1.60 \times 10^{-19}\ \text{J}}{1\ \text{eV}}\right)}{6.64 \times 10^{-27}\ \text{kg}}} = \boxed{1.59 \times 10^7\ \text{m/s}}$$

26. **REASONING AND SOLUTION** The β decay reaction is

$$^{208}_{81}\text{Tl} \rightarrow {}^{208}_{82}\text{Pb} + {}^{0}_{-1}\text{e}$$

so

$$^{A}_{Z}\text{X} \rightarrow {}^{208}_{82}\text{Pb} + {}^{4}_{2}\text{He}$$

gives

$$\boxed{{}^{A}_{Z}\text{X} = {}^{212}_{84}\text{Po}}$$

27. [SSM] **REASONING** Energy is released during the β decay. To find the energy released, we determine how much the mass has decreased because of the decay and then calculate the equivalent energy. The reaction and masses are shown below:

$$\underbrace{{}^{22}_{11}\text{Na}}_{21.994\ 434\ \text{u}} \rightarrow \underbrace{{}^{22}_{10}\text{Ne}}_{21.991\ 383\ \text{u}} + \underbrace{{}^{0}_{+1}\text{e}}_{5.485\ 799 \times 10^{-4}\ \text{u}}$$

SOLUTION The decrease in mass is

$$21.994\ 434\ \text{u} - (21.991\ 383\ \text{u} + 5.485\ 799 \times 10^{-4}\ \text{u} + 5.485\ 799 \times 10^{-4}\ \text{u}) = 0.001\ 954\ \text{u}$$

where the extra electron mass takes into account the fact that the atomic mass for sodium includes the mass of 11 electrons, whereas the atomic mass for neon includes the mass of only 10 electrons.

Since 1 u is equivalent to 931.5 MeV, the released energy is

$$(0.001\,954 \text{ u})\left(\frac{931.5 \text{ MeV}}{1 \text{ u}}\right) = \boxed{1.82 \text{ MeV}}$$

28. **REASONING** As Section 7.2 discusses, the principle of conservation of linear momentum indicates that the total momentum of an isolated system is conserved. Assuming that no external net force acts on the beryllium nucleus, it is isolated and the emission of the gamma ray satisfies this principle. The magnitude of the momentum of the gamma ray photon is h/λ, according to Equation 29.6, where h is Planck's constant and λ is the wavelength. We take the direction of the photon as the positive direction. Then, using Equation 7.2, we write the momentum of the recoiling beryllium nucleus as $-mv_{Be}$, where m is the mass and v_{Be} is the speed. The conservation principle indicates that the total momentum after the emission must be equal to the total momentum before the emission, so we have

$$\underbrace{-mv_{Be}}_{\substack{\text{Momentum of nucleus}\\\text{after emission}}} + \underbrace{\frac{h}{\lambda}}_{\substack{\text{Momentum of gamma}\\\text{ray photon}}} = \underbrace{0}_{\substack{\text{Total momentum}\\\text{before emission}}}$$

This expression can be solved for the wavelength.

SOLUTION Solving the conservation-of-momentum expression, we obtain

$$\lambda = \frac{h}{mv_{Be}}$$

Before we can use this equation, the mass of the beryllium must be converted from atomic mass units to kilograms:

$$m_{Be} = (7.017 \text{ u})\left(\frac{1.6605 \times 10^{-27} \text{ kg}}{1 \text{ u}}\right) = 1.165 \times 10^{-26} \text{ kg}$$

The wavelength, then, is

$$\lambda = \frac{h}{mv_{Be}} = \frac{6.63 \times 10^{-34} \text{ J} \cdot \text{s}}{(1.165 \times 10^{-26} \text{ kg})(2.19 \times 10^4 \text{ m/s})} = \boxed{2.60 \times 10^{-12} \text{ m}}$$

29. **REASONING AND SOLUTION** Since one-eighth of the nuclei remain, three half-lives have elapsed. Therefore, $T_{1/2} = (9.0 \text{ days})/3 = \boxed{3.0 \text{ days}}$.

30. **REASONING AND SOLUTION** The number of seconds in 14.28 days is $T_{1/2} = 1.23 \times 10^6$ s. The decay constant is

$$\lambda = \frac{0.693}{T_{1/2}} = \frac{0.693}{1.26 \times 10^6 \text{ s}} = \boxed{5.62 \times 10^{-7} \text{ s}^{-1}}$$

31. [SSM] **REASONING AND SOLUTION** According to Equation 31.5, $N = N_0 e^{-\lambda t}$, the decay constant is

$$\lambda = -\frac{1}{t} \ln\left(\frac{N}{N_0}\right) = -\frac{1}{20 \text{ days}} \ln\left(\frac{8.14 \times 10^{14}}{4.60 \times 10^{15}}\right) = 0.0866 \text{ days}^{-1}$$

The half-life is, from Equation 31.6,

$$T_{1/2} = \frac{0.693}{\lambda} = \frac{0.693}{0.0866 \text{ days}^{-1}} = \boxed{8.00 \text{ days}}$$

32. **REASONING AND SOLUTION** According to Equation 31.5, the fraction of an initial sample remaining after a time t is $N/N_0 = e^{-\lambda t}$, where λ is the decay constant. The decay constant is related to the half-life $T_{1/2}$. According to Equation 31.6, the decay constant is $\lambda = 0.693/T_{1/2}$. Therefore, the fraction remaining is

$$\frac{N}{N_0} = e^{-0.693 t/T_{1/2}} = e^{-0.693[(30.0 \text{ days})/(8.04 \text{ days})]} = 0.0753$$

This fraction corresponds to a percentage of $\boxed{7.53\%}$.

33. **REASONING AND SOLUTION** The amount remaining is 0.0100% = 0.000 100. We know $N/N_0 = e^{-0.693 t/T_{1/2}}$. Therefore, we find

$$t = -\frac{T_{1/2}}{0.693} \ln\left(\frac{N}{N_0}\right) = -\frac{28.5 \text{ yr}}{0.693} \ln(0.000\,100) = \boxed{379 \text{ yr}}$$

34. **REASONING AND SOLUTION** The activity is $A = \lambda N$. The decay constant is

$$\lambda = \frac{0.693}{T_{1/2}} = \frac{0.693}{(5.27 \text{ yr})\left(\frac{3.156 \times 10^7 \text{ s}}{1 \text{ yr}}\right)} = 4.17 \times 10^{-9} \text{ s}^{-1}$$

As discussed in Section 14.1, the number N of nuclei is the number of moles of nuclei times Avogadro's number (which is the number of nuclei per mole). Thus,

$$N = \underbrace{\left(\frac{0.50 \text{ g}}{59.9 \text{ g/mol}}\right)}_{\substack{\text{Number} \\ \text{of moles}}} \underbrace{(6.02 \times 10^{23} \text{ mol}^{-1})}_{\substack{\text{Avogadro's} \\ \text{number}}} = 5.0 \times 10^{21}$$

Therefore, $A = \lambda N = (4.17 \times 10^{-9} \text{ s}^{-1})(5.0 \times 10^{21}) = \boxed{2.1 \times 10^{13} \text{ Bq}}$.

---

35. **SSM REASONING** We can find the decay constant from Equation 31.5, $N = N_0 e^{-\lambda t}$. If we multiply both sides by the decay constant λ, we have

$$\lambda N = \lambda N_0 e^{-\lambda t} \quad \text{or} \quad A = A_0 e^{-\lambda t}$$

where A_0 is the initial activity and A is the activity after a time t. Once the decay constant is known, we can use the same expression to determine the activity after a total of six days.

SOLUTION Solving the expression above for the decay constant λ, we have

$$\lambda = -\frac{1}{t} \ln\left(\frac{A}{A_0}\right) = -\frac{1}{2 \text{ days}} \ln\left(\frac{285 \text{ disintegrations/min}}{398 \text{ disintegrations/min}}\right) = 0.167 \text{ days}^{-1}$$

Then the activity four days after the second day is

$$A = (285 \text{ disintegrations/min}) e^{-(0.167 \text{ days}^{-1})(4.00 \text{ days})} = \boxed{146 \text{ disintegrations/min}}$$

---

36. **REASONING AND SOLUTION** The decay constant is

$$\lambda = (0.693)/T_{1/2} = (0.693)/(1.60 \times 10^3 \text{ yr}) = 4.33 \times 10^{-4} \text{ yr}^{-1}$$

The mass m of the radium is proportional to the number N of atoms present, so $N/N_0 = m/m_0$. Since

1110 NUCLEAR PHYSICS AND RADIOACTIVITY

$$N = N_0 e^{-\lambda t}, \quad \text{then} \quad m = m_0 e^{-\lambda t}$$

The mass remaining after fifty years is

$$m = (1.000 \times 10^{-9} \text{ kg}) e^{-(4.33 \times 10^{-4} \text{ yr}^{-1})(50.0 \text{ yr})} = 0.979 \times 10^{-9} \text{ kg}$$

The difference in mass, which is the mass of radium that has disappeared, is

$$\Delta m = 1.000 \times 10^{-9} \text{ kg} - 0.979 \times 10^{-9} \text{ kg} = \boxed{2.1 \times 10^{-11} \text{ kg}}$$

---

37. **REASONING AND SOLUTION** We know that $A = \lambda N$. The decay constant is

$$\lambda = \frac{0.693}{T_{1/2}} = \frac{0.693}{(1.6 \times 10^3 \text{ yr})\left(\dfrac{3.156 \times 10^7 \text{ s}}{1 \text{ yr}}\right)} = 1.372 \times 10^{-11} \text{ s}^{-1}$$

As discussed in Section 14.1, the number N of Radium nuclei is the number of moles of nuclei times Avogadro's number (which is the number of nuclei per mole). Thus,

$$N = \underbrace{\left(\frac{1.00 \text{ g}}{226 \text{ g/mol}}\right)}_{\text{Number of moles}} \underbrace{(6.02 \times 10^{23} \text{ mol}^{-1})}_{\text{Avogadro's number}} = 2.664 \times 10^{21}$$

Therefore, $A = \lambda N = (1.372 \times 10^{-11} \text{ s}^{-1})(2.664 \times 10^{21}) = \boxed{3.7 \times 10^{10} \text{ Bq}}$.

---

38. **REASONING** The activity of a radioactive sample is given by the magnitude of Equation 31.4 as λN, where λ is the decay constant and N is the number of radioactive nuclei. The decay constant is related to the half life by Equation 31.6 as $\lambda = 0.693/T_{1/2}$. Thus, the activity can be expressed as $0.693 \, N/T_{1/2}$. The number of nuclei can be obtained as discussed in Section 14.1, by multiplying the number of moles by Avogadro's number N_A. The number of moles is the mass m divided by the mass per mole M. Therefore, the number of radioactive nuclei is $N = mN_A/M$, and our expression for the activity becomes

$$\text{Activity} = \frac{0.693 \, N}{T_{1/2}} = \frac{0.693 \, m N_A}{T_{1/2} \, M}$$

We can apply this result to both krypton and xenon and then set the two activities equal to obtain the mass of krypton.

SOLUTION Setting the activity of the krypton equal to the activity of the xenon, we obtain

$$\frac{0.693\, m_{Kr}\, N_A}{(T_{1/2})_{Kr}\, M_{Kr}} = \frac{0.693\, m_{Xe}\, N_A}{(T_{1/2})_{Xe}\, M_{Xe}} \quad \text{or} \quad \frac{m_{Kr}}{(T_{1/2})_{Kr}\, M_{Kr}} = \frac{m_{Xe}}{(T_{1/2})_{Xe}\, M_{Xe}}$$

Solving for m_{Kr} gives

$$m_{Kr} = \frac{m_{Xe}\,(T_{1/2})_{Kr}\, M_{Kr}}{(T_{1/2})_{Xe}\, M_{Xe}} = \frac{(2.00\text{ g})(1.840\text{ s})(91.9\text{ g/mol})}{(13.6\text{ s})(139.9\text{ g/mol})} = \boxed{0.178\text{ g}}$$

---

39. **SSM** **REASONING** According to Equation 31.5, the number of nuclei remaining after a time t is $N = N_0 e^{-\lambda t}$. Using this expression, we find the ratio N_A / N_B as follows:

$$\frac{N_A}{N_B} = \frac{N_{0A}\, e^{-\lambda_A t}}{N_{0B}\, e^{-\lambda_B t}} = e^{-(\lambda_A - \lambda_B)t}$$

where we have used the fact that initially the numbers of the two types of nuclei are equal $(N_{0A} = N_{0B})$. Taking the natural logarithm of both sides of the equation above shows that

$$\ln(N_A / N_B) = -(\lambda_A - \lambda_B)t \quad \text{or} \quad \lambda_A - \lambda_B = \frac{-\ln(N_A / N_B)}{t}$$

SOLUTION Since $N_A / N_B = 3.00$ when $t = 3.00$ days, it follows that

$$\lambda_A - \lambda_B = \frac{-\ln(3.00)}{3.00\text{ days}} = -0.366\text{ days}^{-1}$$

But we need to find the half-life of species B, so we use Equation 31.6, which indicates that $\lambda = 0.693 / T_{1/2}$. With this expression for λ, the result for $\lambda_A - \lambda_B$ becomes

$$0.693\left(\frac{1}{T_{1/2}^A} - \frac{1}{T_{1/2}^B}\right) = -0.366$$

Since $T_{1/2}^B = 1.50$ days, the result above can be solved to show that $\boxed{T_{1/2}^A = 7.23\text{ days}}$.

1112 NUCLEAR PHYSICS AND RADIOACTIVITY

40. **REASONING AND SOLUTION** We know $t = -(1/\lambda) \ln(A/A_0)$ where

$$\lambda = 1.21 \times 10^{-4} \text{ yr}^{-1} \quad \text{and} \quad A/A_0 = 0.21$$

Therefore, $t = -\dfrac{1}{1.21 \times 10^{-4} \text{ yr}^{-1}} \ln(0.21) = \boxed{13\,000 \text{ yr}}$.

---

41. [SSM] **REASONING AND SOLUTION** The answer can be obtained directly from Equation 31.5, combined with Equation 31.6:

$$\dfrac{N}{N_0} = e^{-\lambda t} = e^{-(0.693)t/T_{1/2}} = e^{-(0.693)(41\,000 \text{ yr})/(5730 \text{ yr})} = 0.0070$$

The percent of atoms remaining is $\boxed{0.70\,\%}$.

---

42. **REASONING AND SOLUTION**
   a. According to Equations 31.5 and 31.6, the ratio N/N_0 is given by

   $$\dfrac{N}{N_0} = e^{-0.693\,t/T_{1/2}} = e^{-0.693\left[(5.00\text{ yr})/(5730\text{ yr})\right]} = \boxed{0.999}$$

   b. Similarly, we find

   $$\dfrac{N}{N_0} = e^{-0.693\,t/T_{1/2}} = e^{-0.693\left[(3600\text{ s})/(122.2\text{ s})\right]} = \boxed{1.36 \times 10^{-9}}$$

   c. Similarly, we find

   $$\dfrac{N}{N_0} = e^{-0.693\,t/T_{1/2}} = e^{-0.693\left[(5.00\text{ yr})/(12.33\text{ yr})\right]} = \boxed{0.755}$$

---

43. **REASONING AND SOLUTION**
   a. We know that

   $$t = -\dfrac{T_{1/2}}{0.693} \ln\!\left(\dfrac{A}{A_0}\right) = -\dfrac{5730 \text{ yr}}{0.693} \ln\!\left(\dfrac{0.0061 \text{ Bq}}{0.23 \text{ Bq}}\right) = \boxed{3.0 \times 10^4 \text{ yr}}$$

   b. If A_0 is 40% larger, then $A_0 = (1.4)(0.23 \text{ Bq}) = 0.32 \text{ Bq}$, so that

   $$t = -\dfrac{T_{1/2}}{0.693} \ln\!\left(\dfrac{A}{A_0}\right) = -\dfrac{5730 \text{ yr}}{0.693} \ln\!\left(\dfrac{0.0061 \text{ Bq}}{0.32 \text{ Bq}}\right) = \boxed{3.3 \times 10^4 \text{ yr}}$$

44. ***REASONING AND SOLUTION*** The time during which radioactive decay occurred is $t = 1988 \text{ yr} - 1200 \text{ yr} = 788 \text{ yr}$. According to Equations 31.5 and 31.6, the ratio N/N_0 that corresponds to this time is given by

$$\frac{N}{N_0} = e^{-0.693\,t/T_{1/2}} = e^{-0.693\left[(788\text{ yr})/(5730\text{ yr})\right]} = 0.909$$

The percentage of the original $^{14}_{6}\text{C}$ nuclei remaining, then, is $\boxed{90.9\%}$.

---

45. **SSM** ***REASONING*** According to Equation 31.5, $N = N_0 e^{-\lambda t}$. If we multiply both sides by the decay constant λ, we have

$$\lambda N = \lambda N_0 e^{-\lambda t} \qquad \text{or} \qquad A = A_0 e^{-\lambda t}$$

where A_0 is the initial activity and A is the activity after a time t. The decay constant λ is related to the half-life through Equation 31.6: $\lambda = 0.693/T_{1/2}$. We can find the age of the fossils by solving for the time t. The maximum error can be found by evaluating the limits of the accuracy as given in the problem statement.

SOLUTION The age of the fossils is

$$t = -\frac{T_{1/2}}{0.693} \ln\left(\frac{A}{A_0}\right) = -\frac{5730 \text{ yr}}{0.693} \ln\left(\frac{0.10 \text{ Bq}}{0.23 \text{ Bq}}\right) = \boxed{6900 \text{ yr}}$$

The maximum error can be found as follows:

When there is an error of $+10\%$, $A = 0.10 \text{ Bq} + 0.010 \text{ Bq} = 0.11 \text{ Bq}$, and we have

$$t = -\frac{5730 \text{ yr}}{0.693} \ln\left(\frac{0.11 \text{ Bq}}{0.23 \text{ Bq}}\right) = 6100 \text{ yr}$$

Similarly, when there is an error of -10%, $A = 0.10 \text{ Bq} - 0.010 \text{ Bq} = 0.090 \text{ Bq}$, and we have

$$t = -\frac{5730 \text{ yr}}{0.693} \ln\left(\frac{0.090 \text{ Bq}}{0.23 \text{ Bq}}\right) = 7800 \text{ yr}$$

The maximum error in the age of the fossils is $7800 \text{ yr} - 6900 \text{ yr} = \boxed{900 \text{ yr}}$.

46. **REASONING** The number of radioactive nuclei remaining after a time t is given by Equations 31.5 and 31.6 as

$$N = N_0 e^{-0.693\, t/T_{1/2}}$$

where N_0 is the number of radioactive nuclei present at $t = 0$ s and $T_{1/2}$ is the half life for the decay. The activity A is proportional to the number of radioactive nuclei that a sample contains, so it follows that

$$A = A_0 e^{-0.693\, t/T_{1/2}} \qquad (1)$$

We know that the activity for the ancient carbon in the sample is 0.011 Bq per gram of carbon, whereas for the fresh carbon it is 0.23 Bq per gram of carbon. Knowing the percentage composition of the contaminated sample, we can determine its activity A by using the given percentages with the known activities:

$$A_{\text{Contaminated}} = 0.980\, A_{\text{Ancient}} + 0.020\, A_{\text{Fresh}} \qquad (2)$$

The true age and the apparent age can be obtained by applying Equation (1) to the corresponding activities.

SOLUTION a. Using Equation (1), we find that the true age is

$$A_{\text{Ancient}} = A_0 e^{-0.693\, t/T_{1/2}} \quad \text{or} \quad 0.011 \text{ Bq} = (0.23 \text{ Bq})e^{-0.693\, t/(5730 \text{ yr})}$$

Taking the natural logarithm of both sides of this result gives

$$\ln\left(\frac{0.011 \text{ Bq}}{0.23 \text{ Bq}}\right) = -\frac{0.693\, t}{5730 \text{ yr}}$$

$$t = \frac{-(5730 \text{ yr})\ln\left(\dfrac{0.011 \text{ Bq}}{0.23 \text{ Bq}}\right)}{0.693} = \boxed{25\,000 \text{ yr}}$$

b. Using Equation (2) to find the activity of the contaminated sample, we obtain

$$A_{\text{Contaminated}} = 0.980\, A_{\text{Ancient}} + 0.020\, A_{\text{Fresh}}$$
$$= 0.980(0.011 \text{ Bq}) + 0.020(0.23 \text{ Bq}) = 0.0154 \text{ Bq}$$

Using this activity in Equation (1), we find that the apparent age of the sample is

$$t = \frac{-(5730 \text{ yr})\ln\left(\dfrac{0.0154 \text{ Bq}}{0.23 \text{ Bq}}\right)}{0.693} = \boxed{22\,000 \text{ yr}}$$

47. **REASONING AND SOLUTION** For $^{202}_{80}$Hg the mass of the separated nucleons is

$$m = 80(1.007\,825\text{ u}) + 122(1.008\,665\text{ u}) = 203.683\,130\text{ u}$$

The mass defect is then $\Delta m = 203.683\,130\text{ u} - 201.970\,617\text{ u} = 1.712\,513\text{ u}$.

This corresponds to a total binding energy of $(1.712\,513\text{ u})\left(\dfrac{931.5\text{ MeV}}{1\text{ u}}\right) = 1595\text{ MeV}$ and a binding energy per nucleon of $\dfrac{1595\text{ MeV}}{202\text{ nucleons}} = \boxed{7.90\text{ MeV per nucleon}}$.

48. **REASONING AND SOLUTION** Equation 31.2 gives for the radius of the nucleus

$$r = (1.2 \times 10^{-15}\text{ m})A^{1/3} = (1.2 \times 10^{-15}\text{ m})(48)^{1/3} = \boxed{4.4 \times 10^{-15}\text{ m}}$$

49. **SSM REASONING AND SOLUTION** The number of radioactive nuclei that remains in a sample after a time t is given by Equation 31.5, $N = N_0 e^{-\lambda t}$, where λ is the decay constant. From Equation 31.6, we know that the decay constant is related to the half-life by $T_{1/2} = 0.693/\lambda$; therefore, $\lambda = 0.693/T_{1/2}$ and we can write

$$\frac{N}{N_0} = e^{-(0.693/T_{1/2})t} \quad\text{or}\quad \frac{t}{T_{1/2}} = -\frac{1}{0.693}\ln\left(\frac{N}{N_0}\right)$$

When the number of radioactive nuclei decreases to one-millionth of the initial number, $N/N_0 = 1.00 \times 10^{-6}$; therefore, the number of half-lives is

$$\frac{t}{T_{1/2}} = -\frac{1}{0.693}\ln(1.00 \times 10^{-6}) = \boxed{19.9}$$

50. **REASONING AND SOLUTION** The reaction and the associated masses are as follows:

$$\underbrace{^{3}_{1}\text{H}}_{3.016\,050\text{ u}} \rightarrow \underbrace{^{3}_{2}\text{He}}_{3.016\,030\text{ u}} + ^{\;\;0}_{-1}\text{e}$$

The mass defect Δm for the reaction is

$$\Delta m = 3.016\,050\text{ u} - 3.016\,030\text{ u} = 0.000\,020\text{ u}$$

1116 NUCLEAR PHYSICS AND RADIOACTIVITY

Since 1 u = 931.5 MeV, this mass defect corresponds to a released energy of

$$(0.000\,020\text{ u})\left(\frac{931.5\text{ MeV}}{1\text{ u}}\right) = \boxed{0.019\text{ MeV}}$$

---

51. **REASONING** We can determine the identity of X in each of the decay processes by noting that for each process, the sum of A and Z for the decay products must equal the values of A and Z for the parent nuclei.

SOLUTION

a.
$$^{211}_{82}\text{Pb} \rightarrow \,^{211}_{83}\text{Bi} + \text{X}$$

Using the reasoning discussed above, X must have $A = 0$ and $Z = -1$. Therefore X must be an electron, $^{0}_{-1}e$; $\boxed{\text{X represents a } \beta^- \text{ particle (electron)}}$.

b.
$$^{11}_{6}\text{C} \rightarrow \,^{11}_{5}\text{B} + \text{X}$$

Similar reasoning suggests that X must have $A = 0$ and $Z = +1$. Therefore X must be a positron; $^{0}_{+1}e$; $\boxed{\text{X represents a } \beta^+ \text{ particle (positron)}}$.

c.
$$^{231}_{90}\text{Th}^* \rightarrow \,^{231}_{90}\text{Th} + \text{X}$$

Similar reasoning suggests that X must have $A = 0$ and $Z = 0$. Therefore X must be a gamma ray; $\boxed{\text{X represents a } \gamma \text{ ray}}$.

d.
$$^{210}_{84}\text{Po} \rightarrow \,^{206}_{82}\text{Pb} + \text{X}$$

Using the reasoning discussed above, X must have $A = 4$ and $Z = 2$. Therefore X must be a helium nucleus, $^{4}_{2}\text{He}$; $\boxed{\text{X represents an } \alpha \text{ particle (helium nucleus)}}$.

---

52. **REASONING** According to Equation 31.5, the number of nuclei remaining after a time t is $N = N_0 e^{-\lambda t}$. If we multiply both sides of this equation by the decay constant λ, we have $\lambda N = \lambda N_0 e^{-\lambda t}$. Recognizing that λN is the activity A, we have $A = A_0 e^{-\lambda t}$, where A_0 is the activity at time $t = 0$. A_0 can be determined from the fact that we know the mass of the specimen, and that the activity of one gram of carbon in a living organism is 0.23 Bq. The decay constant λ can be determined from the value of 5730 yr for the half-life of $^{14}_{6}\text{C}$ using

Equation 31.6. With known values for A_0 and λ, the given activity of 1.6 Bq can be used to determine the age t of the specimen.

SOLUTION For $^{14}_{6}C$, the decay constant is

$$\lambda = \frac{0.693}{T_{1/2}} = \frac{0.693}{5730 \text{ yr}} = 1.21 \times 10^{-4} \text{ yr}^{-1}$$

The activity at time $t = 0$ is $A_0 = (9.2 \text{ g})(0.23 \text{ Bq}/\text{g}) = 2.1$ Bq. Since $A = 1.6$ Bq and $A_0 = 2.1$ Bq, the age of the specimen can be determined from

$$A = 1.6 \text{ Bq} = (2.1 \text{ Bq}) e^{-(1.21 \times 10^{-4} \text{ yr}^{-1}) t}$$

Taking the natural logarithm of both sides leads to

$$\ln\left(\frac{1.6 \text{ Bq}}{2.1 \text{ Bq}}\right) = -(1.21 \times 10^{-4} \text{ yr}^{-1}) t$$

Therefore, the age of the specimen is

$$t = \frac{\ln\left(\dfrac{1.6 \text{ Bq}}{2.1 \text{ Bq}}\right)}{-1.21 \times 10^{-4} \text{ yr}^{-1}} = \boxed{2.2 \times 10^3 \text{ yr}}$$

---

53. **SSM WWW** *REASONING AND SOLUTION* As shown in Figure 31.18, if the first dynode produces 3 electrons, the second produces 9 electrons (3^2), the third produces 27 electrons (3^3), so the Nth produces 3^N electrons. The number of electrons that leaves the 14th dynode and strikes the 15th dynode is

$$3^{14} = \boxed{4\ 782\ 969 \text{ electrons}}$$

---

54. *REASONING AND SOLUTION* The conservation of linear momentum applied to the reaction gives

$$m_\alpha v_\alpha + m_T v_T = 0 \tag{1}$$

The energy released in the decay is assumed to be kinetic, so

$$(1/2) m_\alpha v_\alpha^2 + (1/2) m_T v_T^2 = 4.3 \text{ MeV} \tag{2}$$

Solving Equation (1) for v_α, substituting into Equation (2) and rearranging, gives the kinetic energy of the thorium atom.

$$\tfrac{1}{2}m_T v_T^2 = \frac{4.3\text{ MeV}}{1+\dfrac{m_T}{m_\alpha}} = \frac{4.3\text{ MeV}}{1+\dfrac{234.0436\text{ u}}{4.0026\text{ u}}} = \boxed{0.072\text{ MeV}}$$

The kinetic energy of the α particle is, then, 4.3 MeV – 0.072 MeV = $\boxed{4.2\text{ MeV}}$.

---

55. **REASONING AND SOLUTION** Since the isotopes have the same binding energies, they have the same mass defects, so that

$$\underbrace{Zm_h + (A_1 - Z)m_n - M_1}_{\substack{\text{Isotope \#1 mass defect:}\\ A_1 = \text{nucleon number}\\ M_1 = \text{atomic mass}}} = \underbrace{Zm_h + (A_2 - Z)m_n - M_2}_{\substack{\text{Isotope \#2 mass defect:}\\ A_2 = \text{nucleon number}\\ M_2 = \text{atomic mass}}}$$

In this expression, Z is the atomic number of each isotope, m_h is the atomic mass of a hydrogen atom, and m_n is the mass of a neutron. Simplifying the expression shows that $A_1 m_n - M_1 = A_2 m_n - M_2$. We also know that $A_2 = A_1 + 2$. Substituting this in the equation above and simplifying the result shows that

$$M_2 - M_1 = 2m_n = 2(1.008\,665\text{ u}) = \boxed{2.017\,330\text{ u}}$$

---

56. **REASONING AND SOLUTION** $A = A_0 e^{-\lambda t}$ for each of the substances. Equating and rearranging yields

$$A_{0,g} e^{-\lambda_g t} = A_{0,i} e^{-\lambda_i t}$$

But, $A_{0,g} = 5 A_{0,i}$. Substituting this result into the equation above and algebraically rearranging the variables, we find

$$e^{-(\lambda_g - \lambda_i)t} = \tfrac{1}{5}$$

Taking the natural logarithm of each side of this equation results in

$$t = -\frac{\ln\left(\dfrac{1}{5}\right)}{\lambda_g - \lambda_i}$$

Now,

$$\lambda_g = \frac{0.693}{T_{1/2}} = \frac{0.693}{2.69 \text{ days}} = 0.258 \text{ days}^{-1}$$

$$\lambda_i = \frac{0.693}{T_{1/2}} = \frac{0.693}{8.04 \text{ days}} = 0.0862 \text{ days}^{-1}$$

so,

$$t = -\frac{\ln\left(\frac{1}{5}\right)}{0.258 \text{ days}^{-1} - 0.0862 \text{ days}^{-1}} = \boxed{9.37 \text{ days}}$$

---

57. **CONCEPT QUESTIONS** a. According to Coulomb's law the magnitude of the electrostatic force is $F = ke^2/r^2$.

b. The electrostatic force given by Coulomb's law has the least possible magnitude when the two charges are as widely separated as possible, so that the distance r is a great as possible. In the nucleus, this means that the two protons in question must be located at opposite ends of the diameter of the nucleus.

SOLUTION To find the magnitude of the least possible electrostatic force that either of two protons can exert on the other, we need to know the diameter of the gold nucleus. The diameter is twice the radius. Equation 31.2 indicates that the diameter of the gold nucleus is $d = 2(1.2 \times 10^{-15} \text{ m}) A^{1/3}$, where A is the nucleon number. For gold $^{197}_{79}$Au it follows that $A = 197$. Using Coulomb's law with a separation between the protons that equals the diameter of the nucleus, we find

$$F_{\text{Least possible}} = \frac{ke^2}{d^2} = \frac{ke^2}{\left[2(1.2 \times 10^{-15} \text{ m}) A^{1/3}\right]^2}$$

$$= \frac{(8.99 \times 10^9 \text{ N} \cdot \text{m}^2/\text{C}^2)(1.60 \times 10^{-19} \text{ C})^2}{\left[2(1.2 \times 10^{-15} \text{ m})(197)^{1/3}\right]^2} = \boxed{1.2 \text{ N}}$$

---

58. **CONCEPT QUESTIONS** a. The energy E released is related to the mass decrease Δm by Equation 28.5, $E = (\Delta m)c^2$. A mass decrease of one atomic mass unit ($\Delta m = 1\text{u}$) corresponds to a released energy of 931.5 MeV.

b. Since the released energy E is shared among the three particles, we know that

$$E = E_{\text{Daughter}} + E_{\text{Beta particle}} + E_{\text{Antineutrino}} \quad (1)$$

NUCLEAR PHYSICS AND RADIOACTIVITY

The value for $E_{\text{Beta particle}}$ is known. For a given value of E, $E_{\text{Antineutrino}}$ will have its maximum possible value when E_{Daughter} is zero.

SOLUTION We begin by calculating the decrease in mass for the decay process, which is shown as follows along with the given atomic mass values:

$$\underbrace{^{32}_{15}\text{P}}_{31.973\,907\text{ u}} \rightarrow \underbrace{^{32}_{16}\text{S} + ^{\;\;0}_{-1}\text{e}}_{31.972\,070\text{ u}} + \underbrace{\bar{\gamma}}_{\text{Antineutrino}}$$

When the $^{32}_{15}\text{P}$ nucleus of a phosphorus atom is converted into a $^{32}_{16}\text{S}$ nucleus, the number of orbital electrons remains the same, so the resulting sulfur atom is missing one orbital electron. However, the given atomic mass for the sulfur includes all 16 electrons of a neutral atom. In effect, then, the value of 31.972 070 u already includes the mass of the β^- particle. Thus, the mass decrease for the decay is 31.973 907 u – 31.972 070 u = 0.001 837 u. The released energy E is

$$E = (0.001\,837\text{ u})\left(\frac{931.5\text{ MeV}}{1\text{ u}}\right) = 1.711\text{ MeV}$$

Using Equation (1), we can now find the maximum possible energy carried away by the antineutrino

$$E_{\text{Antineutrino}} = E - E_{\text{Daughter}} - E_{\text{Beta particle}}$$

$$= (1.711\text{ MeV}) - (0\text{ MeV}) - (0.90\text{ MeV}) = \boxed{0.81\text{ MeV}}$$

---

59. **CONCEPT QUESTIONS** a. The fraction of nuclei A that decay in a given time period is less than the fraction of nuclei B. The reason is that the half life for decay is greater for nuclei A. A greater half life means that a greater time is required for one half (or any fraction) of the nuclei present initially to decay. In other words, the decay process is slower for the nuclei with the greater half life, so that, in a given time period, a smaller fraction decays.

b. The ratio N_A/N_B of the number of nuclei present at a later time is greater than the ratio $N_{0,A}/N_{0,B}$ of the number present initially. This follows from our answer to question (a). The fraction of nuclei that decay is smaller for A and greater for B. Therefore, a greater fraction of type A and a smaller fraction of type B will remain at the later time, with the result that the ratio N_A/N_B increases relative to $N_{0,A}/N_{0,B}$.

SOLUTION The number of radioactive nuclei remaining after a time t is given by Equations 31.5 and 31.6 as

$$N = N_0 e^{-0.693\, t/T_{1/2}}$$

where N_0 is the number of radioactive nuclei present at $t = 0$ s and $T_{1/2}$ is the half life for the decay. Applying this relation to each type of nucleus, we obtain

$$\frac{N_{Sr}}{N_{Cs}} = \frac{N_{0,\,Sr}\, e^{-0.693\, t/T_{1/2,\,Sr}}}{N_{0,\,Cs}\, e^{-0.693\, t/T_{1/2,\,Cs}}} = (7.80 \times 10^{-3}) \frac{e^{-0.693(15.0\text{ yr})/(28.5\text{ yr})}}{e^{-0.693(15.0\text{ yr})/(2.06\text{ yr})}} = \boxed{0.842}$$

---

60. **CONCEPT QUESTIONS** a. The heat Q needed to melt a mass m of water is given by Equation 12.5 as $Q = mL_f$, where $L_f = 33.5 \times 10^4$ J/kg is the latent heat of fusion for water.

b. The energy E released is related to the mass decrease Δm by Equation 28.5, $E = (\Delta m)c^2$. A mass decrease of one atomic mass unit ($\Delta m = 1\text{u}$) corresponds to a released energy of 931.5 MeV.

c. The total energy released is just the number of disintegrations times the energy for each one, or $\underline{E_{Total} = nE}$.

d. In a time period equal to one half life, the number of radioactive nuclei that decay is $N_0/2$, where N_0 is the number present initially.

SOLUTION According to Equation 12.5, the mass of water melted is $m = Q/L_f$. The heat Q is provided by the total energy from the decay or $Q = E_{Total} = nE$, as discussed in the Concept Questions. The mass of water melted, then, becomes $m = nE/L_f$. To use this expression, we need the energy E that is released by one disintegration.

The disintegration process is

$$\underset{224.020\,186\text{ u}}{{}^{224}_{88}\text{Ra}} \rightarrow \underbrace{\underset{220.011\,368\text{ u}}{{}^{220}_{86}\text{Rn}} + \underset{4.002\,603\text{ u}}{{}^{4}_{2}\text{He}}}_{224.013\,971\text{ u}}$$

As usual, the masses are atomic masses and include the mass of the orbital electrons. This causes no error here, however, because the same total number of electrons is included on the left and right sides of the arrow in the process above. The decrease in mass, then, is

$$224.020\,186\text{ u} - 224.013\,971 = 0.006215\text{ u}$$

NUCLEAR PHYSICS AND RADIOACTIVITY

Since 931.5 MeV of energy corresponds to 1 u and since 1 eV = 1.60×10^{-19} J, the energy release by one disintegration is

$$E = (0.006215 \text{ u})\left(\frac{931.5 \times 10^6 \text{ eV}}{1 \text{ u}}\right)\left(\frac{1.60 \times 10^{-19} \text{ J}}{1 \text{ eV}}\right) = 9.26 \times 10^{-13} \text{ J}$$

Since the time period being considered is equal to one half life, the number n of disintegrations that occur is one half the number of radioactive nuclei initially present. We can now calculate the mass of ice melted as follows:

$$m = \frac{nE}{L_f} = \frac{\frac{1}{2}(2.69 \times 10^{21})(9.26 \times 10^{-13} \text{ J})}{33.5 \times 10^4 \text{ J/kg}} = \boxed{3720 \text{ kg}}$$

CHAPTER 32 | IONIZING RADIATION, NUCLEAR ENERGY, AND ELEMENTARY PARTICLES

PROBLEMS

1. **SSM** *REASONING AND SOLUTION* The exposure in roentgens is given by Equation 32.1:

$$\text{Exposure (in roentgens)} = \left(\frac{1}{2.58 \times 10^{-4}}\right)\frac{q}{m}$$

The total charge q in the beam is equal to the number of ions in the beam multiplied by the charge on each ion, namely $e = 1.6 \times 10^{-19}$ C. Therefore, the exposure is

$$\text{Exposure} = \left(\frac{1}{2.58 \times 10^{-4}}\right)\frac{(1.7 \times 10^{12})(1.6 \times 10^{-19} \text{ C})}{4.0 \times 10^{-3} \text{ kg}} = \boxed{0.26 \text{ roentgens}}$$

2. *REASONING AND SOLUTION* According to Equation 32.4, the biologically equivalent dose (BED) is the product of the absorbed dose (AD) and the relative biological effectiveness (RBE): BED = AD × RBE. For the protons, BED = (60 rad)(10) = 600 rem. For the α-particles to have the same BED, the absorbed dose is

$$AD = BED/RBE = (600 \text{ rem})/(20) = \boxed{30 \text{ rad}}$$

3. *REASONING AND SOLUTION* The absorbed dose (AD) is given by Equation 32.2,

$$AD = \frac{\text{Energy absorbed}}{\text{Mass of absorbing material}}$$

so

$$\text{Energy absorbed} = AD \times \text{Mass} = (2.5 \times 10^{-5} \text{ Gy})(65 \text{ kg}) = \boxed{1.6 \times 10^{-3} \text{ J}}$$

4. **REASONING AND SOLUTION**
   a. The biologically equivalent dose is equal to the product of the absorbed dose and the RBE (see Equation 32.4):

   $$\text{Biologically equivalent dose} = (\text{Absorbed dose})(\text{RBE}) = (38\text{ rad})(12) = \boxed{460\text{ rem}}$$

   b. According to the discussion at the end of Section 32.1, a person exposed to a whole-body, single dose of 460 rem has $\boxed{\text{a 50\% chance of dying}}$.

---

5. **SSM WWW REASONING AND SOLUTION** The energy absorbed is given by the absorbed dose in grays times the mass of the person in kilograms, according to Equation 32.2. To find the absorbed dose from the biologically equivalent dose (BED), we turn to Equation 32.4:

   $$\text{Absorbed dose in rad} = \frac{\text{BED}}{\text{RBE}} = \frac{45 \times 10^{-3}\text{ rem}}{12}$$

   Using the fact that 0.01 Gy = 1 rad, we find that

   $$\text{Absorbed dose in grays} = \left(\frac{45 \times 10^{-3}}{12}\text{ rad}\right)\left(\frac{0.01\text{ Gy}}{1\text{ rad}}\right)$$

   $$\text{Energy} = (\text{Absorbed dose in grays})(\text{Mass}) = \left(\frac{45 \times 10^{-5}}{12}\text{ Gy}\right)(75\text{ kg}) = \boxed{2.8 \times 10^{-3}\text{ J}}$$

---

6. **REASONING AND SOLUTION** According to Equation 32.2, the absorbed dose (AD) is equal to the energy absorbed by the tumor divided by its mass:

   $$\text{AD} = \frac{\text{Energy absorbed}}{\text{Mass}} = \frac{(25\text{ s})(1.6 \times 10^{10}\text{ s}^{-1})(4.0 \times 10^6\text{ eV})\left(\frac{1.60 \times 10^{-19}\text{ J}}{1\text{ eV}}\right)}{0.015\text{ kg}}$$

   $$= 1.7 \times 10^1\text{ Gy} = 1.7 \times 10^3\text{ rd}$$

   The biologically equivalent dose (BED) is equal to the product of the absorbed dose (AD) and the RBE (see Equation 32.4):

   $$\text{BED} = \text{AD} \times \text{RBE} = (1.7 \times 10^3\text{ rd})(14) = \boxed{2.4 \times 10^4\text{ rem}} \qquad (32.4)$$

7. **SSM** *REASONING* According to Equation 32.2, the absorbed dose is the energy absorbed divided by the mass of absorbing material:

$$\text{Absorbed dose} = \frac{\text{Energy absorbed}}{\text{Mass of absorbing material}}$$

The energy absorbed in this case is the sum of three terms: (1) the heat needed to melt a mass m of ice at 0.0 °C into liquid water at 0.0 °C, which is mL_f, according to the definition of the latent heat of fusion L_f (see Section 12.8); (2) the heat needed to raise the temperature of liquid water by an amount ΔT, which is $cm\Delta T$, where c is the specific heat capacity and ΔT is the change in temperature from 0.0 to 100.0 °C, according to Equation 12.4; (3) the heat needed to vaporize liquid water at 100.0 °C into steam at 100.0 °C, which is mL_v, according to the definition of the latent heat of vaporization L_v (see Section 12.8). Once the energy absorbed is determined, the absorbed dose can be determined using Equation 32.2.

SOLUTION Using the value of $c = 4186$ J/(kg·C°) for liquid water from Table 12.2 and the values of $L_f = 33.5 \times 10^4$ J/kg and $L_v = 22.6 \times 10^5$ J/kg from Table 12.3, we find that

$$\frac{\text{Absorbed dose}}{\text{in grays}} = \frac{\text{Energy}}{\text{mass}} = \frac{mL_f + cm\Delta T + mL_v}{m} = L_f + c\Delta T + L_v$$

$$= 33.5 \times 10^4 \text{ J/kg} + \left[4186 \text{ J/(kg·C°)}\right](100.0 \text{ C°}) + 22.6 \times 10^5 \text{ J/kg}$$

$$= 3.01 \times 10^6 \text{ J/kg}$$

Using the fact that 0.01 Gy = 1 rd, we find that

$$\text{Absorbed dose} = \left(3.01 \times 10^6 \text{ Gy}\right)\left(\frac{1 \text{ rd}}{0.01 \text{ Gy}}\right) = \boxed{3.01 \times 10^8 \text{ rd}}$$

8. *REASONING* The number of nuclei in the beam is equal to the energy absorbed by the tumor divided by the energy per nucleus (130 MeV). According to Equation 32.2, the energy (in joules) absorbed by the tumor is equal to the absorbed dose (expressed in grays) times the mass of the tumor. The absorbed dose (expressed in rads) is equal to the biologically equivalent dose divided by the RBE of the radiation (see Equation 32.4). We can use these concepts to determine the number of nuclei in the beam.

SOLUTION The number N of nuclei in the beam is equal to the energy E absorbed by the tumor divided by the energy per nucleus. Since the energy absorbed is equal to the absorbed dose (in Gy) times the mass m (see Equation 32.2) we have

$$N = \frac{E}{\text{Energy per nucleus}} = \frac{[\text{Absorbed dose (in Gy)}]m}{\text{Energy per nucleus}}$$

We can express the absorbed dose in terms of rad units, rather than Gy units, by noting that 1 rad = 0.01 Gy. Therefore,

$$\text{Absorbed dose (in Gy)} = \text{Absorbed dose (in rad)}\left(\frac{0.01 \text{ Gy}}{1 \text{ rad}}\right)$$

The number of nuclei can now be written as

$$N = \frac{E}{\text{Energy per nucleus}} = \frac{[\text{Absorbed dose (in rad)}]\left(\frac{0.01 \text{ Gy}}{1 \text{ rad}}\right)m}{\text{Energy per nucleus}}$$

We know from Equation 32.4 that the Absorbed dose (in rad) is equal to the Biologically equivalent dose divided by the RBE, so that

$$N = \frac{E}{\text{Energy per nucleus}} = \frac{\left[\frac{\text{Biologically equivalent dose}}{\text{RBE}}\right]\left(\frac{0.01 \text{ Gy}}{1 \text{ rad}}\right)m}{\text{Energy per nucleus}}$$

$$= \frac{\left[\frac{180 \text{ rem}}{16}\right]\left(\frac{0.01 \text{ Gy}}{1 \text{ rad}}\right)(0.17 \text{ kg})}{(130 \times 10^6 \text{ eV})\left(\frac{1.60 \times 10^{-19} \text{ J}}{1 \text{ eV}}\right)} = \boxed{9.2 \times 10^8}$$

9. **REASONING AND SOLUTION** According to Equation 32.2, the energy absorbed is equal to the product of the absorbed dose (AD) and the mass of the tumor:

$$\text{Energy} = \text{AD} \times \text{Mass} = (12 \text{ Gy})(2.0 \text{ kg}) = 24 \text{ J}$$

This energy is carried by ΔN particles in time Δt, so that

$$\text{Energy} = (\Delta N/\Delta t)(850 \text{ s})(0.40 \times 10^6 \text{ eV})(1.60 \times 10^{-19} \text{ J})/(1 \text{ eV})$$

Therefore,

$$\frac{\Delta N}{\Delta t} = \boxed{4.4 \times 10^{11} \text{ s}^{-1}}$$

10. **REASONING AND SOLUTION** The reaction can be written as

$$^{10}_{5}B + ^{4}_{2}He \rightarrow ^{1}_{1}H + ^{13}_{Z}X$$

Since electric charge is conserved, $5 + 2 = 1 + Z$. Therefore, we have

$\boxed{Z = 6}$, and $\boxed{X = \text{carbon}}$

---

11. [SSM] **REASONING AND SOLUTION** When a nitrogen $^{14}_{7}N$ nucleus absorbs a deuterium $^{2}_{1}H$ nucleus during a nuclear reaction, the compound nucleus formed will have $A = 14 + 2 = 16$, and $Z = 7 + 1 = 8$. Reference to the periodic table reveals that the compound nucleus must, therefore, be

$\boxed{\text{oxygen } ^{16}_{8}O}$

---

12. **REASONING AND SOLUTION** The reaction can be can be written as

$$^{27}_{13}Al + ^{4}_{2}He \rightarrow ^{A}_{15}P + ^{1}_{0}n$$

We therefore see that

$$27 + 4 = A + 1 \quad \text{or} \quad A = \boxed{30}$$

---

13. **REASONING AND SOLUTION**
  a. Since the total number of nucleons is conserved, we have $43 + 4 = A + 46$ or $A = 1$. Since the total electric charge is conserved, we have $20 + 2 = Z + 21$ or $Z = 1$. The unknown quantity must, therefore, be a $\boxed{\text{proton, } ^{1}_{1}H}$.

  b. In a fashion similar to that in part a, we have $A = 12 + 1 - 9 = 4$ and $Z = 6 - 4 = 2$. The unknown quantity must, therefore, be an $\boxed{\text{alpha particle, } ^{4}_{2}He}$.

  c. Similarly, $A = 9 + 1 - 4 = 6$ and $Z = 4 + 1 - 2 = 3$. The unknown quantity must, therefore, be $\boxed{\text{lithium, } ^{6}_{3}Li}$.

  d. The nucleon number is $A = 17 + 1 - 4 = 14$. The atomic number is $Z = 8 + 1 - 2 = 7$. The unknown quantity must, therefore, be $\boxed{\text{nitrogen, } ^{14}_{7}N}$.

1128 IONIZING RADIATION, NUCLEAR ENERGY, AND ELEMENTARY PARTICLES

e. The nucleon number is $A = 55 + 1 = 56$. The atomic number is $Z = 25$. The unknown quantity must, therefore, be $\boxed{\text{manganese, } ^{56}_{25}\text{Mn}}$.

---

14. **REASONING AND SOLUTION**

   a. We note that ^1_0n is a neutron (n) and ^1_1H is a proton (p), so the reaction can be written as $\boxed{^{14}_7\text{N}(n, p)\,^{14}_6\text{C}}$.

   b. This reaction can be written as $\boxed{^{238}_{92}\text{U}(n, \gamma)\,^{239}_{92}\text{U}}$.

   c. We note that ^2_1H is a deuteron (d), so the reaction can be written as $\boxed{^{24}_{12}\text{Mg}(n, d)\,^{23}_{11}\text{Na}}$.

---

15. **SSM REASONING** Energy is released from this reaction. Consequently, the combined mass of the daughter nucleus $^{12}_6\text{C}$ and the α particle ^4_2He is less than the combined mass of the parent nucleus $^{14}_7\text{N}$ and ^2_1H. The mass defect is equivalent to the energy released. We proceed by determining the difference in mass in atomic mass units and then use the fact that 1 u is equivalent to 931.5 MeV (see Section 31.3).

   SOLUTION The reaction and the atomic masses are as follows:

   $$\underbrace{^2_1\text{H}}_{2.014\,102\,\text{u}} + \underbrace{^{14}_7\text{N}}_{14.003\,074\,\text{u}} \rightarrow \underbrace{^{12}_6\text{C}}_{12.000\,000\,\text{u}} + \underbrace{^4_2\text{He}}_{4.002\,603\,\text{u}}$$

   The mass defect Δm for this reaction is

   $$\Delta m = 2.014\,102\text{ u} + 14.003\,074\text{ u} - 12.000\,000\text{ u} - 4.002\,603\text{ u} = 0.014\,573\text{ u}$$

   Since 1 u = 931.5 MeV, the energy released is

   $$(0.014\,573\text{ u})\left(\frac{931.5\text{ MeV}}{1\text{ u}}\right) = \boxed{13.6\text{ MeV}}$$

16. **REASONING AND SOLUTION** The reaction can be written as

$$^A_Z X + ^{63}_{29}Cu \rightarrow ^{62}_{29}Cu + ^1_1H + ^1_0n$$

Therefore, $A + 63 = 62 + 1 + 1$, so that we find $A = 1$. In addition, $Z + 29 = 29 + 1$, so that $Z = 1$. Thus, the unknown particle $^A_Z X$ is a proton ^1_1H. Therefore, the compound nucleus has

$$Z = 29 + 1 = 30 \quad \text{and} \quad A = 63 + 1 = 64$$

The compound nucleus is $\boxed{\text{zinc, } ^{64}_{30}Zn}$.

---

17. **REASONING AND SOLUTION** During a fission reaction, neutrons are produced in addition to the other fission products. We can see from the reaction that the missing nucleon number is

$$A = 235 + 1 - 93 - 141 = 2$$

Thus, the nucleons produced are $\boxed{2 \text{ neutrons}}$.

---

18. **REASONING AND SOLUTION** To find the number of neutrons we use the fact that the total number of nucleons is conserved. Therefore,

$$235 + 1 = 133 + 99 + A \quad \text{or} \quad A = 4$$

Since each neutron has $A = 1$, we conclude that the number of neutrons released is $\boxed{4}$.

---

19. **SSM REASONING** The rest energy of the uranium nucleus can be found by taking the atomic mass of the $^{235}_{92}U$ atom, subtracting the mass of the 92 electrons, and then using the fact that 1 u is equivalent to 931.5 MeV (see Section 31.3). According to Table 31.1, the mass of an electron is $5.485\,799 \times 10^{-4}$ u. Once the rest energy of the uranium nucleus is found, the desired ratio can be calculated.

SOLUTION The mass of $^{235}_{92}U$ is 235.043 924 u. Therefore, subtracting the mass of the 92 electrons, we have

Mass of $^{235}_{92}U$ nucleus $= 235.043\,924$ u $- 92(5.485\,799 \times 10^{-4}$ u$) = 234.993\,455$ u

The energy equivalent of this mass is

$$(234.993\ 455\ \text{u}) \left(\frac{931.5\ \text{MeV}}{1\ \text{u}} \right) = 2.189 \times 10^5\ \text{MeV}$$

Therefore, the ratio is

$$\frac{200\ \text{MeV}}{2.189 \times 10^5\ \text{MeV}} = \boxed{9.0 \times 10^{-4}}$$

---

20. **REASONING AND SOLUTION** In order to find the mass of the two fragments we need to know the equivalent mass of 225.0 MeV of energy. We know that 1 u = 931.5 MeV. Therefore, the mass equivalent of 225.0 MeV is

$$(225.0\ \text{MeV})(1\ \text{u})/(931.5\ \text{MeV}) = 0.2415\ \text{u}$$

By balancing each side of the fission reaction we can find the mass of the fragments, i.e.,

$$m(\text{fragments}) = m(\text{U-235}) + m(\text{neutron}) - m(3\ \text{neutrons}) - m_{\text{equiv}}$$

$$m(\text{fragments}) = 235.043\ 924\ \text{u} - 2(1.008\ 665\ \text{u}) - 0.2415\ \text{u} = \boxed{232.7851\ \text{u}}$$

---

21. **REASONING AND SOLUTION** The energy of the neutron after the first collision is $(1.5 \times 10^6\ \text{eV})(0.65)$. After the second collision it is $(1.5 \times 10^6\ \text{eV})(0.65)(0.65)$. Thus, after the nth collision it is

$$(1.5 \times 10^6\ \text{eV})(0.65)^n = 0.040\ \text{eV}$$

Solving for n gives

$$n \log(0.65) = \log\left(\frac{0.040\ \text{eV}}{1.5 \times 10^6\ \text{eV}} \right) = -7.57 \quad \text{or} \quad n = \frac{-7.57}{\log(0.65)} = 40.4$$

Therefore, 40 collisions will reduce the energy to something slightly greater than 0.040 eV, and to reduce the energy to at least 0.040 eV, $\boxed{41}$ collisions are needed.

---

22. **REASONING** The conservation of nucleon number states that the total number of nucleons (protons plus neutrons) before the reaction occurs must be equal to the total number of nucleons after the reaction. This conservation law will allow us to find the atomic mass number A of the unknown nucleus. The conservation of electric charge states that the net electric charge of the particles before the reaction must be equal to the net charge after the reaction. This conservation law will allow us to find the atomic number Z of the unknown nucleus. With a knowledge of the atomic number, we can use the periodic table to identify the element.

SOLUTION

a. The total number of nucleons before the reaction is 1 + 232 = 233. The total number of nucleons after the reaction is A. Setting these two numbers equal to each other yields $\boxed{A = 233}$. The net electric charge before the reaction is 0 + 90 = 90. The net electric charge after the reaction is Z. Setting these two numbers equal to each other yields $\boxed{Z = 90}$. A check of the periodic table shows that this element is $\boxed{\text{Thorium } (^{233}_{90}\text{Th})}$.

b. The $^{233}_{90}\text{Th}$ nucleus subsequently undergoes β^- decay $\left(^{\,0}_{-1}\text{e}\right)$, as does its daughter. The first reaction is $^{233}_{90}\text{Th} \rightarrow ^A_Z\text{X} + ^{\,0}_{-1}\text{e}$. By employing an analysis similar to that used in part (a), the unknown nucleus is found to be $^{233}_{91}\text{Pa}$. This daughter nucleus also undergoes β^- decay according to the reaction $^{233}_{91}\text{Pa} \rightarrow ^A_Z\text{X} + ^{\,0}_{-1}\text{e}$. Using the analysis of part (a) again, we see that the final unknown nucleus is $\boxed{^{233}_{92}\text{U}}$.

---

23. **SSM REASONING** We first determine the energy released by 1.0 kg of $^{235}_{92}\text{U}$. Using the data given in the problem statement, we can then determine the number of kilograms of coal that must be burned to produce the same energy.

SOLUTION The energy equivalent of one atomic mass unit is given in the text (see Section 31.3) as

$$1 \text{ u} = 1.4924 \times 10^{-10} \text{ J} = 931.5 \text{ MeV}$$

Therefore, the energy released in the fission of 1.0 kg of $^{235}_{92}\text{U}$ is

$$(1.0 \text{ kg of } ^{235}_{92}\text{U}) \left(\frac{1.0 \times 10^3 \text{ g/kg}}{235 \text{ g/mol}}\right) \left(\frac{6.022 \times 10^{23} \text{ nuclei}}{1.0 \text{ mol}}\right)$$

$$\times \left(\frac{2.0 \times 10^2 \text{ MeV}}{\text{nuclei}}\right) \left(\frac{1.4924 \times 10^{-10} \text{ J}}{931.5 \text{ MeV}}\right) = 8.2 \times 10^{13} \text{ J}$$

When 1.0 kg of coal is burned, about 3.0×10^7 J is released; therefore the number of kilograms of coal that must be burned to produce an energy of 8.2×10^{13} J is

$$m_{\text{coal}} = \left(8.2 \times 10^{13} \text{ J}\right) \left(\frac{1.0 \text{ kg}}{3.0 \times 10^7 \text{ J}}\right) = \boxed{2.7 \times 10^6 \text{ kg}}$$

24. **REASONING AND SOLUTION** In one kilogram of U-235 there are

$$(6.02 \times 10^{23}/\text{mole})(1000 \text{ g})/(235 \text{ g/mole}) = 2.56 \times 10^{24} \text{ nuclei}$$

Each nucleus produces 2.0×10^2 MeV of energy, so the energy liberated by 1 kg of U-235 is

$$E = (2.56 \times 10^{24} \text{ nuclei})(2.0 \times 10^8 \text{ eV/nucleus})(1.6 \times 10^{-19} \text{ J/1 eV}) = 8.2 \times 10^{13} \text{ J}$$

The amount of U-235 needed to produce 9.3×10^{19} J is, therefore,

$$m = (9.3 \times 10^{19} \text{ J})(1 \text{ kg})/(8.2 \times 10^{13} \text{ J}) = \boxed{1.1 \times 10^6 \text{ kg}}$$

---

25. **REASONING AND SOLUTION** The binding energy per nucleon for a nucleus with $A = 239$ is about 7.6 MeV per nucleon, according to Figure 32.8. The nucleus fragments into two pieces of mass ratio 0.32 : 0.68. These fragments thus have nucleon numbers

$$A_1 = (0.32)(A) = 76 \quad \text{and} \quad A_2 = (0.68)(A) = 163$$

Using Figure 32.8 we can estimate the binding energy per nucleon for A_1 and A_2. We find that the binding energy per nucleon for A_1 is about 8.8 MeV, representing an increase of 8.8 MeV – 7.6 MeV = 1.2 MeV per nucleon. Since there are 76 nucleons present, the energy released for A_1 is

$$76 \times 1.2 \text{ MeV} = 91 \text{ MeV}$$

Similarly, for A_2, we see that the binding energy increases to 8.0 MeV per nucleon, the difference being 8.0 MeV – 7.6 MeV = 0.4 MeV per nucleon. Since there are 163 nucleons present, the energy released for A_2 is

$$163 \times 0.4 \text{ MeV} = 70 \text{ MeV}$$

The energy released per fission is $E = 91 \text{ MeV} + 70 \text{ MeV} = \boxed{160 \text{ MeV}}$.

---

26. **REASONING AND SOLUTION**
a. We know that
$$(2.0 \times 10^8 \text{ eV})(1.6 \times 10^{-19} \text{ J/1 eV}) = 3.2 \times 10^{-11} \text{ J}$$

is released for each nucleus. The number N of nuclei needed to produce the equivalent of 20.0 kilotons of TNT is, therefore,

$$N = (20.0 \text{ kilotons})[(5.0 \times 10^{12} \text{ J})/(1 \text{ kiloton})]/(3.2 \times 10^{-11} \text{ J}) = \boxed{3.1 \times 10^{24}}$$

b. We know there are 6.02×10^{23} nuclei in 235 g of U-235. Therefore, 3.1×10^{24} nuclei have a mass of

$$m = (3.1 \times 10^{24})(235 \text{ g/mole})/(6.02 \times 10^{23}/\text{mole}) = \boxed{1.2 \times 10^3 \text{ g}}$$

c. The equivalent mass of the bomb's energy can be obtained from $E_0 = mc^2$ as follows:

$$m = E_0/c^2 = (1.0 \times 10^{14} \text{ J})/(3.0 \times 10^8 \text{ m/s})^2 = 1.1 \times 10^{-3} \text{ kg} = \boxed{1.1 \text{ g}}$$

---

27. **SSM WWW** *REASONING* We first determine the total power generated (used and wasted) by the plant. Energy is power times the time, according to Equation 6.10, and given the energy, we can determine how many kilograms of $^{235}_{92}\text{U}$ are fissioned to produce this energy.

SOLUTION Since the power plant produces energy at a rate of 8.0×10^8 W when operating at 25 % efficiency, the total power produced by the power plant is

$$(8.0 \times 10^8 \text{ W})4 = 3.2 \times 10^9 \text{ W}$$

The energy equivalent of one atomic mass unit is given in the text (see Section 31.3) as

$$1 \text{ u} = 1.4924 \times 10^{-10} \text{ J} = 931.5 \text{ MeV}$$

Since each fission produces 2.0×10^2 MeV of energy, the total mass of $^{235}_{92}\text{U}$ required to generate 3.2×10^9 W for a year $(3.156 \times 10^7 \text{ s})$ is

$$\underbrace{(3.2 \times 10^9 \text{ J/s})(3.156 \times 10^7 \text{ s})}_{\text{Power times time gives energy in joules}}$$

$$\times \underbrace{\left(\frac{931.5 \text{ MeV}}{1.4924 \times 10^{-10} \text{ J}}\right)}_{\text{Converts joules to MeV}} \underbrace{\left(\frac{1.0 \; ^{235}_{92}\text{U nucleus}}{2.0 \times 10^2 \text{ MeV}}\right)}_{\text{Converts MeV to number of nuclei}} \underbrace{\left(\frac{0.235 \text{ kg}}{6.022 \times 10^{23} \; ^{235}_{92}\text{U nuclei}}\right)}_{\text{Converts number of nuclei to kilograms}} = \boxed{1200 \text{ kg}}$$

1134 IONIZING RADIATION, NUCLEAR ENERGY, AND ELEMENTARY PARTICLES

28. **REASONING** The conservation of nucleon number states that the total number of nucleons (protons plus neutrons) before the reaction occurs must be equal to the total number of nucleons after the reaction. This conservation law will allow us to find the atomic mass number A of the unknown particle Y. The conservation of electric charge states that the net electric charge of the particles before the reaction must be equal to the net charge after the reaction. This conservation law will allow us to find the atomic number Z of the unknown particle X.

SOLUTION

a. The total number of nucleons before the reaction is $1 + A$. The total number of nucleons after the reaction is 3. Setting these two numbers equal to each other yields $\boxed{A = 2}$. The net electric charge before the reaction is $Z + 1$. The net electric charge after the reaction is 1. Setting these two numbers equal to each other yields $\boxed{Z = 0}$. The nucleon $^1_Z X = ^1_0 n$ is a $\boxed{\text{neutron}}$. The nucleon $^A_1 Y = ^2_1 H$ is a $\boxed{\text{hydrogen nucleus}}$.

b. The sum of the atomic masses before the reaction is $1.0087 \text{ u} + 2.0141 \text{ u} = 3.0228 \text{ u}$. The sum of the atomic mass after the reaction is 3.0161 u. The difference between the sums is $3.0228 \text{ u} - 3.0161 \text{ u} = 0.0067 \text{ u}$. This mass difference is equivalent to an energy of

$$(0.0067 \text{ u}) \left(\frac{931.5 \text{ MeV}}{1 \text{ u}} \right) = \boxed{6.2 \text{ MeV}}$$

---

29. **REASONING AND SOLUTION** The mass defect Δm for the fusion process is

$$\Delta m = 2.0141 \text{ u} + 2.0141 \text{ u} - 3.0161 \text{ u} - 1.0078 \text{ u} = 0.0043 \text{ u}$$

Since $1 \text{ u} = 931.5 \text{ MeV}$, the energy released is

$$(0.0043 \text{ u})(931.5 \text{ MeV})/(1 \text{ u}) = \boxed{4.0 \text{ MeV}}$$

---

30. **REASONING AND SOLUTION** The energy released in the reaction is found from the mass defect between 2 deuterium nuclei and a helium nucleus plus neutron;

$$\Delta m = 2(2.0141 \text{ u}) - (3.0160 \text{ u} + 1.0087 \text{ u}) = 0.0035 \text{ u}$$

The energy is, therefore,

$$E = (0.0035 \text{ u})(931.5 \text{ MeV})/(1 \text{ u}) = \boxed{3.3 \text{ MeV}}$$

31. **SSM** *REASONING* To find the energy released per reaction, we follow the usual procedure of determining how much the mass has decreased because of the fusion process. Once the energy released per reaction is determined, we can determine the mass of lithium $^{6}_{3}\text{Li}$ needed to produce 3.8×10^{10} J.

SOLUTION The reaction and the masses are shown below:

$$\underbrace{^{2}_{1}\text{H}}_{2.014 \text{ u}} + \underbrace{^{6}_{3}\text{Li}}_{6.015 \text{ u}} \rightarrow \underbrace{2\ ^{4}_{2}\text{He}}_{2(4.003 \text{ u})}$$

The mass defect is, therefore, $2.014 \text{ u} + 6.015 \text{ u} - 2(4.003 \text{ u}) = 0.023 \text{ u}$. Since 1 u is equivalent to 931.5 MeV, the released energy is 21 MeV, or since the energy equivalent of one atomic mass unit is given in Section 31.3 as $1 \text{ u} = 1.4924 \times 10^{-10} \text{ J} = 931.5 \text{ MeV}$,

$$(21 \text{ MeV})\left(\frac{1.4924 \times 10^{-10} \text{ J}}{931.5 \text{ MeV}}\right) = 3.4 \times 10^{-12} \text{ J}$$

In 1.0 kg of lithium $^{6}_{3}\text{Li}$, there are

$$(1.0 \text{ kg of } ^{6}_{3}\text{Li})\left(\frac{1.0 \times 10^{3} \text{ g}}{1.0 \text{ kg}}\right)\left(\frac{6.022 \times 10^{23} \text{ nuclei/mol}}{6.015 \text{ g/mol}}\right) = 1.0 \times 10^{26} \text{ nuclei}$$

Therefore, 1.0 kg of lithium $^{6}_{3}\text{Li}$ would produce an energy of

$$(3.4 \times 10^{-12} \text{ J/nuclei})(1.0 \times 10^{26} \text{ nuclei}) = 3.4 \times 10^{14} \text{ J}$$

If the energy needs of one household for a year is estimated to be 3.8×10^{10} J, then the amount of lithium required is

$$\frac{3.8 \times 10^{10} \text{ J}}{3.4 \times 10^{14} \text{ J/kg}} = \boxed{1.1 \times 10^{-4} \text{ kg}}$$

32. *REASONING AND SOLUTION*
   a. The number n of atoms of hydrogen and its isotopes in 1 kg of water (molecular mass = 18 u) is

   $$n = 2(6.02 \times 10^{23}/\text{mol})(1.00 \times 10^{3} \text{ g})/(18.015 \text{ g/mol}) = 6.68 \times 10^{25}$$

If deuterium makes up 0.015% of hydrogen in this number of atoms, the number N of deuterium atoms is

$$N = (1.5 \times 10^{-4})(6.68 \times 10^{25}) = \boxed{1.0 \times 10^{22}}$$

b. Each deuterium nucleus provides 7.2 MeV of energy, so the energy from 1 kg of water is

$$E = (1.0 \times 10^{22})(7.2 \times 10^{6} \text{ eV})(1.6 \times 10^{-19} \text{ J})/(1 \text{ eV}) = 1.15 \times 10^{10} \text{ J}$$

To supply 9.3×10^{19} J of energy, we would need

$$m = (9.3 \times 10^{19} \text{ J})/(1.15 \times 10^{10} \text{ J/kg}) = \boxed{8.1 \times 10^{9} \text{ kg}}$$

---

33. **REASONING AND SOLUTION** We will use the atomic mass of hydrogen $^{1}_{1}\text{H}$ (1.007 825 u), deuterium $^{2}_{1}\text{H}$ (2.014 102 u), helium $^{3}_{2}\text{He}$ (3.016 030 u), helium $^{4}_{2}\text{He}$ (4.002 603 u), a positron $^{0}_{1}\text{e}$ (0.000 549 u), and an electron (0.000 549 u). The mass defect Δm for two reactions of type (1) is

$$\Delta m = 2\,(1.007\,825 \text{ u} + 1.007\,825 \text{ u} - 2.014\,102 \text{ u} - 0.000\,549 \text{ u} - 0.000\,549 \text{ u}) = 0.000\,900 \text{ u}$$

In this result, one of the two values of 0.000 549 u accounts for the positron. The other is present because the two hydrogen atoms contain a total of two electrons, whereas a single deuterium atom contains only one electron. The mass defect for two reactions of type (2) is

$$\Delta m = 2\,(1.007\,825 \text{ u} + 2.014\,102 \text{ u} - 3.016\,030 \text{ u}) = 0.011\,794 \text{ u}$$

The mass defect for one reaction of type (3) is

$$\Delta m = 3.016\,030 \text{ u} + 3.016\,030 \text{ u} - 4.002\,603 \text{ u} - 1.007\,825 \text{ u} - 1.007\,825 \text{ u} = 0.013\,807 \text{ u}$$

The total mass defect for the proton-proton cycle is the sum of three values just calculated:

$$\Delta m_{\text{total}} = 0.000\,900 \text{ u} + 0.011\,794 \text{ u} + 0.013\,807 \text{ u} = 0.026\,501 \text{ u}$$

Since 1 u = 931.5 MeV, the energy released is

$$(0.026\,501 \text{ u})[(931.5 \text{ MeV})/(1 \text{ u})] = \boxed{24.7 \text{ MeV}}$$

34. **REASONING** The energy released in the decay is the difference between the rest energy of the π^- and the sum of the rest energies of the μ^- and $\overline{\nu}_\mu$. Each of these energies can be found in Table 32.3

SOLUTION The rest energies of the particles are π^- (139.6 MeV), μ^- (105.7 MeV) and $\overline{\nu}_\mu$ (≈ 0 MeV). The energy released is 139.6 MeV − 105.7 MeV = $\boxed{33.9 \text{ MeV}}$.

---

35. **SSM REASONING AND SOLUTION** The lambda particle contains three different quarks, one of which is the up quark u, and contains no antiquarks. Therefore, the remaining two quarks must be selected from the down quark d, the strange quark s, the charmed quark c, the top quark t, and the bottom quark b. Since the lambda particle has an electric charge of zero and since u has a charge of $+2e/3$, the charges of the remaining two quarks must add up to a total charge of $−2e/3$. This eliminates the quarks c and t as choices, because they each have a charge of $+2e/3$. We are left, then, with d, s, and b as choices for the remaining two quarks in the lambda particle. The three possibilities are as follows:

$$\boxed{(1) \; u,d,s \qquad (2) \; u,d,b \qquad (3) \; u,s,b}$$

---

36. **REASONING AND SOLUTION**
a. The electron and its antiparticle (the positron) both have the same mass. Therefore, each electron produces one gamma ray of the same energy.

$$E_0 = mc^2 = (9.11 \times 10^{-31} \text{ kg})(3.00 \times 10^8 \text{ m/s})^2 = 8.20 \times 10^{-14} \text{ J} \qquad (28.5)$$

or

$$E_0 = (8.20 \times 10^{-14} \text{ J})/(1.60 \times 10^{-19} \text{ J})/(1 \text{ eV}) = \boxed{0.513 \text{ MeV}}$$

b. The wavelength can be found by combining Equations 29.2 ($E_0 = hf$) and 16.1 ($\lambda = c/f$) so that $\lambda = c/f = hc/E_0$:

$$\lambda = hc/E_0 = (6.63 \times 10^{-34} \text{ J·s})(3.00 \times 10^8 \text{ m/s})/(8.20 \times 10^{-14} \text{ J}) = \boxed{2.43 \times 10^{-12} \text{ m}}$$

c. The magnitude of the momentum of each photon is

$$p = h/\lambda = (6.63 \times 10^{-34} \text{ J·s})/(2.43 \times 10^{-12} \text{ m}) = \boxed{2.73 \times 10^{-22} \text{ kg·m/s}} \qquad (29.8)$$

37. ***REASONING AND SOLUTION*** The proton is constructed from the quarks d, u, and u. Therefore, the antiproton is constructed from $\boxed{\bar{d}, \bar{u}, \text{ and } \bar{u}}$.

---

38. ***REASONING AND SOLUTION*** Let us consider the difference in rest energies between the two incident protons and the outgoing proton, neutron and π^+. This would be equivalent to the minimum energy of the incident proton. We have

$$E = (1.0087 \text{ u})(931.5 \text{ MeV})/(1 \text{ u}) + 139.6 \text{ MeV} - (1.0073 \text{ u})(931.5 \text{ MeV})/(1 \text{ u})$$

$$E = 939.6 \text{ MeV} + 139.6 \text{ MeV} - 938.3 \text{ MeV} = \boxed{140.9 \text{ MeV}}$$

---

39. **SSM WWW** ***REASONING*** The momentum of a photon is given in the text as $p = E/c$ (see the discussion leading to Equation 29.6). This expression applies to any massless particle that travels at the speed of light. In particular, if the neutrino has no mass and travels at the speed of light, it applies to the neutrino. Once the momentum of the neutrino is determined, the de Broglie wavelength can be calculated from Equation 29.6 ($p = h/\lambda$).

SOLUTION
a. The momentum of the neutrino is, therefore,

$$p = \frac{E}{c} = \left(\frac{35 \text{ MeV}}{3.00 \times 10^8 \text{ m/s}}\right)\left(\frac{1.4924 \times 10^{-10} \text{ J}}{931.5 \text{ MeV}}\right) = \boxed{1.9 \times 10^{-20} \text{ kg} \cdot \text{m/s}}$$

where we have used the fact that 1.4924×10^{-10} J = 931.5 MeV (see Section 31.3).

b. According to Equation 29.6, the de Broglie wavelength of the neutrino is

$$\lambda = \frac{h}{p} = \frac{6.63 \times 10^{-34} \text{ J} \times \text{s}}{1.9 \times 10^{-20} \text{ kg} \cdot \text{m/s}} = \boxed{3.5 \times 10^{-14} \text{ m}}$$

40. **REASONING AND SOLUTION** In order for the proton to come within a distance r of the second proton, it must overcome the Coulomb repulsive force. Therefore, the kinetic energy of the incoming proton must equal the Coulombic potential energy of the system.

$$KE = EPE = ke^2/r$$

$$KE = (8.99 \times 10^9 \text{ N·m}^2/\text{C}^2)(1.6 \times 10^{-19} \text{ C})^2/(8.0 \times 10^{-15} \text{ m}) = 2.9 \times 10^{-14} \text{ J}$$

$$KE = (2.9 \times 10^{-14} \text{ J})(1 \text{ eV})/(1.6 \times 10^{-19} \text{ J}) = 1.8 \times 10^5 \text{ eV} = \boxed{0.18 \text{ MeV}}$$

---

41. **REASONING AND SOLUTION** The absorbed dose (AD) is

$$AD = (3.1 \times 10^{-5} \text{ Gy/s})(0.10 \text{ s}) = 3.1 \times 10^{-6} \text{ Gy}$$

The energy absorbed is given by

$$\text{Energy absorbed} = AD \times (\text{Mass of absorbing material}) \qquad (32.2)$$

$$= (3.1 \times 10^{-6} \text{ Gy})(1.2 \text{ kg}) = \boxed{3.7 \times 10^{-6} \text{ J}}$$

---

42. **REASONING AND SOLUTION** The additional matter comes from the kinetic energy of the incoming protons. We can find the amount by considering the difference in rest energies as follows:

$$\Delta E = (1.007\ 276 \text{ u})(931.5 \text{ MeV})/(1 \text{ u}) + 139.6 \text{ MeV} + 1116 \text{ MeV} + 497.7 \text{ MeV}$$

$$- 2(1.007\ 276 \text{ u})(931.5 \text{ MeV})/(1 \text{ u}) = \boxed{815 \text{ MeV}}$$

---

43. **SSM** **WWW** **REASONING** The reaction given in the problem statement is written in the shorthand form: $^{17}_{8}\text{O}\,(\gamma, \alpha n)\,^{12}_{6}\text{C}$. The first and last symbols represent the initial and final nuclei, respectively. The symbols inside the parentheses denote the incident particles or rays (left side of the comma) and the emitted particles or rays (right side of the comma).

SOLUTION Using the reasoning above and noting that an α particle is a helium nucleus, $^{4}_{2}\text{He}$, we have

$$\boxed{\gamma + {}^{17}_{8}\text{O} \rightarrow {}^{12}_{6}\text{C} + {}^{4}_{2}\text{He} + {}^{1}_{0}\text{n}}$$

44. **REASONING AND SOLUTION** The energy liberated can be found from the mass defect of the reaction, which is

$$\Delta m = 235.044 \text{ u} - 2(1.009 \text{ u}) - 140.914 \text{ u} - 91.926 \text{ u} = 0.186 \text{ u}$$

The equivalent energy is

$$\Delta E = (0.186 \text{ u})(931.5 \text{ MeV})/(1 \text{ u}) = \boxed{173 \text{ MeV}}$$

---

45. **SSM REASONING** To find the energy released per reaction, we follow the usual procedure of determining how much the mass has decreased because of the fusion process. Once the energy released per reaction is determined, we can determine the amount of gasoline that must be burned to produce the same amount of energy.

SOLUTION The reaction and the masses are shown below:

$$3\,{}^{2}_{1}\text{H} \rightarrow {}^{4}_{2}\text{He} + {}^{1}_{1}\text{H} + {}^{1}_{0}\text{n}$$

$$\underbrace{\phantom{3\,{}^{2}_{1}\text{H}}}_{3(2.0141 \text{ u})} \quad \underbrace{\phantom{{}^{4}_{2}\text{He}}}_{4.0026 \text{ u}} \quad \underbrace{\phantom{{}^{1}_{1}\text{H}}}_{1.0078 \text{ u}} \quad \underbrace{\phantom{{}^{1}_{0}\text{n}}}_{1.0087 \text{ u}}$$

The mass defect is, therefore, $3(2.0141 \text{ u}) - 4.0026 \text{ u} - 1.0078 \text{ u} - 1.0087 \text{ u} = 0.0232 \text{ u}$. Since 1 u is equivalent to 931.5 MeV, the released energy is 21.6 MeV, or since it is shown in Section 31.3 that 931.5 MeV = 1.4924×10^{-10} J, the energy released per reaction is

$$(21.6 \text{ MeV}) \left(\frac{1.4924 \times 10^{-10} \text{ J}}{931.5 \text{ MeV}} \right) = 3.46 \times 10^{-12} \text{ J}$$

To find the total energy released by all the deuterium fuel, we need to know the number of deuterium nuclei present. The number of deuterium nuclei is

$$(6.1 \times 10^{-3} \text{ g}) \left(\frac{6.022 \times 10^{23} \text{ nuclei/mol}}{2.0141 \text{ g/mol}} \right) = 1.8 \times 10^{21} \text{ nuclei}$$

Since each reaction consumes three deuterium nuclei, the total energy released by the deuterium fuel is

$$\tfrac{1}{3}(3.46 \times 10^{-12} \text{ J/nuclei})(1.8 \times 10^{21} \text{ nuclei}) = 2.1 \times 10^{9} \text{ J}$$

If one gallon of gasoline produces 2.1×10^{9} J of energy, then the number of gallons of gasoline that would have to be burned to equal the energy released by all the deuterium fuel is

$$(2.1 \times 10^{9} \text{ J}) \left(\frac{1.0 \text{ gal}}{2.1 \times 10^{9} \text{ J}} \right) = \boxed{1.0 \text{ gal}}$$

46. **REASONING AND SOLUTION** The absorbed dose is

$$750 \text{ rad} = 7.5 \text{ Gy} = 7.5 \text{ J/kg} = Q/m$$

The rise in temperature of the water is determined from $Q = cm\Delta T$, Equation 12.4, so that the temperature rise is

$$\Delta T = (Q/m)/c = (7.5 \text{ J/kg})/[4186 \text{ J/(kg·C°)}] = \boxed{1.8 \times 10^{-3} \text{ C°}}$$

47. **REASONING AND SOLUTION** We have $Q = cm\Delta T$, Equation 12.4. The power is the energy per unit time, so $P = Q/t$. The mass of the water passing through the core each second is, therefore,

$$\frac{m}{t} = \frac{\frac{Q}{t}}{c\Delta T} = \frac{5.6 \times 10^9 \text{ J/s}}{[4420 \text{ J/(kg·C°)}](287\,°\text{C} - 216\,°\text{C})} = \boxed{1.8 \times 10^4 \text{ kg/s}}$$

48. **REASONING AND SOLUTION**
a. The number of nuclei in one gram of U-235 can be obtained as follows:

$$1 \text{ gram of U-235} = [(1/235) \text{ mol}] \times (6.02 \times 10^{23} \text{ g/mol}) = 2.56 \times 10^{21} \text{ nuclei}$$

Each nucleus yields 2.0×10^2 MeV of energy, so we have

$$E = (2.0 \times 10^8 \text{ eV})[(1.60 \times 10^{-19} \text{ J})/1 \text{ eV}](2.56 \times 10^{21}) = \boxed{8.2 \times 10^{10} \text{ J}}$$

b. If 30.0 kWh of energy are used per day, the total energy use per year is

$$E_{\text{total}} = (30.0 \text{ kWh/d})[(3.60 \times 10^6 \text{ J})/(1 \text{ kWh})][(365 \text{ d})/(1 \text{ yr})] = 3.94 \times 10^{10} \text{ J/yr}$$

The amount of U-235 needed in a year, then, is

$$m = (3.94 \times 10^{10} \text{ J})/(8.2 \times 10^{10} \text{ J/g}) = \boxed{0.48 \text{ g}}$$

1142   IONIZING RADIATION, NUCLEAR ENERGY, AND ELEMENTARY PARTICLES

49. **REASONING AND SOLUTION** Using Equation 32.1,

$$\text{Exposure} = \frac{q}{(2.58 \times 10^{-4})m}$$

we can get the amount of charge produced in 1 kilogram of dry air when exposed to 1.0 R of X-rays as follows:

$$q = (2.58 \times 10^{-4})(1.0 \text{ R})(1 \text{ kg}) = 2.58 \times 10^{-4} \text{ C}$$

Alternatively, we know that 1.0 R deposits 8.3×10^{-3} J of energy in 1 kg of dry air. Thus, 8.3×10^{-3} J of energy produces 2.58×10^{-4} C. To find how much energy E is required to produce 1.60×10^{-19} C, we set up a ratio as follows:

$$\frac{E}{8.3 \times 10^{-3} \text{ J}} = \frac{1.60 \times 10^{-19} \text{ C}}{2.58 \times 10^{-4} \text{ C}} \quad \text{or} \quad E = 5.1 \times 10^{-18} \text{ J}$$

Converting to eV, we find

$$E = (5.1 \times 10^{-18} \text{ J})(1 \text{ eV})/(1.6 \times 10^{-19} \text{ J}) = \boxed{32 \text{ eV}}$$

---

50. **CONCEPT QUESTIONS** According to Equation 32.4, the biologically equivalent dose is equal to the product of the absorbed dose and the RBE (relative biological effectiveness).

a. If the absorbed doses are the same, then the radiation with the larger RBE produces the greater biological effect.

b. If the two types of radiation have the same RBE, then the radiation that produces the greater absorbed dose produces the greater biological effect.

SOLUTION Using Equation 32.4, the biologically equivalent dose can be determined for each type of radiation. The rankings are given in the last column of the table.

Radiation	Biologically Equivalent Dose = (Absorbed dose)(RBE)	Ranking
γ rays	$(20 \times 10^{-3} \text{ rad})(1) = 20 \times 10^{-3}$ rem	3
Electrons	$(30 \times 10^{-3} \text{ rad})(1) = 30 \times 10^{-3}$ rem	2
Protons	$(5 \times 10^{-3} \text{ rad})(10) = 50 \times 10^{-3}$ rem	1
Slow neutrons	$(5 \times 10^{-3} \text{ rad})(2) = 10 \times 10^{-3}$ rem	4

51. **CONCEPT QUESTION** The relation between the rad and gray units is presented in Section 32.1 as 1 rad = 0.01 gray. If, for instance, we wanted to convert an absorbed dose of 2.5 grays into rads, we would use the conversion procedure:

$$(2.5 \text{ Gy})\left(\frac{1 \text{ rad}}{0.01 \text{ Gy}}\right) = 250 \text{ rad}$$

In general, the conversion relation is

$$\text{Absorbed dose (in rad)} = \left[\text{Absorbed dose (in Gy)}\right]\left(\frac{1 \text{ rad}}{0.01 \text{ Gy}}\right) \quad (1)$$

SOLUTION
According to Equation 32.4, the relative biological effectiveness (RBE) is given by

$$\text{RBE} = \frac{\text{Biologically equivalent dose}}{\text{Absorbed dose (in rad)}}$$

The absorbed dose (in rad) is related to the absorbed dose (in Gy) by Equation (1), so the RBE can be expressed as

$$\text{RBE} = \frac{\text{Biologically equivalent dose}}{\left[\text{Absorbed dose (in Gy)}\right]\left(\frac{1 \text{ rad}}{0.01 \text{ Gy}}\right)}$$

The absorbed dose (in Gy) is equal to the energy E absorbed by the tissue divided by its mass m (Equation 32.2), so the RBE can be written as

$$\text{RBE} = \frac{\text{Biologically equivalent dose}}{\left(\frac{E}{m}\right)\left(\frac{1 \text{ rad}}{0.01 \text{ Gy}}\right)} = \frac{2.5 \times 10^{-2} \text{ rem}}{\left(\frac{6.2 \times 10^{-3} \text{ J}}{21 \text{ kg}}\right)\left(\frac{1 \text{ rad}}{0.01 \text{ Gy}}\right)} = \boxed{0.85}$$

---

52. **CONCEPT QUESTIONS**
   a. According to the discussion in Section 31.1, the proton and the neutron are nucleons.

   b. The statement means that the total number of nucleons (protons and neutrons) of the particles before the reaction is equal to the total number of nucleons after the reaction.

   c. Only the proton and the electron have electrical charge. The neutron and the γ-ray photon are electrically neutral.

d. The statement means that the total electric charge of the particles before the reaction is equal to the total electric charge of particles after the reaction.

SOLUTION

a. The total number of nucleons before the reaction is $A + 14$. The total number of nucleons after the reaction is $1 + 17$. Setting these two numbers equal to each other yields $A = 4$. The net electric charge before the reaction is $Z + 7$. The net electric charge after the reaction is $1 + 8$. Setting these two numbers equal to each other yields $Z = 2$. The unknown particle is $\boxed{{}_{Z}^{A}X = {}_{2}^{4}He}$.

b. The total number of nucleons before the reaction is $15 + A$. The total number of nucleons after the reaction is $12 + 4$. Setting these two numbers equal to each other yields $A = 1$. The net electric charge before the reaction is $7 + Z$. The net electric charge after the reaction is $6 + 2$. Setting these two numbers equal to each other yields $Z = 1$. The unknown particle is $\boxed{{}_{Z}^{A}X = {}_{1}^{1}H}$.

c. The total number of nucleons before the reaction is $1 + 27$. The total number of nucleons after the reaction is $A + 1$. Setting these two numbers equal to each other yields $A = 27$. The net electric charge before the reaction is $1 + 13$. The net electric charge after the reaction is $Z + 0$. Setting these two numbers equal to each other yields $Z = 14$. The unknown particle is $\boxed{{}_{Z}^{A}X = {}_{14}^{27}Si}$.

d. The total number of nucleons before the reaction is $7 + 1$. The total number of nucleons after the reaction is $4 + A$. Setting these two numbers equal to each other yields $A = 4$. The net electric charge before the reaction is $3 + 1$. The net electric charge after the reaction is $2 + Z$. Setting these two numbers equal to each other yields $Z = 2$. The unknown particle is $\boxed{{}_{Z}^{A}X = {}_{2}^{4}He}$.

---

53. CONCEPT QUESTIONS

a. If energy is released during a reaction, the total rest energy of the particles after the reaction must be less than the total rest energy before the reaction. Since energy and mass are equivalent, the total mass of the particles after the reaction is less than the total mass of the particles before the reaction.

b. According to Einstein's relation between mass and energy, Equation 28.5, the difference Δm in total masses is related to the energy ΔE_0 released by the reaction by $\Delta m = \Delta E_0/c^2$, where c is the speed of light.

SOLUTION The difference between the total mass before the reaction and the total mass after the reaction is

$$\Delta m = \underbrace{1.0078 \text{ u} + 1.0087 \text{ u}}_{\text{Total mass before reaction}} - \underbrace{2.0141 \text{ u}}_{\substack{\text{Total mass} \\ \text{after reaction}}} = 0.0024 \text{ u}$$

Since 1 u = 931.5 MeV, the energy released is

$$E = (0.0024 \text{ u})\left(\frac{931.5 \text{ MeV}}{1 \text{ u}}\right) = \boxed{2.2 \text{ MeV}}$$

CHAPTER 18 | ELECTRIC FORCES AND ELECTRIC FIELDS

CONCEPTUAL QUESTIONS

1. **REASONING AND SOLUTION** In Figure 18.8, the grounding wire is removed first, followed by the rod, and the sphere is left with a positive charge. If the rod were removed first, followed by the grounding wire, the sphere would not be left with a charge. Once the rod is removed, the repulsive force caused by the presence of the rubber rod is no longer present. Since the wire is still attached, free electrons will enter the sphere from the ground until the sphere is once again neutral.

2. **REASONING AND SOLUTION** A metallic rod is given a positive charge by the process of induction as illustrated in Figure 18.8.

    a. The metallic object becomes positive because, during the induction process, electrons are forced from the object to the earth. The mass of the object will decrease by an amount equal to the mass of the electrons that left the metallic object.

    b. The metallic object becomes negative because, during the induction process, electrons are pulled onto the object from the earth. The mass of the object will increase by an amount that is equal to the mass of the "excess" electrons that are pulled onto the object.

3. **REASONING AND SOLUTION** When the charged insulating rod is brought near to (but not touching) the sphere, the free electrons in the sphere will move. If the rod is negatively charged, the free electrons will move to the side of the sphere that is opposite to the side where the rod is; if the rod is positively charged, the free electrons will migrate to the side of the sphere where the rod is. In either case, the region of the sphere near the vicinity of the rod will acquire a charge that has the opposite sign as the charge on the rod.

    a. Since oppositely charged objects always attract each other, the rod and sphere will always experience a mutual attraction.

    b. Since the side of the sphere in the vicinity of the rod will always have charge that is opposite in sign to the charge on the rod, the rod and the sphere will always attract each other. They never repel each other.

4. **REASONING AND SOLUTION** On a dry day, just after washing your hair to remove natural oils and drying it thoroughly, you run a plastic comb through it. As the surface of the comb rubs against your hair, the comb becomes electrically charged. If the comb is brought near small bits of paper, the charge on the comb causes a separation of charge on the bits of paper. Since the paper is neutral, it contains equal amounts of positive and negative charge.

The charge on the comb causes the regions of the bits of paper that are closest to the comb to become oppositely charged; therefore, the bits of paper are attracted to the comb. This situation is analogous to that in Figure 18.9.

---

5. ***REASONING AND SOLUTION*** A balloon is blown up and rubbed against a person's shirt a number of times. The balloon is then touched to a ceiling. Upon being released, the balloon remains stuck to the ceiling. The balloon is charged by contact. The ceiling is neutral. The charged balloon will induce a slight surface charge on the ceiling that is opposite in sign to the charge on the balloon. Since the charge on the balloon and the ceiling are opposite in sign, they will attract each other. Since both the balloon and the ceiling are insulators, charge cannot flow from one to the other. The charge on the balloon remains fixed on the balloon, while the charge on the ceiling remains fixed on the ceiling. The electrostatic force that the ceiling exerts on the balloon is sufficient to hold the balloon in place.

---

6. ***REASONING AND SOLUTION*** A proton and electron are held in place on the x axis. The proton is at $x = -d$, while the electron is at $x = +d$. They are released simultaneously, and the only force that affects their motions is the electrostatic force of attraction that each applies to the other. According to Newton's third law, the force F_{pe} exerted on the proton by the electron is equal in magnitude and opposite in direction to the force F_{ep} exerted on the electron by the proton. In other words, $F_{pe} = -F_{ep}$. According to Newton's second law, this equation can be written

$$m_p a_p = -m_e a_e \qquad (1)$$

where m_p and m_e are the respective masses and a_p and a_e are the respective accelerations of the proton and the electron. Since the mass of the electron is considerably smaller than the mass of the proton, the acceleration of the electron at any instant must be considerably greater than the acceleration of the proton at that instant in order for Equation (1) to hold. Since the electron has a much greater acceleration than the proton, it will attain greater velocities than the proton and, therefore, reach the origin first.

---

7. ***REASONING AND SOLUTION*** In order for a particle to execute simple harmonic motion, it must obey a force law of the form of Equation 10.2, $F = -kx$, where x is the displacement of the object from its equilibrium position, and k is the spring constant. The force described by Equation 10.2 is a restoring force in the sense that it always pulls the particle toward its equilibrium position.

The force of repulsion that a charge q_1 feels when it is pushed toward another charge q_2 of the same polarity is given by Coulomb's law, Equation 18.1, $F = kq_1q_2/r^2$. Clearly, Coulomb's law is an inverse square law. It does not have the same mathematical form as Equation 10.2. Therefore, a charged particle that is pushed toward another charged particle of the same polarity that is fixed in position, will not exhibit simple harmonic motion when it is released. Coulomb's law does *not* describe a restoring force. When q_1 is released, it simply "flies away" from q_2 and never returns.

---

8. **REASONING AND SOLUTION** Identical point charges are fixed to opposite corners of a square, as shown in the figure at the right.

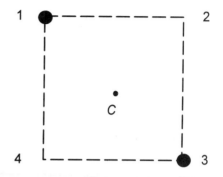

There exists an electric field at each point in space in the vicinity of this configuration. The electric field at a given point is the resultant of the electric field at that point due to the charge at corner 1 and the charge at corner 3.

The magnitude of the electric field due to a single point charge of strength q is given by Equation 18.3: $E = kq/r^2$, where r is the distance from the charge to the point in question.

Since the point charges at corners 1 and 3 are identical, they each have the same value of q. Furthermore, they are equidistant from the center point C. Therefore, at the center of the square, each charge gives rise to an electric field that is equal in magnitude and opposite in direction. The resultant electric field at the center of the square is, therefore, zero. Note that this result is independent of the polarity of the charges.

The distance from either empty corner, 2 or 4, to either of the charges is the same. Since the charges are equidistant from either empty corner, each charge gives rise to an electric field that is equal in magnitude. Since the direction of the electric field due to a point charge is radial (radially inward for negative charges and radially outward for positive charges), we see that the electric fields due to each of the two point charges will be mutually perpendicular. Their resultant can be found by using the Pythagorean theorem.

The magnitude of the force experienced by a third point charge placed in this system is, from Equation 18.2: $F = q_0 E$, where q_0 is the magnitude of the charge and E is the magnitude of the electric field at the location of the charge. Since the electric field is zero at the center of the square, a third point charge will experience no force there. Thus, a third point charge will experience the greater force at one of the empty corners of the square.

---

9. **REASONING AND SOLUTION** The figure shows a thin, nonconducting rod, on which positive charges are spread evenly, so that there is the same amount of charge per unit length at every point. We wish to deduce the direction of the electric field at the point P, which is directly above the midpoint of the rod.

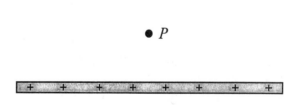

We first imagine dividing the rod up into a large number of small length elements. Since the charge per unit length is the same along the rod, each element will contain an equal amount of charge Δq. If the elements are chosen small enough, then the amount of charge on each may be treated as a point charge. That is, the charge on each element gives rise to an electric field at the point P given by $\Delta E = k\Delta q / r^2$, where r is the distance between that particular element and the point P. The electric field at P is found by adding, vectorially, the effects of all the elements. Since the rod is positively charged, the direction of $\Delta \mathbf{E}$ is away from the rod and along the line that connects the point P to the particular element of charge. In general, each element Δq is a different distance r from the point P, and, therefore, gives rise to a different electric field $\Delta \mathbf{E}$ at the point P. However, every element on the left half of the rod can be paired with an element on the right half of the rod that is the same distance r from the point P. Thus, each element on the left can be paired with an element on the right that gives rise to an electric field at P that has the same magnitude ΔE.

We notice that each $\Delta \mathbf{E}$ has both x and y components; however, for charge elements on the right, the x component points in the negative x direction, while for elements on the left, the x component points in the positive x direction, as shown below.

Thus, the x field components from the charges on the left cancel pair-wise with the x field components from the charges on the right. The y field components from all of the charge elements point in the positive y direction. Therefore, the net electric field at the point P will point only in the positive y direction, that is, perpendicularly away from the rod.

On another identical rod, positive charges are spread evenly over only the left half of the rod, and the same amount of negative charges are spread evenly over the right half. Again, we wish to deduce the direction of the electric field at the point P, which is located above the midpoint of the rod.

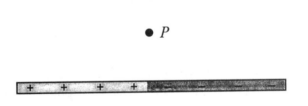

We proceed as we did previously. Since the left half of the rod is positively charged, the direction of $\Delta \mathbf{E}$ is away from the rod and along the line that connects the point P to the particular element of charge on the left half. Conversely, the right half of the rod is negatively charged. Therefore, the direction of $\Delta \mathbf{E}$ is toward the rod and along the line that connects the point P to the particular element of charge on the right half. As with the first rod, each element of charge Δq gives rise to a different electric field $\Delta \mathbf{E}$ at the point P. But once again, each element on the left half of the rod can be paired with an element on the right half of the rod that is the same distance r from the point P and, therefore, gives rise to an electric field at P of the same magnitude ΔE.

Each $\Delta \mathbf{E}$ has both x and y components; however, for charge elements on the right, the y component points in the negative y direction, while for charge elements on the left, the y component points in the positive y direction, as shown in the following drawings.

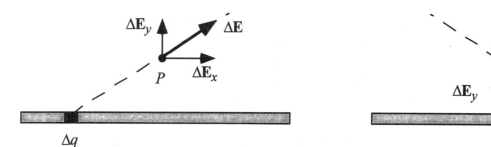

Thus, the y field components that arise from the charges on the left cancel pair-wise with the y field components that arise from the charges on the right. The x field components from all of the charge elements point in the positive x direction. Therefore, the net electric field at the point P will point only in the positive x direction, that is, parallel to the rod, pointing from the positive toward the negative side.

---

10. **REASONING AND SOLUTION** From Equation 18.2, we see that the electric field **E** that exists at a point in space is the force per unit charge, $\mathbf{F}/q_0$, experienced by a small test charge q_0 placed at that point. When a positive or negative charge is placed at a point where there is an electric field, the charge will experience a force given by $\mathbf{F} = q_0 \mathbf{E}$. If the charge q_0 is positive, then **F** and **E** are in the same direction; that is, a positive charge experiences a force in the direction of the electric field. Similarly, if the charge q_0 is negative, then **F** and **E** are in the opposite direction; a negative charge experiences a force that is opposite to the direction of the electric field.

If two different charges experience forces that differ in both magnitude and direction when they are separately placed at the same point in space, we can conclude that the two charges differ in both magnitude and polarity.

---

11. **REASONING AND SOLUTION** Three point charges are fixed to the corners of a square (side length L and diagonal length d), one to a corner, in such a way that the net electric field at the empty corner is zero.

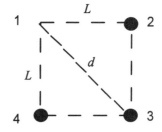

a. Each charge produces an electric field that is radially toward (positive charges) or radially away from (negative charges) the charge. The net electric field at corner 1 is the vector sum of the electric fields due to each of the charges at corners 2, 3 and 4. Since the electric field at corner 1 is zero, then the charges in corners 2 and 4 must have the same sign and have the opposite sign to the charge in corner 3. Suppose, for example, that the charge in corner 3 is positive. Then the electric field at corner 1 due to the charge in corner 3 must point away from corner 3 along the diagonal. In order for the vector sum of the fields due to all three charges to vanish at corner 1, the electric field due to the charge in corner 2 must cancel the x component of the electric field due to the charge in corner 3. Likewise, the electric field due to the charge in corner 4 must cancel the y component of the electric field due to the

104   ELECTRIC FORCES AND ELECTRIC FIELDS

charge in corner 3. Thus, at corner 1, the electric field due to the charge at corner 2 must point to the right, while the electric field due to the charge in corner 4 must point toward the bottom of the page. The charges in corners 2 and 4, therefore, must be negative. Similar arguments apply if the charge in corner 3 is negative; we can then deduce that the charges in corners 2 and 4 must be positive. We conclude, therefore, that the three charges *cannot* all have the same sign.

b. Let all three charges have magnitude Q. Then the magnitude of the electric field at corner 1 due to the charge in corner 3 is given by Equation 18.3:

$$E_{13} = \frac{kQ}{d^2} = \frac{kQ}{2L^2}$$

As discussed in part (a), this electric field vector points along the diagonal. Therefore, its x and y components have the following magnitudes

$$(E_{13})_x = \frac{kQ}{2L^2} \cos 45° = \frac{kQ}{2L^2}\left(\frac{\sqrt{2}}{2}\right) = \frac{kQ}{(2\sqrt{2})L^2}$$

$$(E_{13})_y = \frac{kQ}{2L^2} \sin 45° = \frac{kQ}{2L^2}\left(\frac{\sqrt{2}}{2}\right) = \frac{kQ}{(2\sqrt{2})L^2}$$

In order for the net electric field at corner 1 to be zero, the charges in corners 2 and 4 must have the opposite sign to the charge in corner 3 [as discussed in part (a)], and the electric field at corner 1 due to the charge in corner 2 must be equal to $-(E_{13})_x$. Similarly, the electric field at corner 1 due to the charge in corner 4 must be equal to $-(E_{13})_y$. If the charges in all three corners have the same magnitude, however, the electric field at corner 1 due to the charge at corner 2 and the electric field at corner 1 due to the charge at corner 4 have magnitudes given by Equation 18.3 as

$$E_{12} = E_{14} = \frac{kQ}{L^2}$$

Clearly, $E_{12} \neq (E_{13})_x$ and $E_{14} \neq (E_{13})_y$; therefore, all three charges *cannot* have the same magnitude.

---

12. **REASONING AND SOLUTION** The charges at corners 1 and 3 are both $+q$ while the charges at corners 2 and 4 are both $-q$. Let's concentrate first on the electric field produced at the center C of the rectangle by the positive charges. The positive charge at corner 1 produces an electric field at C that points toward corner 3. In contrast, the positive charge at corner 3 produces an electric field at C that points toward corner 1. Thus, the two fields have opposite directions. The magnitudes of the fields are identical because the charges have the

same magnitude and are equally far from the center. Therefore, the fields from the two positive charges cancel. In a similar manner, the fields due to the negative charges on corners 2 and 4 also cancel. Consequently, the net electric field at the center C due to all four charges is zero.

---

13. **REASONING AND SOLUTION** We can think of dividing the line into three distinct regions. Region I to the left of the positive charge, region II between the charges, and region III to the right of the negative charge. The electric field $\mathbf{E}_+$ from the positive charge always points away from the positive charge, while the electric field $\mathbf{E}_-$ always points toward the negative charge.

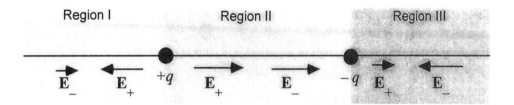

In order for the electric field to be zero, the resultant of the electric field due to the positive charge $\mathbf{E}_+$ and the electric field due to the negative charge $\mathbf{E}_-$ must be zero. This can only occur in regions I or III since those are the only two regions where the vectors $\mathbf{E}_+$ and $\mathbf{E}_-$ point in opposite directions.

The magnitude of the electric field at a distance r from either charge is given by Equation 18.3: $E = kq/r^2$. When the two charges have equal magnitudes, the point on the line where the magnitude of $\mathbf{E}_+$ is equal to the magnitude of $\mathbf{E}_-$ is midway between the charges, since that point is equidistant from either charge (q and r are the same for both $\mathbf{E}_+$ and $\mathbf{E}_-$). The field will not be zero, however, because both $\mathbf{E}_+$ and $\mathbf{E}_-$ point to the right. If the magnitude of the negative charge is greater than the magnitude of the positive charge, then, the position where the magnitude of $\mathbf{E}_+$ is equal to the magnitude of $\mathbf{E}_-$ will occur closer to the positive charge. There will be a point in region I where the magnitudes of $\mathbf{E}_+$ and $\mathbf{E}_-$ are the same. Therefore, a point of zero electric field can occur on the line in region I.

---

14. **REASONING AND SOLUTION** In general, electric field lines are always directed away from positive charges and toward negative charges. At a point in space, the direction of the electric field is tangent to the electric field line that passes through that point. The magnitude of the electric field is strongest in regions where the field lines are closest together. Furthermore, the text states that electric field lines created by a positive point charge are directed radially outward, while those created by a negative point charge are directed radially inward.

a. *In both regions I and II the electric field is the same everywhere.* This statement is false. By inspection, we see that the field lines in II get closer together as we proceed from left to right in the drawing; hence, the magnitude of the electric field is getting stronger. The direction of the electric field at any point along a field line can be found by constructing a line tangent to the field line. Clearly, the direction of the electric field in II changes as we proceed from left to right in the drawing.

b. *As you move from left to right in each case, the electric field becomes stronger.* This statement is false. While this is true for the field in II by reasons that are discussed in part (a) above, we can see by inspection that the field lines in I remain equally spaced and point in the same direction as we go from left to right. Thus, the field in I remains constant in both magnitude and direction.

c. *The electric field in I is the same everywhere but becomes stronger in II as you move from left to right.* This statement is true for the reasons discussed in parts (a) and (b) above.

d. *The electric fields in both I and II could be created by negative charges located somewhere on the left and positive charges located somewhere on the right.* Since electric field lines are always directed away from positive charges and toward negative charges, this statement is false.

e. *Both I and II arise from a single point charge located somewhere on the left.* Since the electric field lines are radial in the vicinity of a single point charge, this statement is false.

---

15. **REASONING AND SOLUTION** The effects of air resistance and gravity are ignored throughout this discussion. As shown in Figure 18.25, the electric field lines between the plates of a parallel plate capacitor are parallel and equally spaced, except near the edges. Ignoring edge effects, then, we can treat the electric field between the plates as a constant electric field. Since the force on the charge is given by Equation 18.2 as $\mathbf{F} = q_0 \mathbf{E}$, we can deduce that the force on the positive charge, and therefore its acceleration, is constant in magnitude and is directed toward the negative plate. Since the charge enters the region with a horizontal velocity and the acceleration is downward, we can conclude that the horizontal component of the velocity will remain constant while the vertical component of the velocity will increase in the downward direction as the particle moves from left to right. This situation is analogous to that of a projectile fired horizontally near the surface of the earth. We can deduce, therefore, that the charged particle will follow a parabolic trajectory.

a. Two possible trajectories for the particle are shown below. For case 1, the particle hits the bottom plate and never emerges from the capacitor. The point at which the particle hits the bottom plate (when indeed it does) depends on the initial speed v of the particle when it enters the capacitor and the magnitude E of the electric field between the plates. For certain values of v and E, the particle may exit from the right side of the capacitor without striking the bottom plate. This is shown in case 2. Note that after the particle leaves the capacitor, it will move in a straight line at constant speed according to Newton's first law (assuming that no other forces are present).

Case 1

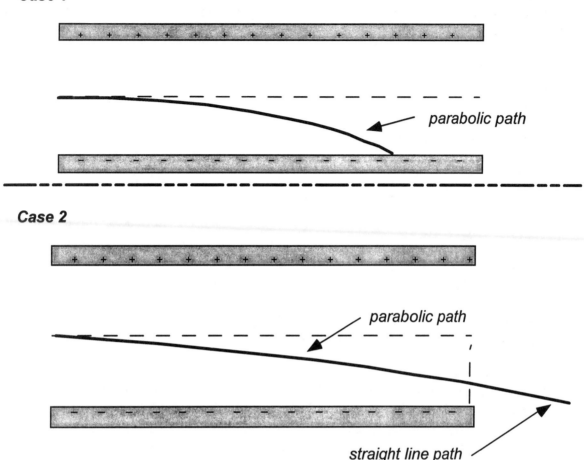

Case 2

b. The direction of the particle's instantaneous displacement vector and velocity vector are in the instantaneous direction of motion of the particle. Furthermore, the direction of the linear momentum vector is always in the same direction as the particle's velocity, and, therefore, in the instantaneous direction of motion. Since the particle travels in a parabolic path as discussed in part (a), we know that the directions of the particle's displacement vector, velocity vector, and linear momentum vector are always tangent to the path shown in part (a). They are not, therefore, parallel to the electric field inside the capacitor. From Newton's second law, we know that the acceleration vector of the particle is in the direction of the net force that acts on the particle. Since the particle is positively charged, we know that the net force on the particle at any point is in the same direction as the direction of the electric field at that point. Therefore, the acceleration vector is parallel to the electric field in the capacitor.

16. **REASONING AND SOLUTION** Refer to Figure 18.27 which shows the electric field lines for two identical positive point charges. Imagine a plane that is perpendicular to the line that passes through the charges and is half into and half out of the paper. The plane is midway between the charges. The electric flux through this plane is given by Equation 18.6: $\Phi_E = \sum (E \cos \phi) \Delta A$, where E is the magnitude of the electric field at a point on the plane, ϕ is the angle that the electric field makes with the normal to the plane at that point, ΔA is a

108   ELECTRIC FORCES AND ELECTRIC FIELDS

tiny area that just surrounds the point, and the summation is carried out over the entire plane. Midway between the charges, the electric field is zero (see Figure 18.27). Therefore, the electric field is zero for all points on the plane. The electric flux through the plane is, therefore, also zero.

---

17. **REASONING AND SOLUTION** Two charges $+q$ and $-q$ are inside a Gaussian surface. Since the net charge inside the Gaussian surface is zero, Gauss' law states that the electric flux Φ_E through the surface is also zero. However, the fact that Φ_E is zero does not imply that **E** is zero at any point on the Gaussian surface.

The flux through a Gaussian surface of any shape that surrounds the two point charges is given by text Equation 18.6: $\Phi_E = \sum (E \cos \phi) \Delta A$, where E is the magnitude of the electric field at a point on the surface, ϕ is the angle that the electric field makes with the normal to the surface at that point, ΔA is a tiny area that surrounds the point, and the summation is carried out over the entire Gaussian surface. The individual terms $(E \cos \phi) \Delta A$ can be positive or negative, depending on the sign of the factor $(\cos \phi)$.

We can see from Figure 18.26 that the electric field in the vicinity of the two point charges changes from point to point in both magnitude and direction. Furthermore, as Figure 18.26 indicates, the electric field lines are directed away from the positive charge and toward the negative charge. In general, the value of E is non-zero at every point on any arbitrary Gaussian surface that surrounds the charges. The magnitude and direction of **E** at each point is such that when the summation in Equation 18.6 is carried out over the entire Gaussian surface, the electric flux through the entire surface is zero.

---

18. **REASONING AND SOLUTION** Gauss' law states that the electric flux Φ_E through a Gaussian surface is equal to Q/ε_0, where Q is the net charge enclosed by the surface and ε_0 is the permittivity of free space.

The drawing shows three charges, labeled q_1, q_2, and q_3.

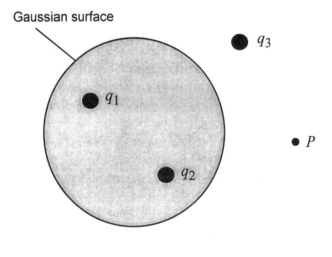

a. Since the charges labeled q_1 and q_2 lie within the Gaussian surface, they determine the electric flux through the surface. From Gauss' law, the flux through the surface is

$$\Phi_E = (q_1 + q_2)/\varepsilon_0$$

b. Each of the three charges produces its own electric field. The magnitude of the field at the point P due to any one of the charges is given by Equation 18.3: $E = kq/r^2$, where q is the magnitude of the charge and r is the distance between the charge and the point P. The net electric field at the point P is the resultant of the electric fields that are produced by each charge individually. Thus, all three charges contribute to the electric field at the point P.

---

19. ***REASONING AND SOLUTION*** Gauss' law states that the electric flux Φ_E through a Gaussian surface is equal to Q/ε_0, where Q is the net charge enclosed by the surface and ε_0 is the permittivity of free space. Gauss' law make no reference to position or location other than to specify that Q refers to the net charged *enclosed* by the Gaussian surface.

Suppose that a charge $+q$ is placed inside a spherical Gaussian surface. The charge is positioned so that it is *not* at the center of the sphere.

a. In order for Gauss' law to tell us anything about the charge within the Gaussian surface, the electric flux through the surface must be known. Since Gauss' law makes no reference to the position of the net charge Q, Gauss' law cannot tell us where the charge is located. If the flux through the surface is known, Gauss' law can only tell us that a net amount of charge $Q = +q$ is located somewhere within the surface.

b. If we know that the net charge enclosed by the sphere is $Q = +q$, then, by the statement of Gauss' law, we know that the flux through the surface is $\Phi_E = +q/\varepsilon_0$ (regardless of where the enclosed charges are positioned relative to the center of the sphere).

CHAPTER 19 | ELECTRIC POTENTIAL ENERGY AND THE ELECTRIC POTENTIAL

CONCEPTUAL QUESTIONS

1. **REASONING AND SOLUTION** The work done in moving a charge q_0 from A to B is given by Equation 19.4: $W_{AB} = q_0(V_A - V_B)$. For the cases in the drawing, we have

    Case 1: $W_{AB} = q_0(150\text{ V} - 100\text{ V}) = q_0(50\text{ V})$

    Case 2: $W_{AB} = q_0[25\text{ V} - (-25\text{ V})] = q_0(50\text{ V})$

    Case 3: $W_{AB} = q_0[-10\text{ V} - (-60\text{ V})] = q_0(50\text{ V})$

    The work done on the charge by the electric force is the same in each case.

2. **REASONING AND SOLUTION** The potential at a point in space that is a distance r from a point charge q is given by Equation 19.6: $V = kq/r$. When more than one point charge is present, the total potential at any location is the algebraic sum of the individual potentials created by each charge at that location.

    A positive point charge and a negative point charge have equal magnitudes. One of the charges is fixed to one corner of a square. If the other charge is placed opposite to the first charge along the diagonal of the square, then each charge will be the same distance L from the empty corners. The potential at each of the empty corners will be

    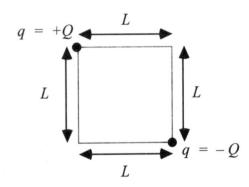

    $$V = \frac{k(+Q)}{L} + \frac{k(-Q)}{L} = 0$$

    Therefore, if the potential at each empty corner is to be the same, then the charges must be placed at diagonally opposite corners as shown in the figure.

3. **REASONING AND SOLUTION** The potential at a point in space that is a distance r from a point charge q is given by Equation 19.6: $V = kq/r$. When more than one point charge is present, the total potential at any location is the algebraic sum of the individual potentials created by each charge at that location.

Three point charges have identical magnitudes, but two of the charges are positive and one is negative. These charges are fixed to the corners of a square, one to a corner, as shown in the figure.

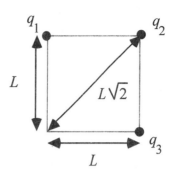

The potential at the empty corner is given by

$$V = \frac{q_1}{L} + \frac{q_2}{L\sqrt{2}} + \frac{q_3}{L}$$

Using q to denote the magnitude of each charge, we have the following possibilities:

q_1 and q_2 are positive: $V = \frac{kq}{L} + \frac{kq}{L\sqrt{2}} - \frac{kq}{L} = \frac{kq}{L\sqrt{2}}$

q_1 and q_3 are positive: $V = \frac{kq}{L} - \frac{kq}{L\sqrt{2}} + \frac{kq}{L} = \frac{kq}{L}\left(2 - \frac{1}{\sqrt{2}}\right)$

q_2 and q_3 are positive: $V = -\frac{kq}{L} + \frac{kq}{L\sqrt{2}} + \frac{kq}{L} = \frac{kq}{L\sqrt{2}}$

In each case, the potential at the empty corner is positive.

---

4. ***REASONING AND SOLUTION*** Four point charges of equal magnitude are placed at the corners of a square as shown in the figure at the right.

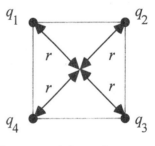

The electric field at the center of the square is the *vector sum* of the electric field at the center due to each of the charges individually. The potential at the center of the square is equal to the *algebraic sum* of the potentials at the center due to each of the charges individually.

All four charges are equidistant from the center (a distance r in the figure). If two diagonal charges have the same magnitude and sign, then the electric field at the center due to these two charges have equal magnitude and opposite directions. Their resultant is, therefore, zero. Thus, the electric field at the center will be zero if each diagonal pair of charges has the same magnitude and sign.

If a diagonal pair of charges has the same magnitude and sign, they will give rise to a non-zero potential at the center. Thus, the potential due to one diagonal pair of charges must cancel the potential due to the other diagonal pair of charges. This will be the case if the two pairs of diagonal charges have opposite signs.

5. **REASONING AND SOLUTION**
Positive charge is spread uniformly around a circular ring, as shown in the figure, with q representing the total amount of charge on the ring. Equation 19.6 gives the correct potential at points along the line perpendicular to the plane of the ring at its center. The equation does not, however, give the correct potential at points that do not lie on this line.

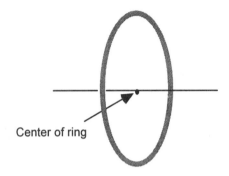
Center of ring

We can consider the ring to be composed of point charges of magnitude Δq spread uniformly around the ring so that $q = \sum \Delta q$. According to Equation 19.6, each "point charge" gives rise to a potential given by $\Delta V = k\Delta q/r$ at a distance r from the "point charge." The potential due to the ring at any point in space is the algebraic sum of the potentials of all the "point charges". For any point along the line perpendicular to the plane of the ring at its center, the distance r is the same for all "point charges" on the ring. Therefore, for any point on that line, $V = \sum \Delta V = \sum k\Delta q/r = kq/r$; thus, we obtain Equation 19.6. For points that are not on the line, the distance r will, in general, be different for each "point charge" that makes up the total charge q; therefore, the potential V at that point will not be given by Equation 19.6 with a single value for r.

6. **REASONING AND SOLUTION** The electric field at a single location is zero. This does not necessarily mean that the electric potential at the same place is zero. According to Equation 19.7, $E = -\Delta V/\Delta s$. Therefore, the fact that **E** = 0 means that the potential gradient, $\Delta V/\Delta s$, is zero, and the potential difference does not change with distance. In other words, in the vicinity of a point where **E** is zero, the electric potential is constant, but not necessarily zero

To illustrate this fact, consider the situation of two identical point charges. The electric field at each location in the vicinity of the charges is the resultant of the electric field at that location due to each of the two charges. The magnitude of the field due to each charge is given by Equation 18.3: $E = kq/r^2$. At the point exactly halfway between the charges, along the line that joins them, r is the same for the field contribution from each charge; therefore, the magnitudes of the field contributions at this point are the same. Since the charges are identical, the directions of the field contributions at this point are in opposite directions. The following figure shows the situation for both positive and negative charges. The resultant of the two field contributions is zero. Thus, the electric field is zero exactly halfway between the charges on the line that joins them. The electric potential at this point, however, is the algebraic sum of the potential at that point due to each individual charge. Therefore, the potential at the point exactly halfway between the charges is

Positive point charges

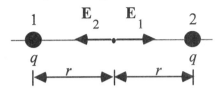

$$V = \frac{kq}{r} + \frac{kq}{r} = \frac{2kq}{r}$$

Negative point charges

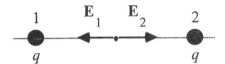

which is clearly non-zero.

---

7. **REASONING AND SOLUTION** An electric potential energy exists when two protons are separated by a certain distance. It is equal to the work that must be done by an external agent to assemble the configuration. Suppose that we imagine assembling the system, one particle at a time. If there are no other charges in the region, there are no existing electric fields; therefore, no work is required to put the first proton in place. That proton, however, gives rise to an electric field that fills the region. Its magnitude at a distance r from the proton is given by Equation 18.3, $E = ke/r^2$, where $+e$ is the magnitude of the charge on the proton. Since the region contains an electric field due to the first proton, there also exists an electric potential, and the external agent must do work to place the second proton at a distance d from the first proton. The electric potential energy of the final configuration is equal to the work that must be done to bring the second proton from infinity and place it at a distance d from the first proton. The electric potential at a distance d from the first proton is $V_{proton} = +ke/d$ (Equation 19.6). According to Equation 19.3, the electric potential energy of the final configuration is therefore

$$\text{EPE} = V_{proton}(+e) = +\frac{ke^2}{d}$$

a. If both protons are replaced by electrons, similar arguments apply. However, since the electron carries a negative charge $(-e)$, the electric potential at a distance d from the first electron is $V_{electron} = -ke/d$. The electric potential energy of the final configuration is now given by

$$\text{EPE} = V_{electron}(-e) = -\frac{ke}{d}(-e) = +\frac{ke^2}{d}$$

Therefore, if both protons are replaced by electrons, the electric potential energy *remains the same*.

b. When only one of the protons is replaced by an electron, we find that

$$\text{EPE} = V_{\text{proton}}(-e) = \left(+\frac{ke}{d}\right)(-e) = -\frac{ke^2}{d}$$

Thus, when only one of the protons is replaced by an electron, the electric potential energy *decreases* from $+ke^2/d$ to $-ke^2/d$.

---

8. **REASONING AND SOLUTION** A proton (charge = +e) is fixed in place. An electron (charge = –e) is released from rest and allowed to collide with the proton. Since the electron is released from rest, its initial kinetic energy is zero. Since the electron is in the vicinity of the proton, it has electric potential energy. The initial total energy of the electron is, therefore, electric potential energy. After the electron is released, the potential energy decreases, while the kinetic energy increases. Assuming that the charges are point charges and that r_i is the initial separation between the charges, while r_f is the separation just before the collision, the conservation of energy gives

$$E_{\text{initial}} = E_{\text{final}}$$

$$\frac{-ke^2}{r_i} = \frac{-ke^2}{r_f} + \tfrac{1}{2} m_{\text{electron}} v^2$$

$$ke^2\left(\frac{1}{r_f} - \frac{1}{r_i}\right) = \Delta(\text{EPE}) = \tfrac{1}{2} m_{\text{electron}} v^2$$

where $\Delta(\text{EPE}) = ke^2\left(\dfrac{1}{r_f} - \dfrac{1}{r_i}\right)$ is the change in the electric potential energy.

Therefore, the speed of the electron when it collides with the proton is

$$v = \sqrt{\frac{2\Delta(\text{EPE})}{m_{\text{electron}}}} \tag{1}$$

When the roles of the electron and proton are interchanged and the same experiment repeated, similar arguments apply.

It can be shown that the speed of the proton when it collides with the electron is

$$v = \sqrt{\frac{2\Delta(\text{EPE})}{m_{\text{proton}}}} \tag{2}$$

Since the mass of the electron is much less than the mass of the proton, we can conclude from Equations 1 and 2 that the electron is traveling faster when the collision occurs.

---

9. ***REASONING AND SOLUTION*** The electric potential is constant throughout a given region of space. Since the potential does not change from point to point in this region, the potential gradient, $\Delta V/\Delta s$, is zero throughout this region. From Equation 19.7, we see that the electric field is $E = -\Delta V/\Delta s$ and must also be zero throughout this region.

10. ***REASONING AND SOLUTION*** In a region where the electric field is constant, we see from Equation 19.7, $E = -\Delta V/\Delta s$, that the potential gradient is constant. In other words, the rate at which the electric potential changes with distance in the region is constant. The potential, therefore, is not zero, but varies from point to point in a linear fashion with distance.

11. ***REASONING AND SOLUTION*** A positive charge is placed in an electric field. At each location in the electric field, the field is perpendicular to an equipotential surface. Therefore, in order for the charge to experience a constant electric potential, it must be moved along a path that is perpendicular to the electric field.

12. ***REASONING AND SOLUTION*** The electric potential at any location in the vicinity of the charges is given by

$$V = \frac{k(+q)}{r_+} + \frac{k(-q)}{r_-}$$

where r_+ is the distance from the positive charge to the point in question and r_- is the distance from the negative charge to the point in question. All points for which $r_+ = r_-$ correspond to points where the potential is zero, according to the above expression. Such points would lie on a plane that is midway between the two charges. Therefore, we can conclude that line B in the figure is the edge-on view of an equipotential surface.

As stated in the text, the electric field created by a group of charges is everywhere perpendicular to the associated equipotential surfaces. Lines A and C can be ruled out as possible edge-on views of equipotential surfaces, because they are not perpendicular to the electric field lines that would arise from the dipole configuration (see Figure 19.15).

13. ***REASONING AND SOLUTION*** A positive test charge is moved along the line between two identical point charges.

a. If the two identical point charges are positive, then the test charge experiences a repulsive force from each of the two fixed charges. Halfway between the charges, the resultant of these two repulsive forces will be zero since the distance from the test charge to either fixed charge is the same. As the test charge is moved from the midpoint, it will feel a net repulsive force which tends to push the charge back to the midpoint. The work done by this repulsive force in pushing the positive charge back to the midpoint is positive. According to the discussion in Section 19.4, the charge therefore goes from a higher potential (away from the midpoint) toward a lower potential (at the midpoint). Thus, with regard to the electric potential, the midpoint is analogous to the bottom of a valley.

b. If the two identical point charges are negative, then the test charge experiences an attractive force from each of the two fixed charges. Halfway between the charges, the resultant of these two attractive forces will be zero since the distance from the test charge to either fixed charge is the same. As the test charge is moved from the midpoint, it will feel a net attractive force which tends to pull the charge away from the midpoint. The work done by this attractive force in pulling the positive charge away from the midpoint is positive. According to the discussion in Section 19.4, the charge therefore goes from a higher potential (at the midpoint) toward a lower potential (away from the midpoint). Thus, with regard to the electric potential, the midpoint is analogous to the top of a hill.

14. *REASONING AND SOLUTION* A negative test charge is moved along the line between two identical point charges.

a. If the two identical point charges are positive, then the test charge experiences an attractive force from each of the two fixed charges. Halfway between the charges, the resultant of these two attractive forces will be zero since the distance from the test charge to either fixed charge is the same. As the test charge is moved from the midpoint, it will feel a net attractive force which tends to pull the charge away from the midpoint. The work done by this attractive force in pulling the test charge away from the midpoint is positive. According to the discussion in Section 19.4, the charge therefore goes from a higher potential (at the midpoint) toward a lower potential (away from the midpoint). Thus, with regard to the electric potential, the midpoint is analogous to the top of a hill.

b. If the two identical point charges are negative, then the test charge experiences a repulsive force from each of the two fixed charges. Halfway between the charges, the resultant of these two repulsive forces will be zero since the distance from the test charge to either fixed charge is the same. As the test charge is moved from the midpoint, it will feel a net repulsive force which tends to push the charge toward the midpoint. The work done by this repulsive force in pushing the test charge back to the midpoint is positive. According to the discussion in Section 19.4, the charge therefore goes from a higher potential (away from the midpoint) toward a lower potential (at the midpoint). Thus, with regard to the electric potential, the midpoint is analogous to the bottom of a valley.

15. *REASONING AND SOLUTION* According to Equation 19.3, the electric potential energy is EPE = $q_0 V$, where V is the potential relative to a zero value at infinity and q_0 is the charge. For a given value of the potential, the electric potential energy depends on the value of the charge. Thus, the electric potential energy is not the same for every charge placed at the point in question.

16. **REASONING AND SOLUTION** Since both particles are released from rest, their initial kinetic energies are zero. They both have electric potential energy by virtue of their respective positions in the electric field between the plates. Since the particles are oppositely charged, they move in opposite directions toward opposite plates of the capacitor. As they move toward the plates, the particles gain kinetic energy and lose potential energy. Using (EPE)₀ and (EPE)_f to denote the initial and final electric potential energies of the particle, respectively, we find from energy conservation that

$$(EPE)_0 = \tfrac{1}{2} m_{particle} v_f^2 + (EPE)_f$$

The final speed of each particle is given by

$$v_f = \sqrt{\frac{2\left[(EPE)_0 - (EPE)_f\right]}{m_{particle}}}$$

Since both particles travel through the same distance between the plates of the capacitor, the change in the electric potential energy is the same for both particles. Since the mass of the electron is smaller than the mass of the proton, the final speed of the electron will be greater than that of the proton. Therefore, the electron travels faster than the proton as the particles move toward the respective plates. The electron, therefore, strikes the capacitor plate first.

---

17. **REASONING AND SOLUTION** A parallel plate capacitor is charged up by a battery. The battery is then disconnected, but the charge remains on the plates. The plates are then pulled apart.

a. The capacitance of the system is given by Equation 19.10 with $\kappa = 1$, $C = \varepsilon_0 A / d$, where A is the area of either plate and d is the distance between the plates. Since d increases when the plates are pulled apart and A remains the same, the capacitance *decreases*.

b. According to Equation 19.8, the potential difference between the plates of the capacitor is given by $V = q/C$. Since q remains the same and C decreases, the potential difference V must *increase*.

c. The magnitude of the electric field between the plates is given by $E = V/d$. Since $V = q/C$, and $C = \varepsilon_0 A / d$, the electric field between the plates is given by

$$E = \frac{(q/C)}{d} = \frac{q(d/\varepsilon_0 A)}{d} = \frac{q}{\varepsilon_0 A}$$

Since both q and A remain the same, the electric field *remains the same*.

d. The energy stored in the capacitor is given by Equation 19.11, Energy $=(1/2)CV^2$. Since $V = q/C$, we have

$$\text{Energy} = \tfrac{1}{2}CV^2 = \tfrac{1}{2}C\left(\frac{q}{C}\right)^2 = \tfrac{1}{2}\frac{q^2}{C}$$

Since q remains the same and C decreases, the electric potential energy stored by the capacitor *increases*. This increase will be equal to the work that is done by the external agent in pulling the plates apart.

CHAPTER 20 | ELECTRIC CIRCUITS

CONCEPTUAL QUESTIONS

1. **REASONING AND SOLUTION**
   a. When S_1 is set to position A, current can flow from the generator only along the path that requires S_2 to be set to position A. Hence, the light will be on when S_2 is in position A.

   b. When S_1 is set to position B, current can flow from the generator only along the path that requires S_2 to be set to position B. Hence, the light will be on when S_2 is in position B.

2. **REASONING AND SOLUTION** When an incandescent light bulb is turned on, the tungsten filament becomes white hot. Since the voltage is constant, the power delivered to the light bulb is given by Equation 20.6c: $P = V^2/R$. From Equation 20.5, $R = R_0[1 + \alpha(T - T_0)]$, where α is the temperature coefficient of resistivity and is a positive number. Thus, as the filament temperature increases, the resistance of the wire increases, and as the filament heats up, the power delivered to the bulb decreases.

3. **REASONING AND SOLUTION** Two materials have different resistivities. Two wires of the same length are made, one from each of the materials. The resistance of each wire is given by Equation 20.3: $R = \rho L/A$, where ρ is the resistivity of the wire material, and L and A are, respectively, the length and cross-sectional area of the wire. Even when the wires have the same length, they may have the same resistance, if the cross-sectional areas of the wires are chosen so that the ratio $\rho L/A$ is the same for each.

4. **REASONING AND SOLUTION** The resistance of a wire is given by Equation 20.3: $R = \rho L/A$, where ρ is the resistivity of the wire material, L is the length of the wire, and A is its cross-sectional area. Since the cross-sectional area is proportional to the square of the diameter, a doubling of the diameter causes the cross-sectional area to be increased four-fold. From Equation 20.3, we see that doubling both the diameter and length causes the resistance of the wire to be reduced by a factor of 2.

5. **REASONING AND SOLUTION** One electrical appliance operates with a voltage of 120 V, while another operates with a voltage of 240 V. The power used by either appliance is given by Equation 20.6c: $P = V^2/R$. Without knowing the resistance R of each appliance, no conclusion can be reached as to which appliance, if either, uses more power.

120 ELECTRIC CIRCUITS

6. ***REASONING AND SOLUTION*** Two light bulbs are designed for use at 120 V and are rated at 75 W and 150 W. The power used by either bulb is given by Equation 20.6c: $P = V^2/R$. We see from Equation 20.6c that, at constant voltage, the power used by a bulb is inversely proportional to the resistance of the filament. Therefore, the filament resistance is greater for the 75-W bulb.

---

7. ***REASONING AND SOLUTION*** Rather than state how many watts of power an appliance uses, appliance instructions often give statements such as "10 A, 120 V" instead. This statement gives the current used by the appliance for a specific voltage. It provides all the information necessary to determine the power usage of the appliance from Equation 20.6a: $P = IV$. Thus, the "current-voltage" rating given in the instructions is equivalent to a statement of the power consumption of the appliance.

---

8. ***REASONING AND SOLUTION*** When the switch is initially closed, a current appears in the circuit, because charges flow through the heater wire, the bimetallic strip, the contact point, and the light bulb. The bulb glows in response. As charges flow through the heater wire, it becomes hot thereby heating the bimetallic strip. As the strip is heated, the brass, having a larger coefficient of thermal expansion, expands more than the steel. Thus, the bimetallic strip bends into an arc away from the contact point, and the electric current drops to zero because the charges no longer have a continuous path along which to flow. The bulb ceases to glow. Since there is no longer a current, the resistance wire, and hence, the bimetallic strip begin to cool down to room temperature. As the bimetallic strip cools, it bends back to its initial position, as shown in the drawing in the text. When it reaches room temperature it will again touch the contact point, and a current will begin to flow again. The wire and the bimetallic strip will become hot and the bimetallic strip will bend away from the contact point and the current will again drop to zero. This cycle will continue as long as the switch remains closed. Therefore, as long as the switch remains closed, the bulb will flash on and off.

---

9. ***REASONING AND SOLUTION*** The power rating of a 1000-W heater specifies the power consumed when the heater is connected to an ac voltage of 120 V. The power consumed by the heater is given by Equation 20.6c: $P_1 = V^2/R$. When two of these heaters are connected in series, the equivalent resistance of the combination is $R + R = 2R$. The power consumed by two of the heaters connected in series is, therefore, $P_2 = V^2/(2R) = P_1/2 = (1000 \text{ W})/2 = 500 \text{ W}$.

---

10. ***REASONING AND SOLUTION*** A number of light bulbs are to be connected to a single electrical outlet. The bulbs will provide more brightness when they are connected in such a way that their power output is greatest. Since the voltage at the outlet is constant, the power delivered by the bulbs is given by Equation 20.6c: $P = V^2/R$, where R is the equivalent resistance of the combination of light bulbs. When the light bulbs are connected in series, their equivalent resistance is given by Equation 20.16: $R_S = R_1 + R_2 + R_3 + ...$ When the

light bulbs are connected in parallel, their equivalent resistance is given by Equation 20.17: $\frac{1}{R_P} = \frac{1}{R_1} + \frac{1}{R_2} + \frac{1}{R_3} + \ldots$ Clearly, the equivalent resistance will be less when the bulbs are connected in parallel. From Equation 20.6c, we can conclude that the power output will be greatest, and the light bulbs will provide more brightness, when they are connected in parallel.

11. ***REASONING AND SOLUTION*** A car has two headlights. The filament of one burns out. However, the other headlight stays on. We can immediately conclude that the bulbs are *not* connected in series. The figure below shows the series arrangement of two such bulbs.

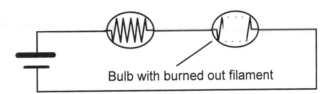

When the bulbs are connected in series, charges must flow through the filaments of *both* lights in order to have a complete circuit. Since the filament of the second bulb is burned out, charges will not be able to flow around the circuit, and neither headlight will stay on.

On the other hand, if the bulbs are connected in parallel, as shown at the right, the current will split at the junction *J*. Charges will be able to flow through the branch of the circuit that contains the good bulb, and that headlight will stay on. Notice that the order of the bulbs does not matter in either case. The results are the same.

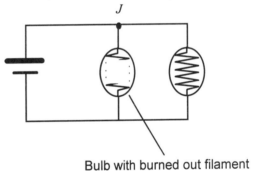

12. ***REASONING AND SOLUTION*** When two or more circuit elements are connected in series, they are connected such that the same electric current flows through each element. When two or more circuit elements are connected in parallel, they are connected such that the same voltage is applied across each element.

The circuit in Figure (*a*) can be shown to be a combination of series and parallel arrangements of resistors. The circuit can be redrawn as shown below.

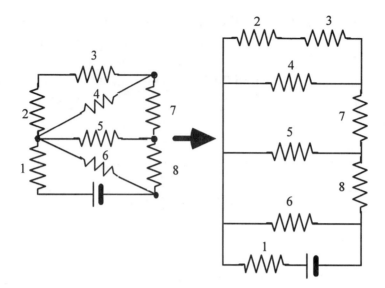

We can see in the redrawn figure that the current through resistors 2 and 3 is the same; therefore, resistors 2 and 3 are in series and can be represented by an equivalent resistance 23 as shown in the following drawing.

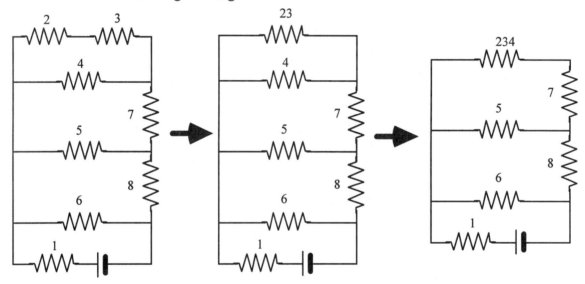

The voltage across resistance 23 and resistor 4 is the same, so these two resistances are in parallel; they can be represented by an equivalent resistance 234. The current through resistance 234 is the same as that through resistor 7, so resistance 234 is in series with resistor 7; they can be represented by an equivalent resistance 2347 as shown in the figure below.

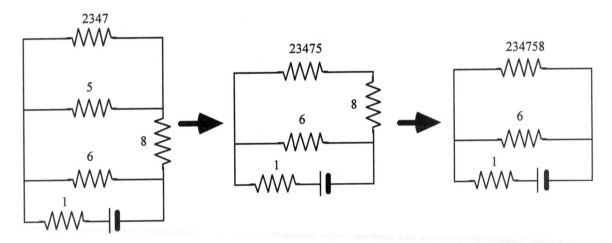

The voltage across 2347 is the same as that across resistor 5; therefore, resistance 2347 is in parallel with resistor 5. They can be represented by an equivalent resistance 23475. Similarly, resistance 23475 is in series with resistor 8, giving an equivalent resistance 234758. Resistance 234758 is in parallel with resistor 6, giving an equivalent resistance 2347586.

Finally, the current through resistor 1 and resistance 2347586 is the same, so they are in series as shown at the right.

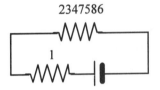

The circuit in Figure (b) can also be shown to be a combination of series and parallel arrangements of resistors. Since both ends of resistors 2 and 3 are connected, the voltage across resistors 2 and 3 is the same. The same statement can be made for resistors 4 and 5, and resistors 6 and 7. Therefore, resistor 2 is in parallel with the resistor 3 to give an equivalent resistance labeled 23. Resistor 4 is in parallel with resistor 5 to give an equivalent resistance 45, and resistor 6 is in parallel with resistor 7 to give an equivalent resistance 67. From the right-hand portion of the drawing below, it is clear that the resistances 23, 45, and 67 are in series with resistor 1.

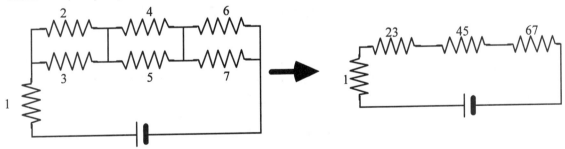

The drawing at the right shows the circuit in Figure (c). No such simplifying arguments can be made for this circuit. No two resistors carry the same current; thus, no two of the resistors are in series. Furthermore, no two resistors have the same voltage applied across them; thus, no two of the resistors are in parallel. Circuit (c) contains resistors that are neither in series nor in parallel.

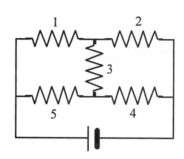

13. **REASONING AND SOLUTION** One way is to make two combinations that consist of two resistors in parallel. The resistance of each combination is

$$R_p = [(1/R) + (1/R)]^{-1} = R/2$$

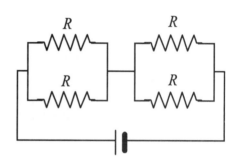

These two combinations can then be placed in series, as shown at the right. The equivalent resistance of the series combination is

$$R_s = (R/2) + (R/2) = R$$

A second way is to make two combinations that consist of two resistors in series. The resistance of each combination is

$$R_s = R + R = 2R$$

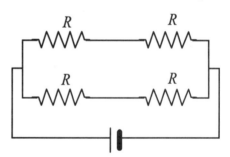

These two combinations can then be placed in parallel, as shown at the right. The equivalent resistance of the parallel combination is

$$R_p = [1/(2R) + 1/(2R)]^{-1} = R$$

14. **REASONING AND SOLUTION** An ammeter is used to measure the current in a particular branch of a circuit, while a voltmeter is used to measure the voltage across two points in a circuit. An ideal ammeter must have a low resistance so that its presence in the circuit does not affect the current measurement. Conversely, an ideal voltmeter must have a large resistance so that it does not draw current and change the voltage between the two points in question.

15. **REASONING AND SOLUTION** A voltmeter is inadvertently mistaken for an ammeter and placed in a circuit. A voltmeter is a high resistance instrument. Since an ammeter is placed in series with the circuit, we would be placing a large resistance in series with the circuit, and the resistance of the circuit would greatly increase. From Ohm's law, $V = IR$, the current in the circuit would drop markedly.

---

16. **REASONING AND SOLUTION**
From Ohm's law, $V = IR$, 1 ohm = 1 volt/ampere = 1 volt/(coulomb/second). From the definition of capacitance, $q = CV$, 1 farad = 1 coulomb/volt. Then

$$1 \text{ ohm} \cdot \text{farad} = \frac{1 \text{ volt}}{(\text{coulomb/second})} \cdot \frac{1 \text{ coulomb}}{\text{volt}} = \frac{1 \text{ volt} \cdot \text{second}}{\text{coulomb}} \cdot \frac{1 \text{ coulomb}}{\text{volt}} = 1 \text{ second}$$

CHAPTER 21 | MAGNETIC FORCES AND MAGNETIC FIELDS

CONCEPTUAL QUESTIONS

1. **REASONING AND SOLUTION** Magnetic field lines, like electric field lines, never intersect. When a moving test charge is placed in a magnetic field so that its velocity vector has a component perpendicular to the field, the particle will experience a force. That force is perpendicular to both the direction of the field and the direction of the velocity. If it were possible for magnetic field lines to intersect, then there would be a different force associated with each of the two intersecting field lines; the particle could be pushed in two directions. Since the force on a particle always has a unique direction, we can conclude that magnetic field lines can never cross.

2. **REASONING AND SOLUTION** If you accidentally use your left hand, instead of your right hand, to determine the direction of the magnetic force on a positive charge moving in a magnetic field, the direction that you determine will be exactly opposite to the correct direction.

3. **REASONING AND SOLUTION** A charged particle, passing through a certain region of space, has a velocity whose magnitude and direction remain constant.

    a. If it is known that the external magnetic field is zero everywhere in the region, we can conclude that the electric field is also zero. Any charged particle placed in an electric field will experience a force given by $\mathbf{F} = q\mathbf{E}$, where q is the charge and $\mathbf{E}$ is the electric field. If the magnitude and direction of the velocity of the particle are constant, then the particle has zero acceleration. From Newton's second law, we know that the net force on the particle is zero. But there is no magnetic field and, hence, no magnetic force. Therefore, the net force is the electric force. Since the electric force is zero, the electric field must be zero.

    b. If it is known that the external electric field is zero everywhere, we *cannot* conclude that the external magnetic field is also zero. In order for a moving charged particle to experience a magnetic force when it is placed in a magnetic field, the velocity of the moving charge must have a component that is perpendicular to the direction of the magnetic field. If the moving charged particle enters the region such that its velocity is parallel or antiparallel to the magnetic field, it will experience no magnetic force, even though a magnetic field is present. In the absence of an external electric field, there is no electric force either. Thus, there is no net force, and the velocity vector will not change in any way.

4. **REASONING AND SOLUTION** Suppose that the positive charge in Figure 21.10b were launched from the south pole toward the north pole, in a direction opposite to the magnetic field. Regardless of the strength of the magnetic field, the particle will always reach the

north pole. Since the charge is launched directly opposite to the magnetic field, its velocity will be antiparallel to the field and have no component perpendicular to the field. Therefore, there will be no magnetic force on the particle. The particle will move at constant velocity from the south pole of the magnet to the north pole.

5. **REASONING AND SOLUTION** Since the paths of the particles are perpendicular to the magnetic field, we know that the velocities of the particles are perpendicular to the field. Since the velocity of particle #2 is perpendicular to the magnetic field and it passes through the field undeflected, we can conclude that particle #2 is neutral.

Particles #1 and #3 move in circular paths. The figure at the right shows the direction of the (centripetal) magnetic force that acts on the particles. If the fingers of the right hand are pointed into the page so that the thumb points in the direction of motion of particle #1, the palm of the hand points toward the center of the circular path traversed by the particle. We can conclude, therefore, from RHR-1 that particle #1 is positively charged. If the fingers of the right hand are pointed into the page so that the thumb points in the direction of motion of particle #3, the palm of the hand points away from the center of the circular path traversed by the particle. We conclude, therefore, from RHR-1 that particle #3 is negatively charged.

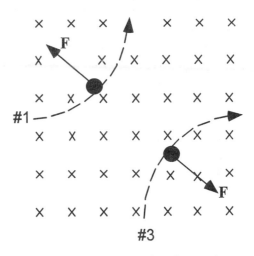

6. **REASONING AND SOLUTION** Three particles have identical charges and masses. They enter a constant magnetic field and follow the paths as shown. The magnitude of the magnetic force on each particle is given by Equation 21.1 as $F = q_0 (v \sin \theta) B$, where q_0 is the magnitude of the charge on the particle, v is the speed of the particle and B is the magnitude of the magnetic field. Since the particle paths are perpendicular to the magnetic field, $\theta = 90°$ and $\sin \theta = 1$, so that $F = q_0 vB$. This magnetic

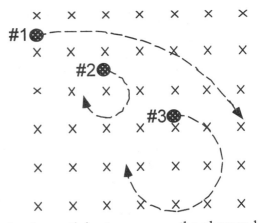

force supplies the centripetal force that is necessary for the particles to move on the observed circular paths, so $q_0 vB = mv^2/r$, where m is the mass of the particle and r is the radius of its circular path. Solving for v, we find $v = q_0 Br/m$. Since all three particles have the same charge and mass, we can conclude that the speed of the particle is directly proportional to the radius of the particle's path. Therefore, particle #1 is traveling the fastest, while particle #2 is traveling the slowest.

7. **REASONING AND SOLUTION** A proton follows the path shown in Figure 21.12. The magnitude of the magnetic force on the proton is given by Equation 21.1 as $F = q_0(v \sin\theta)B$, where q_0 is the magnitude of the charge on the proton, v is the speed of the proton and B is the magnitude of the magnetic field. Since the proton's path is perpendicular to the magnetic field, $\theta = 90°$ and $\sin\theta = 1$, so that $F = q_0 vB$. This magnetic force supplies the centripetal force that is necessary for the proton to move on the observed circular path, so $q_0 vB = m_{proton} v^2 / r$, where m is the mass of the particle and r is the radius of its circular path. The magnitude of the magnetic field is, therefore, $B = m_{proton} v / (q_0 r)$.

If we want an electron to follow exactly the same path, we must adjust the magnetic field. The magnitude of the charge on the electron is the same as that on the proton; however, the electron is negatively charged and the proton is positively charged. As stated in the text, the direction of the force on a negative charge is *opposite* to that predicted by RHR-1 for a positive charge. Therefore, the direction of the magnetic field must be reversed. In order for the electron to travel with the same speed v in a circular path of the same radius r, the magnitude of the magnetic field must be changed to the value $B = m_{electron} v / (q_0 r)$.

Thus, the electron will travel in the same path as the proton in Figure 21.12 if the direction of the magnetic field is reversed and the magnitude of the magnetic field is reduced by a factor $(m_{electron} / m_{proton})$.

---

8. **REASONING AND SOLUTION** The drawing shows a top view of four interconnected chambers. A negative charge is fired into chamber 1. By turning on separate magnetic fields in each chamber, the charge is made to exit from chamber 4.

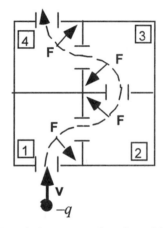

a. In each chamber the path of the particle is one-quarter of a circle. The drawing at the right also shows the direction of the centripetal force that must act on the particle in each chamber in order for the particle to traverse the path. The charged particle can be made to move in a circular path by launching it into a region in which there exists a magnetic field that is perpendicular to the velocity of the particle.

Using RHR-1, we see that if the palm of the right hand were facing in the direction of **F** in chamber 1 so that the thumb points along the path of the particle, the fingers of the right hand must point out of the page. This is the direction that the magnetic field must have to make a *positive* charge move along the path shown in chamber 1. Since the particle is *negatively* charged, the field must point opposite to that direction or into the page. Similar reasoning using RHR-1, and remembering that the particle is negatively charged, leads to the following conclusions: in region 2 the field must point out of the page, in region 3 the field must point out of the page, and in region 4 the field must point into the page.

b. If the speed of the particle is *v* when it enters chamber 1, it will emerge from chamber 4 with the same speed *v*. The magnetic force is always perpendicular to the velocity of the particle; therefore, it cannot do work on the particle and cannot change the kinetic energy of the particle, according to the work-energy theorem. Since the kinetic energy is unchanged, the speed remains constant.

---

9. **REASONING AND SOLUTION** A positive charge moves along a circular path under the influence of a magnetic field. The magnetic field is perpendicular to the plane of the circle, as in Figure 21.12. If the velocity of the particle is reversed at some point along the path, the particle will *not* retrace its path. If the velocity of the particle is suddenly reversed, then from RHR-1 we see that the force on the particle reverses direction. The particle will travel on a different circle that intersects the point where the direction of the velocity changes. The direction of motion of the particle (clockwise or counterclockwise) will be the same as that in the original circle. This is suggested in the figure below.

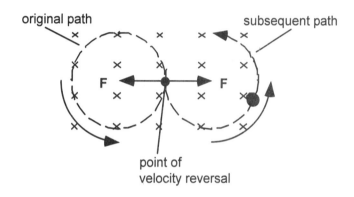

---

10. **REASONING AND SOLUTION** A television tube consists of an evacuated tube that contains an electron gun that sends a narrow beam of high-speed electrons toward the screen of the tube (see text Figure 21.37). The inner surface of the screen is covered with a phosphor coating, and when the electrons strike it, they generate a spot of visible light. The electron beam is deflected by the magnetic fields produced by electromagnets placed around the neck of the tube, between the electron gun and the screen. The magnetic fields produced by the electromagnets exert forces on the moving electrons, causing their trajectories to bend and reach different points on the screen, thereby producing the picture. If one end of a bar magnet is placed near a TV screen, the magnetic field of the bar magnet alters the trajectories of the electrons. As a result, the picture becomes distorted.

---

11. **REASONING AND SOLUTION** When the particle is launched in the *x, y* plane, its initial velocity will be perpendicular to the magnetic field; therefore, the particle will travel on a circular path in the *x, y* plane. In order for the charged particle to hit the target, the target must lie on the circular path of the moving particle. This will occur if the particle moves in a counterclockwise circle that passes through the third quadrant of the coordinate system. RHR-1 can be used to determine possible trajectories in the following way. Place the fingers of the right hand into the page (direction of the magnetic field) and orient the thumb along one of the coordinate axes (direction of the particle's initial velocity). The palm of the right

130   MAGNETIC FORCES AND MAGNETIC FIELDS

hand will face the direction in which the force on the charged particle is directed. Since the particle travels on a circle, the direction of the force will point toward the center of the particle's trajectory. The figures below show the results for all four possible cases.

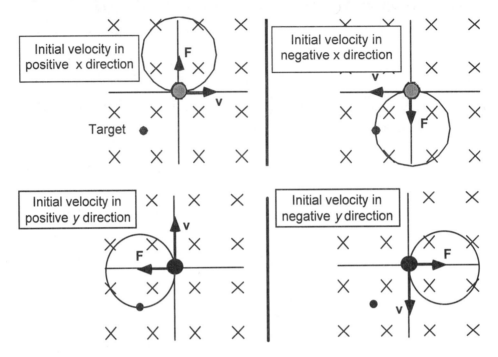

Clearly, the charged particle can hit the target only if the initial velocity of the particle points either in the negative x direction or the positive y direction.

---

12. **REASONING AND SOLUTION**   Refer to Figure 21.17. We can use RHR-1 with the modification that the direction of the velocity of a positive charge is the same as the direction of the conventional current I.

a. If the direction of the current I in the wire is reversed, the thumb of the right hand (direction of I) will be directed into the page, and the palm of the hand (direction of **F**) will face the left side of the page. Therefore, the wire will be pushed to the left.

b. If *both* the current and the magnetic poles are reversed, then the fingers of the right hand must point toward the top of the page (direction of **B**), while the thumb will be directed into the page (direction of I). The palm of the hand will face the right (direction of **F**); therefore, the wire will be pushed to the right.

---

13. **REASONING AND SOLUTION**   The bulb flashes like a turn signal on a car. When current appears in the circuit after the switch is closed, the bulb glows. Adjacent coils of the helix have current in them in the same direction. We have seen that parallel wires carrying current in the same direction attract one another due to magnetic forces. These forces squeeze the coils together, shortening the helix in the process. The bottom end of the wire is withdrawn from the mercury, thus interrupting the current in the circuit, and the bulb goes out. Without

the current, there are no magnetic forces to squeeze the coils together, and the helix, acting like a spring, expands. The bottom end of the wire dips back into the mercury, and current reappears in the circuit, causing the bulb to glow again. As this process repeats itself, the bulb flashes on and off.

---

14. **REASONING AND SOLUTION** In Figure 21.28, assume that current I_1 is larger than current I_2. Consider three regions: Region I lies to the left of both wires, region II lies between the wires, and region III lies to the right of both wires. Let **B**$_1$ represent the magnetic field due to current I_1, and let **B**$_2$ represent the magnetic field due to current I_2.

    a. *The currents point in opposite directions.* From RHR-2, the vectors **B**$_1$ and **B**$_2$ point in the same direction in region II. Therefore, regardless of the magnitudes of **B**$_1$ and **B**$_2$, the resultant of the fields can never be zero between the wires. The fields **B**$_1$ and **B**$_2$ point in opposite directions in regions I and III. Therefore, the resultant of **B**$_1$ and **B**$_2$ could be zero only in these two regions. The resultant of **B**$_1$ and **B**$_2$ will be zero at the point where their magnitudes are equal. The magnitude of the magnetic field at a distance r from a long straight wire is given by Equation 21.5: $B = \mu_0 I / (2\pi r)$, where I is the magnitude of the current in the wire and μ_0 is the permeability of free space. If the current I_1 is larger than the current I_2, the magnitude of **B**$_1$ will be equal to the magnitude of **B**$_2$ somewhere in region III. This is because region III is closer to the smaller current I_2, and the smaller value for r in Equation 21.5 allows the effect of I_2 to offset the effect of the greater and more distant current I_1. Therefore, there is a point to the right of both wires where the total magnetic field is zero.

    b. *The currents point in the same direction.* From RHR-2, the vectors **B**$_1$ and **B**$_2$ point in the same direction in both regions I and III; therefore, the resultant of **B**$_1$ and **B**$_2$ cannot be zero in these regions. In region II, the vectors **B**$_1$ and **B**$_2$ point in opposite directions; therefore, there is a place where the total magnetic field is zero between the wires. It is located closer to the wire carrying the smaller current I_2.

---

15. **REASONING AND SOLUTION** The drawing shows an end-on view of three parallel wires that are perpendicular to the plane of the paper. In two of the wires, the current is directed into the paper, while in the remaining wire the current is directed  out of the paper. The two outermost wires are held rigidly in place. From Example 8, we know that two parallel currents that point in the same direction attract each other while two parallel currents that point in opposite directions repel each other. Therefore, the middle wire will be attracted to the wire on the left and repelled from the wire on the right. Since each of the fixed wires exerts a force to the left on the middle wire, the net force on the middle wire will be to the left. Thus, the middle wire will move to the left.

132 MAGNETIC FORCES AND MAGNETIC FIELDS

16. **REASONING AND SOLUTION** The figure below shows the arrangements of electromagnets and magnets.

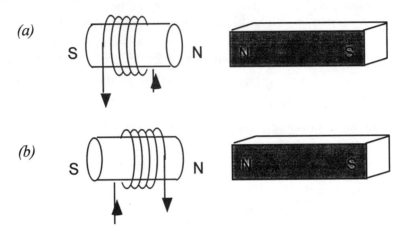

We can determine the polarity of the electromagnets by using RHR-2. Imagine holding the current-carrying wire of the electromagnet in the right hand as the wire begins to coil around the iron core. The thumb points in the direction of the current. For the electromagnet in figure (a), the fingers of the right hand wrap around the wire on the left end so that they point, inside the coil, toward the right end. Thus, the right end of the coil must be a north pole. Similar reasoning can be used to identify the north and south poles of the electromagnet in figure (b). The results are shown in the figure above. Since the like poles of two different magnets repel each other and the dissimilar poles of two different magnets attract each other, we can conclude that in both arrangements, the electromagnet is repelled from the permanent magnet at the right.

17. **REASONING AND SOLUTION** The figure below shows the arrangements of electromagnets.

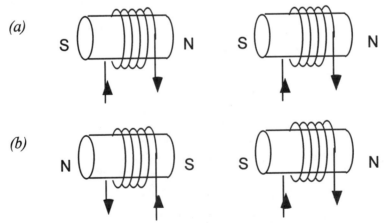

We can determine the polarity of each electromagnet by using RHR-2. Imagine holding the current-carrying wire of the electromagnet in the right hand as the wire begins to coil around the iron core. The thumb points in the direction of the current. For the electromagnet on the left in figure (*a*), the fingers of the right hand wrap around the wire on the left end so that they point, inside the coil, toward the right end. Thus, the right end of the coil must be a north pole. Similar reasoning can be used to identify the north and south poles of the other remaining electromagnets. The results are shown in the figure above. Since the like poles of two different magnets repel each other and the dissimilar poles of two different magnets attract each other, we can conclude that only the arrangement shown in (*a*) results in attraction. The electromagnets shown in arrangement (*b*) result in repulsion.

---

18. **REASONING AND SOLUTION** Refer to Figure 21.5. If the earth's magnetism is assumed to originate from a large circular loop of current within the earth, the plane of the current loop must be perpendicular to the magnetic axis of the earth, as suggested in the figure at the right. Using RHR-2, the current must flow clockwise when viewed looking down at the loop from the north magnetic pole.

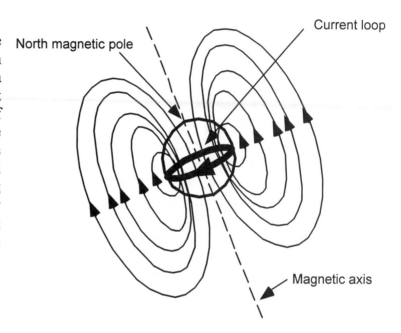

---

19. **REASONING AND SOLUTION** The total magnetic field at the point *P* is the resultant of the magnetic field at *P* due to each individual wire. If the current in all four wires is directed into the page, then from RHR-2, the magnetic field at *P* due to the current in wire 1 must point toward wire 3, the magnetic field at *P* due to the current in wire 2 must point toward wire 1, the magnetic field at *P* due to the current in wire 3 must point toward wire 4, and the magnetic field at *P* due to the current in wire 4 must point toward wire 2, as shown at the right.

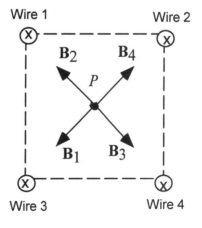

The magnetic field at a distance *r* from a long straight wire that carries a current *I* is $B = \mu_0 I / (2\pi r)$. Since the current in all four wires has the same magnitude and all four wires are equidistant from the point *P*, each wire gives rise to a magnetic field at *P* of the same magnitude. When current is flowing through all four wires, the total magnetic field at *P* is zero. If the current in any single wire is turned off, the total magnetic field will point toward one of the corners. For example, if the current in wire 1 is turned off, the resultant of $\mathbf{B}_2$ and $\mathbf{B}_3$ is still zero and the total magnetic field is $\mathbf{B}_4$. If the current in wire 2 is turned off, then the total magnetic field is $\mathbf{B}_3$, and so on.

We could have achieved similar results if the current in all four wires was directed out of the page. If fact, as long as opposite wires along the diagonal have currents in the same direction (for example, wires 1 and 4 with outward currents and wires 2 and 3 with inward currents) the total magnetic field will point toward one of the corners when one of the currents is turned off.

---

20. ***REASONING AND SOLUTION*** You have two bars, one of which is a permanent magnet and the other of which is not a magnet, but is made from a ferromagnetic material like iron. The two bars look exactly alike.

a. A third bar (the test bar) that is a permanent magnet can be used to distinguish which of the look-alike bars is the permanent magnet and which is the ferromagnetic bar. Both ends of each of the look-alike bars, in succession, should be brought next to one end of the test bar. The look-alike permanent magnet will be attracted to the test bar when opposite poles are brought together, while it will be repelled from the test bar when like poles are brought together. Both ends of the ferromagnetic bar, however, will be attracted to the test bar. The magnetic field of the test bar will always induce a pole that is opposite in polarity to the pole of the magnet. Ferromagnetism always results in attraction.

b. The identities of the look-alike bars can be determined from a third bar that is not a magnet, but is made from a ferromagnetic material. Either end of the look-alike permanent magnet will be attracted to both ends of the ferromagnetic test bar, while the look-alike ferromagnetic bar will have no effect on the ferromagnetic test bar.

---

21. ***REASONING AND SOLUTION*** A strong electromagnet picks up a delivery truck carrying cans of soda pop. Inside the truck cans of the soft drink are seen to fly upward and stick to the roof just beneath the electromagnet. We can conclude that the cans must be made from a ferromagnetic material, at least in part. Since aluminum is a nonferromagnetic material, we know that the cans are not entirely made of aluminum.

CHAPTER 22 | ELECTROMAGNETIC INDUCTION

CONCEPTUAL QUESTIONS

1. **REASONING AND SOLUTION** If the coil and the magnet in Figure 22.1a were each moving with the same velocity relative to the earth, there would be no relative motion between the magnet and the coil. The magnetic flux through the coil due to the bar magnet would be constant and, therefore, the combined motion of the bar magnet and the coil would *not* result in an induced current in the coil. We are ignoring here any effect due to the earth's magnetic field.

2. **REASONING AND SOLUTION** In the discussion concerning Figure 22.5, we saw that a force of 0.086 N from an external agent was required to keep the rod moving at constant speed. Suppose the light bulb in the figure is unscrewed from its socket. Once the light bulb is unscrewed, the conducting rails and the rod are no longer part of a complete circuit (the resistance of the empty socket is infinite). Therefore, even though there will be a charge separation and a motional emf in the rod given by *vBL*, there will be no current in the rod. Since there is no current, there is no magnetic force to resist the motion of the rod. From Newton's first law, the rod will continue to move with constant velocity **v** without the application of an external force.

3. **REASONING AND SOLUTION** A metal sheet moves to the right at a velocity **v** in a magnetic field **B** that is directed into the sheet. At the instant shown in the figure, the magnetic field only extends over half of the sheet. An induced emf leads to the eddy current shown.

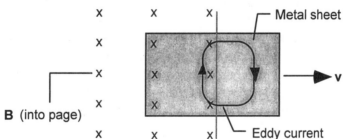

We can apply RHR-1 (modified for currents) to the portion of the eddy current that exists in the portion of the sheet that is in the magnetic field. With the thumb of the right hand pointing toward the top of the page (direction of *I*), and the fingers of the right hand pointing into the page (direction of **B**), the palm of the right hand faces the left (direction of **F**). Thus, there is a retarding magnetic force **F** that acts on the sheet due to the interaction of the eddy current with the magnetic field. Hence, the eddy current causes the sheet to slow down.

4. **REASONING AND SOLUTION** A magnetic field **B** is necessary if there is to be a magnetic flux Φ passing through a coil of wire. Yet, just because there is a magnetic field does not mean that a magnetic flux will pass through a coil.

   The general expression for magnetic flux is given by Equation 22.2: $\Phi = (B \cos \phi)A$, where B is the magnitude of the magnetic field, A is the cross-sectional area of the coil, and ϕ is the angle between the magnetic field **B** and the normal to the surface of the coil. Equation 22.2 shows that the flux depends only on the component of the magnetic field that is perpendicular to the surface of the coil. As shown in Example 4 and in Figure 22.11, when the coil is oriented so that it is parallel to the field, $\phi = 90°$, **B** has no component normal to the surface of the coil, and the magnetic flux through the coil is zero. Therefore, the magnetic flux through a coil can be zero even though there is a magnetic field present.

5. **REASONING AND SOLUTION** It is known that the magnetic flux through a 1-m² flat surface is 2 Wb. From this data alone, it is possible to determine certain information about the average magnetic field at the surface, but *not the magnitude and direction of the field.*

   The general expression for the magnetic flux through the surface is given by Equation 22.2: $\Phi = (B \cos \phi)A$, where B is the magnitude of the magnetic field, A is the cross-sectional area of the surface, and ϕ is the angle between the magnetic field **B** and the normal to the surface. Equation 22.2 shows that the flux depends only on the component of the magnetic field that is perpendicular to the surface. Therefore, from the information given, we can only ascertain that the component of the magnetic field that is perpendicular to the surface has an average magnitude of 2 T.

6. **REASONING AND SOLUTION** Initially, before the switch is closed, neither the conducting rails nor the rod carries a current. When the switch is closed, a conventional current will flow along the conducting rails from the positive toward the negative terminal of the battery. Since the rod is a conducting rod, current will flow through the rod, from top to bottom. According to RHR-1, there will be a force that points to the right on the conducting rod due to the magnetic field; therefore, the rod will be pushed and accelerate to the right. As the rod moves to the right, the area bound by the "loop" increases, thereby increasing the magnetic flux through the loop. As the magnetic flux increases, an induced emf appears around the "loop." According to Lenz's law, the induced emf that appears, will appear in such a way so as to oppose the increase in the magnetic flux. This will occur if the induced emf opposes the battery emf, with the result that the current in the rod begins to decrease and reaches zero when the induced emf exactly offsets the battery emf. With no current in the rod, there is no longer a magnetic force applied to the rod. With no force, there is zero acceleration. In other words, from this point on, the rod moves with a constant velocity.

7. **REASONING AND SOLUTION** A bolt of lightning contains moving charges, and hence, is a current. This current is surrounded by its magnetic field. Since the charges in a bolt of lightning move erratically, the current is a time-dependent and gives rise to a time-dependent magnetic field. Many household appliances contain coils. If the time-dependent magnetic field passes through these coils, the magnetic flux through the coils will be time-dependent. From Faraday's law, there will be an induced emf and, hence, an induced current in the coils

of the appliance. This will result in an induced current in the circuit that contains the appliance. Therefore, a bolt of lightning can produce a current in the circuit of an electrical appliance, even when the lightning does not directly strike the appliance.

---

8. **REASONING AND SOLUTION** A solenoid is connected to an ac source. A copper ring is placed inside the solenoid, with the normal to the ring parallel to the axis of the solenoid.

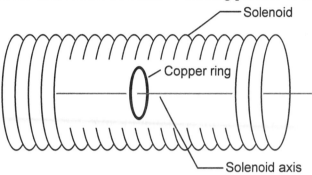

When a current passes through the coils of the solenoid, a magnetic field appears inside the solenoid. Near the center of the solenoid, the magnetic field lines are nearly parallel to the solenoid axis. Since the normal of the copper ring is parallel to the axis, there is a flux through the copper ring when the solenoid carries a current. When the solenoid is connected to an ac source, the current alternates direction, giving rise to a magnetic field inside the solenoid that continuously reverses direction. Therefore, when the solenoid is connected to an ac source, the flux through the copper ring changes with time. From Faraday's law, we know that the changing flux results in an induced emf and, therefore, an induced current around the ring. Since the copper ring has a finite resistance, heat will be generated in the ring at a rate of I^2R, where I is the magnitude of the induced current and R is the resistance of the ring. Therefore, the copper ring will get hot even though nothing touches it.

---

9. **REASONING AND SOLUTION** The cable under the floor carries a steady current I; therefore, it will be surrounded by a magnetic field **B** that does not change with time. The magnitude of the field is given by Equation 21.5, $B = \mu_0 I/(2\pi r)$, where r is the distance from the center of the cable. Since the robot's sensor coil is close to and parallel to the floor, magnetic field lines from the current-carrying cable will pass through the coil's cross-section. As long as the robot moves parallel to the cable, the distance from each point on the coil's cross section and the center of the cable remains constant, and the magnetic flux through the coil does not change. If the robot deviates from its parallel path, the distance from the coil to the cable will change. The magnitude of the magnetic field through the coil will change as the distance r changes. Therefore, the flux through the coil will change, and, from Faraday's law, there will be an induced emf in the coil.

---

10. **REASONING AND SOLUTION** The generator-like action of the alternator in a car occurs while the engine is running and keeps the battery fully charged. The headlights would discharge an old and failing battery quickly if it were not for the alternator action. When a battery is in bad shape, the engine and the alternator must do more work to keep the battery

138   ELECTROMAGNETIC INDUCTION

charged when the headlights are on. Therefore, when the battery is failing, the engine of a parked car runs more "quietly" when the headlights are off than when they are on.

---

11. **REASONING AND SOLUTION** Lenz's Law states that the polarity of an induced emf is such that the induced current produces an induced magnetic field that opposes the change in the flux that causes the emf. In Figure 22.3 a coil of wire is being stretched.

a. As the coil of wire is stretched, the cross-sectional area of the coil decreases. Therefore, the magnetic flux that penetrates the coil decreases. In order to oppose this decrease in flux, a magnetic field must be induced so as to enhance the existing magnetic field. The induced magnetic field, therefore, must point into the page in the region that is surrounded by the coil. RHR-2 can be applied to determine the direction of the induced current. If the wire is held in the right hand with the fingers pointing into the page in the interior of the coil (direction of induced **B**), then the thumb of the right hand (direction of induced I) points so that the induced current is clockwise. This is the direction shown in the drawing.

b. If the direction of the magnetic field were reversed, then using the same reasoning as in part (a), we can reason that the induced magnetic field must point out of the page to oppose the change in flux through the coil as the coil is stretched. Using RHR-2 we deduce that the induced current must be counterclockwise around the loop.

---

12. **REASONING AND SOLUTION** The figure shows the set-up.

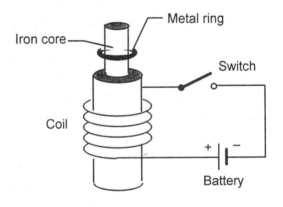

a. When the switch is open, there is no current in the coil and no magnetic field is present. When the switch in the circuit is closed, a current is established in the coil. From RHR-2, the magnetic field associated with the current in the coil leaves the bottom of the coil and enters the top of the coil (see drawing below). In other words, when there is a current in the coil, the bottom of the coil acts like the north pole of a bar magnet, and the top of the coil acts like the south pole of a bar magnet. The magnetic field of the coil causes a magnetic flux through the metal ring.

In the very short time that it takes for the current to rise from zero to its steady value, the magnetic field associated with the current also rises from zero to its steady value. As the magnetic field increases, the magnetic flux through the metal ring changes, and there is an induced voltage and an induced current around the ring.

From Lenz's law, the polarity of the induced voltage will be such that it opposes the changing flux through the metal ring. The induced magnetic field of the metal ring will be directed upward through the ring to subtract from the magnetic field of the coil that points downward through the ring. This is illustrated in the figure at the right. To preserve clarity, the iron core and a portion of the circuit have not been shown. The induced magnetic field of the metal ring is similar to that of a bar magnet, with the north pole above the ring, and the south pole below the ring. Thus, the situation is similar to that of two bar magnets with their south poles next to each other. The two south poles repel, and the ring "jumps" upward.

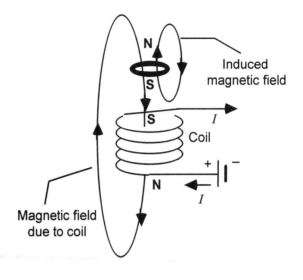

b. If the polarity of the battery were reversed, the ring still "jumps" upward when the switch is closed. When the polarity of the battery is reversed, and the switch is closed, the current moves through the coil in the opposite direction.

The magnetic field associated with the current points in a direction that is opposite to the direction of the magnetic field in part (a). In other words, the top of the coil acts like the north pole of a bar magnet, and the bottom of the coil acts like the south pole of a bar magnet.

When the switch is closed, the magnetic flux through the metal ring changes, and there is an induced voltage and an induced current around the ring.

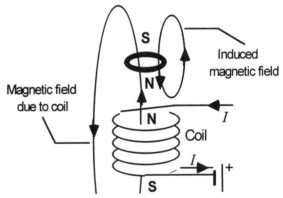

In accord with Lenz's law, the induced magnetic field of the metal ring will be directed downward through the ring to subtract from the magnetic field of the coil that points upward through the ring. The induced magnetic field of the metal ring is similar to that of a bar magnet, with the south pole above the ring, and the north pole below the ring. The two north poles repel, and the ring "jumps" upward.

---

13. **REASONING AND SOLUTION** The string of an electric guitar is made of a magnetizable material. As shown in Figure 22.18, the presence of a permanent magnet inside the pickup coil causes the string to become magnetized with north and south poles. When the magnetized string is plucked, it oscillates above the pickup coil, thereby changing the magnetic flux that passes through the coil. The guitar string vibrates in a standing wave pattern that

consists of nodes and antinodes. The magnetic flux will change most rapidly where the oscillations of the magnetized string are the largest; therefore, the electromagnetic pickup should be placed at an antinode to produce a maximum emf.

14. ***REASONING AND SOLUTION*** The current drawn by an electric motor is given by Equation 22.5: $I = (V - \text{emf})/R$, where V is the applied emf, "emf" is the back emf developed because the coil of the motor is rotating, and R is the resistance of the wire in the coil.

An electric motor in a hair drier is running at normal speed; therefore, the back emf is comparable to the applied emf, and the motor is drawing a relatively small current. If the shaft of the motor is prevented from turning, the back emf is suddenly reduced to zero. We see from Equation 22.5 that the current drawn by the motor will increase to the value V/R. Heat will be produced in the coil at a rate of $I^2 R = (V/R)^2 R = V^2/R$, and the temperature of the coil will increase.

15. ***REASONING AND SOLUTION*** One transformer is a step-up device, while another is step-down. These two units have the same voltage across and the same current in their primary coils.

If we ignore any heat loss within the transformers, then, according to the principle of energy conservation, the average power delivered to the circuit attached to the secondary coil is the same for either transformer.

This conclusion can be verified using Equations 20.15a, and Equation 22.13. According to Equation 20.15a, the average power delivered to the secondary coil, and therefore, the average power that the secondary coil delivers to the circuit is $\overline{P}_s = I_s V_s$. Solving Equation 22.13 for I_s and substituting the resulting expression into Equation 20.15a, we have

$$\overline{P}_s = I_s V_s = \left(\frac{I_p V_p}{V_s}\right) V_s = I_p V_p$$

This result holds for both the step-up, as well as, the step-down device. Since both devices have the same voltage across and the same current in their primary coils, the average power delivered to the secondary coil, and therefore, the average power delivered to the circuit attached to the secondary coil, are the same.

CHAPTER 23 ALTERNATING CURRENT CIRCUITS

CONCEPTUAL QUESTIONS

1. **REASONING AND SOLUTION** A light bulb and a parallel plate capacitor (including a dielectric material between the plates) are connected in series to the 60-Hz ac voltage present at a wall outlet. Since the capacitor and the light bulb are in series, the rms current at any instant is the same through each element, and is given by Equation 23.6: $I_{rms} = V_{rms}/Z$, where Z is the impedance of the circuit. The impedance of the circuit is given by Equation 23.7 with $X_L = 0$ (since there is no inductance in the circuit): $Z = \sqrt{R^2 + (-X_C)^2} = \sqrt{R^2 + X_C^2}$. According to Equation 19.10, if the dielectric between the plates of the capacitor is removed, the capacitance decreases by a factor of κ, where κ is the dielectric constant. From Equation 23.2, $X_C = 1/(2\pi f C)$, we see that decreasing the capacitance increases the capacitive reactance X_C. Therefore, the impedance of the circuit, $Z = \sqrt{R^2 + X_C^2}$, increases. The rms current is $I_{rms} = V_{rms}/Z$ and will, therefore, be *less* than it was before the dielectric was removed. Thus, the brightness will *decrease*.

2. **REASONING AND SOLUTION** The ends of a long straight wire are connected to the terminals of an ac generator, and the current is measured. The wire is then disconnected, wound into the shape of a multiple-turn coil, and reconnected to the generator. After the wire is wound into a coil, it has a greater inductance. When the generator is turned on, the coil will develop a voltage that opposes a change in the current according to Faraday's law. Since the induced voltage opposes the rise in current, the rms current in the circuit will be less than it was before the wire was wound into a coil.

   This can also be seen by considering Equations 23.6 and 23.7. Before the wire was wound into a coil, its primary property was that of resistance, and the current through the wire was given by $I_{rms} = V_{rms}/Z$ with $Z = R$, or $I_{rms} = V_{rms}/R$. After the wire is wound into a coil, the wire possesses both resistance and inductance. Now the current in the coil is given by $I_{rms} = V_{rms}/Z$, where $Z = \sqrt{R^2 + X_L^2}$. Since Z is necessarily greater than R, the rms current is less when the wire is wound into a coil.

3. **REASONING AND SOLUTION** An air-core inductor is connected in series with a light bulb of resistance R. This circuit is plugged into an electrical outlet. The current in the circuit is given by Equation 23.6, $I_{rms} = V_{rms}/Z$, where, from Equation 23.7, $Z = \sqrt{R^2 + X_L^2}$. When a piece of iron is inserted in the inductor, the magnetic field in the inductor is enhanced relative to that in air, and the inductance increases. Equation 23.4, $X_L = 2\pi f L$, shows that when the inductance L increases, the inductive reactance X_L also

increases. The impedance of the circuit, therefore, increases, and the current in the circuit, $I_{rms} = V_{rms}/Z$, decreases. Thus, the brightness of the bulb decreases.

---

4. ***REASONING AND SOLUTION*** It is possible that the current in the simplified circuit has the same rms value as that in the original circuit. This would be the case if the original circuit were at resonance. Then $X_C = X_L$, and Equation 23.7 for the impedance would simplify to $Z = R$. As a result, the currents in the original and the simplified circuits would each be $I_{rms} = V_{rms}/R$.

---

5. ***REASONING AND SOLUTION*** Consider the series RCL circuit shown in Figure 23.9. The impedance of the circuit is given by Equation 23.7: $Z = \sqrt{R^2 + (X_L - X_C)^2}$, where the capacitive reactance is given by Equation 23.2 [$X_C = 1/(2\pi f C)$] and the inductive reactance is given by Equation 23.4 [$X_L = 2\pi f L$]. One example of this expression for Z is plotted in Figure 23.15 (see the blue curve). The vertical axis in this figure gives the impedance, while the horizontal axis gives the frequency. A horizontal line drawn to intersect the vertical axis above the minimum in the curve will intersect the curve at two places. These places correspond to two different frequencies on the horizontal axis. Thus, the circuit in Figure 23.9 can have the same impedance at two different frequencies.

---

6. ***REASONING AND SOLUTION*** An inductor and a capacitor are connected in parallel across the terminals of a generator.

a. *The frequency of the generator becomes very large.* From Equation 23.2 (or Figure 23.2) we see that, in the high frequency limit, the capacitive reactance approaches zero. Therefore, in the high frequency limit, the capacitor behaves as if it were replaced by a wire with zero resistance, and the generator delivers a very large current to the capacitor. From Equation 23.4 (or Figure 23.6) we see that, in the high frequency limit, the inductive reactance becomes very large. In this limit, the inductor behaves as if it has been cut out of the circuit, leaving a gap in the wire, and the generator delivers no current to the inductor. Therefore, when the frequency of the generator becomes very large, the current through the capacitor becomes large, and no current flows through the inductor. The total current delivered by the generator is large.

b. *The frequency of the generator becomes very small.* From Equation 23.2 we see that, in the low frequency limit, the capacitive reactance approaches infinity. The capacitor acts as if it has been cut from the circuit, leaving a gap in the wire. As a result, the generator delivers no current to the capacitor. From Equation 23.4 we see that, in the low frequency limit, the inductive reactance approaches zero. In this limit, the inductor acts as if it has been replaced by a wire with zero resistance, and the generator delivers a very large current to the inductor. Therefore, when the frequency of the generator becomes very small, the current through the inductor becomes very large and the current through the capacitor is zero. The total current delivered by the generator is again large.

7. **REASONING AND SOLUTION** The rms current is given by Equation 23.6: $I_{rms} = V_{rms}/Z$, where V_{rms} is the rms voltage of the generator and Z is the impedance of the circuit. Since the rms voltage of the generator is the same in both cases, the greater current is delivered to the circuit with the smaller impedance. We see from Equation 23.2 $[X_C = 1/(2\pi f C)]$ and Equation 23.4 $[X_L = 2\pi f L]$ that in the high frequency limit, the capacitive reactance is nearly zero and the inductive reactance is essentially infinite. In other words, in the high frequency limit, the capacitors behave as if they have been replaced by wires of zero resistance and the inductors behave as if they have been cut from the circuit leaving gaps in the wires. The figure below shows the two circuits in this limit.

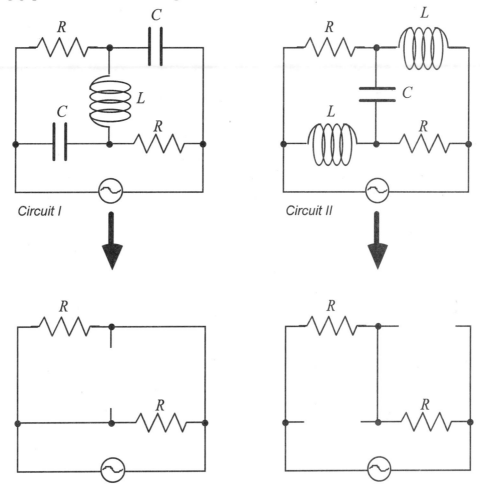

Circuit I behaves as if it consists only of two identical resistors in parallel; therefore, the impedance of circuit I is $1/Z = 1/R + 1/R$, or $Z = R/2$. Circuit II behaves as if it consists only of two identical resistors in series; therefore, the impedance of circuit II is $Z = 2R$. At a very high frequency, then, circuit I has the smaller impedance, and, therefore, its generator supplies the greater rms current.

144 ALTERNATING CURRENT CIRCUITS

8. **REASONING AND SOLUTION** The phase angle between the current in and the voltage across a series RCL combination is the angle ϕ between the current phasor I_0 and the voltage phasor V_0 in Figure 23.11. According to Equation 23.8, the tangent of this angle is related to the resistance R, the inductive reactance X_L, and the capacitive reactance X_C, according to the relation

$$\tan \phi = \frac{X_L - X_C}{R}$$

At resonance, $X_L = X_C$, and so $\tan \phi = 0$. Therefore, $\phi = \tan^{-1}(0) = 0$; in other words, the phase angle between the current phasor and the voltage phasor is zero. Since the phase angle between the current phasor and the voltage phasor is zero, we can conclude that the current is in phase with the voltage.

---

9. **REASONING AND SOLUTION** The resonant frequency of a series RCL circuit is given by Equation 23.10:

$$f_0 = \frac{1}{2\pi\sqrt{LC}}$$

a. Since the resonant frequency of a RCL circuit does not depend on the value of R, it is possible for two series RCL circuits to have the same resonant frequencies and yet have different R values.

b. The resonant frequency of a series RCL circuit is inversely proportional to $\sqrt{LC}$. It is possible for two series RCL circuits to have the same resonant frequencies and yet have different values of C and L, provided that the product LC is the same in both circuits.

---

10. **REASONING AND SOLUTION** The generator connected to a series RCL circuit has a frequency that is greater than the resonant frequency of the circuit. Suppose that it is necessary to match the resonant frequency of the circuit to the frequency of the generator. To accomplish this, a second capacitor will be added to the circuit. In order to match the resonant frequency of the circuit to the frequency of the generator, the resonant frequency of the circuit must be increased. The resonant frequency of a series RCL circuit is given by Equation 23.10: $f_0 = 1/(2\pi\sqrt{LC})$. Since the resonant frequency is inversely proportional to $\sqrt{LC}$, the capacitor must be added to the circuit so that it *decreases* the capacitance. This can be accomplished by placing the second capacitor *in series* with the first capacitor (see Equation 20.19). With the proper choice of the value of C for the second capacitor, the capacitance of the series combination will be smaller than the capacitance of the first capacitor alone, and the resonant frequency of the circuit will increase to match the generator frequency.

11. **REASONING AND SOLUTION** The drawing at the right shows a full-wave rectifier circuit, in which the direction of the current through the load resistor R is the same for both positive and negative halves of the generator's voltage cycle. The points A and B are connected directly to the top and bottom of the generator, respectively. The diodes are labeled a, b, c, and d. The direction of the current through the circuit can be found by recalling the fact that charge flows through a diode only when the diode is in a forward bias condition. When the diode is in a forward bias condition, the side of the diode symbol that contains the arrowhead has a positive potential relative to the other side.

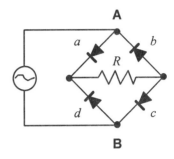

a. When the top of the generator is positive and the bottom is negative, point A in the figure has the positive potential and point B the negative potential. Since point A has the positive potential and is connected directly to the arrowhead for diode a, diode a is in a forward bias condition. In contrast, we see that for diode b, the side opposite the arrowhead is connected to point A, so diode b is in a reverse bias condition. Similarly, since point B is negative, and is connected directly to the side of diode c opposite the arrowhead, the arrowhead must be positive, and diode c is in a forward bias condition. In contrast, the arrowhead of diode d is negative, so diode d is in a reverse bias condition. Thus, when the top of the generator is positive and the bottom is negative, only diodes a and c are in the forward bias condition and allow charge to flow through the resistance R. Drawing 1 below shows the path taken by the current under these circumstances.

b. When the top of the generator is negative and the bottom is positive, reasoning similar to that given above in part (a) indicates that diodes b and d are now forward biased, while diodes a and c are in reverse bias. Furthermore, the forward bias diodes allow charge to flow from left to right through the resistance R. Drawing 2 below shows the path taken by the current under these circumstances.

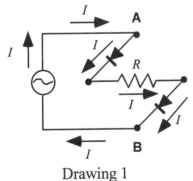

Drawing 1

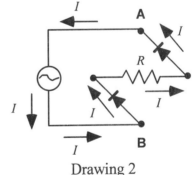

Drawing 2

CHAPTER 24 | ELECTROMAGNETIC WAVES

CONCEPTUAL QUESTIONS

1. **REASONING AND SOLUTION**
   a. The intensity of a wave refers to the power P carried by the wave that passes perpendicularly through a surface in the path of the wave, divided by the area A of the surface. It is given for sound waves by Equation 16.8 and for electromagnetic waves by Equation 24.4. The concept of intensity applies to both sound waves and electromagnetic waves, because both types of wave transfer energy and, therefore, power away from their respective sources.

   b. In a transverse wave, the particles of the medium vibrate perpendicular to the direction of propagation of the wave. In the case of electromagnetic waves, there are no particle vibrations, but rather the electric and magnetic fields oscillate perpendicular to the direction of propagation of the wave. When a wave is produced or altered so that the vibrations or oscillations take place in a particular direction perpendicular to the wave velocity, the wave is said to be *polarized*.
   In a longitudinal wave, the notion of polarization has no meaning, as discussed in Section 24.6 of the text. Therefore, *transverse waves can be polarized*, while *longitudinal waves cannot be polarized*.
   Since sound waves are longitudinal waves, while electromagnetic waves are transverse waves, the concept of polarization applies only to electromagnetic waves.

2. **REASONING AND SOLUTION** Refer to Figure 24.2. Between the times indicated in parts c and d in the drawing, negative charges have moved to the top of the antenna, leaving a net positive charge of equal magnitude on the bottom of the antenna. Therefore, as the negative charges flow, the conventional current points toward the bottom of the antenna. Using RHR-2, the magnetic field for the electromagnetic wave at P must point out of the page.

3. **REASONING AND SOLUTION** A transmitting antenna is located at the origin of an x, y, z axis system, and broadcasts an electromagnetic wave whose electric field oscillates along the y axis. The wave travels along the $+x$ axis. Three possible loops can be used with an LC-tuned circuit to detect the wave: One loop lies in the x, y plane, another in the x, z plane, and the third in the y, z plane.
   The loop that will detect the electromagnetic wave must be oriented so that the normal to the loop is parallel to the magnetic field. Then, as the wave passes by the loop, the changing magnetic field penetrates the loop and results in an induced emf and current, as predicted by Faraday's law. Since the electromagnetic wave travels along the $+x$ direction, and the electric field oscillations of the electromagnetic wave are along the y axis, the magnetic field oscillations will be along the z axis. For optimum reception, therefore, the loop should lie in the x, y plane so that the normal to the loop is in the z direction.

4. ***REASONING AND SOLUTION*** When a straight-wire antenna is used as the receiving antenna to detect electromagnetic waves, the wires must be oriented parallel to the electric field, as shown in Figure 24.5. The electric field exerts a force on the electrons in the wire and causes them to oscillate up and down along the length of the antenna. Since the electric field drives the electrons directly, the peak emf that causes the ac current in the receiving antenna depends on the amplitude of the electric field. It does not depend on the frequency of the wave.

When a loop antenna such as that shown in Figure 24.6, is used as the receiving antenna, the induced emf around the loop depends on the rate at which the magnetic flux changes with time, $\Delta\Phi/\Delta t$. If B represents the instantaneous magnitude of the magnetic field and A represents the cross-sectional area of the loop, and the normal of the loop points in the same direction as **B**, then $\Phi = BA$. Then $\Delta\Phi/\Delta t = \Delta(BA)/\Delta t$. Since the area of the loop, A, does not change, it follows that $\Delta\Phi/\Delta t = (\Delta B/\Delta t)A$. Hence, the induced emf in the receiving loop depends on the rate at which the magnetic field of the wave changes with time, and therefore, on the frequency of the electromagnetic wave. Thus, the peak value of the emf induced in a loop antenna depends on the frequency of the wave.

5. ***REASONING AND SOLUTION*** The electric field of an electromagnetic wave is related to the magnetic field through the relation $E = cB$ (Equation 24.3). Therefore, if the electric field of a wave decreases in magnitude, the magnetic field will also decrease.

This conclusion can also be reached by considering the fact that in an electromagnetic wave propagating through air or vacuum, the electric field and the magnetic field carry equal amounts of energy per unit volume of space. If the electric field of a wave decreases in magnitude, the electric energy per unit volume must decrease. The magnetic energy per unit volume must decrease to the same amount; therefore, the magnitude of the associated magnetic field must also decrease.

6. ***REASONING AND SOLUTION*** The same Doppler effect arises for electromagnetic waves when either the source or the observer of the waves moves; it is only the relative motion of the source and the observer with respect to one another that is important (see Section 24.5). Therefore, when an astronomer measures the Doppler change in frequency for the light reaching earth from a distant star, the astronomer *cannot* tell whether the star is moving away from the earth or whether the earth is moving away from the star.

7. ***REASONING AND SOLUTION*** The only real difference between a polarizer and an analyzer is the purpose for which they are used. Both consist of a piece of polarizing material. When the piece of polarizing material is used to produce a desired polarization direction, it is referred to as a *polarizer*. When the piece of polarizing material is used to change the polarization direction and to adjust the intensity so that the polarization direction of the incident light can be determined, it is referred to as an *analyzer*. The same piece of polarizing material can be used as the polarizer in one situation and the analyzer in another.

8. **REASONING AND SOLUTION** Malus' law applies to the setup in Figure 24.20, which shows the analyzer rotated through an angle θ and the polarizer held fixed. When the analyzer is held fixed and the polarizer is rotated, Malus' law still applies. Malus' law states that the average intensity $\bar{S}$ depends on the angle θ between the polarizer and analyzer. It does not matter whether the polarizer is fixed and the analyzer is rotated, or vice-versa.

9. **REASONING AND SOLUTION** In Example 7, we saw that, when the angle between the polarizer and analyzer is 63.4°, the intensity of the transmitted light drops to one-tenth of that of the incident unpolarized light. The light intensity that is not transmitted is absorbed by both the polarizer and the analyzer. The polarizer absorbs one-half of the incident intensity. The analyzer absorbs four-tenths, or two-fifths, of the original incident light. This absorbed energy results in an increase in the temperature of the polarizer and analyzer.

10. **REASONING AND SOLUTION** Light is incident from the left on two pieces of polarizing material, 1 and 2. As part a of the following drawing illustrates, the transmission axis of material 1 is along the vertical direction, while that of material 2 makes an angle of θ with respect to the vertical. In part (b) of the drawing, the two polarizing materials are interchanged.

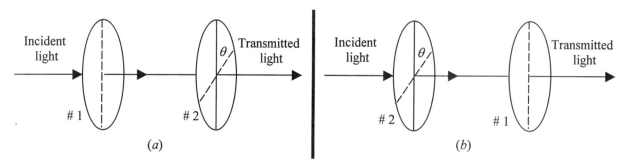

a. If the incident light is unpolarized and has an average intensity $\bar{I}$, then the average intensity of the light that leaves polarizer #1 and strikes polarizer #2 in Figure (a) is $\bar{S}_0 = \bar{I}/2$. According to Malus' law, the average intensity of the light leaving polarizer #2 is $\bar{S} = \bar{S}_0 \cos^2 \theta = (\bar{I} \cos^2 \theta)/2$.

Similarly, the average intensity of the light that leaves polarizer #2 and strikes polarizer #1 in Figure (b) is $\bar{S}_0 = \bar{I}/2$. The angle between the transmission axis of polarizer #2 and polarizer #1 is still θ, so that the average intensity of the light leaving polarizer #1 in (b) is $\bar{S} = \bar{S}_0 \cos^2 \theta = (\bar{I} \cos^2 \theta)/2$. Therefore, the transmitted light in Figure (a) is the same as that in Figure (b).

b. If the incident light in Figure (a) is polarized along the vertical direction with average intensity $\bar{I}$, then, since the polarization direction of the light is the same as the transmission axis of the polarizer, all of the incident light is transmitted. The average intensity of the light that leaves polarizer #1 and strikes polarizer #2 is $\bar{S}_0 = \bar{I}$. According to Malus' law, the average intensity of the light leaving polarizer #2 is $\bar{S} = \bar{S}_0 \cos^2 \theta = \bar{I} \cos^2 \theta$.

In Figure (b) we have from Malus' law that the average intensity of the light that leaves polarizer #2 and strikes polarizer #1 is $\overline{S}_0 = \overline{I}\cos^2\theta$. The angle between the transmission axis of polarizer #2 and polarizer #1 in Figure (b) is θ. Therefore, according to Malus' law, the average intensity of the light leaving polarizer #1 in Figure (b) is $\overline{S} = \overline{S}_0 \cos^2\theta = (\overline{I}\cos^2\theta)\cos^2\theta = \overline{I}\cos^4\theta$. Therefore, the intensity of the transmitted light in Figure (a) is greater than the intensity of the transmitted light in Figure (b).

---

11. **REASONING AND SOLUTION** Light from the sun is unpolarized; however, when the sunlight is reflected from horizontal surfaces such as the surface of a swimming pool, lake, or ocean, the reflected light is partially polarized in the horizontal direction. Polaroid sunglasses are constructed with lenses made of Polaroid (a polarizing material) with the transmission axis oriented vertically. Thus, the horizontally polarized light that is reflected from horizontal surfaces is blocked from the eyes.

    Suppose you are sitting on the beach near a lake on a sunny day, wearing Polaroid sunglasses. When you are sitting upright, the horizontally polarized light that is reflected from the lake is blocked from your eyes, as discussed above. When you lay down on your side, facing the lake, the transmission axis of the Polaroid sunglasses is now oriented in a nearly horizontal direction. Most of the horizontally polarized light that is reflected from the lake is transmitted through the sunglasses and reaches your eyes. Therefore, the sunglasses don't work as well as they did when you were sitting in an upright position.

CHAPTER 25 THE REFLECTION OF LIGHT: MIRRORS

CONCEPTUAL QUESTIONS

1. **REASONING AND SOLUTION** A sign painted on a store window is reversed when viewed from inside the store. A person inside the store views the sign in a plane mirror. As discussed in the text, the image of an object formed in a plane mirror is upright, has the same size as the object, is located as far behind the mirror as the object is in front of it, and is reversed left to right. Therefore, since the image is reversed left to right, the image of the sign painted on the store window, when viewed in a plane mirror in the store, will appear as it does when viewed from outside the store.

2. **REASONING AND SOLUTION** As discussed in the text, the image of an object formed in a plane mirror is reversed left to right. If a clock is held in front of a mirror, its image is reversed left to right, but not up and down. In order to understand the appearance of the image of the second hand, as viewed by a person looking into the mirror, imagine replacing the second hand by a rotating vector that always points away from its rotation axis. At any instant, this vector can be resolved into horizontal and vertical components. Since the image is reversed left to right, the image of the horizontal component will be reversed in direction. In contrast, since the image is *not reversed* up and down, the image of the vertical component will point in the same direction as the object. Therefore, when the horizontal and vertical components of the image are combined, the resultant will rotate *counterclockwise*. Thus, from the point of view of a person looking into the mirror, the image of the second hand of the clock rotates in the counterclockwise direction.

3. **REASONING AND SOLUTION** When parallel rays of light strike a concave mirror, they are reflected; these reflected rays converge at the focal point of the mirror. When parallel rays of light strike a convex mirror, they are also reflected; these reflected rays diverge from the mirror's surface and appear to originate from the focal point located behind the mirror.

    a. The earth-sun distance is very large; therefore, when rays of light from the sun reach the earth, they are essentially parallel. If it is desired to start a fire with sunlight, it is necessary to focus the parallel light rays from the sun on a very small area, preferably a point, on the piece of paper. Since a concave mirror reflects parallel rays so that they converge in front of the mirror, a concave mirror, rather than a convex mirror, should be used.

    b. For best results, the piece of paper should be placed at the focal point of the mirror since this is the location where the rays converge to a point and the heating would be greatest.

4. **REASONING AND SOLUTION** The photograph in the text shows an experimental device at Sandia National laboratories in New Mexico. The device is a mirror that focuses sunlight to heat sodium to a boil.

a. The earth-sun distance is very large; therefore, when rays of sunlight reach the earth, they are essentially parallel. A concave mirror reflects parallel rays of light so that the reflected rays converge at the focal point of the mirror. It is reasonable to conclude, therefore, that the mirror in the photograph is a concave mirror.

b. Since the rays of light converge at the focal point of the mirror, it is reasonable to conclude that the sodium unit is located at the mirror's focal point. Therefore, the distance between the mirror and the sodium unit is equal to the focal length of the mirror.

---

5. **REASONING AND SOLUTION** When parallel rays of light strike a concave mirror, they are reflected. For rays that lie close to the principal axis of the mirror, these rays will converge at a single point, namely the focal point of the mirror. Rays that are far from the principal axis do not converge to a single point after reflection. Each reflected ray obeys the law of reflection; namely, the angle of reflection, as measured with respect to the normal to the surface, is equal to the angle of incidence. Since the reflecting surface is spherical, the direction of the lines that are normal to the surface varies from point to point; however, they are directed radially with respect to the center of the "sphere." Rays that are farther from the principal axis have greater angles of incidence and greater angles of reflection. Therefore, when such rays reflect, they cross the principal axis to the right of the focal point F in Figure 25.14. The top ray in this drawing could be directed through the focal point if the angle of reflection, and therefore, the angle of incidence were decreased. This can be accomplished by changing the shape of the mirror so that the line normal to the surface at the point of incidence is rotated upward as shown in the figure below.

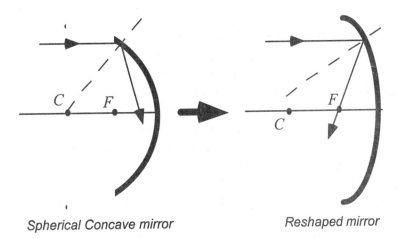

Spherical Concave mirror   Reshaped mirror

Therefore, to bring the top ray closer to the focal point, the mirror must be "opened up" to produce a more gently curving surface.

152   THE REFLECTION OF LIGHT: MIRRORS

6. **_REASONING AND SOLUTION_**
    a. For a real image, light rays actually pass through the points on the image. Such an image can be projected onto a screen directly. This can be accomplished by placing the screen at the location of the image.

    If the object is placed between the focal point and the center of curvature of a concave mirror, a real image is formed beyond the center of curvature. The real image is enlarged and inverted (see Figure 25.19a). If the object is placed beyond the center of curvature, a real image is formed between the center of curvature and the focal point. This real image is reduced in size and inverted with respect to the object (see Figure 25.19b). In these cases, the image is real; therefore, it can be projected directly onto a screen (placed at the location of the image) without the help of other mirrors or lenses.

    If, however, the object is placed between the focal point and the concave mirror, an enlarged, upright *virtual* image is produced (see Figure 25.20a). The image is behind the mirror and cannot be projected directly onto a screen without the help of other optical components.

    b. A convex mirror always produces a virtual image that is behind the mirror; therefore the image cannot be projected directly onto a screen without the help of other optical components.

---

7. **_REASONING AND SOLUTION_**
    a. The back side of a shiny teaspoon acts like a convex mirror. When an object is placed in front of a convex mirror, a virtual image is produced that is reduced in size and upright, as shown in Figure 25.22. Thus, when you look at the back side of a shiny teaspoon held at arms length, you will see yourself upright.

    b. The concave side of a shiny teaspoon acts like a concave mirror. If the teaspoon is held at arm's length, then you (the object) are farther from the reflecting surface than the center of curvature of the surface. The situation is similar to that of an object located beyond the center of curvature of a concave mirror. A real image is formed that is reduced in size and *inverted* with respect to the object, as shown in Figure 25.19b. Therefore, when you look at the concave side of a shiny teaspoon, you will see yourself upside down.

---

8. **_REASONING AND SOLUTION_**   If you stand between two parallel plane mirrors, you see an infinite number of images of yourself. This occurs because an image in one mirror is reflected in the other mirror to produce another image, which is then re-reflected, and so-forth. The multiple images are equally spaced.

    The image produced by a convex mirror is reduced in size relative to the object. Like a plane mirror, the image is virtual and lies behind the mirror; however, the virtual image in a convex mirror is closer to the mirror than it would be if the mirror were planar. If you stand facing a convex mirror, with a plane mirror behind you, you will see an infinite number of images of yourself, as you do in the case of two parallel plane mirrors. The reason is that the first image in the convex mirror is reflected in the plane mirror, to produce an identical image behind it. The image behind the plane mirror is re-reflected in the convex mirror to produce another smaller image closer to the focal point of the convex mirror. This second image in the convex mirror is smaller than the first one, because the image behind the plane mirror is farther from the convex mirror than you are. Further reflections and re-reflections occur, leading to a series of images in the convex mirror that decrease in size as they occur closer and closer to the focal point. The images "pile up" so to speak, becoming closer and closer together as they approach the focal point. The size of the image becomes zero at the focal point. Thus, the series of images appears to vanish at the focal point of the convex mirror.

9. **REASONING AND SOLUTION** The microphone arrangement shown in the figure is used to pick up weak sounds. It consists of a "hollowed-out" shell behind the mike. The shell acts like a mirror for sound waves. Therefore, when parallel rays of sound hit the inside surface of the shell, they will be reflected from it. Since the shell is concave, the reflected rays will converge at the focal point of the shell. Presumably, weak sounds will originate far from the microphone; therefore, when the they reach the microphone arrangement, these rays will be essentially parallel to the principal axis of the shell. Since the reflected rays will converge at the focal point, the microphone should be located at the focal point to detect them optimally.

10. **REASONING AND SOLUTION** When you see the image of yourself formed by a mirror, it is because (1) light rays actually coming from a real image enter your eyes or (2) light rays appearing to come from a virtual image enter your eyes. If the light rays from the image do not enter your eyes, you do not see yourself.

    a. If you stand in front of a convex mirror on its principal axis, your virtual image is behind the mirror. The image will be reduced in size and upright regardless of your location. You will be able to see yourself at any location on the principal axis when you are in front of a convex mirror, because rays that appear to come from this image can enter your eyes.

    b. If you stand in front of a concave mirror on its principal axis, and you are beyond the center of curvature of the mirror, you will see a real image of yourself that is inverted and reduced in size relative to your size and orientation (Figure 25.19b). If you stand between the mirror and its focal point, you will see a virtual image of yourself that is upright and enlarged in size (Figure 25.20). In the former case, rays emanating from the real image can reach your eyes. In the latter case, rays that appear to emanate from the virtual image can reach your eyes.
    When you stand on the principal axis of a concave mirror so that you are between the center of curvature and the focal point of the mirror, the image is formed beyond the center of curvature (Figure 25.19a). You cannot see this image, because it is behind you.

11. **REASONING AND SOLUTION** Plane mirrors and convex mirrors form virtual images. With a plane mirror, the image may be infinitely far behind the mirror, depending on where the object is located in front of the mirror.
    For an object in front of a single convex mirror, the greatest distance behind the mirror at which the image can be found is equal to the magnitude of the focal length of the mirror.

This can be confirmed using the mirror equation (Equation 25.3): $1/d_o + 1/d_i = 1/f$, where f is taken to be a negative number because the mirror is a convex mirror. The largest image distance occurs when the object is at infinity. Then $1/d_o = 1/\infty = 0$, and the mirror equation gives $d_i = f$. Since f is a negative number, d_i is a negative number, indicating that the image is behind the mirror. Therefore, the image will never be located beyond the focal point, behind the mirror.

---

12. ***REASONING AND SOLUTION*** As shown in Figure 25.22, an object placed in front of a convex mirror produces a virtual image behind the mirror that is reduced in size and is upright in orientation for all object distances. Therefore, it is *not* possible to use a convex mirror to produce an image that is larger than the object.

---

13. ***REASONING AND SOLUTION*** Suppose you stand in front of a spherical mirror (concave or convex). As shown in Figures 25.19 and 25.20, a concave mirror can form a real, inverted image (Figures 25.19a and 25.19b) or a virtual, upright image (Figure 25.20). As shown in Figure 25.22, a convex mirror can only form a virtual, upright image.

   a. Therefore, it is not possible for your image to be real and upright.

   b. Similarly, it is not possible for your image to be virtual and inverted.

CHAPTER 26 | THE REFRACTION OF LIGHT: LENSES AND OPTICAL INSTRUMENTS

CONCEPTUAL QUESTIONS

1. **REASONING AND SOLUTION** Since the index of refraction of water is greater than that of air, the ray in Figure 26.2a is bent toward the normal at the angle θ_{w1} when it enters the water. According to Snell's law (Equation 26.2), the sine of θ_{w1} is given by

$$\sin\theta_{w1} = \frac{n_{air}\sin\theta_1}{n_{water}} = \frac{\sin\theta_1}{n_{water}} \tag{1}$$

where we have taken $n_{air} = 1.000$. When a layer of oil is added on top of the water, the angle of refraction at the air/oil interface is θ_{oil} and, according to Snell's law, we have

$$\sin\theta_{oil} = \frac{n_{air}\sin\theta_1}{n_{oil}} = \frac{\sin\theta_1}{n_{oil}} \tag{2}$$

But θ_{oil} is also the angle of incidence at the oil/water interface. At this interface the angle of refraction is θ_{w2} and is given by Snell's law as follows:

$$\sin\theta_{w2} = \frac{n_{oil}\sin\theta_{oil}}{n_{water}} = \frac{n_{oil}\sin\theta_1}{n_{water}\,n_{oil}} = \frac{\sin\theta_1}{n_{water}} \tag{3}$$

where we have substituted Equation (2) for $\sin\theta_{oil}$. According to Equation (1), this result is equal to $\sin\theta_{w1}$. Therefore, we can conclude that the angle of refraction as the ray enters the water does *not* change due to the presence of the oil.

2. **REASONING AND SOLUTION** When light travels from a material with refractive index n_1 into a material with refractive index n_2, the angle of refraction θ_2 is related to the angle of incidence θ_1 by Equation 26.2: $n_1 \sin\theta_1 = n_2 \sin\theta_2$ or $\theta_2 = \sin^{-1}[(n_1/n_2)\sin\theta_1]$. When $n_1 < n_2$, the angle of refraction will be less than the angle of incidence. The larger the value of n_2, the smaller the angle of refraction for the same angle of incidence. The angle of refraction is smallest for slab B; therefore, slab B has the greater index of refraction.

3. *REASONING AND SOLUTION* When light travels from a material with refractive index n_1 into a material with refractive index n_2, the angle of refraction θ_2 is related to the angle of incidence θ_1 by Equation 26.2: $n_1 \sin \theta_1 = n_2 \sin \theta_2$. Therefore, $n_1 = n_2 \sin \theta_2 /(\sin \theta_1)$. Both blocks are made from the same material; therefore, n_2 is the same for each system. Furthermore, the angle of incidence is the same for each system; therefore, $\sin \theta_1$ is the same in both liquids. Thus, the liquid with the greater index of refraction will be the one for which $\sin \theta_2$, and therefore θ_2, is the largest. This occurs in liquid A. Therefore, the index of refraction of liquid A is greater than that of liquid B.

4. *REASONING AND SOLUTION* When an observer peers over the edge of a deep empty bowl, he does not see the entire bottom surface, so a small object lying on the bottom is hidden from view. However, when the bowl is filled with water, the object can be seen.

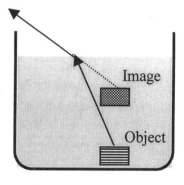

When the object is viewed from the edge of the bowl, light rays from the object pass upward through the water. Since $n_{air} < n_{water}$, the light rays from the object refract away from the normal when they enter air. The refracted rays travel to the observer, as shown in the figure at the right. When the rays entering the air are extended back into the water, they show that the observer sees a virtual image of the object at an apparent depth that is less than the actual depth, as indicated in the drawing. Therefore, the apparent position of the object in the water is in the line of sight of the observer, even though the object could not be seen before the water was added.

5. *REASONING AND SOLUTION* Two identical containers, one filled with water ($n = 1.33$) and the other filled with ethyl alcohol ($n = 1.36$) are viewed from directly above. According to Equation 26.3, when viewed from directly above in a medium of refractive index n_2, the apparent depth d' in a medium of refractive index n_1 is related to the actual depth d by the relation $d' = d(n_2/n_1)$. Assuming that the observer is in air, $n_2 = 1.00$. Since n_1 refers to the refractive index of the liquid in the containers, we see that the apparent depth in each liquid is inversely proportional to the refractive index of the liquid. The index of refraction of water is smaller than that of ethyl alcohol; therefore, the container filled with water appears to have the greater depth of fluid.

6. *REASONING AND SOLUTION* When you look through an aquarium window at a fish, the fish appears to be closer than it actually is. When light from the fish leaves the water and enters the air, it is bent away from the normal as shown below. Therefore, the apparent location of the image is closer to the observer than the actual location of the fish.

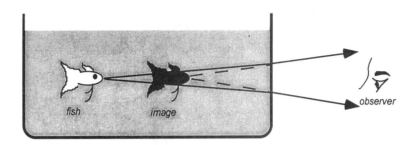

7. **REASONING AND SOLUTION** At night, when it is dark outside and you are standing in a brightly lit room, it is easy to see your reflection in a window. During the day it is not so easy. If we assume that the room is brightly lit by the same amount in both cases, then the light reflected from the window is the same during the day as it is at night. However, during the day, light is coming through the window from the outside. In addition to the reflection, the observer also sees the light that is refracted through the window from the outside. The light from the outside is so intense that it obscures the reflection in the glass.

8. **REASONING AND SOLUTION**
a. The man is using a bow and arrow to shoot a fish. The light from the fish is refracted away from the normal when it enters the air; therefore, the apparent depth of the image of the fish is less than the actual depth of the fish. When the arrow enters the water, it will continue along the same straight line path from the bow. Therefore, in order to strike the fish, the man must aim below the image of the fish. The situation is similar to that shown in Figure 26.5a; we can imagine replacing the boat by a dock and the chest by a fish.

b. Now the man is using a laser gun to shoot the fish. When the laser beam enters the water it will be refracted. From the principle of reversibility, we know that if the laser beam travels along one of the rays of light emerging from the water that originates on the fish, it will follow exactly the same path in the water as that of the ray that originates on the fish. Therefore, in order to hit the fish, the man must aim directly at the image of the fish.

9. **REASONING AND SOLUTION** Two rays of light converge to a point on a screen, as shown below.

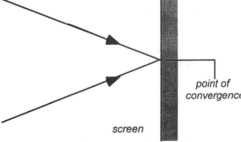

A plane-parallel plate of glass is placed in the path of this converging light, and the glass plate is parallel to the screen, as shown below. As discussed in the text, when a ray of light passes through a pane of glass that has parallel surfaces, and is surrounded by air, the

158   THE REFRACTION OF LIGHT: LENSES AND OPTICAL INSTRUMENTS

emergent ray is parallel to the incident ray, but is laterally displaced from it. The extent of the displacement depends on the angle of incidence, on the thickness, and on the refractive index of the glass.

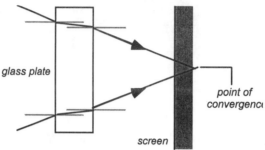

As shown in the scale drawing above, the point of convergence does not remain on the screen. It will move away from the glass as shown.

---

10. **REASONING AND SOLUTION**   Light from the sun is unpolarized; however, when the sunlight is reflected from horizontal surfaces, such as the surface of an ocean, the reflected light is partially polarized in the horizontal direction. Polaroid sunglasses are constructed with lenses made of Polaroid (a polarizing material) with the transmission axis oriented vertically. Thus, the horizontally polarized light that is reflected from horizontal surfaces is blocked from the eyes.

Suppose you are sitting on the beach near a lake on a sunny day, wearing Polaroid sunglasses. When a person is sitting upright, the horizontally polarized light that is reflected from the water is blocked from her eyes, as discussed above, and she notices little discomfort due to the glare from the water. When she lies on her side, the transmission axis of the Polaroid sunglasses is now oriented in a nearly horizontal direction. Most of the horizontally polarized light that is reflected from the water is transmitted through the sunglasses and reaches her eyes. Therefore, when the person lies on her side, she will notice that the glare increases.

---

11. **REASONING AND SOLUTION**   Light from the sun is unpolarized; however, when the sunlight is reflected from horizontal surfaces such as the surface of a swimming pool, lake, or ocean, the reflected light is partially polarized in the horizontal direction. Polaroid sunglasses are constructed with lenses made of Polaroid (a polarizing material) with the transmission axis oriented vertically. Thus, the horizontally polarized light that is reflected from horizontal surfaces is blocked from the eyes.

If you are sitting by the shore of a lake on a sunny and windless day, you will notice that the effectiveness of your Polaroid sunglasses in reducing the glare of the sunlight reflected from the lake varies depending on the time of the day. As the angle of incidence of the sun's rays increases from 0°, the degree of polarization of the rays in the horizontal direction increases. Since Polaroid sunglasses are designed so that the transmission axes are aligned in the vertical direction when they are worn normally, they become more effective as the sun gets lower in the sky. When the angle of incidence is equal to Brewster's angle, the reflected light is completely polarized parallel to the surface, and the sunglasses are most effective. For angles of incidence greater than Brewster's angle, the glasses again become less effective.

12. **REASONING AND SOLUTION** According to the *principle of reversibility* (see Section 25.5), if the direction of a light ray is reversed, the light retraces its original path. While the principle of reversibility was discussed in Section 25.5 in connection with the reflection of light rays, it is equally valid when the light rays are refracted. Imagine constructing a mixture of colored rays by passing a beam of sunlight through a prism in the usual fashion. By orienting a second prism so that the rays of colored light are incident on the second prism with angles of incidence that are equal to their respective angles of refraction as they emerge from the first prism, we have a perfectly symmetric situation. The rays through the second prism will follow the reverse paths of the rays through the first prism, and the light emerging from the second prism will be sunlight.

13. **REASONING AND SOLUTION** For glass (refractive index n_g), the critical angle for the glass/air interface can be determined from Equation 26.4:

$$\sin \theta_c = \frac{1.0}{n_g} \qquad (1)$$

In Figure 26.7 the angle of incidence at the upper glass/air interface is θ_2. Total internal reflection will occur there only if $\theta_2 \geq \theta_c$. But θ_2 is also the angle of refraction at the lower air/glass interface and can be obtained using Snell's law as given in Equation 26.2:

$$(1.0)\sin \theta_1 = n_g \sin \theta_2 \quad \text{or} \quad \sin \theta_2 = \frac{1.0}{n_g} \sin \theta_1$$

Using Equation (1) for $1.0/n_g$, we obtain

$$\sin \theta_2 = \sin \theta_c \, \sin \theta_1 \qquad (2)$$

For all incident angles θ_1 that are less than 90°, Equation (2) indicates that $\sin \theta_2 \geq \sin \theta_c$, since $\sin \theta_1 < 1$. Therefore, $\theta_2 < \theta_c$ and total internal reflection can not occur at the upper glass/air interface.

14. **REASONING AND SOLUTION**
a. When a rainbow is formed, light from the sun enters a spherical water droplet and is refracted by an amount that depends on the refractive index of water for that wavelength. Light that is reflected from the back of the droplet is again refracted at it reenters the air, as suggested in Figure 26.21. Although all colors are refracted for any given droplet, the observer sees only one color, because only one color travels at the proper angle to reach the observer. The observer sees the full spectrum in the rainbow because each color originates from water droplets that lie at different elevation angles.

As shown in Figure 26.21, the sun must be located *behind* the observer, if the observer is to see the rainbow. Therefore, if you want to make a rainbow by spraying water from a garden hose into the air, you must stand with the sun behind you, and adjust the hose so that it sprays a fine mist of water in front of you. The distance between the observer and the droplets is not crucial. The important factor is the angle formed by the intersection of the line that extends from the sun to the droplet with the line that extends from the droplet to the observer. *Remark:* When the distance is only a few meters, as it would be in the case of a "garden-hose rainbow", each eye would receive rays from different parts of the mist. Therefore, the observer could see two rainbows that cross over each other.

b. Each color of light that leaves a given droplet travels in a specific direction that is governed by Snell's law. You can't ever walk under a rainbow, because each color that originates from a single droplet travels in a unique direction. To walk under a rainbow, all the colors would have to be refracted vertically downward, which is not the case. Therefore, you can't walk under a rainbow, because the rays are traveling in the wrong directions to reach the observer's eyes.

---

15. **REASONING AND SOLUTION** A person is floating on an air mattress in the middle of a swimming pool. His friend is sitting on the side of the pool. The person on the air mattress claims that there is a light shining up from the bottom of the pool directly beneath him. His friend insists, however, that she cannot see any light from where she sits on the side.

Rays from a light source on the bottom of the pool will radiate outward from the source in all directions. However, only rays for which the angle of incidence is less than the critical angle will emerge from the water. Rays with an angle of incidence equal to, or greater than, the critical angle will undergo total internal reflection back into the water, as shown in the following figure.

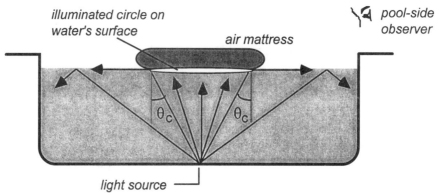

Because of the geometry, the rays that leave the water lie within a cone whose apex lies at the light source. Thus, rays of light that leave the water emerge from within an illuminated circle just above the source. If the mattress is just over the source, it could cover the area through which the light would emerge. A person sitting on the side of the pool would not see any light emerging. Therefore, the statements made by both individuals are correct.

---

16. **REASONING AND SOLUTION** Total internal reflection occurs only when light travels from a higher-index medium (refractive index = n_1) toward a lower-index medium (refractive index = n_2). Total internal reflection does not occur when light propagates from a

lower-index to a higher-index medium. The smallest angle of incidence for which total internal reflection will occur at the higher-index/lower-index interface is called the critical angle and is given by Equation 26.4: $\sin\theta_c = n_2/n_1$ where $n_1 > n_2$.

A beam of blue light is propagating in glass. When the light reaches the boundary between the glass and the surrounding air, the beam is totally reflected back into the glass. However, red light with the same angle of incidence is not totally reflected and some of the light is refracted into the air. According to Table 26.2, the index of refraction of glass is greater for blue light than it is for red light. From Snell's law, therefore, we can conclude that the critical angle is greater for red light than it is for blue light. Therefore, if the angle of incidence is equal to or greater than the critical angle for blue light, but less than the critical angle for red light, blue light will be totally reflected back into the glass, while some of the red light will be refracted into the air.

---

17. **REASONING AND SOLUTION** A beacon light in a lighthouse is to produce a parallel beam of light. The beacon consists of a bulb and a converging lens. As shown in Figure 26.22b, paraxial rays that are parallel to the principal axis converge to the focal point after passing through the lens. From the principle of reversibility, we can deduce that if a point source of light were placed at the focal point, the emitted light would travel in parallel rays after passing through the lens. Therefore, in the construction of the beacon light, the bulb should be placed at the focal point of the lens.

---

18. **REASONING AND SOLUTION** The figure at the right shows a converging lens (in air). The normal to the surface of the lens is shown at five locations on each side of the lens.

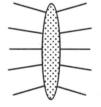

A ray of light bends toward the normal when it travels from a medium with a lower refractive index into a medium with a higher refractive index. Likewise, a ray of light bends away from the normal when it travels from a medium with a higher refractive index into a medium with a lower refractive index. When rays of light traveling in air enter a converging lens, they are bent toward the normal. When these rays leave the right side of the lens, they are bent away from the normal; however, since the normals diverge on the right side of the lens, the rays again converge.

If this lens is surrounded by a medium which has a higher index of refraction than the lens, then when rays of light enter the lens, the rays are bent away from the normal, and, therefore, they diverge. When the rays leave the right side of the lens, they are bent toward the normal; however, since the normals diverge on the right side of the lens, the rays diverge further. Therefore, a converging lens (in air) will behave as a diverging lens when it is surrounded by a medium that has a higher index of refraction than the lens.

---

19. **REASONING AND SOLUTION** A spherical mirror and a lens are immersed in water. The effect of the mirror on rays of light is governed by the law of reflection; namely $\theta_r = \theta_i$. The effect of the lens on rays of light is governed by Snell's law; namely, $n_1 \sin\theta_1 = n_2 \sin\theta_2$. The law of reflection, as it applies to the mirror, does not depend on the index of refraction of the material in which it is immersed. Snell's law, however, as it

applies to the lens, depends on both the index of refraction of the lens and the index of refraction of the material in which it is immersed. Therefore, compared to the way they work in air, the lens will be more affected by the water.

---

20. **REASONING AND SOLUTION** A converging lens is used to project a real image onto a screen, as in Figure 26.27b. A piece of black tape is then placed on the upper half of the lens.

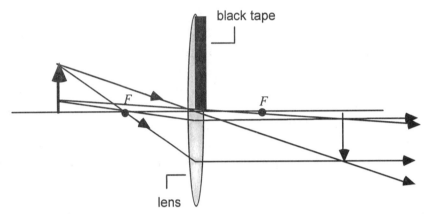

The ray diagram above shows the rays from two points on the object, one point at the top of the object and one point on the lower half of the object. As shown in the diagram, rays from both points converge to form the image on the right side of the lens. Therefore, the entire image will be formed. However, since fewer rays reach the image when the tape is present, the intensity of the image will be less than it would be without the tape.

---

21. **REASONING AND SOLUTION** When light travels from a material with refractive index n_1 into a material with refractive index n_2, the angle of refraction θ_2 is related to the angle of incidence θ_1 by Snell's law (Equation 26.2): $n_1 \sin \theta_1 = n_2 \sin \theta_2$.

A converging lens is made from glass whose index of refraction is n. The lens is surrounded by a fluid whose index of refraction is also n. This situation is known as *index matching* and is discussed in Conceptual Example 7. Since the refractive index of the surrounding fluid is the same as that of the lens, $n_1 = n_2$, and Snell's law reduces to $\sin \theta_1 = \sin \theta_2$. The angle of refraction is equal to the angle of incidence at both surfaces of the lens; the path of light rays is unaffected as the rays travel through the lens. Therefore, this lens cannot form an image, either real or virtual, of an object.

---

22. **REASONING AND SOLUTION** The expert claims that the height of the window can be calculated from only two pieces of information : (1) the measured height on the film, and (2) the focal length of the camera. The expert is not correct. According to the thin-lens equation (Equation 26.6), $1/d_o + 1/d_i = 1/f$, where d_o is the object distance, d_i is the image distance, and f is the focal length of the lens. The magnification equation (Equation 26.7), relates the image and object heights to the image and object distances: $h_o/h_i = -d_o/d_i$. These two equations contain five unknowns. To determine any one of the unknowns, three

of the other unknowns must be known. In this case, all that we know is the height of the image, h_i, and the focal length of the camera, f. Therefore, we do not have enough information given to determine the distance from the ground to the window (the height of the object in this case), h_o. We still need to know either the distance from the photographer to the house (the object distance, d_o), or the distance from the center of the lens to the film (the image distance, d_i). We can conclude, therefore, that the expert is incorrect.

---

23. **REASONING AND SOLUTION** Suppose two people who wear glasses are camping. One is nearsighted, and the other is farsighted. It is desired to start a fire with the sun's rays. A converging lens can be used to focus the nearly parallel rays of the sun on a sheet of paper. If the paper is placed at the focal point of the lens, the sun's rays are concentrated to give a large intensity, so that the paper heats up rapidly and ignites. As shown in Figures 26.36 and 26.37, nearsightedness can be corrected with diverging lenses, and farsightedness can be corrected using converging lenses. Therefore, the glasses of the farsighted person would be useful in starting a fire, while the glasses of the nearsighted person would not be useful.

---

24. **REASONING AND SOLUTION** A 21-year-old with normal vision (near point = 25 cm) is standing in front of a plane mirror. The near point is the point nearest the eye at which an object can be placed and still produce a sharp image on the retina. Therefore, if the 21-year old wants to see himself in focus, he can stand no closer to the mirror than 25 cm from his image. As discussed in Chapter 25, the image in a plane mirror is located as far behind the mirror as the observer is in front of the mirror. If the 21-year-old is 25 cm from his image, he must be 25 cm/2 = 12.5 cm in front of the mirror's surface. Therefore, he can stand no closer than 12.5 cm in front of the mirror and still see himself in focus.

---

25. **REASONING AND SOLUTION** The distance between the lens of the eye and the retina is constant; therefore, the eye has a fixed image distance. The only way for images to be produced on the retina for objects located at different distances is for the focal length of the lens to be adjusted. This is accomplished through with the ciliary muscles. If we read for a long time, our eyes become "tired," because the ciliary muscle must be tensed so that the focal length is shortened enough to bring the print into focus. When the eye looks at a distant object, the ciliary muscle is fully relaxed. Therefore, when your eyes are "tired" from reading, it helps to stop and relax the ciliary muscle by looking at a distant object.

---

26. **REASONING AND SOLUTION** As discussed in the text, for light from an object in air to reach the retina of the eye, it must travel through five different media, each with a different index of refraction. About 70 % of the refraction occurs at the air/cornea interface where the refractive index of air is taken to be unity and the refractive index of the cornea is 1.38.

To a swimmer under water, objects look blurred and out of focus. However, when the swimmer wears goggles that keep the water away from the eyes, the objects appear sharp and in focus. Without the goggles, light from objects must undergo the first refraction at a water/cornea interface. Since the index of refraction of water is 1.33 while that of the cornea is 1.38, the amount of refraction is smaller than it is when the person is in air, and the presence of the water prevents the image from being formed on the retina. Consequently, objects look blurred and out of focus. When the swimmer wears goggles, incoming light passes through the volume of air contained in the goggles before it reaches the eyes of the swimmer. The first refraction of the light in the eye occurs at an air/cornea interface. The refraction occurs to the proper extent, so that the image is formed on the retina. Therefore, when the swimmer wears the goggles, objects appear to be sharp and in focus.

27. **REASONING AND SOLUTION** The refractive power of the lens of the eye is 15 diopters when surrounded by the aqueous and vitreous humors. If this lens is removed from the eye and surrounded by air, its refractive power increases to about 150 diopters. From Snell's law, we know that the effect of the lens on incoming light depends not only on the refractive index of the lens, but also on the refractive index of the materials on either side of the lens. The refractive index of the lens is 1.40, while that of the aqueous humor is 1.33, and that of the vitreous humor is 1.34. Light that leaves the lens has been refracted twice, once when it enters the lens and again when it leaves the lens. Since the refractive indices of these three media are not very different, the amount of refraction at each interface is small. When the lens is surrounded by air, the light is again doubly refracted. In this case, however, the refractive indices at each interface differ substantially, so the amount of refraction at each interface is much larger. Therefore, when the lens is in air, its focal length is much smaller than it is when the lens is in place in the eye. According to Equation 26.8, the refractive power of a lens is equal to $1/f$, where the refractive power is expressed in diopters when the focal length is in meters. The smaller the focal length of the lens, the larger its refractive power. Consequently, the refractive power of the lens is much greater when the lens is surrounded by air.

28. **REASONING AND SOLUTION** A full glass of wine acts, approximately, as a converging lens and focuses the light to a spot on the table. An empty glass consists only of thin glass layers on opposite sides, which do not refract the light enough to act as a lens and produce a focused image.

29. **REASONING AND SOLUTION** The angle θ subtended by the image as measured from the principal axis of the lens of the eye is equal to the angle subtended by the object. This angle is called the angular size of both the image and the object and is given by $\theta \approx h_o/d_o$, where θ is expressed in radians.

Jupiter is the largest planet in our solar system. Yet to the naked eye, it looks smaller than Venus. This occurs because the distance from Earth to Jupiter is about 15 times greater than the distance from Earth to Venus, while the diameter of Jupiter is only about 12 times larger than that of Venus. Consequently, the angular size of Jupiter is about 12/15 or 0.80 times as large as that of Venus. Therefore, Jupiter looks smaller than Venus.

30. **REASONING AND SOLUTON**
a. The figure below is a ray diagram that shows that the eyes of a person wearing glasses appear to be smaller when the glasses use diverging lenses.

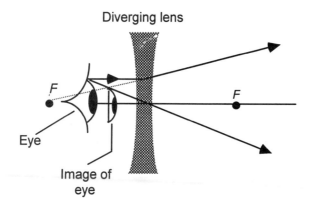

b. The figure below is a ray diagram that shows that the eyes of a person wearing glasses appear to be larger when the glasses use converging lenses.

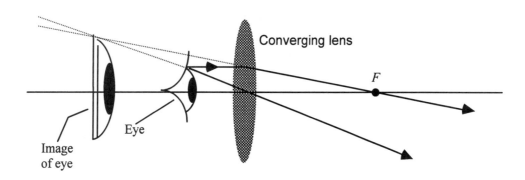

Notice that in both cases, the eye lies between the focal length of the lens and the lens, and that both images are virtual images.

---

31. **REASONING AND SOLUTION** As discussed in the text, regardless of the position of a real object, a diverging lens always forms a virtual image that is upright and smaller relative to the object. The figures below show this for two cases: one in which the object is within the focal point, and the other in which the object is beyond the focal point. In each case, the image is smaller than the object. Therefore, a diverging lens cannot be used as a magnifying glass.

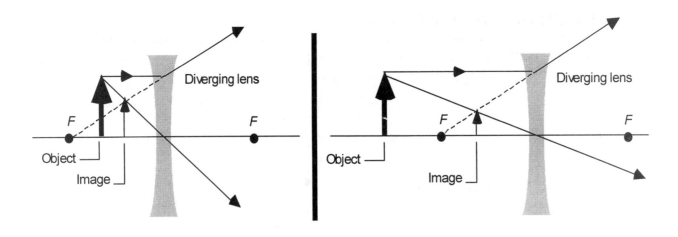

32. **REASONING AND SOLUTION** A person whose near point is 75 cm from the eyes, must hold a printed page at least 75 cm from his eyes in order to see the print without blurring, while a person whose near point is 25 cm can hold the page as close as 25 cm and still find the print in focus. If the size of the print is small, it will be more difficult to see the print at 75 cm than at 25 cm, even though the print is in focus. Therefore, the person whose near point is located 75 cm from the eyes will benefit more by using a magnifying glass.

33. **REASONING AND SOLUTION** The angular magnification of a telescope is given by Equation 26.12: $M \approx -f_o/f_e$, where f_o is the focal length of the objective, and f_e is the focal length of the eyepiece. In order to produce a final image that is magnified, f_o must be greater than f_e. Therefore if two lenses, whose focal lengths are 3.0 and 45 cm are to be used to build a telescope, the lens with the 45 cm focal length should be used for the objective, and the lens with the 3.0 cm focal length should be used for the eyepiece.

34. **REASONING AND SOLUTION** A telescope consists of an objective and an eyepiece. The objective focuses nearly parallel rays of light that enter the telescope from a distant object to form an image just beyond its focal point. The image is real, inverted, and reduced in size relative to the object. The eyepiece acts like a magnifying glass. It is positioned so that the image formed by the objective lies just within the focal point of the eyepiece. The final image formed by the eyepiece is virtual, upright and enlarged.

Two refracting telescopes have identical eyepieces, although one is twice as long as the other. Since the eyepiece is positioned so that the image formed by the objective lies just within the focal point of the eyepiece, the longer telescope has an objective with a longer focal length. The angular magnification of a telescope is given by Equation 26.12: $M \approx -f_o/f_e$, where f_o is the focal length of the objective, and f_e is the focal length of the eyepiece. Both telescopes have the same value for f_e. The longer telescope has the larger value of f_o; therefore, the longer telescope has the greater angular magnification.

35. **REASONING AND SOLUTION** In a telescope the objective forms a first image just beyond the focal point of the objective and just within the focal point of the eyepiece. Thus, as Figure 26.42 shows, the distance between the two converging lenses is $L \approx f_o + f_e$. For the two lenses specified, this would mean that $L \approx 4.5$ cm + 0.60 cm = 5.1 cm. But L is given as $L = 14$ cm, which means that there is a relatively large separation between the focal points of the objective and the eyepiece. This arrangement is like that for a microscope shown in Figure 26.33. Thus, the instrument described in the question is a microscope.

36. **REASONING AND SOLUTION**
   a. A projector produces a real image at the location of the screen.

   b. A camera produces a real image at the location of the film.

   c. A magnifying glass produces a virtual image behind the lens.

   d. Eyeglasses produce virtual images that the eye then sees in focus.

   e. A compound microscope produces a virtual image.

   f. A telescope produces a virtual image.

37. **REASONING AND SOLUTION** Chromatic aberration occurs when the index of refraction of the material from which a lens is made varies with wavelength. Lenses obey Snell's law. If the index of refraction of a lens varies with wavelength, then different colors of light that pass through the lens refract by different amounts. Therefore, different colors come to a focus at different points. Mirrors obey the law of reflection. The angle of reflection depends only on the angle of incidence, regardless of the wavelength of the incident light; therefore, chromatic aberration occurs in lenses, but not in mirrors.

CHAPTER 27 INTERFERENCE AND THE WAVE NATURE OF LIGHT

CONCEPTUAL QUESTIONS

1. **REASONING AND SOLUTION** A radio station broadcasts simultaneously from two transmitting antennas at two different locations. The radio that receives the broadcast could have better reception depending on the location of the receiving antenna. According to the principle of linear superposition, when the electromagnetic waves from the transmitting antenna arrive at the same point, the resultant wave is the sum of the individual waves. The amplitude of the resultant wave depends on the relative phase between the two waves. The relative phase between the waves depends on the path length difference between the two waves. If the two waves arrive at the receiving antenna so that the path length difference between them is equal to an integer number of wavelengths, the two waves will be in phase, they will reinforce each other, and constructive interference will occur. The resulting amplitude of the radio wave will be larger than it would be from either transmitting antenna alone; therefore, the radio reception will be better. If the two waves arrive at the receiving antenna so that the path length difference between them is equal to an odd multiple of half of a wavelength, the two waves will be exactly out of phase, they will mutually cancel, and destructive interference will occur. The receiving antenna will receive little or no signal in this situation. As a result, the radio reception will be very bad. Thus, having two transmitting antennas does not necessarily lead to better reception.

2. **REASONING AND SOLUTION** Constructive interference occurs between waves from two in-phase sources when the difference in travel distances from the sources to the point where the waves combine is the same or equal to an integer number of wavelengths. For a square the distance from any corner to the center is the same. Hence, constructive interference occurs at the center.

3. **REASONING AND SOLUTION**
   a. The light waves coming from *both* slits in a Young's double slit experiment have their phases shifted by an amount equivalent to half a wavelength. Since the light from both slits is changed by the same amount, the relative phase difference between the light from the two slits is zero when the light leaves the slits. When the light reaches the screen, the relative phase difference between light waves from the two slits will be the same as if the phase of the waves had not been shifted at the slits. Therefore, the pattern will be exactly the same as that described in the text.

   b. Light coming from *only one* of the slits in a Young's double slit experiment has its phase shifted by an amount equivalent to half of a wavelength. Now, there is a relative phase difference of one half of a wavelength between the light leaving the slits. When the light reaches the screen, there will be a relative phase difference due to the fact that light from each slit, in general, traveled along different paths. In addition, there will be the initial phase difference that is equivalent to half of a wavelength. Therefore, the pattern will be similar to that described in the text; however, the points of constructive interference will be points of destructive interference, and the points of destructive interference will be points of constructive interference. In other words, the positions of the light and dark fringes are interchanged.

4. ***REASONING AND SOLUTION*** According to Equation 27.1, the bright fringes of a double slit experiment occur at values of θ for which $\sin\theta = m\lambda/d$ where $m = 0, 1, 2, 3, ...$ According to Equation 27.2, the dark fringes occur at values of θ for which $\sin\theta = [m+(1/2)]\lambda/d$ where $m = 0, 1, 2, 3, ...$ The slits S_1 and S_2 in Figure 27.4 are replaced with identical loudspeakers and use the same ac electrical signal to drive them. The two sound waves produced will then be identical, and we have the audio equivalent of Young's double slit experiment. Sound waves that reach the center of the screen have traveled along paths of equal length; therefore, the sound waves from each loudspeaker arrive there in phase and constructive interference occurs. Consequently, at the center point on the screen, the sound will have maximum loudness. As you proceed to walk away from the center (in either direction), the sound will decrease in volume. When the angle θ is such that $\sin\theta = (1/2)\lambda/d$, the sound will decrease to zero amplitude. If you continue to walk beyond this point, the loudness of the sound will increase. It will reach a maximum (a point of constructive interference) when the angle θ is such that $\sin\theta = \lambda/d$. The sound will again decrease in loudness until the next point of destructive interference occurs. This point occurs when $\sin\theta = (3/2)\lambda/d$. In general, there will a pattern of alternating loud and soft sounds heard as you walk away from the center of the screen. The intensity at the points of constructive interference varies in a manner that is similar to that shown in Figure 27.6. The central maximum has the greatest intensity. The sound at each region of constructive interference decreases in intensity as one proceeds away from the center (in either direction).

5. ***REASONING AND SOLUTION*** The angle θ for the interference maximum in Young's double-slit experiment is given by Equation 27.1: $\sin\theta = m\lambda/d$ where $m = 0, 1, 2, 3, ...$ When the wavelength λ of the light is greater than the distance d between the slits, the ratio λ/d is greater than one; however, $\sin\theta$ cannot be greater than one. Therefore, it is *not* possible to see interference fringes when the wavelength of the light is greater than the distance between the slits.

6. ***REASONING AND SOLUTION*** A camera lens is covered with a nonreflective coating that eliminates the reflection of perpendicularly incident green light. If the light were incident on the nonreflective coating at 45°, rather than perpendicularly, it would not be eliminated by the coating and an observer would see it.

Using reasoning similar (but not identical to) that in Example 3, if the refractive index of the film is less than the refractive index of the lens, the minimum nonzero thickness of the nonreflecting coating is given by $t = \lambda_{coating}/4$. This result follows from the fact that the

170 INTERFERENCE AND THE WAVE NATURE OF LIGHT

extra distance traveled by the wave that travels through the coating is $2t$ when the wave strikes the film at normal incidence.

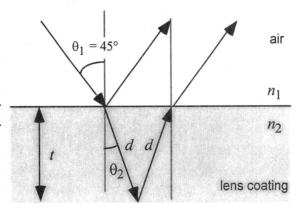

If the wave strikes the thin film at 45°, then the extra distance traveled by the wave that travels through the coating is no longer $2t$. Instead, it is $2d$, where d is the distance shown in the figure at the right. The value of d depends on the value for θ_2, the angle of refraction. This angle, in turn, depends on the refractive index n_1 for air and n_2 for the lens coating, according to Snell's law (Equation 26.2). To eliminate the light that is incident at 45°, the coating would have to have a minimum nonzero thickness such that $d = \lambda_{coating}/4$. Since the coating is designed such that $t = \lambda_{coating}/4$, it does not eliminate the light incident at 45°.

---

7. ***REASONING AND SOLUTION*** When sunlight reflects from a thin film of soapy water, the film appears multicolored, in part because destructive interference removes different wavelengths from the light reflected at different places, depending on the thickness of the film. As the film becomes thinner and thinner, the path length difference between the light reflected from the top of the film and the light reflected from the bottom of the film becomes closer to zero. The light reflected from the top of the film undergoes a phase change equivalent to half of a wavelength. Therefore, as the thickness of the film approaches zero, the light reflected from the top of the film is almost exactly out of phase with the light reflected from the bottom of the film. Therefore, as the film becomes thinner and thinner, it looks darker and darker in reflected light. It appears black just before it breaks, because at that point destructive interference occurs between the light reflected from the top and the light reflected from the bottom of the film.

---

8. ***REASONING AND SOLUTION*** Two pieces of the same glass are covered with thin films of different materials. The thickness t of each film is the same. In reflected sunlight, however, the films have different colors.

This could occur because, in general, different materials have different refractive indices. Using reasoning similar to (but not identical to) that in Example 3, the wavelength λ_{air} that is removed from the reflected light corresponds to a value of λ_{film} in the film that satisfies either of the following conditions:

$$2t = m\lambda_{film}, \quad m = 0, 1, 2, 3, \ldots \qquad n_{film} > n_{glass}, \qquad (1)$$

and

$$2t = \left(m + \tfrac{1}{2}\right)\lambda_{film}, \quad m = 0, 1, 2, 3, \ldots \qquad n_{film} < n_{glass}. \qquad (2)$$

However, the value of the wavelength λ_{film} of the light in the film depends on the index of refraction of the film. According to Equation 27.3, $\lambda_{film} = \lambda_{vacuum}/n$, where n is the index of refraction of the film. If the index of refraction of the two materials are different, then conditions (1) and (2) will be satisfied for two different values of λ_{film}. These values of λ_{film} correspond to different values of λ_{air}, and, therefore, different wavelengths will be removed from the reflected light.

---

9. *REASONING AND SOLUTION* When a spherical surface is put in contact with an optically flat plate, circular interference patterns, called Newton's rings, can be observed in reflected light, as shown in Figure 27.14*b*. There is a dark spot at the center of the pattern of Newton's rings. The dark spot occurs for the following reason. At the center of the plate, the spherical surface is in contact with the plate. Destructive interference occurs there, because the thickness of the "air wedge" is essentially zero at this point. Therefore, the only difference in phase between the light reflected from the bottom of the spherical surface and the light reflected from the top of the optical flat is the half-wavelength phase change due to reflection from the optical flat. This situation is similar to that discussed in part (b) of Example 5 in the text.

---

10. *REASONING AND SOLUTION* A thin film of material is floating on water ($n = 1.33$). When the material has a refractive index of $n = 1.20$, the film looks bright in reflected light as its thickness approaches zero. But when the material has a refractive index of $n = 1.45$, the film looks black in reflected light as its thickness approaches zero.

When the material has a refractive index of $n = 1.20$, both the light reflected from the top of the film and the light reflected from the bottom occur under conditions where light is traveling through a material with a smaller refractive index toward a material with a larger refractive index. Therefore, both the light reflected from the top and the bottom undergoe a phase change upon reflection. Both phase changes are equivalent to half of a wavelength. Therefore, the reflections introduce no net phase difference between the light reflected from the top and bottom. As the thickness of the film approaches zero, the path difference between the light reflected from the top and the bottom of the film approaches zero. Since the path difference is close to zero, and there is no relative phase difference due to reflection, the light reflected from the top will be in phase with the light reflected from the bottom of the film; constructive interference will occur, and the film looks bright in reflected light.

When the material has a refractive index of $n = 1.45$, only the light reflected from the top of the film is traveling through a material with lower refractive index toward a material with a larger refractive index. Therefore, only the light reflected from the top undergoes a phase change upon reflection. This phase change is equivalent to half of a wavelength. Therefore, there is a net phase difference equivalent to half of a wavelength between the light reflected from the top and bottom of the film. As the thickness of the film approaches zero, the path difference between the light reflected from the top and the light reflected from the bottom of the film approaches zero. Since the path difference is close to zero, and there is a relative phase difference due to reflection that is equal to half of a wavelength, the light reflected from the top will be exactly out of phase with the light reflected from the bottom of the film, and destructive interference will occur. Therefore, the film looks black in reflected light.

11. **REASONING AND SOLUTION** A transparent coating is deposited on a glass plate and has a refractive index that is larger than that of the glass. For a certain wavelength within the coating, the thickness of the coating is a quarter wavelength. The coating enhances the reflection of the light corresponding to this wavelength.

As discussed in Example 3, only the light reflected from the top of the coating is traveling in through a material (air) with lower refractive index toward a material coating with a larger refractive index. Therefore, only the light reflected from the top of the coating undergoes a phase change upon reflection. This change in phase is equivalent to half of a wavelength. Therefore, there is a net phase difference equivalent to half of a wavelength between the light reflected from the top and light reflected from the bottom of the coating. Since the thickness of the coating is a quarter wavelength, the light reflected from the bottom of the coating traverses a distance equal to two times a quarter wavelength, or half a wavelength, as it travels through the coating. The overall relative phase difference between the light reflected from the top and the light reflected from the bottom of the coating is equivalent to half of a wavelength due to the reflection from the top of the coating and another half of a wavelength due to the path difference between the light reflected from the top and the light reflected from the bottom. Thus, the overall phase difference is equivalent to a whole wavelength, and the light reflected from the top is in phase with the light reflected from the bottom. Constructive interference occurs, and the coating enhances light corresponding to that wavelength.

12. **REASONING AND SOLUTION** Rayleigh's criterion states that two point objects are just resolved when the first dark fringe in the diffraction pattern of one falls directly on the central bright fringe in the diffraction pattern of the other. Equation 27.6 approximates the minimum angular displacement between the two objects so that they are just resolved by an optical instrument: $\theta_{min} \approx 1.22 \lambda / D$, where λ is the wavelength of light used and D is the diameter of the aperture of the optical instrument. According to Equation 27.6, optical instruments with the highest resolution should have the largest possible diameter D and utilize light of the shortest possible wavelength.

The f-number setting on a camera gives the ratio of the focal length of the camera lens to the diameter of the aperture through which the light enters the camera. Therefore, smaller f-numbers mean larger aperture diameters. According to Equation 27.6, if we wish to resolve two closely spaced objects in a picture, we should adjust the aperture to have the largest possible diameter. As a result, we should use a small f-number setting.

13. **REASONING AND SOLUTION** As discussed in the text, the extent to which a wave bends or diffracts around the edges of an obstacle or opening is determined by the ratio λ/W, where λ is the wavelength of the wave and W is the width of the obstacle or opening. As the ratio λ/W becomes larger, diffraction effects become more pronounced.

The wavelength of light is extremely small and is normally expressed in nanometers (1 nm = 10^{-9} m). Therefore, the ratio λ/W is very small for light, and when light passes through a typical doorway, diffraction effects are not noticeable. Many sound waves encountered in the everyday world have wavelengths that are comparable to the dimensions of a typical doorway. Therefore, the ratio λ/W is nearly unity for these sounds, and when sound passes through a doorway, diffraction effects are noticeable; you can hear around corners, although you cannot see around corners.

---

14. **REASONING AND SOLUTION** The Rayleigh criterion is $\theta_{min} \approx 1.22\, \lambda/D$, where λ is the wavelength of the light, D is the diameter of the opening into the eye, and θ_{min} is the minimum angle that two point objects must subtend at the opening in order to be resolved by the eye. In this case D is the opening through which light from the painting enters the eye. Before squinting, the angle subtended at the eye is greater than θ_{min}, since the dots are distinguishable. Because of the squinting, light reaches the retina through an opening that is smaller than the normal size of the eye's pupil. A smaller value for D, with a given wavelength, means that the value for θ_{min} is larger. Thus, after squinting, the angle subtended at the eye is smaller than the larger value for θ_{min}, and the dots of color can no longer be individually resolved, which gives the painting a more normal look.

---

15. **REASONING AND SOLUTION** Four light bulbs are arranged at the corners of a rectangle that is three times longer than it is wide. You look at this arrangement perpendicular to the plane of the rectangle. From very far away, your eyes cannot resolve the individual bulbs and you see a single rectangular "smear" of light. From close in, you see the individual bulbs. The figures below illustrate what you would see at two positions between these two extremes.

In the first case, you are still far away, but you have moved close enough so that you can begin to resolve the bulbs at opposite ends of the diagonal. Essentially, you see a rectangular "smear" of light with a gap in the middle.

In the second case, you have moved closer still and now can resolve the two bulbs on one of the short sides of the rectangle from the two on the other short side. Essentially, you see two "smears" of light, each smear consisting of the two bulbs at the ends of a short side of the rectangle.

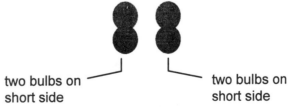

174   INTERFERENCE AND THE WAVE NATURE OF LIGHT

16. **REASONING AND SOLUTION** The resolving power, or the ability to resolve two closely spaced images, through any circular aperture (such as the pupil of your eye) is limited by diffraction effects. The Rayleigh criterion for resolution focuses on these effects and is given by $\theta_{min} \approx 1.22\lambda/D$ (Equation 27.6), where λ is the wavelength of light, D is the diameter of the aperture, and θ_{min} is the minimum angle that can be resolved between two point objects by a circular aperture. A smaller value for θ_{min} or λ/D means a greater resolving power.

a. Suppose the pupil of your eye were elliptical instead of circular in shape, with the long axis of the ellipse oriented in the vertical direction. Since the resolving power depends on the diameter of the aperture and since the ellipse is longer than it is wide, the resolving power of your eye would not be the same in the horizontal and vertical directions.

b. If the pupil of your eye were elliptical instead of circular in shape, with the long axis of the ellipse oriented in the vertical direction, the "diameter" or "width" would be larger in the vertical direction and smaller in the horizontal direction. Consider two "planes" of light rays. Plane 1 is oriented so that neighboring rays are horizontal, and plane 2 is oriented so that neighboring rays are stacked vertically. Since the "diameter" of your eye is largest in the vertical direction, the ratio λ/D is smaller for light in plane 2 than it is for light in plane 1. Thus, diffraction effects will be less noticeable for light rays stacked vertically. Therefore, the resolving power will be greater in the vertical direction.

---

17. **REASONING AND SOLUTION** Rayleigh's criterion states that two point objects are just resolved when the first dark fringe in the diffraction pattern of one falls directly on the central bright fringe in the diffraction pattern of the other. Equation 27.6 approximates the minimum angular displacement θ_{min} between the two objects so that they are just resolved by an optical instrument: $\theta_{min} \approx 1.22\lambda/D$, where λ is the wavelength of light used and D is the diameter of the aperture of the optical instrument. According to Equation 27.6, optical instruments with the highest resolution should have the largest possible diameter D and utilize light of the shortest possible wavelength.

Suppose you were designing an eye and could select the size of the pupil and the wavelength of the electromagnetic waves to which the eye is sensitive. As far as the limitation created by diffraction is concerned, the following list ranks the design choices in order of decreasing resolving power (greatest first): (a) large pupil (large D) and ultraviolet wavelengths (small λ); (c) small pupil (small D) and ultraviolet wavelengths (small λ); (b) small pupil (small D) and infrared wavelength (larger λ).

---

18. **REASONING AND SOLUTION** In our discussion of single-slit diffraction, we ignored the height of the slit, in effect assuming that the height was much larger than the width. Suppose that the height and width were the same size, so that diffraction occurred in both dimensions. The resulting pattern would change in the following way. The height of the horizontal fringes would be reduced. Furthermore, there would be two sets of fringes. One set would be along the horizontal axis, and the other set would be along the vertical axis. The two sets would share a common central maximum. The intensity of the central

maximum will be greatest. The intensity will decrease on either side of the central maximum, in both the horizontal and the vertical directions. The figure below shows the central maximum and the less intense higher order on either side of the central maximum in both dimensions.

19. **REASONING AND SOLUTION** According to Equation 27.7, the angles θ that locate principal maxima of a diffraction grating satisfy the condition $\sin\theta = m\lambda/d$, where $m = 0$, 1, 2, 3, . . .

If the entire interference apparatus of a diffraction grating (light source, grating, and screen) were immersed in water, the wavelength of the light would *decrease* from the value λ_{air} to the value λ_{water}. This is because these values are related by Equation 27.3, $\lambda_{water} = \lambda_{vacuum}/n_{water} = \lambda_{air}/n_{water}$, where we have used the fact that λ_{vacuum} and λ_{air} are nearly the same, since n_{air} is nearly 1. Thus, angles θ that locate principal maxima of a diffraction grating immersed in water will be smaller than when the apparatus is immersed in air. Consequently, the principle maxima will be closer together when the apparatus is immersed in water relative to their location when the entire apparatus is immersed in air.

CHAPTER 28 | SPECIAL RELATIVITY

CONCEPTUAL QUESTIONS

1. **REASONING AND SOLUTION** The speed of light postulate states that the speed of light in *a vacuum*, measured in any inertial reference frame, always has the same value of c, no matter how fast the source of the light and the observer are moving relative to each other.

   The speed of light in water is c/n, where $n = 1.33$ is the refractive index of water. Thus, the speed of light in water is less than c. This does not violate the speed of light postulate, because the postulate refers to the speed of light in a vacuum, not in a physical medium.

2. **REASONING AND SOLUTION** A baseball player at home plate hits a pop fly straight up (the beginning event) that is caught by the catcher at home plate (the ending event).

   a. A spectator in the stands is at rest relative to these events; therefore, this spectator would record the proper time interval between them Δt_0.

   b. A spectator sitting on the couch and watching the game on TV is at rest relative to these events; therefore, this spectator would also record the proper time interval between them Δt_0.

   c. The third baseman running in to cover the play is moving relative to the events; therefore, the third baseman *will not* record the proper time interval between them. He will record a dilated time interval Δt given by the time-dilation equation (Equation 28.1): $\Delta t = \Delta t_0 / \sqrt{1 - v^2/c^2}$, where v is the relative speed between the observer who measures Δt_0 and the observer who measures Δt.

3. **REASONING AND SOLUTION** The earth spins on its axis once each day. To a person viewing the earth from an inertial reference frame in space, a clock at the equator would run more slowly than a clock at the north (magnetic) pole. The observer in the inertial reference frame is not in the rest frame of either clock, so he will measure a dilated time for both clocks. According to the time dilation equation (Equation 28.1): $\Delta t = \Delta t_0 / \sqrt{1 - v^2/c^2}$, where v is the relative speed between the observer who measures Δt_0 and the observer who measures Δt. Both clocks move around the earth's rotation axis with the same angular speed ω as that of the earth. The linear speed of each clock is given by $v = r\omega$, where r is the distance from the clock to the rotation axis. The clock at the equator has the greater value of r, and, therefore, the greater linear speed v. According to Equation 28.1, the observer in the inertial reference frame will see the clock with the larger linear speed v register the longer time interval Δt. Therefore, the clock at the equator will appear to run slower when viewed by the inertial observer.

Chapter 28 Conceptual Questions 177

4. *REASONING AND SOLUTION*
    a. You are standing at a railroad crossing watching a train go by. Both you and a passenger in the train are looking at a clock on the train. The passenger on the train is at rest relative to the clock; therefore, the passenger on the train measures the proper time interval.

    b. A passenger in the train car is at rest relative to the car; therefore, the passenger on the train measures the proper length of the train car.

    c. Since you are standing at the railroad crossing, you are at rest relative to the track; therefore, you measure the proper distance between the railroad ties under the track.

---

5. *REASONING AND SOLUTION*   There are tables that list data for the various particles of matter that physicists have discovered. Often, such tables list the masses of the particles in units of energy, such as MeV, rather than kilograms. This is possible because of the equivalence between mass and energy. The total energy E of a moving object is related to its mass m and its speed v by Equation 28.4: $E = mc^2 / \sqrt{1 - v^2/c^2}$. The mass of a particle listed in the table is actually the total energy of the particle when it is at rest ($v = 0$ m/s); the rest energy of an object of mass m is given by Equation 28.5: $E_0 = mc^2$.

---

6. *REASONING AND SOLUTION*   When an object is accelerated from rest to a speed v, the object acquires kinetic energy in addition to its rest energy. The total energy E of the object is the sum of its rest energy E_0 and its kinetic energy KE. According to Equation 28.6, the kinetic energy of an object of mass m moving at speed v is $KE = E - E_0 = mc^2 \left[\left(1/\sqrt{1 - v^2/c^2} \right) - 1 \right]$. The work required to accelerate an object from rest to speed v is, from the work-energy theorem (Chapter 6), equal to the change in the kinetic energy of the object. Since the object begins at rest, its change in kinetic energy is equal to its final kinetic energy and is given by Equation 28.6.
    According to Equation 28.6, it is easier to accelerate an electron to a speed that is close to the speed of light, compared to accelerating a proton to the same speed. This is because the mass of the electron is much less than that of the proton. Hence, the final kinetic energy of the proton, will be greater than that of the electron at the same speed v. Therefore, the proton will require more work to get it to the same speed as the electron.

---

7. *REASONING AND SOLUTION*   Light travels in water at a speed of 2.26×10^8 m/s. It is possible for a particle that has mass to travel through water at a faster speed than this. But objects that have mass *cannot reach the speed of light in a vacuum*, that is, they cannot attain a speed of $c = 3.00 \times 10^8$ m/s. It is the speed of light in a vacuum that is the ultimate speed, not the speed of light in a material like water.

---

8. ***REASONING AND SOLUTION*** Two positive, electric charges separated by a finite distance have more mass than when they are infinitely far apart. The electric potential energy of the two-particle system is zero when the particles are infinitely far apart. Work is required to bring two positive charges in from infinity, so that there is a finite separation between them; therefore, the electric potential energy and consequently the total energy of the two-particle system increases as they are brought in from infinity. Consistent with the theory of special relativity, any increase in the total energy of the system, including a change in the electric potential energy, is equivalent to an increase in the mass of the system. Therefore, the two-particle system has more mass when the particles have a finite distance between them, than when they are infinitely far apart.

9. ***REASONING AND SOLUTION*** One system consists of two stationary electrons separated by a distance r. Another consists of a positron and an electron, both stationary and separated by the same distance r. A positron has the same mass as an electron, but it has a positive electric charge $+e$. Since the charges of unlike sign attract each other, and the charges of like sign repel each other, more work is required to bring the charges from infinity to assemble the system that consists of the two stationary electrons. Since more work is required to assemble the system of two stationary electrons, that system has the greater electric potential energy. Consistent with mass-energy equivalence predicted by the theory of special relativity, the system with the larger total energy will have the greater mass. Therefore, the system that consists of two stationary electrons has the greater mass.

10. ***REASONING AND SOLUTION*** A parallel plate capacitor is initially uncharged. The capacitor is charged up by removing electrons from one plate and placing them on the other plate. Work and, therefore, energy is required to charge the capacitor. It is stored in the electric field between the plates of the charged capacitor in the form of electric potential energy. Since the charged capacitor has more electric potential energy than the uncharged capacitor, its also has more total energy than the uncharged capacitor. Therefore, from the equivalence of mass and energy, we can conclude that the charged capacitor has more mass.

11. ***REASONING AND SOLUTION*** The speed limit on many interstate highways is 65 miles per hour. If the speed of light were 65 miles per hour, you would never be able to drive at the speed limit. When the speed, v, of an object is equal to the speed of light, c, the term $1/\sqrt{1-v^2/c^2}$ becomes infinitely large. Therefore, according to Equation 28.6, the kinetic energy of the object would be infinitely large. According to the work-energy theorem, an infinite amount of work would be required to increase the speed of the object to $v = c$. Since an infinite amount of work is not available, it would *not* be possible for the car to reach the speed limit.

12. ***REASONING AND SOLUTION*** A person is approaching you in a truck that is traveling very close to the speed of light. This person throws a baseball toward you. Relative to the truck, the ball is thrown with a speed nearly equal to the speed of light, so the person on the truck sees the baseball move away from the truck at a very high speed. Yet you see the baseball move away from the truck very slowly.

The relative velocities in this problem are:

v_{BY} = velocity of the **B**aseball relative to **Y**ou
v_{BT} = velocity of the **B**aseball relative to the **T**ruck
v_{TY} = velocity of the **T**ruck relative to **Y**ou

These velocities are related by the velocity-addition formula (Equation 28.7),

$$v_{BY} = \frac{v_{BT} + v_{TY}}{1 + \frac{v_{BT}v_{TY}}{c^2}}$$

If both v_{BT} and v_{TY} are close to the speed of light, $v_{BY} \approx (c+c)/[1+c^2/c^2] = 2c/2 = c$, and we can conclude that v_{BY} is very close to c. Since both v_{BY} and v_{TY} are measured relative to the *same* inertial reference frame, namely your reference frame, you see the baseball leaving the truck at a relative speed that is the difference between the two speeds, that is, $v_{BY} - v_{TY}$. Since v_{BY} and v_{TY} are unequal, but close to the speed of light, the difference $v_{BY} - v_{TY}$ will be very small. Therefore, you see the baseball move away from the truck very slowly.

---

13. ***REASONING AND SOLUTION*** The quantities listed are (a) the time interval between two events, (b) the length of an object, (c) the speed of light in a vacuum, and (d) the relative speed between the observers.

a. The time interval between two events depends on the reference frame in which the time measurements are made. An observer in the rest frame of the two events will measure the proper time interval. In any other inertial reference frame in motion relative to the rest frame of the events, the time interval will be longer or dilated. Therefore, two observers in relative motion will not, in general, agree on the measured value of the time interval between two events.

b. The length of an object depends on the reference frame in which the length measurement is made. An observer in the rest frame of the object will measure the proper length of the object. In any other inertial reference frame in motion relative to the rest frame of the object, the length will be shorter or contracted in the direction of motion. Therefore, two observers in relative motion will not, in general, agree on the measured value of the length of the object.

c. According to the speed of light postulate, the speed of light in a vacuum will be the same in all inertial reference frames that are in motion relative to each other. Therefore, two observers will always measure the speed of light to be the same, regardless of the relative velocity between them.

d. Since there is no preferred inertial reference frame, two observers in inertial reference frames that are in motion relative to each other will always measure the same relative speed between them, regardless of the relative velocity.

14. *REASONING AND SOLUTION* If the speed of light were infinitely large, instead of 3×10^8 m/s, the effects of time dilation and length contraction would *not* be observable. According to the time dilation equation (Equation 28.1), $\Delta t = \Delta t_0 / \sqrt{1 - v^2/c^2}$, where v is the relative speed between the observer who measures Δt_0 and the observer who measures Δt. According to the length contraction equation (Equation 28.2), $L = L_0 \sqrt{1 - v^2/c^2}$, where v is the relative speed between the observer who measures L_0 and the observer who measures L. If c were infinite, then the factor v^2/c^2 would be equal to zero, $\sqrt{1 - v^2/c^2}$ would be equal to unity, and it would follow that $\Delta t = \Delta t_0$ and $L = L_0$. In other words, the time interval for an event would be the same in all inertial reference frames, regardless of the relative speed between them. Similarly, the length of an object would be the same in all inertial reference frames, regardless of the relative speed between them. Therefore, if c were infinitely large, time dilation and length contraction would not be observable.

CHAPTER 29 | PARTICLES AND WAVES

CONCEPTUAL QUESTIONS

1. **REASONING AND SOLUTION** A monochromatic light source emits photons of a single frequency. According to Equation 29.2, the energy, E, of a single photon is related to its frequency, f, by the relation $E = hf$, where h is Planck's constant.

    The photons emitted by a source of light do *not* all have the same energy. Since the photons do not all have the same energy, then, from Equation 29.2, we can conclude that the photons do not all have the same frequency. Therefore, the source is not monochromatic.

2. **REASONING AND SOLUTION** According to the data given in Example 1, Chapter 24, the frequency of visible light ranges from 4.0×10^{14} Hz (red light) to 7.9×10^{14} Hz (violet light). According to Equation 29.2, the energy, E, of a photon is related to its frequency, f, by the relation $E = hf$, where h is Planck's constant. According to Equation 29.2, the energy of a photon is directly proportional to its frequency.

    a. The red-colored light bulb emits photons with the lowest frequency compared to light bulbs of other colors (orange, yellow, green, or blue); therefore, the red-colored light bulb emits photons with the lowest energy.

    b. The color blue appears next to violet in the continuous visible spectrum; therefore, the frequency of blue light is slightly smaller than that of violet, but greater than the frequency of other colors of the visible spectrum. Thus, the blue-colored light bulb emits photons with the highest frequency compared to the other light bulbs; therefore, the blue-colored light bulb emits photons with the greatest energy.

3. **REASONING AND SOLUTION** A photon emitted by a higher-wattage red light bulb *does not* have more energy than a photon emitted by a lower-wattage red bulb. The wattage of a bulb describes the power output of a bulb. Since average power is defined as energy per unit time, the power output of a light bulb tells us the *rate* at which the light bulb produces energy. According to Equation 29.2, the energy, E, of a photon is related to its frequency, f, by the relation $E = hf$, where h is Planck's constant. Thus, the energy of a photon depends *only* on the frequency of the associated light wave. The frequency of red light is the same, regardless of the rate of energy production; therefore, all photons of red light have the same energy, regardless of the nature of their source.

    Remark: Since the higher-wattage bulb provides more energy per unit time than the lower-wattage bulb, we can conclude that the higher-wattage bulb produces *more photons* per unit time. All of the "red photons," however, have the same energy.

4. **REASONING AND SOLUTION** When a sufficient number of visible light photons strike a piece of photographic film, the film becomes exposed. An X-ray photon is more energetic than a visible light photon. Yet, most photographic films are not exposed by the X-ray machines used at airport security checkpoints. Since a single X-ray photon is more energetic than a single photon of visible light, we can conclude that the number of X-ray photons per unit time emitted by airport security machines is much smaller than the number of visible light photons per unit time produced by normal lighting fixtures.

---

5. **REASONING AND SOLUTION** When radiation strikes a metallic surface, a photon of the radiation can give up its energy to an electron in the metal. If the photon has enough energy to perform the minimum amount of work required to remove the electron from the metal, the electron can be emitted from the metal. The minimum amount of work required to remove an electron from a metal is called its *work function*. If the photon does not have enough energy to remove an electron, the energy of the photon is absorbed by the electron; the increase in energy is manifested as thermal motion.

Radiation of a given wavelength causes electrons to be emitted from the surface of one metal (metal 1) but not from the surface of another metal (metal 2). We can conclude that the individual photons of the radiation have sufficient energy to do the work required to remove the electrons from the surface of metal 1, but not enough energy to remove the electrons from the surface of metal 2. Therefore, the work function of metal 1 must be less than the work function of metal 2.

---

6. **REASONING AND SOLUTION** In a photoelectric effect experiment, the intensity of the light is increased while the frequency is kept constant. The frequency is greater than the minimum frequency f_0, so that photoelectrons are emitted from the negatively charged metal plate (see Figure 29.4).

Light intensity is the energy per second per unit area that crosses a transverse surface. According to Einstein's photon hypothesis (see Equation 29.2), a beam of light is composed of individual photons, each having an energy $E = hf$, where h is Planck's constant and f is the frequency of the light. From this viewpoint, the intensity of the light is equal to the energy of each photon multiplied by the number of photons per second per unit area crossing a transverse surface. An increase in the intensity of the light beam, therefore, corresponds to an increase in the number of photons per second that strikes the negatively charged metal plate. This will result in the emission of an increased number of photoelectrons per second at the negative plate.

a. As the number of photoelectrons emitted per second from the negative plate increases, the number of photoelectrons collected per second at the positive plate increases. Since the current in the phototube is proportional to the number of photoelectrons per second collected at the positive plate, increasing the light intensity will cause the current in the phototube to increase.

b. As discussed above, an increase in intensity will cause the number of photoelectrons emitted per second from the metal surface to increase.

c. According to Equation 29.3, the maximum kinetic energy that a photoelectron could have is given by $KE_{max} = hf - W_0$, where W_0 is the work function of the metal. Since the frequency is kept constant, and W_0 is a property of the metal that is independent of the intensity, an increase in intensity will have no effect on KE_{max}. The maximum kinetic energy that an electron could have remains the same.

d. The kinetic energy of an object of mass m moving non-relativistically with speed v can be expressed in terms of its momentum p as follows:

$$KE = \tfrac{1}{2}mv^2 = \frac{(mv)^2}{2m} = \frac{p^2}{2m}$$

Therefore, treating the electron classically, the maximum momentum that a photoelectron could have is $p_{max} = \sqrt{2m(KE_{max})}$. Since the maximum kinetic energy that an electron could have remains the same as the intensity is increased at constant frequency, the maximum momentum that an electron could have remains the same as well.

e. According to Equation 29.8, the minimum de Broglie wavelength of an object of maximum momentum p_{max} is given by

$$\lambda_{min} = \frac{h}{p_{max}}$$

Since the maximum momentum that a photoelectron could have remains the same as the intensity is increased at constant frequency, we can conclude that the minimum de Broglie wavelength that an electron could have also remains the same.

---

7. **REASONING AND SOLUTION** As the result of a Compton scattering experiment, an electron is accelerated straight ahead in the same direction as that of the incident X-ray photon. Momentum conservation requires that the total initial momentum be equal to the total final momentum. The total initial momentum consists only of the forward momentum of the incoming photon, which we assume to be traveling in the +x direction. The electron is initially at rest. Therefore, the direction of the final total momentum of the recoiling electron and the scattered photon must also point in the +x direction. There can be no component of the final total momentum along the +y or –y direction. We know that the recoiling electron has momentum only in the +x direction. Thus, in order for the y component of the total final momentum to be absent, the scattered photon must move either in the +x or –x direction. To distinguish between these two choices, it is necessary to refer to energy conservation, as well as momentum conservation.

---

8. **REASONING AND SOLUTION** A photon can undergo Compton scattering from a molecule such as nitrogen, just as it does from an electron. However, the change in photon wavelength is much less than when an electron is scattered. To see why, let us examine Equation 29.7 which gives the difference between the wavelength λ' of the scattered photon

and the wavelength λ of the incident photon in terms of the scattering angle θ: $\lambda' - \lambda = [h/(mc)](1 - \cos\theta)$, where h is Planck's constant, m is the mass of the target particle, and c is the speed of light in a vacuum. The mass of a nitrogen molecule is much greater than the mass of an electron. Therefore, the factor $h/(mc)$ will be much smaller if the target particle is a nitrogen molecule. Consequently, the change in the photon wavelength, $\lambda' - \lambda$, is much less than it is when the target particle is an electron.

---

9. ***REASONING AND SOLUTION*** When the speed of a particle with mass doubles, its momentum doubles, and its kinetic energy becomes four times greater. A photon, however, does not behave the same as a particle with mass. The momentum of a photon is $p = E/c$ (see the derivation of Equation 29.6); therefore, the energy of the photon can be written as $E = pc$. Clearly, when the momentum of a photon doubles, its energy also doubles. It does *not* become four times greater.

---

10. ***REASONING AND SOLUTION*** When bright light is incident on the radiometer, photons strike both the black and the shiny surfaces. The photons are absorbed by the black surfaces and are reflected by the shiny surfaces. Linear momentum is transferred to the panels because of the photon collisions. Since the orientation of the light is arbitrary, photons will strike the panels at arbitrary angles of incidence. Only the component of the photon's momentum that is perpendicular to the face of the panel contributes to its motion. As discussed in Conceptual Example 3, the momentum transfer to the panels is a maximum for the shiny surfaces and is twice as large in magnitude than it is for the black surfaces. Following the reasoning of Conceptual Example 3, we can use the impulse-momentum theorem and Newton's third law to deduce that the force on the shiny side is greater than the force on the black side. Therefore, in bright light, the arrangement would spin in the direction from the shiny side toward the black side. The observed spinning is in the opposite direction. Thus, photon collisions with the panels *cannot* be the cause of the spinning.

---

11. ***REASONING AND SOLUTION*** The linear momentum of an object of mass m traveling with velocity **v** is **p** = m**v**. The de Broglie wavelength of an object moving with momentum **p** is given by Equation 29.8: $\lambda = h/p$, where h is Planck's constant.

A stone is dropped from the top of a building. As the stone falls, it is uniformly accelerated, and its velocity increases uniformly in the downward direction. Consequently, the linear momentum of the stone also increases uniformly in the downward direction. According to Equation 29.8, the de Broglie wavelength of an object or particle is inversely proportional to the magnitude of its linear momentum. Therefore, as the stone falls, the de Broglie wavelength of the stone must *decrease*.

12. **REASONING AND SOLUTION** An electron and a neutron have different masses. According to Equation 29.8, $\lambda = h/p$, the de Broglie wavelength of a particle is inversely proportional to the magnitude of its linear momentum. Therefore, if the electron and the neutron have different speeds such that the magnitudes of their respective momenta are the same, they will have the same de Broglie wavelength.

---

13. **REASONING AND SOLUTION** Suppose that in Figure 29.1, the electrons are replaced with protons that have the same speed. The speed of the protons is the same as that of the electrons; however, the mass of a proton is greater than the mass of an electron. Therefore, the magnitude of the linear momentum of a proton is greater than the linear momentum of an electron traveling at the same speed. According to Equation 29.8, $\lambda = h/p$, the de Broglie wavelength of the "proton beam" will be smaller than that of the "electron beam." Equation 27.1 gives the condition that must be satisfied by the angle θ that locates the interference maxima in a Young's double-slit experiment: $\sin\theta = m\lambda/d$, where m takes on integer values. According to Equation 27.1, the sine of the angle θ is directly proportional to the wavelength of the incident beam. Since the proton beam has a smaller wavelength than the electron beam, the values of $\sin\theta$ that correspond to maxima will be smaller. Consequently, the difference in the corresponding values of θ will be smaller, and the angular separation between the bright fringes will *decrease* when the electrons are replaced by protons.

CHAPTER 30 | *THE NATURE OF THE ATOM*

CONCEPTUAL QUESTIONS

1. **REASONING AND SOLUTION** A tube is filled with atomic hydrogen at room temperature. Electromagnetic radiation with a continuous spectrum of wavelengths, including those in the Lyman, Balmer, and Paschen series, enters one end of the tube and leaves the other end. The exiting radiation is found to contain absorption lines.

   At room temperature, most of the atoms of atomic hydrogen contain electrons that are in the ground state ($n = 1$) energy level. Since the radiation contains a continuous spectrum of wavelengths, it contains photons with a wide range of energies ($E = hf = hc/\lambda$). In particular, it will contain photons with energies that are equal to the energy difference between the atomic states in the Lyman series. When the radiation is incident on the atoms in the tube, these photons are absorbed by the electrons. When a photon, whose energy is equal to the energy difference for a transition in the Lyman series, is absorbed by a ground state electron, that electron will make a transition to a higher energy level. Every photon of this energy will be absorbed by a ground state electron and cause a transition. The wavelength of radiation that corresponds to that particular photon energy will, therefore, be removed from the radiation. When the radiation is analyzed, the wavelengths that correspond to transitions in the Lyman series will be absent. Since most of the atoms in the tube are in the ground state ($n = 1$), the electron populations in the $n = 2$ and $n = 3$ states are extremely small. Therefore, any absorption lines resulting from Balmer or Paschen transitions will be extremely weak. When the radiation is analyzed, the only predominant absorption lines in the exiting radiation will correspond to wavelengths in the Lyman series.

2. **REASONING AND SOLUTION** Refer to the situation described in Question 1. Suppose the electrons in the atoms are mostly in excited states. Most of the electrons are in states with $n \geq 2$; therefore, Balmer and Paschen series transitions will occur, and the absorption lines in the exiting radiation will correspond to wavelengths in the Balmer and Paschen series. Since there are relatively few electrons in the ground state, only a relatively few number of photons that correspond to wavelengths in the Lyman series will be absorbed. Most of the "Lyman photons" will remain in the radiation; therefore, the exiting radiation will *not* contain absorption lines that correspond to wavelengths in the Lyman series. Although the absorption lines that correspond to transitions in the Lyman series are not present, there will be more absorption lines in the exiting radiation compared to the situation when the electrons are in the ground state, because absorption lines corresponding to both the Balmer and Paschen series will be present.

3. **REASONING AND SOLUTION** According to Equation 30.13, the energy E_n of the nth atomic state in a Bohr atom is given by $E_n = -(13.6 \text{ eV}) Z^2 / n^2$, where $n = 1, 2, 3, \ldots$, and Z is equal to the number of protons in the nucleus. The energy that must be absorbed by the electron to cause an upward transition from the initial state n_i to the final

state n_f is $\Delta E_{fi} = E_f - E_i$. Using Equation 30.13, we find that the required energy is $\Delta E_{fi} = -(13.6 \text{ eV})Z^2 (1/n_f^2 - 1/n_i^2)$, with n_i, n_f = 1, 2, 3, . . . and $n_f > n_i$. The energy required to ionize the atom from when the outermost electron is in the state n_i can be found by letting the value of n_f approach infinity. The ionization energy is then

$$\Delta E_\infty = -(13.6 \text{ eV})Z^2 (1/\infty - 1/n_i^2) = -(13.6 \text{ eV})Z^2 (0 - 1/n_i^2) = (13.6 \text{ eV})Z^2/n_i^2$$

From this expression, we see that the ionization energy is inversely proportional to the square of the principal quantum number n_i of the initial state of the electron; thus, less energy is required to remove the outermost electron in an atom when the atom is in an excited state. Therefore, when the atom is in an excited state, it is more easily ionized than when it is in the ground state.

4. **REASONING AND SOLUTION** In the Bohr model for the hydrogen atom, the closer the electron is to the nucleus, the smaller is the total energy of the atom. This is not true in the quantum mechanical picture of the hydrogen atom.

   In the Bohr model, the nth orbit is a circle of definite radius r_n; every time that the position of the electron in this orbit is measured, the electron is found exactly a distance r_n from the nucleus. In contrast, according to the quantum mechanical picture, when an atom is in the nth state, the position of the electron is uncertain. Even though the atomic state is well defined by the principal quantum number n, the location of the electron is not definite. When the atom is in the nth state, the electron can sometimes be found close to the nucleus, while, at other times, it can be found far from the nucleus, or at some intermediate position. In the absence of any external magnetic fields, the energy of the state is determined by the value of n, and for a given n, the position of the electron is uncertain. Therefore, it is not correct to say that in the quantum mechanical picture of the hydrogen atom, the closer the electron is to the nucleus, the smaller is the total energy of the atom.

5. **REASONING AND SOLUTION**
   a. Consider two different hydrogen atoms. The electron in each atom is in an excited state. The Bohr model uses the same quantum number n to specify both the energy and the orbital angular momentum. According to the Bohr model, if an atom is in its nth state, the corresponding energy is $E_n = -(13.6 \text{ eV})Z^2/n^2$, where n = 1, 2, 3, . . ., and Z is equal to the number of protons in the nucleus (Equation 30.13). The integer n that specifies the energy also specifies the orbital angular momentum according to $L_n = nh/(2\pi)$, where h is Planck's constant (Equation 30.8). Therefore, according to the Bohr model, it is *not* possible for the electrons to have different energies, but the same orbital angular momentum.

   b. The quantum mechanical model uses two <u>different</u> quantum numbers to specify the energy and the orbital angular momentum of an atomic electron. In the absence of magnetic fields, the energy is determined by the principal quantum number n, while the orbital angular

188 THE NATURE OF THE ATOM

momentum is determined by the orbital quantum number ℓ. According to quantum mechanics, if an atom is in its nth state with orbital quantum number ℓ, the energy is identical to that of the Bohr model, and the magnitude of the orbital angular momentum is $[h/(2\pi)]\sqrt{\ell(\ell+1)}$, where $\ell = 0, 1, 2, \ldots, (n-1)$. Since $\ell = 0, 1, 2, \ldots, (n-1)$, different values of ℓ are compatible with the same value of n. For example, when $n = 3$, the possible values of ℓ are 0, 1, 2, and when $n = 4$, the possible values of ℓ are 0, 1, 2, 3. Thus, the electron in one of the atoms could have $n = 3$, $\ell = 2$, while the electron in the other atom could have $n = 4$, $\ell = 2$. Therefore, according to quantum mechanics, it is possible for the electrons to have different energies but have the same orbital angular momentum.

6. *REASONING AND SOLUTION* In the quantum mechanical picture of the hydrogen atom, the orbital angular momentum of the electron *may* be zero in any of the possible energy states. This is true because, if an atom is in its nth state with orbital quantum number ℓ, the magnitude of the orbital angular momentum is given by $[h/(2\pi)]\sqrt{\ell(\ell+1)}$, where $\ell = 0, 1, 2, \ldots, (n-1)$; therefore, for any value of n, the value of ℓ may be zero. However, when the hydrogenic electron is in the ground state, $n = 1$, and the only possible value for the orbital quantum number is $\ell = 0$. Therefore, when the electron is in the ground state, its orbital angular momentum *must* be zero.

7. *REASONING AND SOLUTION*

a. According to the Pauli exclusion principle, no two electrons in an atom can have the same set of values for the four quantum numbers n, ℓ, m_ℓ, and m_s. For a given value of the orbital quantum number ℓ, the magnetic quantum number m_ℓ, can have $2\ell + 1$ possible values. Each of these values can be combined with two possible values for the spin quantum number m_s. Therefore, the maximum number of electrons that the ℓth subshell can hold is $2(2\ell + 1)$. All of the electrons in a 5g subshell have $n = 5$, and $\ell = 4$. Thus, the maximum number of electrons that can be contained in a 5g subshell is $2(2\ell + 1) = 2[2(4) + 1] = 18$. Therefore, a 5g subshell can contain at most 18 electrons and cannot contain 22 electrons.

b. Since 17 is less than 18, a 5g subshell could contain 17 electrons.

8. *REASONING AND SOLUTION* An electronic configuration for manganese ($Z = 25$) is written as $1s^2\, 2s^2\, 2p^6\, 3s^2\, 3p^6\, 3d^4\, 4s^2\, 4p^1$. According to Figure 30.17, the ground state electronic configuration for manganese is written as $1s^2\, 2s^2\, 2p^6\, 3s^2\, 3p^6\, 3d^5\, 4s^2$. Clearly, the electronic configuration given in the question corresponds to *an excited state* in which the last electron in the 3d subshell has been excited to the 4p subshell.

9. *REASONING AND SOLUTION* Characteristic X-rays are produced when an electron with enough energy strikes the target atom and knocks one of the K-shell ($n = 1$) electrons

entirely out of the atom. An electron in one of the outer shells ($n \geq 2$) then drops into the K-shell and emits an X-ray photon in the process.

The ground state of hydrogen contains one electron in the $n = 1$ state, whereas, the ground state of helium contains two electrons in the $n = 1$ state. In both atoms, the ground state configuration contains only K-shell electrons. Even if one of the K-shell electrons were knocked out by an impinging electron, there are no electrons in higher levels to fall into the K-shell vacancy. Therefore, we would not expect hydrogen and helium atoms in their ground state to emit characteristic X-rays.

---

10. *REASONING AND SOLUTION* The characteristic lines in the X-ray spectrum depend on the target material. They occur at the wavelength value that corresponds to the wavelength of the X-ray photon that was emitted when an electron from the outer shell of the atom in the target material drops down to the K-shell. Since all identical atoms of the same element contain the same energy levels, the K_α and K_β lines will occur at the same wavelength values for all X-ray targets made of those atoms, regardless of the voltage applied across the X-ray tube.

The cutoff wavelength λ_0, however, depends on the voltage V across the X-ray tube. According to Equation 30.17, $\lambda_0 = (hc)/(eV)$; the cutoff wavelength is inversely proportional to the voltage across the tube. Therefore, when a smaller voltage is used to operate the tube, the cutoff wavelength increases.

We can conclude, therefore, that when the tube is operated at a smaller voltage, the characteristic lines should occur at the same wavelengths as they did at a higher voltage, and the cutoff wavelength should increase (i.e., shift to the right on the wavelength axis). The drawing in the text, shows the X-ray spectra produced by an X-ray tube when the tube is operated at two different voltages. The features of the drawing are consistent with the conclusions reached above.

---

11. *REASONING AND SOLUTION* Bremsstrahlung X-rays are produced when the electrons decelerate upon hitting the target. In the production of X-rays, it is possible to create Bremsstrahlung X-rays without producing the characteristic X-rays. The wavelengths of the characteristic lines depend on the target material; the cutoff wavelength, however, is independent of the target material and depends only on the energy of the impinging electrons. In order to produce Bremsstrahlung X-rays without producing the characteristic X-rays, the potential difference V applied to the tube must be chosen so that the cutoff wavelength is greater than the wavelengths of the characteristic X-rays for the target material. Therefore, using Equation 30.17, the value of V must be chosen so that $V = (hc)/(e\lambda_0)$, where $\lambda_0 > \lambda_{K_\alpha}$.

---

12. *REASONING AND SOLUTION* The short side of X-ray spectra ends abruptly at a cutoff wavelength λ_0. This cutoff wavelength depends only on the energy of the impinging electrons. Regardless of the composition of the target, the incident electron cannot give up more than all of its kinetic energy when it is decelerated by the target. The kinetic energy that the electron has acquired upon reaching the target is given by eV (see Section 19.2),

190 THE NATURE OF THE ATOM

where V is the potential difference that is applied to the X-ray tube. Therefore, the cutoff wavelength depends on the value of V, and not on the composition of the target material used in the X-ray tube.

13. **REASONING AND SOLUTION** A laser beam focused to a small spot can cut through a piece of metal. Light rays leaving the laser are nearly parallel, so the light can be focused to a very small spot. The intensity is the electromagnetic energy per unit time per unit area (see Equations 16.8 and 24.4). Therefore, when the beam is focused to a spot with a very small area, the intensity delivered by the beam is very large. For example, a pulsed ruby laser with a peak power of 10^8 W can be focused to a peak intensity of 10^{20} W/cm^2. This intensity is large enough to ionize air or burn a hole in a piece of metal.

14. **REASONING AND SOLUTION** The energy of a photon of light of frequency f is given by Equation 29.2: $E = hf$. Since $c = \lambda f$ for light, the energy of a photon can be written in terms of the wavelength λ as $E = hc/\lambda$. According to Table 26.2, the wavelength of green light is smaller than the wavelength of red light; therefore, the laser that produces green light emits photons that are more energetic than those emitted by the helium/neon laser.

CHAPTER 31 NUCLEAR PHYSICS AND RADIOACTIVITY

CONCEPTUAL QUESTIONS

1. **REASONING AND SOLUTION** Isotopes are nuclei that contain the same number of protons, but a different number of neutrons. A material is known to be an isotope of lead, although the particular isotope is not known.

   a. The atomic number Z of an atom is equal to the number of protons in its nucleus. It is different for different elements, and all isotopes of the same element have the same atomic number. Therefore, if we know that a material is an isotope of lead, we can specify its atomic number ($Z = 82$).

   b. The various isotopes of the same element differ in the number of neutrons in the nucleus. Since we do not know which particular isotope of lead makes up the material, we cannot specify the neutron number.

   c. The atomic mass number is equal to the total number of protons and neutrons in the nucleus. Since we do not know which isotope of lead makes up the material, we do not know how many neutrons are in the nucleus. Therefore, we cannot specify the atomic mass number.

2. **REASONING AND SOLUTION** Two nuclei have different nucleon numbers A_1 and A_2. The two nuclei may or may not be isotopes of the same element. Two isotopes of the same element have the same number of protons in the nucleus, but differ in the number of neutrons. The nucleon number is equal to the total number of protons and neutrons in the nucleus. If the nucleon numbers differ only because the neutron numbers differ, then the two nuclei are isotopes of the same element. However, if the nucleon numbers differ because the nuclei contain a different number of protons, then the two nuclei are *not* isotopes of the same element (regardless of their respective neutron numbers). Therefore, two nuclei with different nucleon numbers are *not* necessarily isotopes of the same element.

3. **REASONING AND SOLUTION** Two nuclei that contain different numbers of protons and different numbers of neutrons can have the same radius. According to Equation 31.2, the approximate radius of the nucleus depends on the atomic mass number A and is given by $r \approx (1.2 \times 10^{-15} \text{ m}) A^{1/3}$. The atomic mass number A is equal to the total number of protons and neutrons. Therefore, if two nuclei have different numbers of protons and different numbers of neutrons, but the sum of the protons and neutrons is the same for each nucleus, then the two nuclei have the same radius.

4. **REASONING AND SOLUTION** Using Figure 31.6, the following nuclei have been ranked in ascending order (smallest first) according to the binding energy per nucleon: $^{232}_{90}\text{Th}$, $^{184}_{74}\text{W}$, $^{31}_{15}\text{P}$, $^{59}_{27}\text{Co}$.

---

5. **REASONING AND SOLUTION** The general form for α decay is

$$\underbrace{^{A}_{Z}\text{P}}_{\text{Parent nucleus}} \rightarrow \underbrace{^{A-4}_{Z-2}\text{D}}_{\text{Daughter nucleus}} + \underbrace{^{4}_{2}\text{He}}_{\substack{\alpha \text{ particle} \\ (\text{helium nucleus})}}$$

while the general form for the two types of beta decay are

[β⁻ decay]
$$\underbrace{^{A}_{Z}\text{P}}_{\text{Parent nucleus}} \rightarrow \underbrace{^{A}_{Z+1}\text{D}}_{\text{Daughter nucleus}} + \underbrace{^{0}_{-1}\text{e}}_{\substack{\beta^- \text{ particle} \\ (\text{electron})}}$$

and

[β⁺ decay]
$$\underbrace{^{A}_{Z}\text{P}}_{\text{Parent nucleus}} \rightarrow \underbrace{^{A}_{Z-1}\text{D}}_{\text{Daughter nucleus}} + \underbrace{^{0}_{1}\text{e}}_{\substack{\beta^+ \text{ particle} \\ (\text{positron})}}$$

According to Equation 31.2, the approximate radius of the nucleus depends on the atomic mass number A and is given by $r \approx (1.2 \times 10^{-15} \text{ m}) A^{1/3}$.

a. For α decay, the mass number of the daughter nucleus is four nucleons smaller than that of the parent nucleus. Therefore, according to Equation 31.2, the daughter nucleus should have a smaller radius than the parent nucleus.

b. For both types of β decay, the mass number of the daughter nucleus is identical to that of the parent nucleus. Therefore, according to Equation 31.2, the radius of the daughter nucleus should be the same as that of the parent nucleus.

---

6. **REASONING AND SOLUTION** The general forms for α decay, β decay, and γ decay are, respectively:

$$\underbrace{^{A}_{Z}\text{P}}_{\text{Parent nucleus}} \rightarrow \underbrace{^{A-4}_{Z-2}\text{D}}_{\text{Daughter nucleus}} + \underbrace{^{4}_{2}\text{He}}_{\alpha \text{ particle}}$$

$$\underbrace{^{A}_{Z}\text{P}}_{\text{Parent nucleus}} \rightarrow \underbrace{^{A}_{Z+1}\text{D}}_{\text{Daughter nucleus}} + \underbrace{^{0}_{-1}\text{e}}_{\beta \text{ particle}}$$

$$\underbrace{^A_Z P^*}_{\text{Excited energy state}} \rightarrow \underbrace{^A_Z P}_{\text{Lower energy state}} + \underbrace{\gamma}_{\gamma\text{ ray}}$$

As discussed in Section 31.1, the identity of an element depends on the number of protons in the nucleus. Clearly, in both α decay and β decay, the number of protons of the parent nucleus changes as the parent nucleus emits a charged particle. In contrast, in the process of γ decay, the parent nucleus emits only energy and the number of protons in the nucleus remains the same. Therefore, α and β decay result in transmutations, while γ decay does not.

7. **REASONING AND SOLUTION** In order for a nuclear decay to occur, it must bring the parent nucleus toward a more stable state by allowing the release of energy. The total mass of the decay products must be less than the mass of the parent nucleus. The difference in mass between the products and the parent nucleus represents an equivalent amount of energy that is given by Equation 28.5: $E_0 = mc^2$. This is the energy that is released during the decay.

As discussed in Example 4, uranium $^{238}_{92}$U decays into thorium $^{234}_{90}$Th by means of α decay. The $^{238}_{92}$U nucleus never decays by just emitting a single proton, instead of an alpha particle. The hypothetical decay scheme for the unobserved decay, is given below along with the pertinent atomic masses:

$$\underbrace{^{238}_{92}U}_{\substack{\text{Uranium}\\238.050\ 78\ u}} \rightarrow \underbrace{^{237}_{91}Pa}_{\substack{\text{Protactinium}\\237.051\ 14\ u}} + \underbrace{^1_1 H}_{\substack{\text{Proton}\\1.007\ 83\ u}}$$

The total mass of the decay products is (237.051 14 u) + (1.007 83 u) = 238.058 97 u. This is greater, not smaller, than the mass of the parent nucleus $^{238}_{92}$U; therefore, the daughter nucleus is less stable than the parent nucleus. Consequently, the hypothetical decay never occurs.

8. **REASONING AND SOLUTION** According to Equation 31.5, if there are N_0 radioactive nuclei present at time $t = 0$, then the number N of radioactive nuclei present at time t is $N = N_0 e^{-\lambda t}$, where the decay constant λ is given by Equation 31.6: $\lambda = (\ln 2)/T_{1/2}$, where $T_{1/2}$ is the half-life of the nuclei. According to Equation 31.6, unstable nuclei with short half-lives have large decay constants. When λ is large, the term $e^{-\lambda t}$ approaches zero even for very short times. Therefore, unstable nuclei with short half-lives typically have only a small or zero abundance after a short period of time.

194   NUCLEAR PHYSICS AND RADIOACTIVITY

9. **REASONING AND SOLUTION** The half-life of a radioactive isotope is the time required for one-half of the nuclei present to disintegrate. The disintegrations do not occur continuously at regular time intervals. Individual disintegrations occur randomly; we cannot predict *a priori* which particular nucleus will decay. Any given nucleus may decay at any time. The time may be much shorter than the half-life or it may be much longer than the half-life. Therefore, even though the half-life of indium $^{115}_{49}$In is 4.41×10^{14} yr, it is possible for any single nucleus in the sample to decay after only one second has passed.

---

10. **REASONING AND SOLUTION** The activity of a radioactive sample is the number of disintegrations per second that occur. In other words, it is the magnitude of $\Delta N / \Delta t$. According to Equation 31.4, $\Delta N / \Delta t = -\lambda N$, where λ is the decay constant that is characteristic of the isotope in question, and N is the number of radioactive nuclei that are present.

    a. While each isotope has its own characteristic decay constant, the activity is proportional to the product λN, according to Equation 31.4. Therefore, two samples with different decay constants λ, and different numbers of radioactive nuclei N, can have identical activities if the product λN is the same for each. Thus, it is possible for two different samples that contain different radioactive elements to have the same activity.

    b. Since each isotope has its own characteristic decay constant λ, two different samples of the same radioactive isotope have the same decay constant. If the two samples contain different numbers of radioactive nuclei, the product λN will be different for each sample. Therefore, they will have different activities.

---

11. **REASONING AND SOLUTION** Radiocarbon dating utilizes the $^{14}_{6}$C isotope that undergoes β^- decay with a half-life of 5730 yr. If the present activity of $^{14}_{6}$C can be measured in a carbon-containing object, the age of the object can be determined. Example 11 shows how to determine the age of an object, once the present activity is known.

    Consider the following objects, each about 1000 yr old: a wooden box, a gold statue, and some plant seeds. Since 1000 yr is of the same order of magnitude as the half-life of $^{14}_{6}$C, the change in activity should be significant, and the present activity should be easily measured. However, the sample must contain $^{14}_{6}$C atoms. Since both the wooden box and the seeds contain $^{14}_{6}$C, radiocarbon dating can be used to determine their ages. If the gold statue is made of pure gold, it will not contain any $^{14}_{6}$C atoms. Therefore, radiocarbon dating could not be applied to the gold statue.

12. ***REASONING AND SOLUTION*** Two isotopes have half-lives that are short relative to the age of the earth. Today, one of these isotopes (isotope A) is found in nature and the other (isotope B) is not. It is reasonable to conclude that the supply of isotope A has been replenished by some natural process that off-sets the radioactive decay, while the supply of isotope B has not been replenished. The replenishing of isotope A could occur, for example, as the result of the interaction of cosmic rays with the earth's upper atmosphere or the radioactive decay of some other isotope.

13. ***REASONING AND SOLUTION*** In radiocarbon dating, the age of the sample is determined from the ratio of the present activity to the original activity. Suppose that there were a greater number of $^{14}_{6}C$ atoms in a plant living 5000 yr ago than is currently believed. According to Equation 31.4: $\Delta N / \Delta t = -\lambda N$, and the original activity would be greater than is currently believed. A longer time would be required for the activity of a larger number of $^{14}_{6}C$ nuclei to decrease to its present value. Therefore, the age of the seeds obtained from radiocarbon dating would be too small.

14. ***REASONING AND SOLUTION*** Tritium is an isotope of hydrogen and undergoes β^- decay with a half-life of 12.33 yr. Like carbon $^{14}_{6}C$, tritium is produced in the atmosphere because of cosmic rays and can be used in a radioactive dating technique. The age cannot be significantly smaller or significantly larger than the half-life of the radioisotope. During a 700 yr period, 700/12.33 or roughly 57 half-lives would occur, and the activity of a given sample would decrease to an immeasurably small fraction of its initial value (on the order of 10^{-17}). Therefore, tritium dating could *not* be used to determine a reliable date for a 700 yr old sample.

15. ***REASONING AND SOLUTION*** Because of radioactive decay, one element can be transmuted into another. Thus, a container of uranium $^{238}_{92}U$ ultimately becomes a container of lead $^{206}_{82}Pb$. According to Figure 31.16, the first step is the transmutation from uranium $^{238}_{92}U$ to thorium $^{234}_{90}Th$. This step has the greatest half-life, with a value of 4.5×10^9 y. Therefore, we can conclude that the complete transmutation from uranium $^{238}_{92}U$ to lead $^{206}_{82}Pb$ would take at least billions of years.

16. **REASONING AND SOLUTION** According to Equation 21.2, the radius r of curvature for a particle with charge q and mass m, that moves perpendicular to a magnetic field of magnitude B, is given by

$$r = \frac{mv}{qB}$$

Thus, the ratio of the radii of curvature for the motion of the α particle and the β^- particle is

$$\frac{r_\beta}{r_\alpha} = \frac{m_\beta v / q_\beta B}{m_\alpha v / q_\alpha B} = \frac{m_\beta / q_\beta}{m_\alpha / q_\alpha} = \frac{m_\beta}{m_\alpha} \frac{q_\alpha}{q_\beta}$$

The mass of the α particle is 1.0026 u, while the mass of a β^- particle (an electron) is

$$(9.109\ 389\ 731 \times 10^{-31}\ \text{kg}) \left(\frac{1.0073\ \text{u}}{1.6726 \times 10^{-27}\ \text{kg}} \right) = 5.486 \times 10^{-4}\ \text{u}$$

Since the magnitudes of the charge on the α and β^- particles are $2e$ and e, respectively, the ratio of the radii of curvature is

$$\frac{r_\beta}{r_\alpha} = \left(\frac{5.486 \times 10^{-4}\ \text{u}}{4.0026\ \text{u}} \right) \left(\frac{2e}{e} \right) = 2.7 \times 10^{-4}$$

Therefore, the radius of curvature of the path of the β^- particle is smaller than the radius of curvature of the path of the α particle. Therefore, the trajectory of the β^- particle has a greater curvature than that of the α particle.

CHAPTER 32 | IONIZING RADIATION, NUCLEAR ENERGY, AND ELEMENTARY PARTICLES

CONCEPTUAL QUESTIONS

1. **REASONING AND SOLUTION** The RBE, or relative biological effectiveness, of a particular type of radiation compares the dose of that radiation needed to produce a certain biological effect to the dose of 200 keV X-rays needed to produce the same biological effect. The RBE depends on the nature of the ionizing radiation and the type of tissue that is being irradiated. Therefore, if two different types of radiation have the same RBE, this *does not* mean that each type delivers the same amount of energy to the tissue that it irradiates. Rather, it implies that the two different types of radiation are equally effective in producing the same biological effect on the same kind of tissue.

2. **REASONING AND SOLUTION** When a dentist X-rays your teeth, a lead apron is placed over your chest and lower body. The purpose of this apron is to protect the tissues on your chest and lower body from the ionizing effect of the X-rays. The thickness of the lead is chosen so that the apron blocks the penetration of the X-rays.

3. **REASONING AND SOLUTION**
   a. The units "rads" and "grays" are used to express the *absorbed dose* of ionizing radiation. According to Equation 32.2, the absorbed dose is equal to the energy absorbed from the radiation per unit mass of the absorbing material. The gray (Gy) is the SI unit for absorbed dose and is equal to 1 J/kg. The relationship between rads and grays is: 1 rad = 0.01 gray.

   b. The units "rads" and "roentgens" are *not* related. The rad is used to express the *absorbed dose*, and 1 rad = 0.01 J/kg. The roentgen is used to express the *exposure*, and 1 roentgen = 2.58×10^{-4} C/kg. The exposure expresses the ionizing abilities of X-rays or γ rays in air, while the absorbed dose specifies the amount of energy absorbed per kilogram of absorbing material.

4. **REASONING AND SOLUTION** In any nuclear reaction, the following quantities are conserved: (1) mass / energy; (2) electric charge, (3) linear momentum, (4) angular momentum, and (5) nucleon number.

   a. $^{60}_{28}\text{Ni}\,(\alpha, p)\,^{62}_{29}\text{Cu}$ corresponds to the reaction

   $$^4_2\text{He} + ^{60}_{28}\text{Ni} \rightarrow ^{62}_{29}\text{Cu} + ^1_1\text{H}$$

198 IONIZING RADIATION, NUCLEAR ENERGY, AND ELEMENTARY PARTICLES

The nucleon number for the reactants is 4 + 60 = 64, while the nucleon number of the products is 62 + 1 = 63. Therefore, this reaction does *not* occur, because it violates the conservation of nucleon number.

b. $^{27}_{13}$Al (n, n) $^{28}_{13}$Al corresponds to the reaction

$$^{1}_{0}n + ^{27}_{13}Al \rightarrow ^{28}_{13}Al + ^{1}_{0}n$$

The nucleon number for the reactants is 1 + 27 = 28, while the nucleon number of the products is 28 + 1 = 29. Therefore, this reaction does *not* occur, because it violates the conservation of nucleon number.

c. $^{39}_{19}$K (p, α) $^{36}_{17}$Cl corresponds to the reaction

$$^{1}_{1}H + ^{39}_{19}K \rightarrow ^{36}_{17}Cl + ^{4}_{2}He$$

The nucleon number for the reactants is 1 + 39 = 40, while the nucleon number of the products is 36 + 4 = 40. Nucleon number is conserved. The proton number on the left is $1 + 19 = 20$, while the proton number on the right is $17 + 2 = 19$. Since each proton carries a unit of charge +e, the total electric charge on the left is +20 e, while the total electric charge on the right is +19 e. Therefore, this reaction does *not* occur because it violates the conservation of electric charge.

---

5. **REASONING AND SOLUTION** Thermal neutrons and thermal protons have nearly the same speed. But thermal neutrons and thermal electrons have very different speeds. As discussed in Section 32.3, the adjective "thermal" applies to particles that have a kinetic energy of about 0.04 eV or less. According to Equation 6.2, the kinetic energy of a particle of mass m traveling with speed v is given by $KE = \frac{1}{2}mv^2$. Solving for v we have $v = \sqrt{2(KE)/m}$. Therefore, for particles that have the same average kinetic energy, the speed of the particle is inversely proportional to the square root of its mass. Since the mass of the neutron is very nearly equal to the mass of a proton, thermal neutrons and thermal protons have nearly the same speed. The mass of the electron is much smaller than the mass of the neutron; therefore, thermal electrons will have a higher average speed than thermal neutrons.

---

6. **REASONING AND SOLUTION** Consider the hypothetical fission of a nucleus of nucleon number $A = 25$ into two fragments of about equal mass ($A \approx 12.5$). The curve in Figure 32.8 shows that the binding energy of a nucleus with $A = 25$ is about 8.0 MeV per nucleon. Further inspection of Figure 32.8 shows that the binding energy of the fragments *has decreased* to about 6.5 MeV per nucleon. The difference between the binding energy per nucleon of the parent and the binding energy per nucleon of the daughter nucleus is about 6.5 MeV − 8.0 MeV = − 1.5 MeV of energy per nucleon, where the minus sign

indicates that the energy is absorbed. Therefore, a release of energy *would not* accompany the hypothetical fission reaction.

---

7. **REASONING AND SOLUTION** In the fission of $^{235}_{92}$U, there are, on the average, 2.5 neutrons released per fission. The fact that the fission of $^{235}_{92}$U releases on the average, 2.5 neutrons, makes it possible for a self-sustaining series of fissions to occur. Each neutron released by the fission process can initiate another fission reaction, resulting in the release of still more neutrons. These neutrons initiate more fission reactions, and the process results in a self-sustaining chain reaction.

    Suppose a *different* element is being fissioned and, on the average, only 1.0 neutron is released per fission. If a small fraction of the thermal neutrons absorbed by the nuclei does *not* produce a fission, a self-sustaining chain reaction *cannot* be produced using this element. Since each fission produces only one neutron, the product of a fission reaction can, at best, initiate only one more fission reaction. However, since a small fraction of the thermal neutrons absorbed by the nuclei does *not* produce a fission, at some point, the absorbed neutron will not produce a fission and, since there are no other neutrons to initiate a fission, the process will eventually stop.

---

8. **REASONING AND SOLUTION** Much more energy is released when a uranium $^{235}_{92}$U nucleus fissions than when a carbon $^{12}_{6}$C atom in coal is combined with oxygen during the coal-burning process. It is true that the mass of the uranium nucleus is about 20 times greater than that of a carbon nucleus (235/12 = 20). But even 20 carbon atoms, when burned, release much less energy than a single $^{235}_{92}$U nucleus releases when it fissions. In fact, they release about 2 million times less energy. Thus, the mass of coal needed is about 2 million times greater than the mass of uranium.

---

9. **REASONING AND SOLUTION** Figure 32.8 shows that the binding energy of a nucleus with $A = 60$ is about 9.0 MeV per nucleon. If two such nuclei fuse, the daughter nucleus would have $A = 120$. According to Figure 32.8, the binding energy of the daughter nucleus *decreases* to about 8.3 MeV per nucleon. Consequently, the difference between the binding energy per nucleon of the initial nuclei and the binding energy per nucleon of the fused nucleus is about 8.3 MeV – 9.0 MeV = – 0.7 MeV of energy per nucleon, where the minus sign indicates that the energy is absorbed. Therefore, the fusion of two nuclei, each with $A = 60$, would *not* release energy.

---

10. **REASONING AND SOLUTION** Nuclear fission occurs when a massive nucleus captures a slow moving neutron and is split into two less-massive fragments that have a greater binding energy per nucleon than the massive parent nucleus. Since the binding energy per nucleon of the daughter nuclei is greater than that of the parent nucleus, energy is released. It appears primarily as kinetic energy of the fission products.

Nuclear fusion occurs when two low-mass nuclei with relatively small binding energies per nucleon combine or "fuse" into a single, more massive nucleus that has a greater binding energy per nucleon. Since the binding energy per nucleon of the more massive nucleus is greater than that of the parent nuclei, a substantial amount of energy is released. The energy released per nucleon in a fusion reaction is greater than that released in a fission reaction. Therefore, for a given mass of fuel, a fusion reaction will yield more energy than a fission reaction.